AN ENVIRONMENTAL HISTORY OF SCOTLAND

A Land Won from Waste

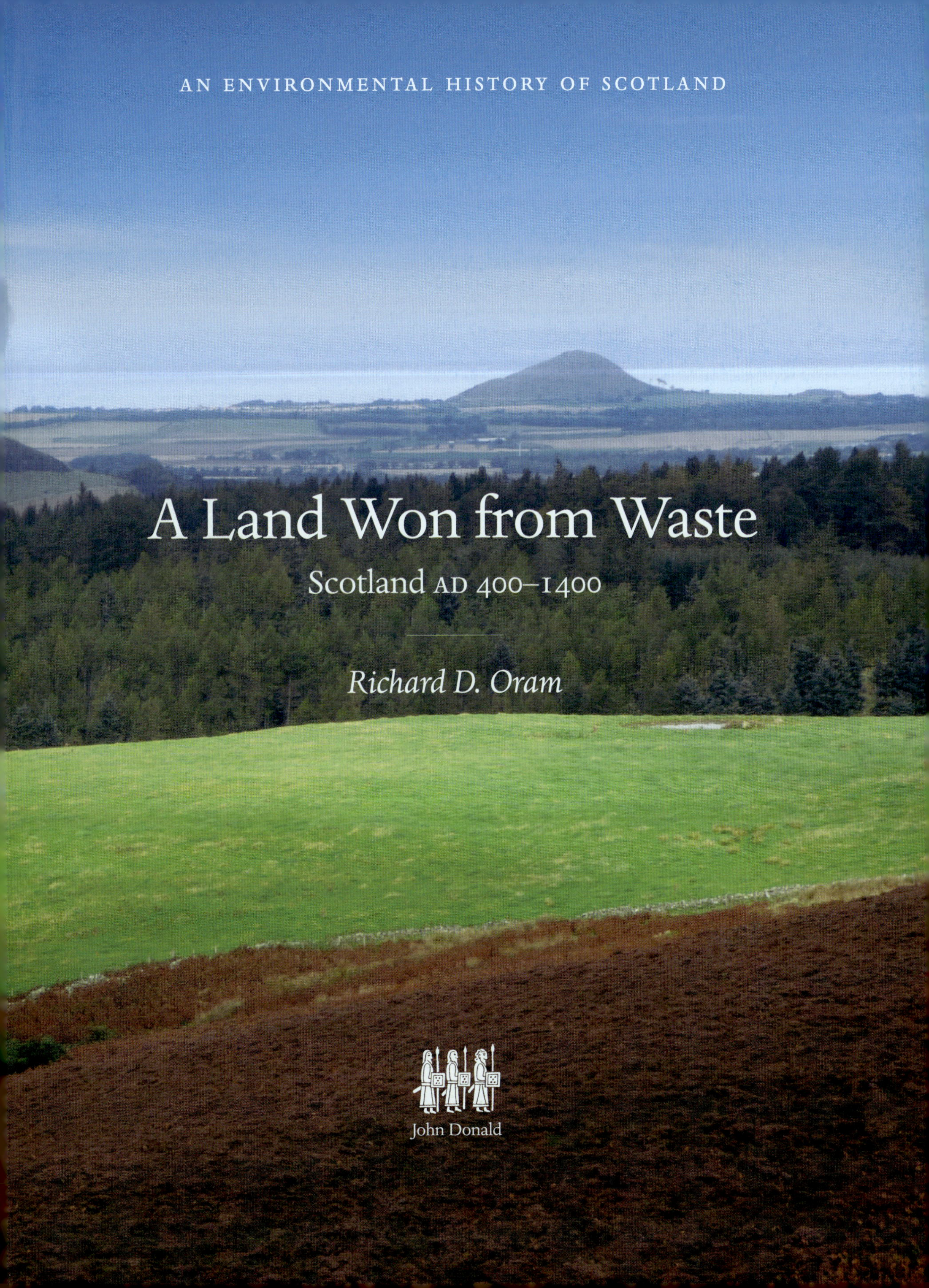

AN ENVIRONMENTAL HISTORY OF SCOTLAND

A Land Won from Waste

Scotland AD 400–1400

Richard D. Oram

John Donald

Title page image:
Traprain Law and North Berwick Law, East Lothian. The probable centre of the Celtic tribe known as the Votadini or the Gododdin, Traprain housed a major concentration of population supported by the rich agricultural land of the Lothian Plain and the seasonal pastures of the Lammermuir Hills to the south.

Opposite:
Dunsinnan Hill, Collace, Perth and Kinross. Traditionally the site of Macbeth's castle, the fort was already nearly two millennia old by the time of his reign. Located on an isolated summit in the Sidlaw Hills, the settlement at Dunsinnan was set amidst an upland landscape of extensive pasture with blocks of plough-cultivated arable land.

Page vi:
Ben Nevis from the east ridge of Carn Eige, Highland. The mountains of the western Highlands took their current form under the icesheet of the Loch Lomond re-advance, some 12,000 to 13,000 years ago.

First published in Great Britain in 2025 by
John Donald, an imprint of Birlinn Ltd

West Newington House
10 Newington Road
Edinburgh, EH9 1QS

www.birlinn.co.uk

ISBN: 978 0 85976 719 4

The publishers gratefully acknowledge the support of
the Scotland Inheritance Fund

towards the publication of this book.

British Library Cataloguing-in-Publication Data
A catalogue record for this book is available on request from the British Library

Designed and typeset by Mark Blackadder

Printed and bound in Britain by Bell and Bain Ltd, Glasgow

To Chris Smout, who started
everything that has led me to this place.
and
To Emma, who has been with me every step of the
journey from planning to completing this odyssey
through Scotland's past, present and future.

Contents

Acknowledgements ix

Abbreviations xi

Introduction: Framing Scotland's Environmental History 1

1 Out of the Mists 37

2 After the 'Great Wood': Trees and Woodland Management to AD 1000 69

3 Riding the Rollercoaster from the Late Antique Little Ice Age to the Medieval Climate Anomaly *c.*500–*c.*1000 101

4 Raiders, Ranchers, Farmers and Fishers through the Great Winters of the Viking Age 131

5 Beasts of the Wild Wood, Mountain and Sea Shore 153

6 An Era of Increase and Abundance? Scotland *c.*1000–1256 179

7 Forests and Hunting Reserves 207

8 Feeding and Fuelling Growth 243

9 Facing into the Coming Storm 287

10 The Worst Disaster? War, Famine, Disease and Death in the Long Fourteenth Century 327

Conclusion – Stepping Back from the Brink? 373

Bibliography 377

Index 393

Acknowledgements

There is a vast array of people that I must thank for helping me through the long process of writing and researching this book and its companion volumes, *Where Men No More May Reap or Sow* and *Standing on the Edge of Being*. The opportunity to sit down and start the process of writing was provided by a long research leave at the end of my term as Dean of Arts and Humanities at the University of Stirling, when I finally had the headroom and time to plan out the structure of a single- volume *Environmental History of Scotland from the Romans to COP26* that, by the end of that year, had morphed into the three volumes of which this is the first. In the two years since the first draft was submitted to Hugh Andrew at Birlinn, much has changed in terms of structure, weighting in the content and span of each volume, and I am eternally grateful for his patience in waiting for a book that he was eager to publish nearly five years ago. As ever, I owe a massive debt of gratitude to Mairi Sutherland and her colleagues, for guiding me through the final stages from re-drafting to publication. To Laura Davey, who has helped to turn some of the clunkier prose into something far more readable, go my undying thanks.

Getting to this point has been aided and speeded by many friends and colleagues. Alison Cathcart, Catherine Mills, Michael Penman and Philip Slavin, all of History, Heritage and Politics at Stirling, read drafts and supported funding applications for research and for publication costs. Annie Tindley at the University of Newcastle, Alex Woolf at the University of St Andrews and Sam Alberti at the National Museum of Scotland also read texts in draft and gave their generous support for grant applications. The ideas that run through these texts have been sparked and nurtured through extended conversations over many years with colleagues past and present at Stirling, especially Paul Adderley, Sarah Bromage, Jane Cameron, Donald Davidson, Alastair Jump, Aedin ni Loinsigh, Greg Mannion, Ben Marsh, Ian Simpson, Phia Steyn, Richard Tipping and Eileen Tisdall, and of course, the much-missed Alasdair Ross. There are huge debts too to my former Master's and PhD students in Environmental History – Gregor Adamson, Kate Buchanan, Sharla Chittick, Anne Dance, Val Dufeu, Ben Fanstone, Laura Hamlet, Sven Leman, Kevin Malloy, Richard Millar, Stuart Morrison, Edward Nelson, Simon Parkin, Verona Helen Shaw, Gilbert Stevenson and David Weinczok – for the conversations that underlie some of the discussion in these volumes.

Much of what is discussed here was supported by many of the staff in the University Library at Stirling, to whom goes my deepest gratitude. Likewise, the guidance and assistance of staff of Historic Environment Scotland (and the former Royal Commission on the Ancient and Historical Monuments of Scotland), Piers Dixon, Strat Halliday, Ewan Hyslop, and staff at the National Records of Scotland, National Library of Scotland and National Museums of Scotland, without whom this book would have been infinitely duller.

To Booboo and Lóki, who have kept me company (and stepped on the keyboard) during the long hours of writing, your distractions have always been welcome reminders that there are other important things in life. To you, Emma, the most important of them all, goes my undying love. To all our children and our first grandchild, Nora Violetta, the message in these books is for you, for it is you who will live with the decisions for the future of our environment that are being made today.

Abbreviations

AHRB	Arts and Humanities Research Board
AMOC	Atlantic Meridional Overturning Circulation or Current
ENSO	El Niño Southern Oscillation
GISP2	Greenland Ice Sheet Project, Phase 2
GRIP	Greenland Ice-core Project
HES	Historic Environment Scotland
LALIA	Late Antique Little Ice Age
LIA	'little ice age'
MCA	Medieval Climate Anomaly
NAO	North Atlantic Oscillation
NLS	National Library of Scotland
NRS	National Records of Scotland
RCAHMS	Royal Commission on the Ancient and Historic Monuments of Scotland
RPS	*The Records of the Parliaments of Scotland to 1707*, ed. K.M. Brown et al. (St Andrews, 2007–24), http://www.rps.ac.uk/
RWP	Roman Warm Period
TEK	traditional environmental knowledge
UNESCO	United Nations Educational Scientific and Cultural Organisation
VEI	Volcanic explosivity index

Inauguration site, Dunadd, Argyll and Bute. Rising now from the grazing-land of the late eighteenth- and nineteenth-century Improvement era fields, the fort was once surrounded by the raised bog of the Moine Mhòr, exploited for fuel, pasture, scrubby woodland and hunting opportunities by Dunadd's inhabitants.

Introduction: Framing Scotland's Environmental History

We listened into the Iceage
And we built up man round the Picts
And the daybreak hammered out warning
To the weak
We dreamed on moor with passion
And on the long lochs bluer than eyes
'Till the mists of bygone ages
Heard our cries[1]

According to the website of VisitScotland, the agency that promotes the wonders of this land to domestic and international tourists alike, our environment – or 'Landscapes and Nature' as it is labelled on the site – is one of the country's greatest assets and attractions.[2] It is a land replete with natural wonders laid out for visitors, of 'Towering mountains, glittering lochs, dense woodlands and miles upon miles of golden beaches' that form 'landscapes and natural scenery [that] really will take your breath away.' You are invited to 'Tour the rolling hills and lush farmland of the Lowlands, or drive through deep glens surrounded by mountains in the Highlands. Uncover magnificent coastal features in the west, and explore the atmospheric, royal settings of the east.' For the adventurous, you can 'Head to one of our Dark Sky Parks or spots where you can marvel at the sparkling night sky, dip your toes into our crisp sea water, wander around one of the many glistening freshwater lochs, or venture out into the wilderness to see if you can catch the Northern Lights dancing in the sky.' Here, in a few sentences, is the delight, wonderful variety, astonishing diversity and seeming timelessness of the land and environment of one small country captured in a handful of breathless lines.

It is an attractive vision, almost idealised in its conjuring of juxtaposed docility and wildness, like

1 Extract from 'The Wire', from Runrig, *Heartland*, Ridge Records RR005, 1985, LP, music and lyrics C. and R. Macdonald, published by Storr Music.

2 VisitScotland, 'Scotland's landscapes, scenery and nature spots', https://www.visitscotland.com/things-to-do/landscapes-nature.

The white sands and aquamarine waters of Scousburgh, South Mainland, Shetland, a scene that captures the spirit of the VisitScotland promotion of the country's natural beauty.

the imagined landscapes of some medieval Italian fresco.[3] Look a little deeper, however, and questions begin to form. Where are the people in this landscape? Apart from the Lowlands' 'lush farmland' or, perhaps, those enigmatic 'atmospheric, royal settings of the east', it is a 'wilderness' with which we are presented, of glens, mountains, lochs and woodlands, empty of other life – human and non-human – and awaiting your discovery as intrepid explorers. There is almost nothing in the pitch to suggest that what you will find is far from being in any sense wild nature or a wilderness, and only the barest of hints that it is the product of around nine millennia of human interaction with, and transformation of, almost every part of the land and waters of what we now know as Scotland. Far from being anything close to pristine nature or in any sense wilderness, all of what we see today is part of a complex cultural construction, layer upon layer of cultural landscapes. Every one of these superimposed levels has been manufactured by millennia of interaction between humans and the wider environment through which they moved and which they altered and adapted to deliver the means to sustain themselves. The complexity of this landscape is staggering in the multiplicity of strata of alteration and input, created by shifting economic and social conditions, population levels, management fashions and, of course, climate change.[4] When there is little recognition of this reality of layering amongst the many stakehold-

3 See, for example, the juxtaposition of the attractive light-coloured and well-ordered fields around the city and the forbidding darkness of the forests and mountains in Ambrogio Lorenzetti's late fourteenth-century allegorical fresco depicting the effects of good and bad government in the council chamber of the Palazzo Pubblico in Siena: J.M. Greenstein, 'The vision of peace: meaning and representation in Ambrogio Lorenzetti's Sala della Pace cityscapes', *Art History*, 11:4 (1988), pp. 492–510.

4 For discussion of 'layered landscapes' in the Scottish Highlands, see H. Deary, 'Restoring wildness to the Scottish Highlands: a landscape of legacies', in M. Hourdequin and D.G. Havlick (eds), *Restoring Layered Landscapes: History, Ecology, and Culture* (Oxford, 2016), pp. 95–111, and other chapters in that volume.

ers who are trying to shape public perceptions of the nature of Scotland's landscape – in some cases, the human presence has been almost literally airbrushed out of the scene to emphasise the apparent emptiness of this 'wild land' – it is perhaps unfair to single out what is, after all, a successful promotional pitch designed for an agency charged with boosting national income from tourist revenues. It nevertheless serves as a useful illustration of the myth-making that for decades if not centuries has moulded perceptions of Scotland's history, of its people and most powerfully of the land itself both at home and abroad. Importantly, these are the myths which as factoids are being repackaged and redeployed by participants in current debate concerning the future shape and content of Scotland's landscapes and environment. Too often, the repetition of some of these pieces of false or misleading data and their loud declaration as 'truth' create a distorted vision of the past, present and future of Scotland's environment. But simple, indeed glib, explanations are seductive.

The airbrushing of humanity's presence and impacts from the picture presented in this marketing material – unconsciously reflecting one consequence of generations of enforced population displacement as a result of the Clearances and the lure of a better life abroad through emigration, and more recently the actual physical removal of such evidence – is a contributor to what has been labelled a 'syndrome of "shifting baselines"'.[5] This 'syndrome' has been explained as our individual internalising of the 'ecological impoverishment in our science, culture and institutions' whereby 'each generation assumes the nature they experienced in their youth to be normal and unwittingly accepts the decline and damage of the generations before', thereby perpetuating a continually moving picture. We are, however, in a dangerous place where steps to reverse that ecological impoverishment result in a cultural impoverishment. It is made worse by that impoverishment occurring through wilful, self-conscious acts of social amnesia that obliterate and then deny past

5 P. Jepson and C. Blythe, *Rewilding: The Radical New Science of Ecological Recovery* (London, 2020), p. 1.

Lubvan ruins, Laggan, Highland. The manufactured emptiness of a landscape cleared first for sheep and then to create a deer forest is for many the epitome of Highland 'wilderness'.

human contributions to the creation of landscape features that evidence our ancestors' struggles to subsist and succeed in places that are often hostile and inhospitable for settlement. A long educational tradition of teaching the history of 'Improvement' and 'Enlightenment' in schools and the ingrained popular knowledge of the Highland Clearances have, however, preconditioned some awareness of moments of significant change and their enduring physical legacies, packaged as 'positive' or 'negative' in their effects. Many people know of the transformation of rural landscapes that started in the eighteenth century and reached its zenith on the eve of the First World War, and whose physical remains still define the patterns into which the land is divided and the uses to which it is put. Similarly, many recognise and understand the consequences of industrialisation / de-industrialisation and urbanisation since 1800. These are all processes that altered the face of the land, creating agricultural landscapes or engulfing them beneath expanding towns and cities, and leaving profound scars that are the unmistakable evidence of massive human interventions of the relatively recent past.

Fewer people perceive the expanses of emptiness found in many upland districts (not only in the Highlands) as anthropogenic constructions of equally recent creation. Yet whether it is the grassy hillsides of the Borders or the heather-clad moors and mountains of the Highlands, the depopulated and often ecologically degraded places we see today are usually cultural products of eighteenth- or nineteenth-century social and economic systems which valued sheep and deer more than people. For many contemporary commentators, the point of transition that delivered these socio-economic systems is the critical baseline, such that Scotland's current environment is understood in terms of division between a pre-Improvement era of nature's abundance and a post-Improvement one of continuous and continuing loss.[6] What few acknowledge, however, is that such a baseline is arbitrary, reflecting neither the millennia

6 For a perceptive, popular exploration of the legacies of the Victorian impact on the environment and human society of the Highlands, see R. Noble, *Castles in the Mist: The Victorian Transformation of the Highlands* (Glasgow, 2016). As Robin Noble says in the first sentence of his book, 'The passing of time alters one's perceptions.'

of subtle and not so subtle shifts that have been powered by climatic, pathogenic and human agencies, nor that the systems which brought about the changes they identify were transitory and gave way to a slow-moving landslide of consequences with variable and still unfolding modern outcomes.[7] That slow-moving landslide is our unconscious experience of 'shifting baselines' and to an extent it excuses our relative blindness to the reality that environments are not static, have no 'pure' or 'perfect' state in which they should persist, but exist in constant flux. Rolling back the flow to some arbitrary, notionally pristine state, as sought for in even the most carefully planned programme of restoration ecology, at best resets the dial on the change process, but how it evolves from there will be determined by wholly different climatic, environmental and anthropogenic variables from those that obtained in the past. To suggest otherwise is to deny the realities of evolution.

Enter a new discipline

While *Standing on the Edge of Being*, the final companion volume to this one, ends with the question of where we go from here in our human relationship with the environment in Scotland, it is how we got to our present state that is the central theme of this book and of the second volume, *Where Men No More May Reap or Sow*. What is offered is an environmental history of Scotland, not a history of Scotland's environment. It is a study of human relationships with the land and the biota in and on it rather than one that focuses on the development over time of the plant life and wildlife to be found here throughout history. Such environmental histories – that is, histories that explore the interrelationship between humans and the environment in which we live, where nature matters[8] – started to emerge as a new strand in historical writing in the third quarter of the twentieth century, around the time we gradually awakened to the immensity of the negative impact we were having on the world. From the atom bomb to overpopulation, pollution and 'forever' chemicals, to resource exhaustion, societal disintegration and ecological collapse, academics in the 1960s saw a horrifying prospect of imminent apocalypse and responded with an outpouring of existential anxiety in a wave of publications that sought to understand how we had arrived at this place and how we might survive the inevitable crises to come.[9] Ironically, given the contribution of the threat of nuclear annihilation during the Cold War era to that existential anxiety, many of the major advances in that understanding were triggered by research programmes in the Arctic funded chiefly but not exclusively by the US military, which had the overarching aim of better understanding the survivability of profound global change – i.e. through nuclear war – but which began to deliver hard evidence for episodes of past global climate change. The most significant advances, however, came with the European Science Foundation-funded Greenland Ice-core Project (or GRIP, as it was known), which ran from 1988 to 1995, and subsequent

7 Deary, 'Restoring wildness to the Scottish Highlands', pp. 98–99.

8 R.C. Hoffmann, *The Catch: An Environmental History of Medieval European Fisheries* (Cambridge, 2023), p. xx.

9 For the bleak vision of humanity's future, see, for example, D.H. Meadows et al., *The Limits to Growth: A Report for the Club of Rome's Project on the Predicament of Mankind* (Falls Church, Virginia, 1972). An earlier response to concerns about overpopulation and the exhaustion of global resources came through the influential but disturbingly misanthropic work of American ecologist Garrett Hardin, whose interest in eugenics and sterilisation programmes and his opposition to a multi-ethnic society are now seen as racist and fascist. See, for example, G. Hardin, 'The tragedy of the commons', *Science*, 162 (1968), pp. 1243–1248. A more nuanced cultural historical perspective is offered by the groundbreaking but now quite dated L. White Jnr, 'The historical roots of our ecologic crisis', *Science*, 155 (1967), pp. 1203–1207.

projects which tested and expanded upon GRIP's findings.[10] Archaeologists and historians, especially those working in the North Atlantic, were quick to waken to the potential of this new form of data to explain in terms of climate change the early successes of ventures like the Norse colonisation of Iceland and Greenland and the near collapse or total failure of those colonies later in the Middle Ages.[11] If we could understand how and why that happened, might we not better understand the kinds of challenges that humanity could face as climate change and, in particular, global warming accelerated? For a country located in an especially climatically sensitive position on the north-western edge of Europe, backed by the North Atlantic and dependent for its relatively benign climatic conditions on the circulation of oceanic waters and atmospheric pressure systems, the precarity of agricultural regimes in Scotland that depended on the continuity of those mechanisms was immediately evident in the new data from Greenland.

It is important to reflect on the fact that until the data from GRIP and the subsequent GISP2 projects started to percolate beyond the climatological community – indeed until around twenty-five years ago – almost all history, not just Scottish history, was written from an entirely human and overwhelmingly male perspective. Archaeology was already moving ahead of historians in its embracing by the 1970s of evidence for ancient environmental conditions – palaeoenvironmental data – that could enrich understanding of lifestyles and practices for which there was otherwise no written record, and was developing methodologies and philosophies that looked beyond the human male experience to better understand how societies in the past had functioned.[12] Regardless of the emerging influence of gendered methodologies, however, archaeology's focus long remained almost wholly anthropocentric. It was concerned with what such evidence might tell us about the arrival and spread of agriculture, prehistoric woodland decline or the diversity of foraged and farmed resources exploited by a community. The emphasis was on the human lived experience. Reflection on the non-human experience and the role of climate change and its potential contribution to some of the

10 For GRIP and its results, see B. Stauffer, 'The Greenland Ice Core Project', *Science*, 260:5115 (1993), pp. 1766–1767, https://doi.org/10.1126/science.260.5115.1766, and J. Anklin et al., 'Climate instability during the last interglacial period recorded in the Greenland Ice-core Project', *Nature*, 364:6434 (1993), pp. 203–207.

11 The 'collapse' thesis in respect of the Norse western colonies in Iceland and Greenland developed strongly through the late 1990s and early 2000s. See, for example, O. Vésteinsson, T.H. McGovern and C. Keller, 'Enduring impacts: social and environmental aspects of Viking age settlement in Iceland and Greenland', *Archaeologia Islandica*, 2 (2002), pp. 98–136; J. Diamond, *Collapse: How Societies Choose to Fail or Succeed* (New York, 2005); A. Dugmore, C. Keller and T. McGovern, 'Norse Greenland settlement: reflections on climate change, trade, and the contrasting fates of human settlements in the North Atlantic islands', *Arctic Anthropology*, 44:1 (2007), pp. 12–36; and A. Dugmore et al., 'Norse Greenland settlement and limits to adaptation', in W.N. Adger, I. Lorenzoni and K. O'Brien (eds), *Adapting to Climate Change: Thresholds, Values, Governance* (Cambridge, 2009), pp. 96–113. More recent research has revised significantly and reframed the stark model as sensationally popularised by Diamond. See, for example, T.H. McGovern et al., 'Landscapes of settlement in northern Iceland: historical ecology of human impact and climate fluctuation on the millennial scale', *American Anthropologist*, new series, 109:1 (2007), pp. 27–51, and S. Hartman et al., 'Medieval Iceland, Greenland, and the New Human Condition: a case study in integrated environmental humanities', *Global and Planetary Change*, 156 (2017), https://doi.org/10.1016/j.gloplacha.2017.04.007.

12 See, for example, the thinking on the potential of such data in J.G. Evans, *An Introduction to Environmental Archaeology* (St Albans, 1978), and, for a reflection on how thinking had developed by the turn of the millennium, K.J. Edwards and I.B.M. Ralston (eds), *Scotland after the Ice Age: Environment, Archaeology and History 8000 BC–AD 1000*, 2nd edition (Edinburgh, 2002).

Glacier, Prince Christian Sound, Greenland. The now fast-melting Greenland ice sheet contains a record of millennia of northern hemisphere climate change. (Image: Anna Ritchie.)

changes evident in the archaeological record developed only gradually through the 1980s. From there, however, application of that alternative 'gaze' evolved rapidly such that by the 1990s perspectives had already swung round, to the point where climate was being proposed by some researchers as one of the most significant influences on human cultural development.[13]

While archaeologists were pursuing these exciting new approaches to their discipline, many academic historians remained oblivious to, or dismissive of, the potential for similar forms of data to expand understanding of processes that they could otherwise barely descry in their documents. That is not to say that there were no developments in historical methodologies, but by the late 1970s, while some scholars adopted a top-down or bottom-up social approach, or attempted to adopt a gendered perspective, historiography still largely followed the human – usually male – gaze. Most research remained driven by textual evidence or, occasionally, oral record, giving only the most cursory of nods to other forms of evidence. For the most part it also considered the world in which we live as a mere backdrop to political events, economic developments or individual human experiences. That world, furthermore, existed in a

13 The contribution of climate studies and ecology is evident, for example, in T. O'Connor and J.G. Evans, *Environmental Archaeology: Principles and Methods*, 1st edition (Stroud, 1999) and 2nd edition (Stroud, 2005). The more cautious view set out in G. Whittington and K.J. Edwards, 'Climate change', in K.J. Edwards and I.B.M. Ralston (eds), *Scotland after the Ice Age: Environment, Archaeology and History 8000 BC–AD 1000*, 2nd edition (Edinburgh, 2002), pp. 11–22, argues for subtler and often short-term events that could have major rapid but transitory impacts on local conditions.

Aonach Mor and the mountains south of Glen Affric from the east ridge of Carn Eige, Highland. Scientific understanding of the effects of the last ice age of the Younger Dryas and the recovery from it in the early stages of the present Holocene era expanded rapidly in the early twentieth century. Research into the sub-arctic area of Scotland's montane regions continues to add to that understanding.

relatively constant physical state that changed only when we humans decided that it needed to be improved upon, or in some way modified to deliver more of what we wanted from it. In that entirely anthropocentric worldview, a capitalised 'Nature' was a relatively passive force that impinged little on human decision-making and, when it did, was there to be subdued to the will of humankind.

This negative perspective needs some tempering, for from the middle of the twentieth century, before even GRIP had been conceived, growing scientific awareness in Europe and North America of a far more complex interplay between humans and the planetary system in which we live had begun to permeate some areas of historical thinking. In the early twentieth century, work by Scandinavian palaeobotanists provided the underpinning data for what was presented as a four-stage development of the northern hemisphere's climate regime across the Holocene era from the end of the last ice age to the present, revised in the middle of the century into a new, tripartite scheme, then revised again more radically in the last quarter of the century. Until the GRIP and GISP2 data began to become available and be analysed and modelled, the view that prevailed for Scotland was of a rapid post-glacial warming and increasing rainfall to the early sixth millennium BC, after which shifts in climate regime were 'relatively minor'.[14] The impact of this climate awareness was strongest in areas like historical geography and climatology, where researchers were attracted by the emergent scientific data and began to advance their own

14 Whittington and Edwards, 'Climate change', p. 14.

interpretations of the way in which larger and more powerful forces – like climate change or pandemic and panzootic disease – might have impacted upon human decision-making and daily life in the past.

At the forefront of this new thinking in Britain was the historical climatologist Hubert Lamb, whose research by the mid 1960s pointed towards episodes of climatic warming and cooling that, although 'relatively minor', had had effects on the largely rural society of the Middle Ages that were likely to have been profound. He identified the most pronounced warming episode as having spanned the period from the later eleventh to the mid thirteenth century, and a major cooling that had started in the fifteenth century and ended in the nineteenth.[15] His theories on the warming episode that he termed 'the medieval warm epoch' struck a chord with English medieval economic historians led by Michael Postan. It was Postan who developed a depressingly neo-Malthusian thesis of population growth outstripping the capacity of the land to feed the expanded numbers, triggering a subsistence crisis and then catastrophic famine- and disease-delivered collapse. He argued that the initial growth was enabled by the benign conditions of what we now style the Medieval Climate Anomaly (MCA), which had stimulated population growth after a long period of stagnation from the middle of the first millennium. Growth of communities created pressures on resources and on individuals' social advancement opportunities – to secure a marriage, young males needed first to acquire land on which to support a family – and so drove peasant land-hunger, expansion into new areas and intensification of exploitation in older settlement landscapes. Postan saw these processes as creating a spiral of ultimately unsustainable growth, which led to overpopulation and soil fertility depletion, leading in turn to yield decline, harvest failure, famine and mass mortalities.[16] It was a compelling model, especially when viewed within the context of 1960s anxieties over global population levels and 'the limits to growth', and it set a trend of enquiry which for the first time looked at the science within the natural processes upon which human society depended. Unfortunately, Postan's use of the extensive surviving southern and Midlands English manorial accounts, which documented expansion of the area under cultivation, population figures, yield levels and stock numbers, coupled with Lamb's analysis of narrative sources for weather trends and his own meteorological models, could not be replicated for northern England, let alone Scotland. Such records simply do not exist for the northern parts of the British Isles. The result was that beyond a disappointingly meagre handful of generalised statements about population trends and agricultural expansion, the revolution in English historical thinking on the interplay of climate and human populations throughout the Middle Ages and beyond failed to take root north of the border.[17]

Largely as a consequence of that perceived lack of equivalent data, as late as the early 1990s the major overview histories of Scotland could still be written with a perspective in which almost every determinant – plague pandemics apart – in the socio-economic and political evolution of the Scottish state was founded in human action.[18] But the steady increase

15 H. Lamb, 'The early medieval warm epoch and its sequel', *Palaeogeography, Palaeoclimatology, Palaeoecology*, 1 (1965), pp. 13–37. His thoughts on the warm episode and the subsequent 'little ice age' are refined considerably in H. Lamb, *Climate, History and the Modern World* (London, 1982).

16 The fully evolved 'thesis' was expounded in M.M. Postan, *The Medieval Economy and Society: An Economic History of Britain in the Middle Age* (London, 1972).

17 See the bland statement of medieval agricultural conditions in R. Mitchison, *A History of Scotland* (London, 1970), and the barely changed comments in the 2nd edition (London, 1982), by which time Lamb's work had circulated widely and Postan's thesis had been articulated.

18 See, for example, M. Lynch, *Scotland: A New History* (London, 1991).

in the availability of scientific data to provide proxies for both climate and weather records and of historical materials such as manorial accounts and yield figures meant that the delayed revolution was imminent. The new movement was spearheaded from the 1960s in social history and from the late 1990s in environmental history by the work of one scholar, Christopher Smout, the pre-eminent social and economic historian of early modern Scotland. In the decade following his retirement from an academic post at the University of St Andrews, a prodigious outpouring of new research and forensic re-examination of much of his older work saw Smout single-handedly introduce Scotland to the potential of environmental history as a tool by which to better understand human relationships and interdependencies with the places in which they lived, the resources available there and the trees, plants and wildlife with which they shared the land. The influence of his work will be most evident in *Where Men No More May Reap or Sow* and *Standing on the Edge of Being*, in the exploration of the period from 1500 to the present.

It was Smout who masterminded the setting up of a joint Centre for Environmental History and Policy at the universities of St Andrews and Stirling, its evolution into the then AHRB-funded Research Centre for Environmental History, and from 2002 the development at Stirling especially of programmes of teaching and research in this field. Over the last twenty-five years, the Stirling Centre became the principal focus of enquiry into the environmental history of medieval and early post-medieval Scotland and the output of several of the scholars associated with it has shaped the themes covered in this volume and the two following volumes that continue its narratives down to the present. Despite that output, however, and despite a sustained effort to communicate research findings in an accessible manner that could carry them far beyond a narrow academic community, popular awareness of the complex histories that underlie many of our current environmental issues remains shallow and fragmentary. Myth and misunderstanding have thrived as new audiences have become engaged and enraged by the ecological loss with which they are everywhere confronted, and the seeming indifference or incapacity of governments, agencies and the public at large to respond effectively to the continuing degradation and destruction. Much of their outrage is based on misconception and misinformation and is often misdirected. Most is fuelled and targeted by the factoids distributed by usually well-meaning but ill-informed commentators and activists, whose perception of the past has been framed by the often vivid but ultimately fantastical word imagery of nineteenth-century authors rather than by the cooler, carefully researched – but sometimes poorly communicated – findings of modern academics.

Modern myth-making, however, has directed its gaze much deeper into the human past in Scotland than the reach of the nation's recorded history, and it is there that the means to dismantle some of the most persistent of these distortions lie. Although this book and its companion volumes cover only the last two millennia, it is vital to recognise that an environmental history of Scotland does not simply span that 'historic' era or commence with the earliest surviving written accounts relating to the geographical area occupied by the modern nation. Using different forms of non-textual record evidence, we can instead take an environmental history far deeper into what is conventionally labelled 'prehistory'. Using climate proxy data (forms of evidence that provide a record of prevailing climatic conditions), such as the annual growth rings of some tree species (whose thicknesses vary according to the tree's sensitivity to temperatures and precipitation), speleothems (the limestone structures that form in caves as a result of the slow percolation of water through the rock strata above, which varies according to annual rainfall or drought), palaeoenvironmental records (such as pollen records or ancient insect and other invertebrate remains) and the physical evidence from archaeological excavation and field survey, it is possible to synthesise an account of the main trends and processes that shaped the environment through which the first settlers in this land and some two hundred generations of their successors moved before the Roman general Agricola led his armies north or his son-in-law the historian Tacitus set pen to parchment.[19]

Weather's drivers

So, what are these primarily non-anthropogenic forces that underlie such fundamental transformations of the environment upon which Scotland's human population depended? In the second half of the twentieth century, but especially from the 1990s as the results of the GRIP project unfolded, there was growing scientific awareness that whilst the post-glacial climate regime might have seen only 'relatively minor changes' there were other factors that affected prevailing weather patterns as opposed to more general climatic conditions. In particular, there was better understanding of the degree to which major oceanic and atmospheric circulation systems drive prevailing weather patterns. The growing research into global circulatory systems, however, indicated that even relatively minor variations can have potentially beneficial or catastrophic consequences for human societies. Such variable outcomes were greatest for societies that had developed economic practices – agriculture, fisheries and other forms of resource management – founded on traditional environmental knowledge, or TEK, shaped during long phases of relatively stable conditions by close observation across generations of the rhythms of the seasons and patterns of weather. In the North Atlantic, the most important of the oceanic circulation systems is that known by the acronym AMOC, otherwise the Atlantic Meridional Overturning Circulation, which is the major system of oceanic currents that carry warm water from the tropics northwards. It works like a conveyor belt that is powered by differences in water density that arise from changes in temperature and salinity. When the water flows north it gradually cools but also sees some evaporation, which causes the salt content of the remaining water to increase. The low temperatures and higher salt content cause the water to become denser. As the density increases, the dense water sinks deep into the ocean – several thousand metres beneath the surface – and begins to spread southwards. There, it begins to be drawn towards the surface, warming as it rises, and so starts a new circulation.

But it is not solely the AMOC which has an effect on the British Isles, and we have become increasingly aware of the impact of the greatest global circulation system in the Pacific Ocean and its influence on weather patterns even in the north-eastern Atlantic. Most people are familiar with the El Niño/La Niña phenomenon by name but less so with what it is and

19 Progressive development in an understanding of the value of such proxies can be traced through, for example, C.J. Proctor et al., 'A thousand year speleothem proxy record of North Atlantic climate from Scotland', *Climate Dynamics*, 16 (2000), pp. 815–820; D.J. Charman et al., 'Paleohydrological records from peat profiles and speleothems in Sutherland, northwest Scotland', *Quaternary Research*, 55:2 (2001), pp. 223–234; A. Crone and C.M. Mills, 'Seeing the wood and the trees: dendrochronological studies in Scotland,' *Antiquity*, 76 (2002), pp. 788–794; C.J. Proctor, A. Baker and W.L. Barnes, 'A three thousand year record of North Atlantic climate', *Climate Dynamics*, 19 (2002), pp. 449–454; C. Mills, 'Historic pine and dendrochronology in Scotland', *Scottish Woodland History Discussion Group: Notes*, 13 (2008), pp. 9–14; A.G. Cage and W.E.N. Austin, 'Marine climate variability during the last millennium: the Loch Sunart record, Scotland, UK', *Quaternary Science Reviews*, 29 (2010), pp. 1633–1647; C.M. Mills and A. Crone, 'Dendrochronological evidence for Scotland's native timber resources over the last 1000 years', *Scottish Forestry*, 66:1 (2012), pp. 18–33; and M. Rydval et al., 'Reconstructing 800 years of summer temperatures in Scotland from tree rings', *Climate Dynamics*, 49:2951 (2017), pp. 2951–2974.

Sunset over the North Atlantic, looking west from near Virda Field, Papa Stour, Shetland. Whilst we have long had a heightened awareness of the way our interrelationship with the surrounding seas has shaped us culturally, our understanding of the influence of oceanic and atmospheric circulation patterns on our environment is still growing.

how it affects us. 'El Niño' means 'the little boy' in Spanish and is derived from the name that fishermen on Peru's Pacific coast used from the 1600s for periods of unusually warm water in the ocean that affected the migration of the shoals on which they depended. Its full name is 'El Niño de Navidad' (the little boy of the Nativity, i.e. the Christ Child), because the warm-water episode usually peaked in December. 'El Niño' is now used to describe this episodic warming of sea surface temperature, which is concentrated normally in the central-east equatorial Pacific. In Met Office terms, an El Niño is declared when 'sea temperatures in the tropical eastern Pacific rise 0.5 °C above the long-term average. El Niño is felt strongly in the tropical eastern Pacific with warmer than average weather'.[20] El Niño's counterpart is La Niña or 'the little girl'. It is the term adopted by climatologists from the traditional name applied to the opposite side of the Pacific's temperature fluctuation. This is episodes of cooler than average sea surface temperature occurring in that same region of the equatorial Pacific. The conditions for declaring a La Niña event differ between various monitoring agencies, but it is generally accepted that during an event surface temperatures might fall to 3–5 °C below average. This is generally experienced as cooler, drier than average weather in the tropical eastern Pacific.

But it is not just the Pacific region that is affected by the El Niño/La Niña cycle. We now know that the cycle has global ramifications, with often severe weather impacts in regions distant from the Pacific. In the northern British Isles, El Niño years are recognised as an important factor with the potential to increase the threat of colder winter weather. El Niño has been seen as a limiter on the development of tropical storms in the North Atlantic. By way of contrast, La Niña can enhance development of storm systems and cause increased rainfall in some regions but droughts or cooler temperatures in others. As a note here, current monitoring at the time of writing suggests that summer and autumn 2024 will see a La Niña event.

Neither the AMOC nor El Niño/La Niña function in isolation; both are accompanied by major switches in atmospheric pressure systems and circulation. For AMOC there is the North Atlantic Oscillation and for El Niño/La Niña there is the El Niño Southern Oscillation, or ENSO, which is an irregular interannual cycle involving the oscillation of atmospheric pressure changes between the east and west tropical Pacific regions and is now known to accompany both El Niño and La Niña episodes. The ENSO name is a reminder of the intimate interaction of atmosphere and ocean that is an essential dimension of the weather-shaping processes they represent. Of the multiple examples of such systems interactions globally, ENSO is the single most dominant phenomenon driving weather variability on year-to-year timescales across much of the planet. From a historical perspective it is essential to recognise that these

20 Met Office, 'What are El Niño and La Niña', https://www.metoffice.gov.uk/weather/learn-about/weather/oceans/el-nino.

systems are also not modern developments, although their increasing dynamism and unpredictability is linked to anthropogenic factors. In recent experience, these events have been associated with widespread disruption of global climate systems that can last several months, sometimes recurring annually, and have been seen to lead to significant impacts on human life, affecting everything from agriculture (good and bad harvests) and fish migration patterns, the physical infrastructure of buildings, roads and bridges, human and animal health and the provision of traditional sources of thermal energy.

In the northern part of the British Isles, most direct weather impacts arise from the North Atlantic Oscillation, or NAO. This is the name applied to the changing relative relationships of the area of atmospheric high pressure centred on the Azores (the so-called Azores high) and low pressure over Iceland (the so-called Icelandic low). The interaction between these systems exerts a strong influence on winter weather and climate patterns in Europe and North America, as it causes changes in the intensity and location of the North Atlantic jet stream, the flow of very fast winds high in the atmosphere that affects the movement of regions of low pressure. The Met Office describes the alternating states of the NAO as follows:

> The positive [or 'high'] NAO phase represents a stronger than usual difference in pressure between the two regions. Winds from the west dominate, bringing with them warm air, while the position of the jet stream enables stronger and more frequent storms to travel across the Atlantic. These support mild, stormy and wet winter conditions in northern Europe and eastern US. Conversely, northern Canada, Greenland and southern Europe are prone to cold and dry winter conditions.
>
> The negative [or 'low'] NAO phase represents the reverse with a weaker than usual difference in pressure. Winds from the east and north-east are more frequent, bringing with them cold air, while the adjusted position of the jet stream leads to weaker and less frequent storms. As a result, Europe and eastern US are more likely to experience cold, calm and dry winters. In contrast, northern Canada and Greenland will tend to be mild and wet.[21]

How we see the effects of these phenomena on Scotland is founded on a combination of modelling based on different forms of climate proxy data, physical evidence from archaeological excavation, principally palaeoenvironmental records, and the often qualitative data of surviving written records.

Introducing climate proxy data

Our evidence for what would have been the lived experience of the physical effects of these various atmospheric and oceanic phenomena is derived mainly from what are referred to as climate proxy records. The most important long-sequence proxy record – that is, one which provides a run of data across a very long chronological span – is from GRIP and GISP2 and most of our temperature and precipitation models for Scotland draw heavily on that ice core data. Obviously, Greenland is geographically distant from the northern British Isles and there would have been considerable buffering of weather effects by the intervening ocean. The development of geographically more proximate proxy sequences has therefore been of huge importance over the last two decades in enabling the construction of more nuanced models. There are currently three main sets of widely available proxy data from Scotland – the Traligill speleothem record (mainly winter precipitation), the Loch Sunart bottom water temperature record and the Cairngorm tree-ring sequence – all of which are linked closest to weather patterns orig-

21 Met Office, 'North Atlantic Oscillation', https://www.metoffice.gov.uk/weather/learn-about/weather/atmosphere/north-atlantic-oscillation.

Weather front and rain squalls over the Ross of Mull, Argyll and Bute. The positive or 'high' NAO state delivers waves of rainy and possibly stormy weather to the western, Atlantic-facing parts of Scotland.

inating in the NAO. Of these proxies, only the Traligill and Loch Sunart records currently extend into the deeper past, with the speleothems offering a three millennia record and the sediments one millennium; the Cairngorm data so far available starts only in the early thirteenth century.

The first of these proxies comes from western Sutherland in the record of the Uamh an Tartair caves in the Traligill basin in Assynt, where the growth rate in the annually deposited laminae of the speleothems – stalagmites and stalactites – reflects the levels of moisture percolating through from the peat overlying the subterranean system. Speleothems offer a well-nuanced proxy record for winter precipitation with strong regional significance. For north-west mainland Scotland at least, the speleothems from Uamh an Tartair have begun to yield good evidence for North Atlantic climate variability over a 3,000-year period and allow the construction of broad models of the winter climate regime experienced in Atlantic-facing northern Scotland. Uamh an Tartair's record has been analysed in most detail for the thirteenth century and later, but the same kind of trends of high/low winter precipitation revealing positive and negative NAO states are identifiable for earlier centuries. The MCA sequence commences with three decades of high winter precipitation to around 1030. This is associated with a high winter NAO characterised by strong westerly airflow and cyclonic rainfall, then plunges to the lowest precipitation – signalling a low winter NAO – recorded for the north-west Highlands until the early seventeenth century, with the dry conditions lasting into the 1080s. Low winter NAO is associated more with cold, dry winters and a northerly or easterly airflow but the depth of the change reflected in the laminar growth of the speleothems at this point perhaps signals the additional influence of an episode of low sunspot activity known as the Oort Solar Minimum.[22]

At Loch Sunart, the long, deep sea loch that penetrates the western Highland coast to the south of the Ardnamurchan peninsula, isotopic analysis of benthic invertebrates – clams, worms and crustaceans – from the bottom waters of the inlet yielded a record of summer coastal temperature anomalies spanning the

22 The role of solar minima in affecting climate and weather on Earth is still deeply contested. What can be said is that there is a marked coincidence of phases of solar minimum sunspot activity and periods of intense climatic disturbance and weather extremes. These phenomena are discussed in greater detail in Chapters 9 and 10 and in the volume *Where Men No More May Reap or Sow*.

Loch Sunart, looking east over the island of Oronsay, Highland. The long, fjord-like sea loch that runs deep inland between Ardnamurchan and Morvern provides ideal conditions for the development of the seabed sediments that record temperature changes over long time spans.

last millennium. Their clearest evidence is for a change in climate regime that commenced around 1400, which the researchers saw as being driven by the large-scale reorganisation of northern hemisphere atmospheric circulation that marked the transition into the 'little ice age' (LIA). Although this climate shift has been classed as minor in comparison to that which heralded the onset of the last true ice age and its worst effects were localised to the North Atlantic and Europe primarily, it was sufficiently significant to deliver profound change in the weather patterns upon which Scotland's agricultural economy was founded. This transition point and the changes that it heralded are the topic of discussion in the opening chapters of *Where Men No More May Reap or Sow*.

In comparison with many other parts of Europe and especially Ireland, Scotland lacks a long run of climate data derived from the annual growth rings of certain tree species. In the 1980s, preliminary research started to indicate that there was potential for the reconstruction of a sequence of historic summer temperatures in the central Highlands, using pine timbers rather than the oak that is the basis for the deep chronologies constructed for Ireland. That work ran into the barrier of the relatively young age of the majority of still growing trees – around 225 years old – and a lack of knowledge of the provenance of surviving structural timbers of greater age. In the early 2000s, however, there was a breakthrough using modern, historical and subfossil pine material obtained from current native pinewoods, from historic structures and from loch-bed sediments. The material recovered spans the last 8,000 years and we are fortunate that the published results of a first stage of analysis, tree-ring sequences from the Cairngorms, commence around 1200. These sequences give us insight into summer conditions in the central and north-east Highland uplands over the last 800 years, taking us through the era from the decline of the MCA to the end of the LIA.

As an illustration of the value of these proxies, let us look at just one short episode in the thirteenth century. We will return to this material later in the volume. From the Cairngorm tree-ring evidence, we have a record of summers in this region having been generally cooler since the start of the century, declining slightly through the 1210s but becoming significantly cooler through the 1220s. The extreme for that decade was reached in 1227, which was the fifth coldest year across the entire 800-year sequence currently available, but worse was to come. The extreme European winter of 1233/4 has been described as 'the most outstanding meteorological episode of the thirteenth century',[23] and Scottish proxy data suggests that this was indeed a time of great cold. At Traligill, the speleothem growth rings indicate that the 1230s were the driest decade of the thirteenth century, albeit with still relatively high levels of precipitation.[24] What this points to is a marked weakening of winter NAO which brought a much colder anticyclonic airflow pulling Arctic air south and across Scotland. This cooling is also present as a sharp dip in the Loch Sunart deep water temperatures – followed by an equally sharp positive movement – perhaps pointing to a combination of cooler ocean surface temperatures and cold run-off from the surrounding land.[25] In the Cairngorms, summer 1232 was the coldest of the last 800 years.[26] Collectively, these fragments of evidence point to the late 1220s and early 1230s as a brief episode of sharply deteriorating summer and winter weather conditions, followed by a bounce back to the warmer conditions of previous decades. Taken together, this data provides some context for the political disturbance of this period, which saw King Alexander II facing challenges from the same northern and western regions that the proxies indicate experienced adverse weather conditions. It is a matter of record that the king's campaigns in the Clyde estuary, Sutherland and Caithness and Galloway were all hindered by adverse summer weather. Alexander, however, was by no means the first and certainly not the last ruler to have his plans upset by circumstances beyond his control. Indeed, both our archaeological and palaeoenvironmental record and our albeit fragmentary historical docu-

23 C. Pfister et al., 'Winter air temperature variations in western Europe during the Early and High Middle Ages (AD 750–1300)', *The Holocene*, 8 (1998), pp. 535–552 at 545.

24 Proctor et al., 'A thousand year speleothem proxy record', p. 818 fig. 4; B.M.S. Campbell, *The Great Transition: Climate, Disease and Society in the Late-Medieval World* (Cambridge, 2016), p. 46 fig. 2.6 (A).

25 Cage and Austin, 'Marine climate variability', p. 1641 fig. 7.

26 Rydval et al., 'Reconstructing 800 years of summer temperatures', p. 2960.

Glen Derry and Glen Luibeg pinewoods, southern Cairngorms, Aberdeenshire. The living pines, 'sub-fossil' remains found in peat or in loch-bed sediments and pine timbers from historic buildings are providing new data for climate change and related weather impacts on tree growth in this region over many centuries.

ments indicate that the human experience in Scotland has been shaped by interactions with climate, weather and the environment since the first human arrival in this land.

History versus mythology

Exactly when those first human settlers arrived in Scotland will probably never be known, as coastal erosion and changes in sea level, millennia of intensive land-use and modern urban sprawl have erased many places that would have been attractive to these first people. It was long believed that no humans could have subsisted in the northern parts of the region that is now the British Isles until several millennia after the last phase of the ice age, the so-called Younger Dryas era. That episode had seen the re-establishment around 12,900 years ago of an ice sheet over north-western Scotland, extending from the Torridon mountains to the south-eastern end of the Loch Lomond basin, in what we know as the Loch Lomond Stadial or Loch Lomond Readvance. The ice developed and spread for over a millennium, until around 11,700 years ago the climate began to warm and the ice contracted rapidly, signalling the start of a new climate era, the Holocene. Archaeologists have found increasing evidence, however, that human populations were present in what is now Scotland from around 12,900 BC and remained within the region on the southern fringes of the ice-covered zone throughout the stadial. They were members of a culture that used the stone tool technology known as 'Hamburgian' (after the north German site locations near the city of Hamburg where encampments were identified), which developed during the period to which the broad label of Late Upper Palaeolithic (Old Stone Age) is applied. This culture, identified from artefacts, art and midden remains as hunters of the reindeer herds that thrived on the tundra at the edge of the ice sheet, persisted through the Loch Lomond Stadial and until around 10,000 BC, when ice probably still lingered on the high summits and in corries.[27] Research into these early people is just commencing and a new 'history' of the earliest human settlement of Scotland starting to be written.

We can be sure that it was the rapidly warming climate at the start of the Holocene that created the conditions which drew many more groups of hunters and gatherers north, users of a new form of stone technology that characterised the period we know as the Mesolithic (Middle Stone Age), starting a 10,000-year interaction between humans, land and environment that shaped the cultural artefact that is our modern landscape and the distinctive ecologies that now exist upon it. Some studies, influenced by the growing perception of human impacts on the environment as wholly negative, sought for evidence to support suggestions that such transformations were as old as the earliest human presence in Scotland. Thus, it was argued that by around 8000 BC significant anthropogenic interventions were already having an impact on the physical character of the land and the vegetation it supported, starting that slow landslide of ecological change that has given rise to some of Scotland's regionally distinctive landscapes. With hindsight, we can recognise that there is an inherent danger in applying a deterministic interpretation to such evidence, where the 'butterfly effect' of an event that is perhaps only indirectly attributable to human action can be overblown into a proposition that from that one action flowed an inevitable trend that resulted in a negative outcome whose legacies remain with us today. In the Flow Country of north-east Sutherland and Caithness, for example, it was proposed in the 1990s that widespread burning of vegetation was already being practised by the Mesolithic hunter-gatherer peoples who moved through the region around 9,000 years ago. That burning, it was suggested, started to impede the development of woodland cover by tree species already at the limit of their geographical range, over four millennia before the arrival of Neolithic pastoralists and cultivators.[28] While some archaeologists and palaeoenvironmental specialists further suggested that this periodic burning might have contributed to the start of the waterlogging that led to the development of the iconic peat wetlands of the Flow Country, others have argued strongly that the two are more likely unrelated and have pointed to the very limited nature of the charcoal spreads that have

27 The key site whose evidence triggered the reappraisal is Howburn Farm in South Lanarkshire. See T.B. Ballin et al., 'An Upper Palaeolithic flint and chert assemblage from Howburn Farm, South Lanarkshire, Scotland: first results', *Oxford Journal of Archaeology*, 29:4 (2010), pp. 323–360; T.B. Ballin and C. Wickham-Jones, 'Searching for the Scottish Late Upper Palaeolithic: a case study from Nethermills Farm, Aberdeenshire', *Journal of Lithic Studies*, 4 (2017), https://doi.org/10.2218/jls.v4i1.1907; and T.B. Ballin et al., *Reindeer Hunters at Howburn Farm, South Lanarkshire: A Late Hamburgian Settlement in Southern Scotland – Its Lithic Artefacts and Natural Environment* (Oxford, 2018). For more recent developments, see K. Hardy and T. Ballin, 'South Cuidrach: a new Late Upper Palaeolithic and Mesolithic site on the Isle of Skye, western Scotland', *Proceedings of the Society of Antiquaries of Scotland*, 153 (2024), https://doi.org/10.9750/PSAS.153.1367, and T.B. Ballin, 'ARO48: A likely late Upper Palaeolithic assemblage from Lunanhead in Angus: re-examination and discussion', Archaeology Reports Online (2021), https://www.archaeologyreportsonline.com/PDF/ARO48_Lunanhead.pdf.

28 D.J. Charman, 'Late-glacial and Holocene vegetation history of the Flow Country, northern Scotland', *New Phytologist*, 127 (1994), pp. 155–168 at 166; C.R. Wickham-Jones, *Scotland's First Settlers* (London, 1994), p. 61.

Looking south-west over the Bog na Gaoithe to the Sutherland hills and the Caithness Flow Country, Highland.

been claimed as evidence of extensive woodland or heathland burning. Most recent studies have rejected anthropogenic causality in peat formation, other than where cultivated methods have reduced natural drainage and triggered waterlogging when rainfall levels began to increase.[29] Rather than being the unwanted end product of a human action, blanket peat spread has been proposed alternatively as 'an inevitable but rapid end-stage to soil development' in what was generally a cold, wet climate conducive to its growth, or as a consequence of past climatic change.[30]

That shifting understanding in respect of one of the most iconic landscapes of northern Scotland, a landscape that at the time of writing has been submitted by the UK Government for designation as a UNESCO World Heritage Site,[31] is representative of a wider series of largely myth-driven controversies concerning the interplay of human and non-human factors in the shaping of Scotland's current environment and cultural landscapes. The most contentious of these controversies, whose influence will be seen recurrently throughout this volume and in both *Where Men No More May Reap or Sow* and *Standing on the Edge of Being*, is that regarding the degree to which the decline of Scotland's woodland cover – where its changing form and the processes affecting its fate have become the principal proxy for exploration of wider environmental loss – has been a consequence of human actions as opposed to successive climatic switches and associated weather trends.[32] From medieval to nineteenth-century efforts to rationalise the disappearance of extensive post-glacial 'climax forest', based on observation of layers of preserved

29 R. Tipping, 'Blanket peat in the Scottish Highlands: timing, cause, spread and the myth of environmental determinism', *Biodiversity and Conservation*, 17 (2008), pp. 2097–2113, where findings from Oliclett in Caithness and four other sites have been used to challenge Charman's model.

30 Ibid., p. 2097.

31 BBC Scotland News, 2 February 2023, 'Formal bid made for Flow Country UNESCO status', https://www.bbc.co.uk/news/uk-scotland-highlands-islands-64497032. Since this text was written, that submission has been successful (July 2024) and the Flow Country inscribed in the list of World Heritage Sites.

32 Current scientific understanding of the extent of woodland cover and the mixture of anthropogenic and climatic factors in its decline is outlined in T.C. Smout, A.R. MacDonald and F. Watson, *A History of the Native Woodlands of Scotland, 1500–1920* (Edinburgh, 2005), pp. 25–34.

Scots pine trees, Allt Mheuran, Glen Etive, Highland. For around two hundred years, fragments of 'Caledonian Forest' of this kind have been misrepresented as remnants of a once vast 'Great Wood of Caledon'.

timbers in peat bogs,[33] to later nineteenth- and earlier twentieth-century scientific understanding of the legacies of past land-use practices that were hostile to woodland regeneration, and even as far as the 1990s, the narrative was almost entirely one of continuous decline caused by human agency. Central to this tradition was the construction of the myth of the 'Great Wood of Caledon', which is discussed in depth in Chapter 1. This myth, in its fullest nineteenth-century development, led to a vision of Scotland as still swathed shore to shore in dense woodland – mainly of Scots pine or pine and oak – that blocked the advance of Roman legions, from whom safe refuge was provided within this 'primeval northern forest'[34] for the Celtic tribesmen who refused to bow to the might of empire.

Although already challenged by the 1860s and comprehensively demolished in a series of major studies in the 1990s and early 2000s, the myth has proven resilient and persists in much popular and populist environmental literature. It is especially prevalent in calls for an acceleration of woodland

33 As recently as the late 1960s, it was still believed that such peat-preserved timbers were of relatively recent date and so evidenced the existence of extensive forests into post-medieval times in what by then were treeless expanses of wetland. See, for example, M.L. Anderson, *A History of Scottish Forestry*, 2 volumes (Edinburgh, 1967), volume 1, pp. 66–71.

34 H. Miles and B. Jackman, *The Great Wood of Caledon* (Lanark, 1991), p. 11. The impact of this beautiful but flawed book is discussed in Chapter 1 and subsequently.

restoration with the express purpose of the recreation of this 'lost' forest in large-scale rewilding projects. Further fuelled by the frequent misused interchangeability of the 'Great Wood of Caledon' of myth and the scientific category 'Caledonian Forest', which lends an air of hard scientific respectability and veracity to the ahistorical fantasy woodland, the written record has been weaponised as the unimpeachable authority for the vision of a blanket of trees that swathed the land until Roman legionaries and later waves of invaders destroyed them. It is understandable that this particular myth has secured such prominence, for it offers a modern parable for the environmental impact of human actions and the possibility of remediation, and consequently a form of conscience-salving atonement and potential redemption for our ancestors' crimes. While this 'putting right the wrongs of the past' approach undoubtedly yields benefits in the heightened public engagement with afforestation and rewilding debates, the distorted picture which it presents of the processes that contributed to woodland decline is unhelpful for a wider understanding of cultural landscapes, historic environments and the interleaving of human and non-human agency in effecting environmental change. Most damagingly, where the myth of the Great Wood is deployed commonly and uncritically, the contribution of climate change to past, current and possible future episodes of woodland loss is omitted or packaged as another aspect of the Holocene's catalogue of anthropogenic destruction.[35]

Since the development of carbon-14 dating in the 1950s, which made it possible to determine the relative antiquity of the ancient timbers present in peat bogs, coupled with the increasing sophistication of pollen analysis, it has been clear that major episodes of large-scale woodland loss occurred at a series of points through Scotland's prehistoric era. A detailed picture of the evolution and decline of woodland over the last ten and a half millennia, based on scientific evidence, was formulated and refined in the early 1990s.[36] The picture is one of gradual evolution over several thousand years after the end of the last glacial episode in Scotland, around 8300 BC, at which time Scotland's only woodlands were probably groves of juniper. Scattered remnants of juniper scrubland that remain in the valleys fringing the Cairngorms are descendants of a genuine 'primeval forest'. This ice age survivor was joined over the next millennium by new arrivals. The first colonists were willow and rowan, then birch and some centuries later hazel. The tree long believed to have constituted the bulk of the mythical 'Great Wood' – Scots pine – had established a substantial presence in Wester Ross by around 7000 BC, possibly spreading from a refugium that had escaped the Loch Lomond ice sheet but has now been inundated by the Atlantic, or from Ireland, which had not been subject to the full extent of ice cover. Pine colonised the north and west Highland area rapidly in the following centuries, some from this initial Wester Ross zone. Other pine populations came from unknown sources to the east or south, perhaps out of the now submerged 'Doggerland' of the North Sea basin.[37] Although dominant through much of the central, western and northern High-

35 An overview is offered in Smout, MacDonald and Watson, *A History of the Native Woodlands*, pp. 20–25.

36 The broad trends set out by Richard Tipping in his comprehensive overview published in 1994 have only been further refined by subsequent, more localised studies: R. Tipping, 'The form and fate of Scotland's woodlands', *Proceedings of the Society of Antiquaries of Scotland*, 124 (1994), pp. 1–54.

37 Ibid., pp. 9–10. For discussion of the Cairngorms Scots pine, which is genetically distinct from those of Wester Ross, and the spread of Scots pine in the post-glacial era, see K.D. Bennett, 'Late Quaternary vegetation dynamics of the Cairngorms', *Botanical Journal of Scotland*, 48 (1996), pp. 51–63, and K.D. Bennett, 'Post-glacial dynamics of pine', in J.R. Aldhous (ed.), *Our Pinewood Heritage* (1996), pp. 23–39.

Looking south over the Cnoc Coig shell midden, Oronsay, towards the north-west coasts of Islay and Jura, Argyll and Bute. Oronsay and its neighbouring islands have yielded rich evidence for the seasonal exploitation of marine resources by nomadic bands of hunter-gatherers.

lands, Scots pine was not the only species present in number, nor did it spread significantly anywhere south of the Highland Boundary Fault. From around 6500–6000 BC, much of southern and eastern Scotland supported oak-dominated woodland with a significant presence of elm, with mixed hazel- and oak-dominated woodland in the central Southern Uplands, Argyll and southern Hebrides, along the Highland/Lowland interface from Breadalbane and through the Angus Glens to the north-east, Moray lowlands and Easter Ross. In the far north mainland, Northern and Western Isles, birch remained dominant around 5,000 years ago, when the forests of Scotland reached probably their maximum extent.[38] That maximum, however, never constituted blanket coverage, with at most 50–54 per cent of the land surface being under some form of woodland; neither did those woods present any semblance of uniformity or stability, being instead in a state of constant flux as wildfire, anthropogenic burning, grazing pressures from wild and domesticated animals, disease and, critically, climatic variation affected the trees and their regeneration. Their constituents, with the probable exception of juniper, were all post-glacial colonists that advanced from their late Pleistocene refugia as the ice retreated and soil formation encouraged tree growth, but they form the group that modern science has misleadingly labelled 'native' species. Any species absent from that list before the large-scale introduction of new types in the sixteenth to later seventeenth centuries AD, including species like sycamore (or plane), beech and hornbeam that were introduced directly from Europe or that had been spreading more slowly northwards through England since the tenth millennium BC, have been stigmatised as 'alien'.

Human colonisation

As we have seen, humans were already moving through the land area of what is now Scotland in nomadic bands of Late Upper Palaeolithic, reindeer-hunting Hamburgian people many millennia before

38 For this distribution, see Tipping, 'The form and fate of Scotland's woodlands', p. 12 illus. 3. See also Smout, MacDonald and Watson, *A History of the Native Woodlands*, p. 26.

Morton Lochs, Tentsmuir, Fife. Now located over 1km from the nearest seashore, the lochs were once an area of coastal lagoon and saltmarsh, where Mesolithic hunter-gatherers exploited overwintering wildfowl and other game and the shellfish resources of the tidal waters.

the pine forests started to grow. Technically more advanced flint, chert and bloodstone tools dating from around 10,000 years ago point to the arrival of new tool-making methods and possibly to a new wave of Mesolithic hunters and gatherers who came in the wake of these Upper Palaeolithic peoples. The discoveries of Mesolithic encampment sites in places like Oronsay in the Inner Hebrides, Lussa on Jura, Kinloch on Rùm or the Morton Lochs in north-east Fife down to the 1980s helped to create a sense of seasonal movement between locations that offered a diverse range of hunting and gathering opportunities at the interface of distinct environmental ranges, mainly in lowland or coastal locations.[39] As long ago as the early 1930s, however, Mesolithic material had been identified from sites in the coastal districts of Ayrshire and penetration inland from there had been postulated by the French-born archaeologist Armand Lacaille.[40] Understanding of where these people travelled across the face of early post-glacial Scotland was limited only by our perceptions of what the land had looked like as new soils developed and vegetation regenerated over the ice-scoured ground.

By the late 1960s, scatter sites of small flint flakes, characteristic of the manufacture of stone cutting edges and points in the Mesolithic period, had been identified at sites like Rink Farm deep in the Southern Uplands at the confluence of the rivers Ettrick and Tweed. The distribution of such sites indicated – as Lacaille had believed – that the river valleys at least had already been penetrated by these nomadic people some 7,000–10,000 years ago.[41] The limited evidence for inland sites took a major leap forward from the 1990s, especially as a result of the reporting of new locations by observant forestry workers and, more recently, in archaeological surveys in advance of windfarm developments. In the Southern Uplands, this has revealed extensive evidence for widespread Mesolithic activity in the upper reaches of the Clyde and its Daer Water tributary, deep in the hilly interior of the region.[42] Further north, archaeologists had identified Mesolithic activity on sites on the coastal strip from Fife round into the Moray Firth and by the 1970s were discovering similar evidence for penetration up the major river valleys, such as the valley of the Dee.[43] Five decades later, fieldwork and excavation in the upper reaches of the

39 Wickham-Jones, *Scotland's First Settlers*. For a summary discussion of one of the Oronsay shell mounds (Caisteal nan Gillean), see C.R. Wickham-Jones, *The Landscape of Scotland: A Hidden History* (Stroud, 2001), pp. 20–21.

40 A.D. Lacaille, 'Mesolithic implements from Ayrshire', *Proceedings of the Society of Antiquaries of Scotland*, 64 (1930), pp. 34–48.

41 See H. Mulholland, 'The microlithic industries of the Tweed Valley', *Transactions of the Dumfriesshire and Galloway Natural History and Antiquarian Society*, 47 (1970), pp. 81–110.

42 See, for example, T. Ward, 'Daer Reservoir (Crawford parish): bastle house, cairns, find-spots, Mesolithic knapping site', *Discovery and Excavation in Scotland*, 1995, p. 87; T. Ward, 'Daer Reservoir (Crawford parish): Mesolithic sites; burnt mounds; cairns', *Discovery and Excavation in Scotland*, 1997, p. 75; T. Ward, 'Daer Reservoir (Crawford parish): Mesolithic flint-knapping site and other lithic scatters', *Discovery and Excavation in Scotland,* new series, 1 (2000), p. 86; T. Ward, 'Weston Farm (Carnwath; Dunsyre parishes): Mesolithic and later prehistoric lithic scatters; medieval brooch', *Discovery and Excavation in Scotland,* new series, 1 (2000), p. 85; T. Ward, 'Daer Valley Project (Crawford parish): excavation; survey', *Discovery and Excavation in Scotland*, 6 (2005), p. 134; and T. Ward, 'Upper Tweed survey – Burnetland Hill: excavation', *Discovery and Excavation in Scotland*, 8 (2007), p. 170.

43 Through fieldwork in the early 1970s, the late James Kenworthy, for example, had found flint scatter sites on the gravel river terrace of the Dee near Crathes: J.B. Kenworthy, 'The prehistory of north east Scotland', in A.M.D. Gemmell (ed.), *Quaternary Studies in North East Scotland* (Aberdeen, 1975), pp. 74–81; J.B. Kenworthy, *Excavation of a Mesolithic Settlement Site at Nethermills Farm, Crathes, near Banchory, Grampian, 1978–80: Interim Statement* (1981), unpublished report.

Dee valley has revealed that hunter-gatherer bands had penetrated the heart of the Cairngorms.[44] It is now beyond doubt that these first human settlers were moving freely across the whole of Scotland, traversing montane districts as well as the coastal plains and river valleys. With that expanding knowledge of their geographical ubiquity, we are also beginning to gain a greater appreciation of how the land, waters and their resources were exploited by the first peoples.

Despite the advancement of hypotheses concerning their use of fire to manage areas of land – for example, to create swards where game would graze and be hunted – there is, as noted in respect of the Caithness Flow Country, no evidence that these first Mesolithic populations had any impact at a level that would have resulted in long-term environmental change. Although the debate over the nature of Mesolithic human impacts and alternative causes for environmental change at landscape scale remains unresolved, there is growing consensus that a combination of moderate climatic cooling, increased storminess and rising levels of precipitation likely had a greater role than human agency in affecting woodland across the four millennia of this era. Those factors are more likely to have halted and then reversed the spread of certain tree species – including Scots pine and oak – in districts that lay at the maximum of their range, especially where deteriorating climatic conditions were unfavourable to tree growth but encouraging for the establishment and spread of blanket mire. While Mesolithic hunting practices might have affected the local availability of certain animal or bird species, or of foraged shellfish, it is unlikely that population levels were anywhere near a size where nut and berry foraging would have affected the composition of shrub or tree cover. Change in that regard only commenced around the time when the nomadic hunter-gatherer culture of the Mesolithic period was giving way to the settled farming culture of the Neolithic.

The ephemeral nature of the campsites of the hunter-gatherer peoples of the Mesolithic is a major reason for the slow growth of the evidence base for their presence in the archaeological landscape. As our understanding of the types of location they favoured and the nature of the remains they have left has grown, so also has our identification of Mesolithic sites in areas that we nowadays regard as remote or inaccessible. Importantly, this growing body of evidence is emphasising that our supposedly 'primeval forest' covered parts of the land that had possibly been hunting grounds for millennia before the first generation of the pine or oak forests of the Southern Uplands and Cairngorms, and that these early human populations had adapted to the new opportunities offered by the changing ground cover as it developed across the period from the Late Glacial Interstadial, through the Loch Lomond Readvance and into the warmer centuries of the Mesolithic. Like the constantly evolving plant communities of their hunting and foraging ranges, the Mesolithic people were static in neither their settlement nor their culture and would have changed their practices and equipment repertoire as the forests arose and spread around them. When the nomadic culture of the ninth to fifth millennia BC gave way to the settled farming communities of the Neolithic, therefore, the landscape had already seen 5,000 years of human exploitation. What changed was not that people had begun to enter a landscape previously empty of human population, but that the transience of the Mesolithic was being replaced by a permanence that required more substantial transformation of the environment to sustain the now settled communities.

44 BBC Scotland News, 7 April 2021, '"Exciting" Stone Age discoveries in the Cairngorms', https://www.bbc.co.uk/news/uk-scotland-north-east-orkney-shetland-56661834; BBC Scotland News, 8 November 2021, 'Hundreds of prehistoric tools found along river', https://www.bbc.co.uk/news/articles/cz88vzkdgyeo.

Farming begins

Amongst the earliest of the major palaeoecological discoveries of the later twentieth century was that a major episode of elm decline across Europe commenced at the beginning of the Neolithic period and lasted for several centuries thereafter. Various theories were advanced for the causes of that decline, most of them revolving around the impact of woodland clearance for agriculture, overzealous pollarding to provide feed for cattle and an associated grazing impact preventing woodland regeneration.[45] In the 1980s, a pathogen or fungal infestation was added to that list as the impact of then current *Ceratocystis* or Dutch elm disease on Britain's woodlands became visible to modern observers.[46] In Scotland, the prehistoric elm decline event appears as a cliff-edge moment around 3500 BC across the southern part of the country, with a sudden and very marked drop in the pollen of this species throughout areas where it had once been widespread if not dominant. It is difficult to read this evidence as marking anything other than a collapse in elm numbers to a very low level that persisted for around a further millennium.[47] Here, it seemed to palaeoenvironmentalists of the time, was evidence for the devastation wrought on parts of Scotland by early agriculturalists, whose slash-and-burn techniques for clearance cut ever-expanding swathes through the climax forests. That there did not seem to be an attendant decline at a similar scale in the pollen of other tree species led to some questioning of the underlying cause of the virtual disappearance of elm alone,[48] but records and observation of forest clearance in colonial America and Africa and knowledge of indigenous slash-and-burn techniques ensured that human agency remained top of the list of suspects. By the early 2000s, however, climate change and pathogens working in combination, perhaps aided by human agricultural practices, were reckoned to have been primary contributors to this cataclysmic episode.[49] And the emphasis has to be on 'primary', for it is becoming increasingly clear for the Neolithic as much as for any subsequent era of history that some human actions helped to accelerate and exacerbate already current negative environmental trends driven by climate change and adverse weather conditions. But it was not simply a question of continuously intensifying human pressure delivering irreversible change, for other environmental processes that

45 For example, as discussed in A.G. Smith and J.R. Pilcher, 'Radiocarbon dates and vegetational history of the British Isles', *New Phytologist*, 72:4 (1973), pp. 903–914; P. Rasmusen, 'Leaf-foddering of livestock in the Neolithic: archaeobotanical evidence from Weier, Switzerland', *Journal of Danish Archaeology*, 8:1 (1989), pp. 51–71; and D. Moe and O. Rackham, 'Pollarding and a possible explanation of the Neolithic elmfall', *Vegetation History and Archaeobotany*, 1 (1992), pp. 63–68.

46 M.A. Girling and J. Greig, 'A first fossil record for *Scolytus scolytus* (F.) (elm bark beetle): its occurrence in elm decline deposits from London and the implications for neolithic elm disease', *Journal of Archaeological Science*, 12:5 (1985), pp. 347–351; I. Perry and P.D. Moore, 'Dutch elm disease as an analogue of Neolithic elm decline', *Nature*, 326 (1987), pp. 72–73.

47 Tipping, 'The form and fate of Scotland's woodlands', pp. 18–22.

48 K.J. Edwards and G. Whittington, 'Vegetation change', in K.J. Edwards and I.B.M. Ralston (eds), *Scotland after the Ice Age: Environment, Archaeology and History 8000 BC–AD 1000*, 2nd edition (Edinburgh, 2002), pp. 63–82 at 73; G.J. Barclay, 'The Neolithic', in K.J. Edwards and I.B.M. Ralston (eds), *Scotland after the Ice Age: Environment, Archaeology and History 8000 BC–AD 1000*, 2nd edition (Edinburgh, 2002), pp. 127–149 at 143.

49 A.G. Parker et al., 'A review of mid-Holocene elm decline in the British Isles', *Progress in Physical Geography: Earth and Environment*, 26:1 (2002), pp. 1–45.

Elm trees, Airthrey, Stirling. Elms were amongst the most common trees in the British landscape to the late 1960s, when the *Ophiostoma novo-ulmi* strain of Dutch elm disease first arrived in these islands. By the late 1970s, it had reached Scotland. Over 25 million trees were lost throughout the UK in that outbreak. Was a similar pathogen perhaps responsible for the Neolithic collapse in elm numbers?

included pandemic disease acted to slow, if not reverse, the clearing effects of settled farming through episodic high mortality.[50]

The tendency to apply broad labels as descriptors to prehistoric cultures has helped to create a distorted public perception of the way in which different ancient peoples interacted with land and waters and exploited their resources. This issue is particularly evident in respect of the Neolithic cultures that moved into and evolved within Scotland. Emphasis is rightly placed on their settled, communal social structures and arable and pastoral subsistence basis, which marked a genuine revolution in the human–environmental interrelationship. Yet the adoption of farming using domesticated plants – many of which were introduced into these islands by farmer-colonists from the European mainland – and livestock did not mean that they stopped hunting, fishing and foraging for the wild resources which abounded beyond their cleared fields and herd-browsed grassland. Where Mesolithic peoples had hunted and gathered seasonally, so too did the peoples of the Neolithic era, as revealed by the abundant evidence from their excavated settlements and from the pathology of the bones from their communal burials.[51] These settlements and the associated arable and pasture, moreover, were not distributed evenly across the land, and with population numbers likely to be only in the thousands across the whole of Scotland, their direct impacts were very localised. Settled agriculture is likely to have clustered thickest in regions where soils were most suitable for the cultivation technologies available at the time. Those were mainly the lighter and more friable types that included the brown forest soils that occur most commonly in the river valleys of south-eastern, central and north-eastern Scotland and that supported oak-, hazel- and elm-dominated woodland where elm decline was most marked.[52] At no Scottish location as yet investigated, however, is there evidence that the decline in elm was accompanied by an increase in the pollen of cultivated plants or the weeds of cultivation. Analysis of soil from the Bowmont valley in the northern Cheviot Hills has suggested that, rather than being converted into arable land, the gap left by elm was swiftly occupied by other tree species, including lime.[53] In this regard it is also important to understand that in the early Neolithic era settlement and agriculture was 'permanent' only in that it lasted perhaps for some decades before abandonment and relocation, with woodland regeneration potentially occurring on the now derelict fields.[54] Rather than laying down an indelible marker of decisive change (other than where there

50 See, for example, N. Rascovan et al., 'Emergence and spread of basal lineages of *Yersinia pestis* during the Neolithic decline', *Cell*, 176:1–2 (2019), pp. 295–305, and A. Andrades Valtueña et al., 'Stone Age *Yersinia pestis* genomes shed light on the early evolution, diversity, and ecology of plague', *Proceedings of the National Academy of Sciences*, 119:17 (2022), https://doi.org/10.1073/pnas.2116722119.

51 Taking just two sites as a broad representation, see, for example, J.W. Hedges, *Isbister: A Chambered Tomb in Orkney*, British Archaeological Reports, British Series 115 (Oxford, 1983), and J. Hunter with J.M. Bond and A.N. Smith, *Investigations in Sanday, Orkney*, volume 1: *Excavations at Pool, Sanday: A Multi-period Settlement from Neolithic to Late Norse Times* (Kirkwall, 2007). The former contains analysis of muscle development and joint damage in the skeletal remains, which have been interpreted as consequences of foraging for sea bird eggs and chicks on the sea cliffs, while at the latter, the midden remains catalogue the farmed produce and wild foods repertoire of Neolithic Orkney.

52 For modern agricultural soil classifications, see J.S. Bibby et al., *Land Capability Classification for Agriculture* (Aberdeen, 1982). See also the James Hutton Institute online resource 'Soil data and maps', https://www.hutton.ac.uk/soil-data-and-maps/.

53 R. Tipping, *Bowmont: An Environmental History of the Bowmont Valley and the Northern Cheviot Hills, 10,000 BC–AD 2000* (Edinburgh, 2010), pp. 166–168; Edwards and Whittington, 'Vegetation change', pp. 73–74.

54 Smout, MacDonald and Watson, *A History of the Native Woodlands*, p. 31.

was a cultural focus on the construction of monumental ritual or funerary structures of the sort we can still see in Orkney, Caithness or parts of Argyll and the Western Isles), the first farmers' activities were transitory. Farming communities, it seems, constituted temporary islands in an ever-shifting, restless and well-forested landscape. When the farmers moved to new places, their abandoned fields were speedily recolonised by the still dominant woodland. There is, therefore, a certain ephemerality in the evidence for the extent and distribution pattern of inhabited locations and their associated cultivated zones.

With our modern mindsets preconditioned by the graphic imagery of forest clearance and its legacies of ravaged and degraded land in Madagascar or Amazonia, it is perhaps difficult to reconcile the possible ephemerality and transitory nature of Neolithic agriculture with a deeply ingrained perception of these first farmers as systematic and permanent destroyers of Scotland's post-glacial climax forests.[55] Indeed, a prominent strand in much of the more populist rewilding literature presents the development of societies based on agriculture as the beginning of the end for our now lost Arcadia. That is a view that has fixated on 'the onslaught of environmental degradation' unleashed by the social, cultural and technological advances of the Neolithic that archaeologists and palaeoecologists have long recognised.[56] While it is likely that Neolithic peoples accelerated some processes that were already in train, however, it is doubtful if their numbers, their technologies or their husbandry practices were the sole drivers of the substantial losses to Scotland's climax ecology rather than being among several contributory agents.[57] In the valleys of the northern Cheviots, for example, there is little evidence from the early Neolithic that anything more than small patches of farmland existed within what remained for millennia a wooded region.[58] By the middle of the third millennium BC, human communities there were clustered in the well-drained locations and were growing barley crops in woodland gaps. They were, however, apparently neither felling nor burning on any scale sufficient to create those gaps. Nor were they keeping flocks and herds in numbers sufficient to clear the trees more generally and then prevent regeneration through grazing pressures.[59]

Pine's decline

While the fourth millennium BC had seen the great wave of elm decline, the middle of the third millennium saw the start of a major retreat of Scots pine around the time of the transition into the new metal-using technologies of the early Bronze Age.[60] The peak of pine cover had apparently been achieved before *c.*3000 BC, when the species was present as far north as the northern coast of Caithness and had spread from its western and central Highland core territory to colonise islands from Lewis in the north-west to Skye and Rùm in the south.[61] Like that of the elm, the decline of the pine from the maximum reach was rapid and uniform, bringing the distribution of the species down to something like its

55 For a summary of models of agriculture and settlement in this era, see Barclay, 'The Neolithic', pp. 141–143.

56 Deary, 'Restoring wildness to the Scottish Highlands', p. 97.

57 Ibid. For some of the theories concerning why Neolithic forests seem to have remained dynamic and recovered from the elm decline, see Edwards and Whittington, 'Vegetation change', pp. 73–74.

58 R.J. Mercer and R.M. Tipping, 'The prehistory of soil erosion in the northern and eastern Cheviot Hills, Anglo-Scottish Borders', in S. Foster and T.C. Smout (eds), *The History of Soils and Field Systems* (Aberdeen, 1996), pp. 1–25.

59 Tipping, *Bowmont*, p. 171.

60 Tipping, 'The form and fate of Scotland's woodlands', pp. 26–27.

61 Smout, MacDonald and Watson, *A History of the Native Woodlands*, p. 29.

Eroded blanket peat moss, Gleann an Dubh Choirein, Perth and Kinross. The rapid growth of peat in the late second and first millennia BC might be the result of the interaction between human land management practices and changing weather patterns.

modern extent. The reasons for this decline are still debated, but as with the elm it has proven impossible to attribute it to purely or even largely anthropogenic factors. As in the Southern Uplands, Neolithic communities were present and active as cultivators, both at the advance front of the woodland expansion in Caithness and the Hebrides and also in the heartlands of Scots pine territory in the central Highlands. Strathspey, whose tributary valleys are still home to some of the most extensive remaining Scots pine woodland, contains almost equally extensive evidence for Neolithic settlement in the ritual and funerary monuments distributed from Inveravon south to Newtonmore.[62] Moreover, as argued for the woodland of southern Scotland, exploitation of the natural resources of this region was not limited to what could be grown on cultivated plots, but is likely to have seen hunting and foraging that routinely penetrated the valley heads and corries and onto the montane plateaux flanking the main strath. While the impact of Neolithic settlements may have been as transitory as that posited in the oak and hazel woods outside the Highlands, their presence added another layer to the already layered cultural landscape that formed with the first Mesolithic exploitation of the wildlife and plants of the Cairngorm

62 For the burial cairns, see A.S. Henshall and J.N.G. Ritchie, *The Chambered Cairns of the Central Highlands: An Inventory of Their Structures and Their Contents* (Edinburgh, 2001).

region. We are, however, only just beginning to glimpse the ephemerality of evidence for Neolithic activities in such upland districts away from the still visible megalithic structures that signpost the main foci of their activities.[63]

Although the scale of any impact must remain questionable, it is likely that settled human presence in these pinewood districts from the fourth millennium BC started to modify their structure. Colonists practising agriculture introduced new dynamics into the mix of factors that helped to shape their ever-changing character and determine their extent. It has been suggested that what occurred was more an opening up of the afforested area rather than its clearance,[64] creating the kind of heather and herb understorey with more widely dispersed trees that is still visible in the current woodlands around Loch Garten, Rothiemurchus or the Glen Lui, Glen Derry and Glen Quoich pinewoods of the south-eastern Cairngorms. Where the woods began to fail, however, they were not succeeded by areas of arable cultivation but in almost every case by the growth of blanket peat. It can be argued that woodland clearance, especially after *c.*2500 BC and the introduction of bronze tools which speeded and eased the process, might have helped to accelerate peat growth in northern and western parts of Scotland by exposing the underlying soils to the physical impact of increasing precipitation. It was, however, the rainfall that contributed most to the waterlogging in which the ideal conditions occurred for the growth of sphagnum moss, whose partly decomposed remains form the basic component of peat. Evidence from across the northern hemisphere indicates that a major episode of climate transition was underway around this time, bringing with it a shift to higher, heavier and more prolonged rainfall, increasing wind strength moving eastwards across the Atlantic and the rising frequency and intensity of storms. It was a devastating mix that the standards of the northern Highland and Hebridean pinewoods could not withstand. When human pressures were added, the woods at the limits of species range began to retreat at an accelerating rate along the outer edge of this new weather-defined limit.

More people, more pressure

Those human pressures increased significantly in the Bronze Age. It is widely understood that rising population levels and changes in social organisation as well as in the technological toolkit became increasingly powerful drivers in the process of environmental change.[65] By the end of the third millennium BC, the general trend in terms of woodland cover was towards decline across much of Scotland, with brief episodes of limited regeneration towards the end of the first millennium slowing an accelerating shift towards a more open landscape. Archaeological survey, most frequently in advance of the large-scale afforestation projects that commenced in the later 1980s, has revealed the presence of extensive agricultural landscapes evidenced by mounds of cleared field rubble, field-edge banks, enclosures and the footings of buildings. Excavation has underscored that these visible remains are the result of the cumulative growth of such indicators of agricultural intensification through slow intakes of newly cleared land over centuries rather than in an 'explosive expansion' of settlement.[66] The location of most of these surviving Bronze Age landscapes in upland and more

63 Tipping, *Bowmont*, pp. 171–172, section 8.7i.

64 Smout, MacDonald and Watson, *A History of the Native Woodlands*, p. 29.

65 Edwards and Whittington, 'Vegetation change', p. 75.

66 T.G. Cowie and I.A.G. Shepherd, 'The Bronze Age', in K.J. Edwards and I.B.M. Ralston (eds), *Scotland after the Ice Age: Environment, Archaeology and History 8000 BC–AD 1000*, 2nd edition (Edinburgh, 2002), pp. 151–168 at 165.

marginal districts might be an artefact of the archaeological record, skewed by the loss of evidence in the more intensively worked and built-over lowlands. It could also reflect a colonisation of less desirable areas driven by population pressure, possibly associated with a cultural trend towards greater territorialisation of human communities and competition over control. Territorialisation is also possibly reflected in the increasing shift from 'open' farm settlements to more clearly defensive or monumentally imposing enclosed ones, with the monumentality evidencing the ability of an individual or of some 'elite' group within the community to command the labour and resources to construct them. Programmes of aerial reconnaissance, however, have also revealed the cropmark presence of similar extensive landscapes of settlement and agriculture in lowland areas, where post-medieval agricultural practices have resulted in the obliteration of their upstanding physical remains.[67]

Although it was not a synchronous process across the whole of Scotland, the Bronze Age seems to have seen an infilling of the settlement pattern that had become established by the end of the Neolithic in both lowland and upland districts. Much of this infilling occurred in areas that were suboptimal for arable cultivation using Bronze Age technologies and are still regarded as at best marginal today. The vulnerabilities which marginality brought were quite likely compounded in such areas by the rollercoaster of climate change and, importantly, by deficiencies in farming practice that resulted in soil impoverishment and erosion, or the overgrazing of pasture.[68] Debate over the causes of what was the ultimate abandonment of such areas is ongoing. The more cautious argue for a complex interplay of social, economic and environmental factors, but others point to possible cataclysmic events as the trigger for desertion. An episode of profound climatic transition and associated weather deterioration in the late first millennium BC, for example, has been linked by some researchers to the eruption in 1159 BC of the Icelandic volcano Hekla.[69] As we shall explore in later chapters and in *Where Men No More May Reap or Sow*, volcanic forcing of climate change – as the consequences on weather patterns of the release of aerosol particulates of volcanic ash into the high atmosphere are termed – is recognised as a contributory factor in the deepening or prolonging of some of the major episodes of climatic instability and generally adverse weather conditions. Few scholars nowadays, however, would accept that they were the sole or even primary cause, with change dynamics already being in progress before the volcanic events. The cooling and dampening of the prevailing weather conditions in the last two centuries of the second millennium BC, which saw an acceleration of peat growth, a further sharp decline in the range of certain tree species and the permanent abandonment of large tracts of land that had been cleared with great effort for arable exploitation across the preceding millennium, might have been intensified by the Hekla eruption, but the volcano was not their cause. Indeed, continuing advances in the study of ancient DNA are showing the extent to which Bronze Age populations, like their Neolithic antecedents and

67 For examples of this extensive survey work, see Royal Commission on the Ancient and Historical Monuments of Scotland, *Mar Lodge Estate, Grampian: An Archaeological Survey* (Edinburgh, 1995); Royal Commission on the Ancient and Historical Monuments of Scotland, *South-East Perth: An Archaeological Landscape* (Edinburgh, 1994); and Royal Commission on the Ancient and Historical Monuments of Scotland, *North-East Perth: An Archaeological Landscape* (Edinburgh, 1990).

68 Cowie and Shepherd, 'The Bronze Age', p. 166.

69 M.G.L. Baillie, 'Do Irish bog oaks date the Shang dynasty', *Current Archaeology*, 117 (1989), pp. 310–313; C. Burgess, 'Volcanoes, catastrophe and the global crisis of the late second millennium BC', *Current Archaeology*, 117 (1989), pp. 325–329.

(a) Above. Nether Largie North, early Bronze Age cairn, Kilmartin, Argyll and Bute. The pasture and scattered woodland of the modern landscape of Kilmartin Glen contains the remains of one of the most remarkable late Neolithic and early Bronze Age ritual landscapes in Scotland. What we now see as an open landscape with its stone circles and burial sites intervisible was densely wooded at the time these monuments were reared, challenging the notion of the relentless prehistoric clearance of the land for agriculture. Nether Largie North Cairn and the Glebe Cairn beyond it were reared over cist burials.
(b) Left. On the underside of the raised lid of Nether Largie North's cist, carved representations of bronze flat axes can be seen.

Early Historic descendants, were exposed to the ravages of plague pandemics and cycles of epidemic that had the potential to arrest and reverse growth in their numbers and at least temporarily relieve the pressures that were contributing to the expansion of settlement and agriculture.[70] While we can never know the exact degree to which such diseases affected the populations of prehistoric Scotland, recognition of their presence and potential scale of impact requires us to abandon the traditional linearity of our old models for the expansion of settlement and agriculture. Rather than relentlessly upward trajectories in population levels and the concomitant intensifying impacts on woodland and grassland landscapes, there is likely to have been a series of steps, checks, reversals and readvances over the centuries.

Into history

When stability returned to weather patterns in the second quarter of the first millennium BC, there was no stampede to return to much of the land that had been abandoned during the preceding era of poorer conditions. That failure to resettle these areas perhaps speaks to a degree to the level of population loss that might have been associated with a late Bronze Age pandemic. We must be careful not to overstate the likely degree of reversion that occurred, however, for the lifting of one pressure does not necessarily mean that there will be a return to the previous state. While some woodland had regenerated, it is unlikely that the wider climatic regime had been encouraging of tree growth, and in fact the wetter conditions that saw an acceleration in peat development were not conducive to the expansion of woodland. In some areas of the north and west, colder, wetter and stormier conditions probably hastened the decline of already marginal woodlands – as was happening in many parts of the Hebrides – or delivered a drop downslope in the altitudinal range of tree growth in more exposed montane locations. Furthermore, abandonment of settlement does not necessarily mean an end to the exploitation of the land around it: a transition to more extensive rather than intensive use might have occurred, from cultivation to pasture, which would still have kept woodland regeneration suppressed. The balances of woodland, pasture and cultivation that had become established by the close of the first millennium BC continued to change in the centuries that followed, but their composition remained relatively stable for a further millennium and a half. As we shall see in *Where Men No More May Reap or Sow*, although there had been incremental changes since the later Middle Ages, it was from the seventeenth century AD that our ancestors began to make significant transformative moves to reshape the land at a scale and for purposes that had hitherto been unimaginable. It was then that they began to move from an adversarial approach to their environment, where they struggled constantly against hostile conditions to subsist on what they could win from the land and sea, to one of perceived mastery, where they worked to 'improve' it to ensure delivery of the yields required to meet the demands of a culturally advanced society based on reason, science and civility.

It is thus into an already radically transformed cultural landscape that we step at the start of this exploration of Scotland's environmental history, but one that was yet to see some of the more profound changes that laid down the patterns still broadly recognisable today. While it was not so sharply compartmentalised – divided by hedges, dykes and

70 S. Rasmussen et al., 'Early divergent strains of *Yersinia pestis* in Eurasia 5,000 years ago', *Cell*, 163:3 (2015), pp. 571–582; P. Slavin and F. Sebbane, 'Emergence and spread of ancestral *Yersinia pestis* in Late-Neolithic and Bronze-Age Eurasia, *ca.* 5,000 to 2,500 y B.P.', *Proceedings of the National Academy of Sciences*, 119:21 (2022), https://doi.org/10.1073/pnas.2204044119.

Dun Dornaigil, Hope, Highland. Once thought of as the fortified homes of an Iron Age social elite, the numbers and close siting of these tower-like structures make that interpretation problematical. One thing that does seem clear from their massive stone construction and limited use of large structural timbers is that mature woodland was long gone from the landscape in which they stood.

fences – as it had become by the middle of the nineteenth century, at the time of the first Roman penetration of what we would recognise as Scotland in the third quarter of the first century AD it was an open landscape of arable fields and pasture with limited stands of woodland that prevailed over much of the country. Denser forest persisted in some remoter and thinly populated districts, but, as we shall see, those areas were also subject to modes of exploitation that rendered them at best 'semi-natural' and most likely comprised regenerated growth on land that had experienced periodic grazing and possibly cultivation; the post-glacial wildwoods and the abundance of bird and mammal life they had sustained were long gone. Some of the largest non-human predator species that had moved into this land in the wake of the ice – such as lynx and bears – were extinct or nearly so, with the only remaining large predator, the wolf, already persecuted for the threat it was believed to pose to domesticated livestock and the competition it offered in the hunting of game, as it transitioned from being a hunter of the forests to a denizen of the open landscape. Herbivores large and small had also become more limited in their diversity. The reindeer of the Upper Palaeolithic were long gone and the remaining large, mainly woodland browsers were under pressure. Those larger herbivores were perhaps worst affected but smaller species like beavers also saw their habitats transformed by greater human exploitation of riverside tree species for material needs. The remaining semi-natural woods also still bore the distinctive imprint of past human management in their form and in the constitution of understorey herbage, where species that thrived on disturbance or were spread through the actions of grazing domesticated animals had replaced the flora that had co-evolved with the woodland. And across the seemingly empty wastes in which these woods stood as islands of trees, the spread of blanket peat or expanses of impoverished grassland often signalled legacies from past efforts to stimulate pasture regeneration through burning or some other form of anthropogenic intervention. Some of those attempts at improvement had instead delivered unforeseen outcomes as weather conditions brought waterlogging or helped to accelerate acidification and podzolisation of the soil.

The brief exploration of the millennia of change that form Scotland's prehistory offered in this introduction has highlighted the interplay of cultural and natural factors within environmental change and the continuously shifting landscape which that interplay helped to create. It has provided a necessary backdrop for the deeper exploration of the dynamics of environmental reconfiguration across the last 2,000 years that is the purpose of this volume and those that follow. With recognition of the continuum that is climate and wider environmental change, and how it interacts with and influences human behaviours, we can demystify and demythologise the manifold actions and processes that have brought us to our present condition. Through the clarity achieved by that stripping away of myth, we will arrive at a better understanding of the possible futures that lie ahead.

CHAPTER 1

Out of the Mists

They create a desolation and call it peace.[1]

Times of first encounter are the moments when impressions are made and perceptions formed that, rightly or wrongly, colour future understanding of the event and the context within which it occurred. Such times are often presented as defining episodes in historical development, as triggers for state formation and as tipping points in struggles for survival or dominance, giving rise to origin myths that take a tenacious hold in the popular imagination. The influence of the historical record in this respect is enduring and can be pernicious in respect of the way the parties involved are portrayed and, consequently, how this frames and constrains the attitudes fostered by the language of the record. In twenty-first-century Britain, we are familiar with the movement towards the 'de-colonisation' of history and attempts to generate a balanced narrative that avoids the binary opposites of 'us' and 'them' that have dominated national histories for centuries. Yet gaining such a balanced view is impossible when one party in the encounter lacks any 'voice', where the record is written entirely from the perspective of one side – to the point where imagined words are placed in the mouths of those on the opposing side – and where even the land in which the encounter takes place is seen through the eyes of an alien culture from a wholly different geographical, topographical and environmental context.

Such was the fate of that part of northern mainland Britain that became Scotland, when it was first entered by the armies of imperial Rome in the second half of the first century AD. Widespread familiarity with the Latin narrative accounts of that encounter has won them general acceptance as basic fact and, other than where the language of the speeches attributed to the Celtic tribal leaders is now understood to be self-reflective commentary on Roman culture, was long seen as an accurate representation of the peoples and lands of northern Britain. Visions of dense forests and swamps, swirling mists and towering mountains, through which intrepid legionaries fought until their foes turned at last to meet them in battle, captured the imaginations of generations of school children who marched with them through

1 From the speech put into the mouth of the Caledonian chieftain Calgacus at the Battle of Mons Graupius by the Roman author and historian Cornelius Tacitus: Tacitus, *The Agricola and the Germania*, trans. H. Mattingly, rev. S.A. Handford (Harmondsworth, 1970), p. 81.

Us and Them. The triumph of Roman civility over the naked barbarism of the Celtic tribes of the North, as depicted on the end panel of the Bridgeness Slab found at the eastern end of the Antonine Wall near Bo'ness, West Lothian, and now displayed in the National Museum of Scotland in Edinburgh. This representation of the triumph of civilisation over a primitive people resonated with the classically educated Scots who were extending the British overseas empire and who, in turn, fed their experience of colonial interaction with indigenous people back into an understanding of Roman and Celtic clashes in the first and second centuries AD.

the Latin texts of Cornelius Tacitus, in ways that the open grassland and arable fields over which the great battle raged could not have done. This Roman account was reinforced in twentieth-century popular tradition through the vivid descriptions of Rosemary Sutcliff's novels and again in the early 2000s through films like *The Eagle* and *The Centurion*. In these depictions, Romans and Celts acquired the mid twentieth-century stereotypes of Hollywood westerns, while the inhospitably empty terrain and harsh environment were presented as adversaries as formidable as the barbarians who inhabited them. Weather, land, wildlife and people were all portrayed as implacably hostile to civilised, Romanised people.

Where did this representation begin? In AD 80, we are told, the armies of imperial Rome advanced from bases on the Tyne in what is now Northumberland and the Eden in modern Cumbria, penetrating as far north as the Tay before the end of the first campaigning season. The accounts of that expedition and its aftermath, written nearly twenty years later by the Roman author Cornelius Tacitus as a celebration of the life and achievements of his father-in-law, Cnaeus Julius Agricola, the general who commanded that northward advance, constitute our first written record of the lands that became Scotland.[2] The debate over whether or not Tacitus appropriated for his much-admired father-in-law the successes of a previous campaign by Quintus Petillius Cerialis in AD 69,[3] while of great importance for academic interpretation of the archaeology of Roman Scotland, makes little material difference to wider understanding of interactions between the late Iron Age Celtic tribes of northern Britain and the southern invaders, or of their impact on the lands they sought to add to Roman territory. With the coming of the Romans, whatever the exact year, there commences Scotland's Historic era, the period for which there is an externally composed written record upon which the one-sided Roman narrative accounts of the political, social, cultural and economic structures of the northern part of the British Isles in the first half millennium AD – however sketchy – can be constructed.[4] For the next three and a half centuries, the history of the territories north of the Tyne–Solway isthmus, over which the Roman Empire exerted fluctuating degrees of control and attempted intermittent episodes of full military conquest, was linked intimately with that of the imperial province of Britannia to the south.

An 'icy winter wasteland'?[5]

For all that the expansion of the power of imperial Rome and its literate military clerks and officers into the northern parts of mainland Britain increased written knowledge of the lands beyond the northern frontier, the surviving record is relatively uninformative about the climate and weather of those territories. None of the surviving Roman writings concerned with the British Isles provide any meaningful

2 The principal but deeply problematical written source for the Roman advance is Tacitus, *The Agricola*. For a summary discussion of this and other sources, see D.J. Breeze, *Roman Scotland: Frontier Country*, new edition (London, 2006), chapter 1.

3 Summarised in J.E. Fraser, *From Caledonia to Pictland: Scotland to 795* (Edinburgh, 2009), pp. 17–20, and Breeze, *Roman Scotland*, pp. 33–35.

4 For example, the first volume of the New History of Scotland series, that was produced in the 1980s, commences with the Romans' northward march: A.P. Smyth, *Warlords and Holy Men: Scotland AD 80–1000* (London, 1984). The first volume of the New Edinburgh History of Scotland series outlines the debate over the AD 69 or AD 80 dates: Fraser, *From Caledonia to Pictland*, chapter 1.

5 The quotation is from A.G. Dawson, *So Foul and Fair a Day: A History of Scotland's Climate and Weather* (Edinburgh, 2009), p. 93.

accounts of conditions encountered in the north, beyond the generic and somewhat dismissive early second-century description produced by Cornelius Tacitus. He, however, never visited Britain and drafted his account entirely from second-hand reports. 'The climate', he asserted, 'is wretched, with its frequent rains and mists, but there is no extreme cold', adding that most crops grew well on account of the moist conditions but ripened slowly for the same reason.[6] We have no explicit descriptions of 'normal' summer conditions after Tacitus, but a more intimate experience of the reality of some winters on the northern frontier might be represented by a letter sent by an anonymous near contemporary of Tacitus to a likewise anonymous friend or family member based with the army there. This individual composed one of a remarkable group of late first-century letters on inked wood tablets found in excavations of the Roman civilian settlement at the gates of the garrison fort of Vindolanda in Northumberland. It was sent to a soldier in the fort's garrison and mentions various items of clothing enclosed with it, including socks (*udones*) and underpants (*subligaria* or *subligaculum*), which were possibly not standard-issue components of military dress but special requests by the recipient.[7] It is dangerous to speculate on the basis of a single letter, but there is a suggestion of inadequate underclothing and footwear for the more northerly climes where his unit was based and a lack of preparedness for winter cold. If the soldier was a member of the ninth cohort of Batavians from the area of the present-day Netherlands, however, which formed the garrison of Vindolanda around that time, it is unlikely that he was wholly unfamiliar with the extremes of winter cold to be experienced around the North Sea. We might rather be looking at the failure of a Roman quartermaster to secure proper winter gear for his colleagues. Whatever triggered the request, the response is a unique reflection on daily experience, for much of the textual evidence that survives from this era belongs to the literary convention of educated Romans using stock descriptions of a land beyond the limits of their civilised world. It is in that light that we should read, for example, the fourth- and early fifth-century panegyric references to the emperor Flavius Constantius campaigning in the misty swamps and forests of the north or to encampments amidst the snows of Caledonia made in the time of the emperor Honorius (384–423).[8] If we are to get closer to the reality of life beyond the northern frontier of the empire, we must look to archaeological, palaeoenvironmental and climate proxy data.

There is broad agreement amongst climatologists that Rome's northwards spread coincided with a climate episode spanning approximately the last two centuries BC and the first four or five centuries AD – the range varies geographically – referred to as the Roman Warm Period (RWP) or Roman Climatic Optimum. Persistent high summer temperatures around the Mediterranean basin – averaging 2°C warmer than in the sixth century AD and representing the consistently warmest episode there before the rapid temperature rises of the late twentieth century – have recently been argued to have helped to produce agricultural surpluses that underpinned the empire's military expansion and provided political stability in Rome itself.[9] As the RWP slowly declined

6 Tacitus, *The Agricola*, pp. 62, 63.

7 A. Bowman and D. Thomas, *Vindolanda: The Latin Writing Tablets* (London, 1983), p. 50.

8 *In Praise of Later Roman Emperors: The Panegyrici Latini*, ed. and trans. C.E.V. Nixon and B.S. Rodgers (Berkeley, 1994), Panegyric 6, pp. 211–253; Claudius Claudianus, *Panegyricus de Quarto Consulatu Honorii Augusti*, ed. and trans. W. Barr (Liverpool, 1981).

9 G. Margaritelli et al., 'Persistent warm Mediterranean surface waters during the Roman period', *Scientific Reports*, 10:10431 (2020), https://doi.org/10.1038/s41598-020-67281-2.

On the edge of the civilised world. Looking north from Stroness over the site of the first-century Roman fort at Fendoch to the Sma' Glen and the Highland Boundary Fault. The retired military men who shaped our perception of Roman army life on the northern frontier on the basis of their own experience of harsh Crimean and Afghan winters understood postings to such locations as having been dreaded by soldiers who had grown up in milder Mediterranean climes.

towards the middle of the first millennium AD, so too did the stability of the central regions of the empire, as competition for its less abundant resources intensified. Research on the northern and north-western edges of the empire has confirmed similar episodes of summer warmth and conditions that could have favoured Rome's advance, with glacial contraction in the Alps in the second to fifth centuries, optimal plant-growth conditions in the southern Biscay region and warmer oceanic waters extending as far north as Iceland into the mid second century.[10] While these are very generalised results from geographically remote datasets, collectively they suggest that the Atlantic-facing regions of the British Isles at least were experiencing a more benign climate regime than had been the case in the later Iron Age or than occurred in the second half of the first millennium AD.

Summer warmth, however, does not always mean good weather for the British Isles generally or Scotland specifically. There might indeed be more reality in Tacitus's reference to rain, mists and mildness than its superficial generality might suggest. From western Sutherland, the climate proxy record of the Uamh an Tartair caves in the Traligill basin in Assynt, where the growth rate of the speleothems

10 S. Desprat, M.F.S. Goñi and M.-F. Loutre, 'Revealing climatic variability of the last three millennia in northwestern Iberia using pollen influx data', *Earth and Planetary Science Letters*, 213:1–2 (2003), pp. 63–78; W.P. Patterson et al., 'Two millennia of North Atlantic seasonality and implications for Norse colonies', *Proceedings of the National Academy of Sciences*, 107:12 (2010), pp. 5306–5310.

The climate regime of the Roman Warm Period was dominated by strongly positive NAO conditions, which brought drier summers and milder winters, but with episodes of intense rainfall moving in from the Atlantic. It is probably these conditions that are reflected in Tacitus's description of British weather in the late first century AD.

in the limestone caves reflects the levels of moisture seeping through from the peat overlying the subterranean system, shows that drier conditions persisted in this region from the last quarter of the first century BC through to the end of the sixth century AD.[11] Drier conditions do not mean an unbroken era of relative drought, however, for the speleothems at Traligill continued to grow but at a slower rate. Some phases within this period were significantly drier or wetter. There was, for example, a spike of particularly dry conditions in the middle of the first century AD, a half century of wetter conditions centred on 200, then a peak of dryness for the two centuries (and beyond) after *c*.290, but again with wetter troughs in the 410s–420s and 460s, perhaps reflecting more unstable conditions towards the end of the RWP. The dryness is associated with a strongly positive summer NAO, where the relative strengths of the Azores high pressure cell and weakness of the Iceland low influence the strength and direction of westerly airflow over the ocean and the position of the oceanic storm tracks. In summer, this generally means higher temperatures and less frequent and intense rainfall for the British Isles, especially in western districts. In winter positive NAO conditions, temperatures are also higher but with greater rainfall. In negative NAO conditions, lower temperatures but higher precipitation is the summer norm, while winter sees colder temperatures and normally lower precipitation as the atmospheric conditions draw in colder but drier Arctic air from the north and north-east. It is in this seasonality that we might see the reality of Tacitus's wet but mild description for British winters and general positivity about the climate's crop-producing capacity.

Although Uamh an Tartair's proxy record indicates that the north-west of Scotland saw relatively dry conditions persist into the later sixth century, elsewhere we can see clearer indications of a general climatic deterioration as the long era of the RWP drew to a close. There is possibly corroboration of a suggested extended episode of deterioration into greater instability on the Atlantic-facing western edge of the country from around the middle of the fourth century, to be found in evidence for significant sand-drift events in the dune systems of the Western Isles commencing around that time.[12] The dune systems and machair on the west side of the island chain were

11 A. Baker et al., 'A composite annual-resolution stalagmite record of North Atlantic climate over the last three millennia', *Scientific Reports*, 5 (2015), https://doi.org/10.1038/srep10307, fig. 4; D.J. Charman et al., 'Paleohydrological records'.

12 D.D. Gilbertson et al., 'Sand-drift and soil formation along an exposed Atlantic coastline: 14,000 years of diverse geomorphological climatic and human impacts', *Journal of Archaeological Science*, 26 (1999), pp. 439–469 at 465.

destabilised as damage to the root mats of the grasses that bound them together, from cultivation, foraging, overgrazing and wave erosion, exposed the sand below to the winnowing action of the westerly winds. The result was deflation – the winnowing out of the sand by wind action – and its redeposition over the land to the east in a deep layer. Sand movement on this scale and over a spread of sites almost the length of the islands from Harris down to Barra and its satellite islands is strongly suggestive of a general regional deterioration with strengthening westerlies to intensify the dune deflation and redeposition events. A slide into wetter conditions over mainland districts as the RWP fizzled out in the decline into the Late Antique Little Ice Age (LALIA) might also be seen in south-eastern Scotland, with upland mosses in the Pentland Hills in Lothian showing evidence for growth arising from increasing rainfall, possibly reflecting a weakening of the NAO condition that had prevailed for much of the preceding five centuries, consistent with that fifth-century instability that is visible in the Uamh an Tartair speleothem evidence.[13]

All told, it is as yet a limited set of data from which to build a picture of climate and weather trends across a period of half a millennium, and we are largely left to conjecture how the inhabitants of these lands responded to the opportunities and challenges with which they were confronted. As we shall see, some palaeoenvironmental data helps to illustrate how the warmer mean annual temperatures possibly allowed agriculture to extend to higher altitudes in upland districts, creating greater arable surpluses and enabling larger populations to be sustained. Equally, however, it has shed light on the extent to which even the reduced rainfall of this era affected soils and, where clearance of woodland and scrub to create fields on sloping ground was being pursued, how the agricultural techniques of the time combined with weather trends to increase soil erosion and sediment transport in watercourses. But the expansion of settled agriculture speaks of prolonged and relatively stable conditions, politically as much as environmentally, and the creation of opportunities to trade and generate wealth. These developments had already commenced in the centuries before the first incursion of the Romans, but it was the eruption of Roman power northwards through mainland Britain that was perhaps the greatest galvaniser of change as the indigenous Celtic

13 Baker et al., 'A composite annual-resolution stalagmite record', fig. 4; P.G. Langdon, K.E. Barber and P.D.M. Hughes, 'A 7500-year peat-based palaeoclimatic reconstruction and evidence for an 1100-year cyclicity in bog surface wetness from Temple Hill Moss, Pentland Hills, southeast Scotland', *Quaternary Science Reviews*, 22:2–4 (2003), pp. 259–274.

Eildon Hill North from Eildon Middle Hill, Scottish Borders. The ramparts of the late Iron Age hillfort are visible encircling the upper slopes of the hill, enclosing a settlement whose population probably numbered in the thousands at the start of the Roman advance. The presence of such major concentrations of population speaks of intensive agricultural exploitation of a wide area around the settlement producing the surpluses needed to sustain them.

tribes positioned themselves to confront or collaborate with the intruder. Rome's interventions could destabilise and disrupt old patterns of exploitation or foster and advance the agricultural expansion that the imperial armies found already underway in the territories they entered, but it was most likely the continuing general climatic stability that enabled the northern tribes to respond in a manner that brought relative peace for much of the first century of the encounter. Even when changes in Roman policy brought friction and finally warfare to the lands beyond the frontier in the late second and early third centuries, the prevailing climate regime favoured the growth of stable tribal communities that were slowly coalescing into the more hierarchical societies from which the Early Historic era kingdoms would emerge in the sixth and seventh centuries. We do not know if greater climatic instability in the fourth and early fifth centuries stimulated the raiding culture that looked to the Roman province to the south to provide through plundering the wealth that could no longer be secured through Roman gift giving and trade. Although the plunge into the extremes of the LALIA still lay several decades into the future, as the presence of Rome faded from the minds of the inhabitants of northern Britain in the fifth century, it might have seemed that the largely favourable weather of the preceding half millennium had departed with the legions.

A forested land? The myth of the *Caledonia silva*

According to one vision of Early Historic Scotland that has been presented to us by generations of antiquarians, historians, conservationists and, more recently, restoration ecologists and 'rewilders', Roman legionaries had marched into a land still swathed beneath a blanket of trees. It is there, we are told, in the written record left by the invaders and in the folk memory handed down over generations, but the identity of the invader responsible for the depredations changes with each retelling. We have been embarrassed with a veritable smorgasbord of suspects: Romans first, then the Norse Vikings, followed by Edward I of England, cruel Scots noblemen in the fourteenth century, Oliver Cromwell and English ironmasters in the late sixteenth and seventeenth centuries, English, lowland Scottish and Irish tanbark and charcoal producers through the eighteenth century, government soldiers during the Jacobite era, sheep farmers and their herds in the late eighteenth and nineteenth centuries, deer forest owners in the later nineteenth and twentieth centuries and overzealous loggers in both world wars.[14] Although all can be seen as perpetrators of deforestation, the vision of a seemingly endless 'blue-green canopy' of pine forest that once 'roofed the glens' in the modern accounts of their excesses all flows from a primary notion of a *Caledonia silva* or 'Great Wood of Caledon' drawn from the written legacy of the Roman empire. It is now over a century and a half since Cosmo Innes questioned the notion, 125 years since Hugh Boyd Watt queried traditional ideas of its extent, twenty-five years since that 'Great Myth of Caledon' was once more debunked by David Breeze and over twenty since Chris Smout comprehensively dismantled the myth's components, yet the idea and the romantic and inspirational imagery it can generate still maintain a powerful hold in popular and, sadly, poorly researched academic literature.[15] Evocative photography and a narrative of human abuse of the environment, party political publications and mainly environmental science and ecological restoration literature, often drawing on much older, tendentious and quite polemical historical discussions, have combined to keep the myth not just alive but spreading its roots deeper into the public consciousness.[16] Yet as the palynological and palaeoenvironmental research discussed in this book's introduction suggests, it is unlikely that much by way of extensive pine woodland – mixed pine,

14 T.C. Smout, *Nature Contested: Environmental History in Scotland and Northern England since 1600* (Edinburgh, 2000), pp. 37–46. A study which attempts to tread the path between myth and reality is Jim Crumley's *The Great Wood: The Ancient Forest of Caledon* (Edinburgh, 2011). The fullest (and an utterly fanciful) catalogue of destruction and its perpetrators was compiled in the late nineteenth century by David Nairne in 'Notes on Highland woods, ancient and modern', *Transactions of the Gaelic Society of Inverness*, 17 (1892), pp. 170–221.

15 C. Innes, 'Some account of early planting in Scotland', *Transactions of the Royal Highland and Agricultural Society*, new series, 9 (1861), pp. 40–53; H.B. Watt, 'Scottish forests and woodland in early historic times', *Annals of the Andersonian Institute*, 2 (1900), pp. 89–107; D.J. Breeze, 'The Great Myth of Caledon', in T.C. Smout (ed.), *Scottish Woodland History* (Edinburgh, 1997), pp. 47–51; Smout, *Nature Contested*, pp. 37–46; Smout, MacDonald and Watson, *A History of the Native Woodlands*, p. 24; T.C. Smout, 'The history and the myth of Scots pine', *Scottish Forestry*, 68:1 (2014), pp. 1–4.

16 The nineteenth-century and more recent promotion of the mythical Caledonian Forest – and the purposes behind that promotion – is discussed in the second and third volumes in this series. For examples, see Miles and Jackman, *The Great Wood of Caledon*, published to accompany a documentary film of the same name, and Scottish Green Party, *A Rural Manifesto for the Highlands: Creating the Second Great Wood of Caledon* (Inverness, 1989). The repetition of the myth in influential sources like F. Fraser Darling, 'History of the Scottish forests', *Scottish Geographical Magazine*, 65:3 (1949), pp. 132–137, and Anderson, *A History of Scottish Forestry*, the latter maintaining a position as the 'go-to' text on Scotland's woodland history into the early 2000s, helped to cement the presence of the Great Wood of Caledon in some modern academic literature. For the importance of forest as an element in national identity-building, see K.L. Syse, 'The ebb and flow of trees and farmland: symbols of nationhood in Scotland and Norway', *Journal of the North Atlantic*, special issue 4 (2013), pp. 219–228.

Scots pine plantation, Muir o' Fauld, Perth and Kinross. Belief that first-century Scotland was swathed in impenetrable forests such as these was based on a combination of the eighteenth-century British experience of the North American forests and the importation of mythical imagery concerning the darkness of the central European conifer-dominated forests, redolent with the folk traditions of the Brothers Grimm.

elm, oak, hazel and birch predominated in varying permutations in different areas – remained in any Southern Upland districts or the Western or Northern Isles by the first century AD, but with the greater survival of significant wooded areas in the mainland north of the Forth–Clyde line it is possible that up to 25 per cent of the land surface was still under some form of tree cover.[17] While that figure is greater than current levels (around 17–18 per cent), it is still far from representing the kind of forest canopy covering the land that is projected by proponents of the Great Wood of Caledon.[18]

Where does the idea of the all-covering forest originate? By the mid nineteenth century, scholars were agreed in tracing the notion to the writings of the Roman author Pliny the Elder, who in his *Natural History*, completed in AD 77, commented that the empire's legionaries had not yet extended knowledge of the island of Britain beyond the neighbourhood of the Caledonian Forest.[19] Although Pliny referred for details of British geography to the authority of earlier Graeco-Roman writers including the third-century-BC Greek geographer Pytheas of Massalia, none of their surviving works refer to such a forest and the immediate source of Pliny's information is unknown.[20] What is certain is that his forest lay somewhere in the parts of northern Britain beyond the Forth–Clyde isthmus, for this is where the country labelled 'Caledonia' and inhabited by a tribe called 'Caledones' by first- and early second-century writers such as Tacitus is located.[21] But Scotland north of the Forth–Clyde isthmus is a huge and topographically diverse region and none of these writers suggest at any point that it was a tree-filled wilderness. Even Pliny the Elder suggests that a forest was just one element on the land in this territory, not its defining feature. Even the overly casual modern equation of Caledonia with all of Scotland or, in a slightly more limited sense, encompassing the whole of the Highland massif and its lowland fringes, which has then been presented as the territorial extent of the Caledonian Forest, again distorts what Pliny the Elder wrote. He simply referred to unnamed districts in the vicinity of the forest and gave no label to the unexplored regions of northern Britain. Even Tacitus is studiously vague, referring to Caledonia only as an area beyond the zone of Roman military occupation that under the Flavian dynasty of emperors had extended north to a line of forts parallel with the edge of the mountains; he does not label the whole area beyond that line as 'Caledonia' or describe it as blanketed by forest.[22]

It is the map and texts produced by the second-century geographer Ptolemy that are seen as beginning to offer a better sense of where the land of the Caledones (or 'Kaledonioi', as he termed them in his native Greek) was located. Ptolemy's evidence gives the lie to the post-Roman equivalence of Caledonia with the whole country north of the Forth–Clyde isthmus and confirms that the Caledones were one

17 Smout, MacDonald and Watson, *A History of the Native Woodlands*, p. 34.

18 I. Armit and I. Ralston, 'The coming of iron', in T.C. Smout (ed.), *People and Woods in Scotland: A History* (Edinburgh, 2003), pp. 40–59 at 41.

19 Smout, MacDonald and Watson, *A History of the Native Woodlands*, p. 24. For Pliny's text, see Pliny the Elder, *The Natural History of Pliny*, ed. and trans. J. Bostock and H.T. Riley, volume 1 (London, 1855), p. 350. Bostock and Riley followed what seems to have already been established convention by the mid nineteenth century and suggested, with no rationale offered, that this forested region was in the area of the Grampian Mountains.

20 Pliny, *The Natural History*, p. 350.

21 Tacitus, *The Agricola*, p. 60.

22 The notion that 'Caledonia' was the Roman term for what is now Scotland is highlighted and challenged in A. Bruford, 'What happened to the Caledonians?', in E.J. Cowan and R.A. McDonald (eds), *Alba: Celtic Scotland in the Medieval Era* (East Linton, 2000), pp. 43–68 at 44–44.

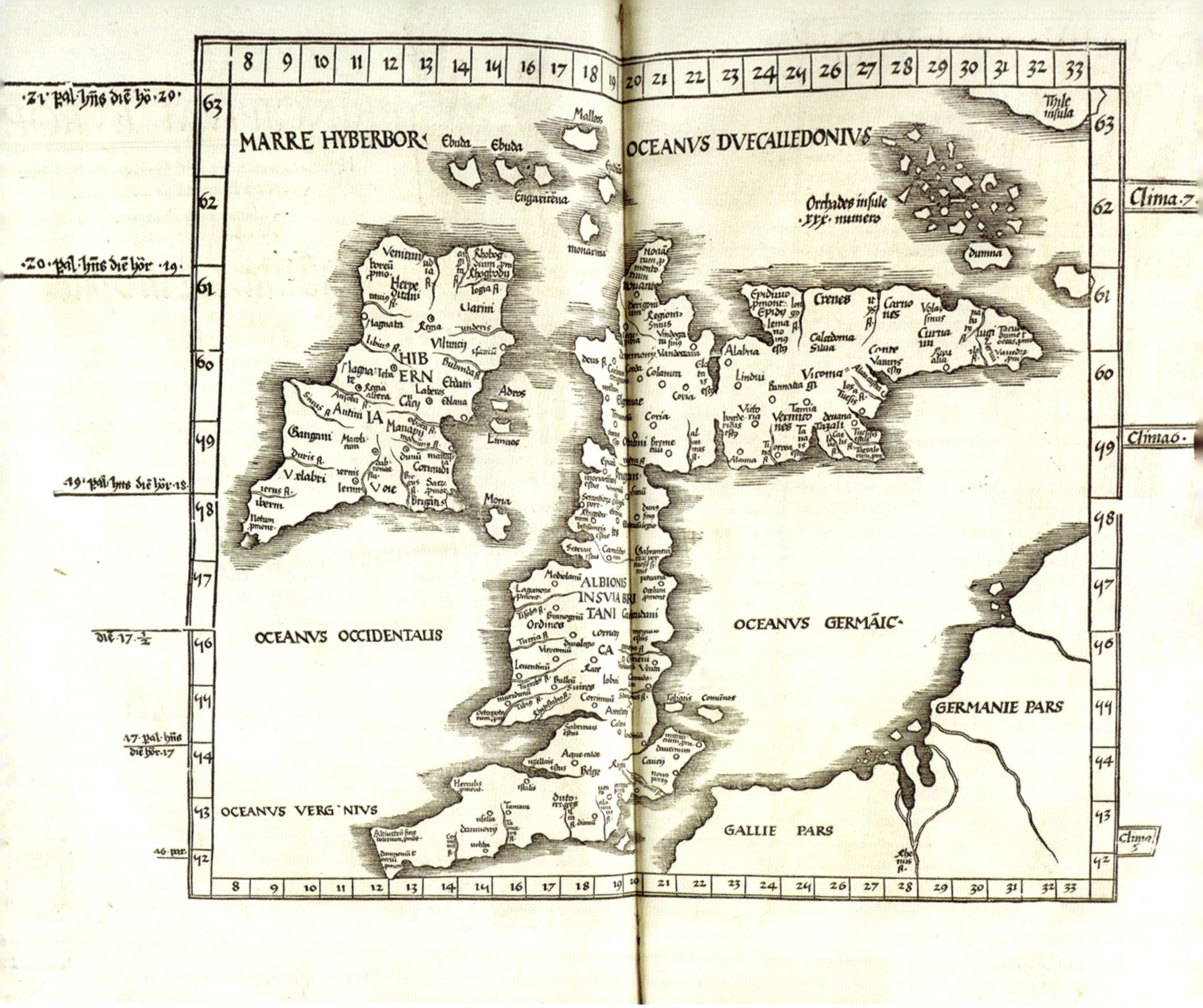

Claudius Ptolemy, 'Europae Tabula 1', from *Claudii Ptolemaei Geographicae Enarrationis Libri Octo* (Strasbourg, 1525). The detail (opposite) shows the location of the *Caledonia silva* in the central / West Highland area.

of possibly twelve tribes then living in the northern districts of Scotland rather than its sole inhabitants.[23] His locational detail, however, is questionable, for he places them as occupants of a zone that stretched from the sea lochs of southern Argyll to the rivers feeding into the head of the Beauly Firth, extending approximately along the length of the Great Glen, with the *Caledonia silva* running parallel with their northern and western limit.[24] Important though the Great Glen was as a major routeway through the Highlands from early prehistory onwards, limited archaeological remains suggest that it was not a

23 For discussion of the northern Scottish names in Ptolemy's *Geography*, see A.L.F. Rivet and C. Smith, *The Place Names of Roman Britain* (London, 1979), pp. 140–141. James Fraser, in *From Caledonia to Pictland*, offers a frank assessment of the development of the labelling and provides some suggestions on how best to view it.

24 Breeze, 'The Great Myth of Caledon', pp. 47–48.

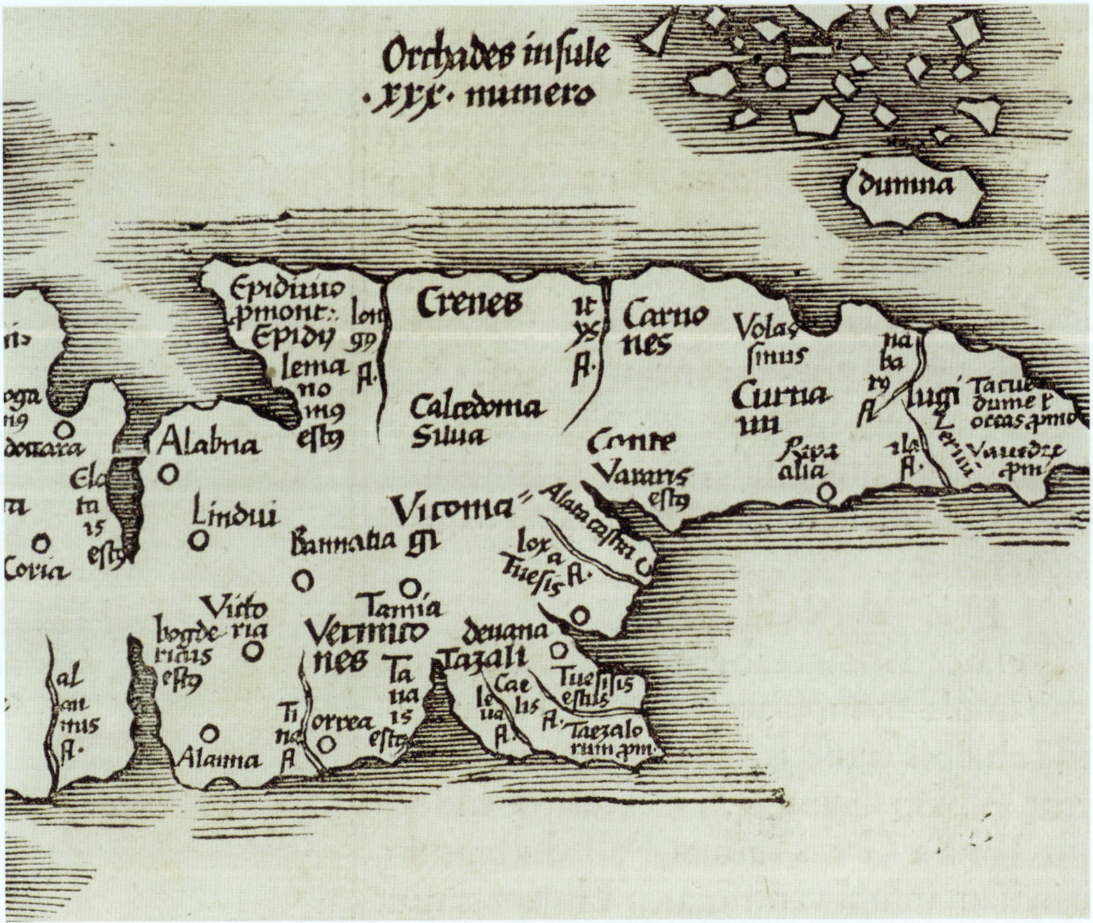

major settlement zone and its steep flanks and limited valley-bottom areas do not seem to have supported a significant population other than at its northern and southern extremities. This relatively lightly settled zone seems an unlikely location for the heartlands of the supposedly most important northern tribe, a possibly confederacy-leading people, as Alan Bruford has suggested.[25] Rather than in the Great Glen, place names containing the Caledon element – Dunkeld, Rohallion, Schiehallion – are all clustered in the Tay–Tummel basin, a zone of quite dense Iron Age settlement and a district more probably reflecting at least one main zone of the Caledones' heartlands.[26] There is now consensus that their tribal territory probably spanned Drumochter, giving them possession of the main zone of the Tay–Tummel basin and at least the upper part of the Spey catchment, which had been home to substantial agricultural communities since the Neolithic era, over 3,000 years earlier.[27] If we take this zone as reflecting

25 Bruford, 'What happened to the Caledonians?', p. 44.

26 Breeze, 'The Great Myth of Caledon', p. 48; Bruford, 'What happened to the Caledonians?', p. 44.

27 The varying extents of an 'inner' and 'outer' Caledonia are discussed in Fraser, *From Caledonia to Pictland*, p. 20.

Section from Conrad Cichorius's casts of Trajan's Column in Rome (scene 92), showing legionary pioneers felling woodland to clear a path for a road. It was scenes like this on well-known Roman monuments that helped to fix in the minds of late nineteenth-century antiquarians the idea of the imperial army carving its way into the heart of 'Caledonia' through the 'Great Wood'.

Ptolemy's possible intention of presenting the Caledonians as a central Highland tribe and his *Caledonia silva* as flanking their western extremity, this reduces a possible woodland expanse to a more likely spread in the upper reaches of the valley systems flowing out from the massif to north and south.[28] That is already beginning to look like the distribution of known afforested areas later in the Middle Ages.

While Pliny, Tacitus and probably Ptolemy drew their information from reports written by men who had served in Agricola's northern campaigns and their locational descriptions have been viewed as precise, they are in fact vague, generalised and slightly contradictory. Tacitus, the one writer who provides more detailed information on the topography of northern Britain as seen through Roman eyes, does not mention a *Caledonia silva* or, indeed, any extensive Caledonian wood or forest at all. He does mention,

28 The alternatives are discussed in Breeze, 'The Great Myth of Caledon', p. 48.

Central Tweeddale looking west from Tinnis Castle, Scottish Borders. This district is central to the myth of Coit Celidon and the poet-mage Myrddin or Merlin losing his mind in the forest. Yet it was already a heavily settled and cultivated area by the late Iron Age, with its remaining woodland probably surviving in carefully managed expanses of coppice-with-standards.

however, that Agricola's army encountered smaller patches of woodland where the defeated Britons sought cover in the aftermath of the failed attack on the Ninth Legion and the subsequent Battle of Mons Graupius.[29] Rather than forest, Tacitus's description speaks primarily of arable land, open plains, rough ground and bare hillsides, a cleared landscape of extensive, populous settlement in which chariot warfare could be conducted.[30] David Breeze, furthermore, has pointed to the manner in which most Roman writers after Ptolemy seem to have used the Caledonian label simply to mean somewhere in the extreme north-west of the world, beyond the limits of the empire, and noted that references to Caledonian 'forests' occur in contexts that again suggest generic application to any and all forested land that

29 Tacitus, *The Agricola*, pp. 77, 85, 89.

30 Ibid., pp. 79, 81, 86, 88–90. Tacitus refers to the burning houses scattered across the land in the aftermath of the battle. See also Breeze, 'The Great Myth of Caledon', pp. 49–50.

they believed to exist as a natural condition in all parts of the remote and barbaric north.[31] Caledonian forest had become a Roman literary trope before the end of the second century, just as descriptions of legionaries toiling through swamps and forests had become a standard means of conveying the hardships of military life by the second century BC. Like the progressively more fantastic descriptions of the land and people of northern Britain given in Cassius Dio's and Herodian's mid-century accounts of the early third-century campaigns of Septimius Severus or the panegyrics celebrating the early fourth-century campaigns of Flavius Constantius,[32] none of these sources can be used as evidence for the existence of the 'blue-green canopy' swathing the Highland glens so prominent in modern fantasies.

The literary mobility of the woodland to which the Caledonian label was applied had by the ninth century become so elastic that it was used by the author of the *Historia Brittonum* as the name of one of the mythical battles fought against the Saxon invaders by King Arthur, Cat Coit Celidon (Battle of the Caledonian Wood), which he claimed took place somewhere in the central Southern Uplands of Scotland.[33] This appropriation of the name was part of an early medieval Welsh literary convention that relocated the mythical woodland to somewhere in the zone between Annandale and Tweeddale and identified it as the wild land in which the maddened bard-magician Merlin found his shelter and met his end and where Arthur defeated his ubiquitous Saxon enemies. Like its Highland counterpart, this ninth-century description of a battle fought in a forested landscape has spawned an equally mythical 'great wood' swathing the Border Hills at the end of the Roman period in what archaeology and palaeoenvironmental studies indicate was a largely treeless zone that had been intensively settled and cultivated for millennia before Cerialis or Agricola marched north. The fanciful notion of a Southern Uplands location for the Great Wood eventually climaxed in the first quarter of the nineteenth century, when Sir John Sinclair of Ulbster spoke eloquently of the historical evidence for the emperor Septimius Severus losing some 50,000 of his army to enemy ambushes within the forest,[34] which he had intended to cut down and burn to deprive the native peoples of its cover and resources. For Sinclair's informants, the surviving and evidently ancient woodland of the Torwood between Stirling and Falkirk and Callendar Park east and south of Falkirk represented the northern outpost remnants of a formerly vast forest whose canopy had once stretched unbroken from the carse-land of the Forth to the Hadrianic frontier line. This early nineteenth-century tale of Roman losses in impenetrable forests has strong echoes of the recorded slaughter of three legions and upwards of 16,000 men of the army led into the German forests of the Teutoburger Wald by Publius Quinctilius Varus in AD 9 and seems to be an appropriation of that narrative to explain the Romans' failure to subdue 'Caledonia' in the same terms as the emperor Augustus's admonition to his heirs to cease future efforts to expand the empire east of the Rhine. As we shall explore further in this volume and return to in *Standing on the Edge of Being*, this notion of a southern Scottish forest during the Roman era is as mythical as the tradition of a vast pine forest spanning the Highlands, yet it plays an equally powerful role in contemporary enthusiasm for a regeneration

31 Breeze, 'The Great Myth of Caledon', pp. 48–49.

32 Smout, *Nature Contested*, pp. 41–42; Breeze, 'The Great Myth of Caledon', p. 49.

33 Breeze, 'The Great Myth of Caledon', p. 49; *Historia Brittonvm cvm Additamentis Nennii*, in *Monumenta Germaniae Historica, Auctorum Antiquissimi* 13: *Chronica Minora*, volume 3, ed. T. Mommsen (Berlin, 1898), pp. 111–222 at 199.

34 J. Sinclair (ed.), *General Report of the Agricultural State and Political Circumstances of Scotland*, volume 4 (Edinburgh, 1814), pp. 465–467.

of woodlands within the Southern Uplands. Any reality of a Great Wood was already long gone centuries, if not millennia, before the cluster of minor settlements on the low hills above the marshy banks of the Tiber had evolved into the first small city-state of Rome.

Of 'Celtic cowboys' and cultivators

Exploitation of woodland resources across the first millennium of the Historic era is the subject of Chapter 2. Here, we look at the management of the land for agricultural purposes in the five centuries that saw engagement with the Roman-controlled territories in the southern parts of mainland Britain. As for the woodland areas, perceptions regarding the wider interaction of the people of Early Historic Scotland with the natural resources of the land and surrounding waters have undergone considerable revision over the last sixty years. It is now emphasised that, in many respects, there was continuity from deeper prehistory into the medieval period. Exploitation of aquatic resources, for example, continued to employ techniques from an earlier age and the limited marine technologies of the pre-Norse period saw no advance in sea-fishing practices. Fishbone assemblages from near-adjacent sites like Dun Vulan and Bornais in South Uist, for example, where settlement spanned a millennium, are almost identical in composition and point to inshore fisheries operating seasonally from beaches or rock perches.[35] Deep-water species are largely absent until the colonisation of the region by Scandinavian peoples in the ninth century, who brought with them a different and more intensive culture of marine resource exploitation and the sea-faring technology needed to pursue it. The practices and preferences evident at Dun Vulan and Bornais, where catching and processing was on a domestic scale and conducted in the shallower coastal waters, are replicated at sites dispersed around Scotland's coast from Galloway and the Hebrides to Shetland and the Lothian coast. Fishing was certainly an important part of the food-production regime of the early centuries AD, but it was not being practised on the scale and intensity of later centuries and the calorific intake of most of Scotland's people at this time likely came largely from terrestrial produce supplemented by freshwater fish. As evidenced by representations of salmon carved on Pictish stones from the seventh and eighth centuries (discussed in Chapter 5), this anadromous species (spawning and juvenile years spent in fresh waters, adult life in salt) was probably already exploited on a significant scale by the peoples who lived close to the great river estuaries and along the valleys of the major spawning rivers. The history of the salmon fishery in the pre-modern era, however, is only just beginning to be researched. It is instead in recognition of the intensity and sophistication of late Iron Age agriculture that the principal shift in our understanding of this era lies.

That shift has been in respect of the managed form, scale and extent of the pastoral expanses and the ubiquity of arable exploitation in all districts capable of sustaining them. For long, framed by visions of the cattle-rearing and cattle-raiding culture presented in the Irish epic tale *Táin bó Cúailnge*,[36] until the third quarter of the twentieth century archaeological discussion of late Iron Age culture in southern

35 R. Cerón-Carrasco and M. Parker Pearson, 'The fish bones', in M. Parker Pearson and N. Sharples (eds), *Between Land and Sea: Excavations at Dun Vulan, South Uist* (Sheffield, 1999), pp. 274–282.

36 The epic is believed to have originated in oral tales dating to the early centuries AD but was first written down only in the seventh century and survives now in mainly twelfth-century versions, so its presentation of a possibly first-century-ad warrior culture has gone through multiple reframings since it was first composed. For a modern translation, see *The Táin: Translated from the Irish Epic Tain Bo Cuailnge*, ed. and trans. F. Kinsella (Dublin, 1969).

Cademuir Hill West, Scottish Borders, looking east along the ridge overlooking the valley of the Tweed. The stones in the foreground and the band across the middle of the view are the tumbled remnants of the ramparts of the Iron Age fort. The stony area in the centre of the photograph, beyond the further rampart, is the remains of a *chevaux de frise* defence, long interpreted as designed to break up the charge of chariots across the grassy uplands of the Celtic cattle pastures.

mainland districts of Scotland focused on their supposedly primarily pastoral economic basis. Arable figured little in generalised representations of Iron Age rural regimes. Stuart Piggott's memorable description of 'Celtic cowboys and shepherds, footloose and unpredictable',[37] practising a largely mobile, non-arable livestock economy that roved across the grassy heights of the Southern Uplands, with great linear earthworks defining grazing ranges, dominated interpretations into the 1980s. As Tacitus's account of chariot warfare conducted by the Caledones implies the existence of open, grassy expanses across which horse-drawn wheeled vehicles could move and manoeuvre at speed, it was perhaps understandable that the notion of a largely pastoral culture whose flocks and herds cropped the land traversed by the Celtic charioteers gained such a firm hold in both the academic and the popular imagination.

The 'Celtic cowboy' was an attractive hypothesis based on the more limited evidence available to

37 S. Piggott, 'Native economies and the Roman occupation of the north', in I.A. Richmond (ed.), *Roman and Native in North Britain* (London, 1958), pp. 1–27.

Piggott in the late 1950s, but it has since been comprehensively dismantled. Instead, fieldwork and excavation has identified more arable-based settlements and the evidence points to a mixed agricultural economy being practised in regions as diverse as Shetland and Galloway, both upland and lowland.[38] Cattle – and sheep – were certainly valued by the Celtic tribes of what is now Scotland and probably constituted an important measure of wealth in the early centuries AD, just as they were throughout the medieval period. Excavated bone assemblages imply that large numbers of cattle were available locally to the Roman forts in southern Scotland, supplying their demand for meat across the four centuries of military occupation.[39] It is also likely that in some areas flocks and herds were present in sufficient numbers to render them important agents of woodland decline, when their grazing on cleared woodland areas prevented regeneration. The greatest cause of woodland decline in this era, however, appears to have remained active anthropogenic clearance for cultivation rather than for pasture. Elsewhere, it has been argued that there was similar grazing pressure on sandy heathland, where management using fire was adopted as a method to stimulate improvement of the pasture resource from at least the early fifth century. On Tentsmuir in north-eastern Fife, for example, pollen and soil analysis has revealed evidence for what was interpreted as the deliberate and repeated burning of the *Calluna*-dominated heath on the sandy district between the estuaries of the Tay and the Eden to stimulate grass growth.[40] Here, however, it has been suggested that the burning led to losses of *Calluna* cover that threatened widespread erosion, but that windblown deposits of soil from adjacent arable districts stabilised the sand areas exposed through the burning and created ideal conditions for grass growth that delivered – accidentally – the improved pasture. The hypothesis of anthropogenic muirburn in this coastal heathland is not without challenge, but the district does seem to have been grazed intensively through the fifth century and into the following centuries as part of a mixed agricultural landscape. Probably much like the neighbouring south-east Angus district where the cultivated zone was flanked by low-lying sandy heath, upland pasture and heather moorland, the grazing on Tentsmuir bordered an intensively cultivated arable zone. In north-eastern Fife, again, it seems that the 'Celtic cowboy' is at least as likely to have been a tiller of the soil as a herdsman.

The earlier discussion of the extent of woodland clearance and expansion of arable that occurred from the last few centuries BC and into the early centuries AD should not be taken as suggesting the ubiquity of cultivated land. Arable practice was still very much limited by the available farming technology, with a restricted range of equipment and, despite the era's labelling as the 'Iron Age', lack of easy access to the forged iron tools that enabled rapid expansion in later centuries. For most of the Early Historic period, large-scale cultivation was undertaken using the ard type of plough, which cut shallow furrows and was of little use in the breaking-in of new cultivated land.[41] Lacking the mould board and coulter of later and heavier ploughs, it was used mainly on lighter

38 I. Armit, 'Life after Hownam: the Iron Age in south-east Scotland', in W. Bevan (ed.), *Northern Exposure: Interpretive Devolution and the Iron Ages in Britain*, Leicester Archaeology Monographs 4 (Leicester, 1999), pp. 65–80.

39 Breeze, *Roman Scotland*, pp. 93–94.

40 G. Whittington and J. McManus, 'Dark Age agricultural practices and environmental change: evidence from Tentsmuir, Fife, eastern Scotland', in C.M. Mills and G. Coles (eds), *Life on the Edge: Human Settlement and Marginality* (Oxford, 1998), pp. 111–119.

41 P. Fowler, *Farming in the First Millennium ad: British Agriculture between Julius Caesar and William the Conqueror* (Cambridge, 2002), chapter 9.

soils and often the fields required cross-ploughing to entirely break up the turf layer to prepare the soil for sowing. In southern parts of Britain, this technique produced the pattern of square 'Celtic' fields familiar to archaeologists, but there is only limited evidence for such fields anywhere in Scotland. That absence, however, is possibly a consequence of the intensity of later agricultural activity, especially since the mid eighteenth century, rather than evidence for different equipment or ploughing techniques. Heavier soils and especially land that was being broken into cultivation for the first time were probably worked using spades, which would have enabled the cultivation of smaller patches amidst rocky outcrops or boggy hollows. What we should think of when visualising this kind of practice is not a zone of continuous, expansive fields such as we see in Scotland's rural areas today. Instead, there was a patchwork of ploughed and spade-worked 'infield' land on the best soils for that technology in more or less continuous cropping, interspersed with an 'outfield' used mainly for pasture but with blocks of occasionally cropped land at first, later given over to more regular cultivation. Beyond that intensive cultivation was an interface zone of grazing land which merged into the wider upland areas that were exploited for hunting, seasonal grazing and other resources. Outwith these zones of intensive and extensive exploitation there was still much that was 'wild' land, whose animal, bird and plant life were all viewed as part of the wider resource base necessary to sustain the human population.

We have already seen that the northern Cheviot valleys experienced an episode of extensive agricultural development around the BC/AD turn. In its effects, the rapid and widespread removal of woodland in Bowmont offers some sense of the impact of agricultural expansion on the eve of the Roman invasion. There, clearance triggered a severe erosion of soils, both of topsoil from the freshly deforested land and of anciently deposited valley-bottom sediments, as watercourses formerly constrained by carr woodland gouged new channels across flood plains.[42] The worst impact seems to have been in the last century BC and to around AD 50, but mineral sediment transport in the streams and rivers continued to the end of the sixth century, suggesting that intensive agricultural activity was maintained on the newly established fields throughout the Roman period. The fact that agricultural activity persisted throughout this erosion episode, however, is strongly indicative of the worst impacts being localised. Hillsides, where woodland loss probably saw increased water-flow in already established gullies and streams, seem to have been relatively stable. Techniques adopted for erosion prevention might have seen the introduction of walled fields at the end of the first millennium BC and in the early centuries AD, and possibly the use of cultivation terraces, both of which limited downslope creep or losses of soil on ploughed land.[43] Spade cultivation was practised in smaller patches of cleared land, apparently mainly in the upland zones – although that distribution is likely an artefact of differential survival rates in zones that have been cultivated for centuries and those only intermittently exploited – and this technique helped to reduce soil loss such as occurred through ploughing. It is likely that similar effects and methods occurred in most districts of the upland margins, where the milder climate of the era perhaps encouraged expansion of arable and the creation of open areas of grassland for pasture.

Along the northern edge of the Cheviots, the lower-lying land cleared of woodland was probably organised into fields for arable cultivation. Pollen analysis indicates that wheat and oats were grown, possibly rye also, maximising opportunities on both the more fertile and well-drained soils of the valley

42 Tipping, *Bowmont*, pp. 183, 188.
43 Ibid., p. 188.

Bowmont Valley, Cheviot Hills, Scottish Borders. Nowadays mainly pasture with some limited arable on the lower slopes and valley bottom, Bowmont saw intensification of agriculture through the late Iron Age and into the Roman period. Woodland clearance and expanded arable cultivation triggered an episode of significant soil erosion.

floors and the wetter, poorer ground of the flanking slopes.[44] Higher up the valley, clearance also triggered soil erosion, which accelerated as exploitation intensified from the early first century BC. Here, there was a greater emphasis on grazing land, but there were also fields of oats with some barley and, again, possibly rye, perhaps only cropped intermittently on ground otherwise lying as heath or grassy grazing. Although there is a question mark against the confident identification of the rye (*Secale*) pollen in material from Bowmont, it is suggested that this cereal – which appears to occur in deposits datable to *c.*220 BC to AD 175 – might have been a late Iron Age introduction to the crop repertoire in northern regions, with charred rye grains found in excavated material from sites in the northern Pennines of England dating to the later Iron Age.[45] Rye has also been identified at the late Iron Age settlement at Bornais in South Uist, but its very low volume suggests that it was not a deliberate cultivar at this date or was grown only in very small quantities, before its large-scale introduction in the ninth century by Norse settlers.[46] In comparison to northern parts of mainland Europe, where it is still a favoured winter-sown crop and is a culturally in-demand cereal type, rye seems never to have been cultivated as extensively as barley or oats in even these more marginal areas of Scotland.

What is striking about the expansion of agriculture in the northern Cheviots is that it was already underway in the two centuries before Petillius Cerialis started the advance into northern Britain and seems to have been maintained throughout the Roman period and on past the fifth century. Agricul-

44 Ibid., p. 183.

45 G. Hillman, 'Reconstructing crop husbandry practices from charred remains of crops', in R. Mercer (ed.), *Farming Practice in British Prehistory* (Edinburgh, 1981), pp. 123–162; M. van der Veen, *Crop Husbandry Regimes: An Archaeological Study of Farming in Northern England 1000 BC–AD 500* (Sheffield, 1992).

46 N. Sharples (ed.), *A Late Iron Age Farmstead in the Outer Hebrides: Excavations at Mound 1, Bornais, South Uist* (Oxford, 2012), pp. 89, 185, 230.

Roundhouse and souterrain, Ardestie, Angus. Located on the south-facing upslope from the Tay estuary, the settlement at Ardestie is one of several excavated sites with souterrains in this rich arable zone. This clustering of such monuments on the better-quality agricultural land has helped to form an interpretation of them as storehouses attached to substantial farm settlements.

tural expansion had been buoyed by rising population, climatic amelioration and greater social cohesion in the later Iron Age and received impetus from, rather than containment by, the intervention of Rome. Is it possible that a further stimulus for the investment in crop growing and cattle raising was the demand and market opportunity from the urbanised civil province of Britannia in southern Britain and the substantial military garrisons in the north of England and, intermittently, south and central Scotland? There is strong evidence from other frontier zones of the Roman Empire, from Germania to North Africa, that a policy of interaction, trade and exchange that promoted stable relations and gave security to the *limes* was pursued to good effect over periods of decades if not centuries. It seems likely that similar efforts at containment and stabilisation were employed for at least some time in respect of the Celtic tribes of northern Britain.

Subsidy, tribute and supply

Much research from the late twentieth century identified a clustering of native settlement in the broad zone close to Hadrian's Wall, which possibly reflected a crystallisation of pre-Roman patterns as a consequence of the enforced stability of the frontier or as a novel response to the new markets created by the Roman garrisons.[47] There was also a suggestion that the development of large and apparently unenclosed agricultural settlements through mainly the east-

47 G. Jobey, 'Between Tyne and Forth: some problems', in P.A.G. Clack and S. Haselgrove (eds), *Rural Settlement in the Rural North* (Durham, 1981), pp. 7–20; G.D.B. Jones and J. Walker, 'Towards a minimalist view of Romano-British agricultural settlement in the north west', in J.C. Chapman and H.C. Mytum (eds), *Settlement in North Britain 1000 BC–AD 1000*, British Archaeological Reports, British Series 118 (Oxford, 1983), pp. 185–204.

central Lowlands and extending up the east coast was likewise a response to either taxation/tribute demands from Rome or increased trade opportunities for the supplying of Roman garrisons during the Flavian (*c*.77–90) and Antonine (139–65) occupation of the region.[48] Ever since the first programme of investigation of such sites in east-central Scotland from the late 1940s to early 1960s, major roundhouses with complex souterrains (subterranean passages) have been interpreted as developing in response to both a need for storage for the grain surpluses delivered by the benign climatic regime of the RWP and a new level of demand for that grain from an external source, i.e. the Roman military. Although a link has been made between the apparently deliberate dismantling and abandonment of souterrains and the military campaigns of the emperors Septimius Severus and Caracalla in their 208–11 war against the people known to the Romans as the Maeatae, reappraisal of the artefact evidence from excavated sites has suggested that the abandonment occurred much earlier. Some analysis indicates that it took place in the aftermath of the Antonine withdrawal from central and southern Scotland in 158.[49] In this scenario, the disappearance of the souterrains in which agricultural surpluses were believed to have been stored was linked to a breakdown in the old relationship of trade and subsidy and a consequent reversion to raiding to secure the luxury goods and portable wealth that were no longer accessible to northern tribal leaders.

This 'souterrain abandonment horizon', however, is contested. Other sites excavated in the later twentieth century as well as more recently – for example, at Shanzie near Alyth in eastern Perthshire – show no evidence for either deliberate dismantling or abandonment and have a settlement continuity extending into the third and possibly fourth centuries.[50] Given that souterrains are distributed from Shetland to Midlothian and that their chronology apparently predates the Flavian invasion of northern Britain and extends beyond the Antonine period in some areas, there is a danger of overreading the evidence for a Roman stimulus to their development. It is very unlikely that the souterrains of Sutherland or Skye were constructed purely as a response to the opportunities for trade with Roman outposts in southern and central mainland Scotland.[51] Nevertheless, the distribution of the largest known examples – Carlungie in Angus is around 40m long and Pitcur in eastern Perthshire has a main gallery 58m long with a subsidiary gallery of 18m[52] – lies at the limit of the attempted permanent zone of military occupation and might reflect an opportunistic response by some of the local social leaders to the potential for wealth generation and status enhancement through trade and diplomatic interaction with the Romans and access to the exotic goods which that afforded. There is a substantial number of such settlements with souterrains clustered in the intensively cultivated districts of eastern Perthshire and Angus, exemplified by the excavated and publicly accessible sites at Ardestie and Carlungie in Angus, between Dundee and Arbroath, the excavated site at Ardownie

48 Breeze, *Roman Scotland*, pp. 92–93.

49 I. Armit, 'The abandonment of souterrains: evolution, catastrophe or dislocation?', *Proceedings of the Society of Antiquaries of Scotland*, 129 (1999), pp. 577–596 at 592–594.

50 R. Coleman et al., 'The excavation of a souterrain at Shanzie Farm, Alyth, Perthshire', *Tayside and Fife Archaeological Journal*, 8 (2002), pp. 77–101.

51 A. Morrison, 'Souterrains in Sutherland', in J.R. Baldwin (ed.), *The Province of Strathnaver* (Edinburgh, 2000), pp. 215–235; R.F. Miket, 'The souterrains of Skye', in B. Ballin Smith and I. Banks (eds), *In the Shadow of the Brochs* (Stroud, 2002), pp. 77–110.

52 HES Canmore, Pitcur, https://canmore.org.uk/site/77193/pitcur.

in the same district, or the now destroyed site at Newmill, north of Perth.[53] These have a floruit from the mid first century to the mid second century, coincidental with the two primary phases of Roman military occupation. Intriguingly, the likely construction of a cluster of datable souterrains in southern Lothian in the later second century has been hypothesised as an attempt by some communities to maintain a trade and exchange relationship such as that which had developed further north to *c.*158 and to capitalise on the reintensified garrison presence in the Southern Uplands after the Antonine withdrawal from the Forth–Clyde frontier.[54] But the same eastern Perthshire and Angus region in which Ardestie and Carlungie lie is also the location of the long-functional Shanzie site and others that provide evidence for continuity of use into the third century or beyond. Rather than a coherent, single episode of abandonment, it looks as if some of the biggest sites – for example, Newmill or Carlungie – were perhaps excessive for the reduced market opportunities of the late second and third centuries, while smaller sites like Shanzie continued to function as they met the requirements of the communities in which they were located. In the absence of historical record, we are left to weigh these various hypotheses.

Debate over the dating of and context for the souterrains will no doubt continue as further archaeological investigation occurs, but what is now reasonably clear is that regardless of political, social and economic drivers they arose as a response to the increased agricultural opportunities of the RWP. Palaeoenvironmental analysis of the excavated deposits from the Ardownie souterrain established that a wide variety of plant material had been stored within it, of which cereals were just one important component.[55] Grasses for animal feed and bedding, heather for kindling, sphagnum moss and a wide range of other plants were apparently kept there, highlighting both the mixed agricultural basis of the settlement economy and the environmental range that was exploited to support the human and animal populations. The heather was brought to the site from open heathland, possibly closer to the coastal strip south of Ardownie or the higher ground to its north. Those districts might have been exploited as range grazing for the flocks and herds of the arable farms in the intermediate zone. The sphagnum, which has a variety of medicinal and practical uses, speaks of wet hollows and areas of peat growth. The volume of cereal and especially barley pollen, however, points to a large arable presence and emphasises the picture we have already seen in the Southern Uplands of the intensification of agriculture and breaking into cultivation of a greater zone than had been exploited in the pre-Roman Iron Age. The charcoal remains from Ardownie also point to a settlement standing in fields within a largely treeless landscape extending from the sandy coastal lowlands to the high ground of the Sidlaws. In the absence of plentiful wood, heather was one of the principal sources of kindling, if not burned as a main fuel. Heather charcoal recovered from hearth, midden and floor deposits represents the possible burning of old heather thatch and bedding material as well as of heather gathered specifically for the fire, but with birch and hazel also well represented. Oak, too, was present, along with pine and alder, but at volumes

53 F.T. Wainwright, *The Souterrains of Southern Pictland* (London, 1963); B. Walker and G. Ritchie, *Fife, Perthshire and Angus*, 2nd edition (Edinburgh, 1995), p. 159, no. 85; S. Anderson et al., 'The excavation of a large double-chambered souterrain at Ardownie Farm Cottages, Monifieth, Angus', *Tayside and Fife Archaeological Journal*, 12 (2006), pp. 14–60; T. Watkins and G. Barclay, 'Excavation of a settlement and souterrain at Newmill, near Bankfoot, Perthshire', *Proceedings of the Society of Antiquaries of Scotland*, 110 (1981), pp. 165–208.

54 Armit, 'The abandonment of souterrains', p. 594.

55 Anderson et al., 'The excavation of a large double-chambered souterrain', p. 54.

Souterrain, Carlungie, Newbigging, Angus. One of the biggest such sites yet located and excavated, Carlungie is likely to have been a key element in a settlement of high status, possibly acting as a collecting place for surplus produce.

that speak of a conserved but probably small mixed woodland resource. As will be explored in the discussion of woodland in Chapter 2, this evidence is suggestive of birch and hazel coppice perhaps surviving as curated stands on poorer and wetter land or at the limits of the cultivated zone.[56] Overall, the picture we have of this part of the east-central Lowlands in the early first millennium AD is one of a landscape of extensive clearance covering most of the soils optimal for tillage, bounded by or interspersed with zones of grassland and grazed heath and with small pockets of woodland and wetland that were exploited for timber, fuel and other materials, such as moss.

Farming at the margins

Is that picture replicated in regions away from the east-central Lowlands? Most of our evidence comes from very site-specific pollen and soil analyses, but there are a number of large-scale regional studies that offer insight into developments across a wider landscape. North and west of the Great Glen, for example, exploitation in Glen Affric, west of Loch Affric, varied across the different soil types available on the valley floor and low on its sides. At Camban, close to the western extremity of the main valley system, there was a revival of fen woodland, indicative of a decline in grazing pressure from the mid first century, accompanied by diminishing evidence for cultivation and pasturing by around the end of the second century.[57] On the better alluvial soils at Carnach Mor further east in the valley, however, grazing intensified from the middle of the second century, with an expansion of arable in the same district from around the turn from the third to the fourth century. These opposite but near synchronous responses, less than 3km apart, suggest that while every opportunity was taken to increase arable

56 Ibid., pp. 46–47.

57 A.L. Davies, 'Upland agriculture and environmental risk: a new model of upland land-use based on high spatial-resolution palynological data from West Affric, NW Scotland', *Journal of Archaeological Science*, 34 (2007), pp. 2053–2063 at 2055 table 2.

West Glen Affric, Highland, looking west from Alltbeithe towards Camban at the head of the glen. Nowadays remote and depopulated, this area of Affric saw pulses of intensive settlement and cultivation through later prehistory and into the Historic era.

production on good soils, demand for cereal crops in this Highland district was never driven in the early first millennium AD by local population pressure. Nor was it driven by access to an external market to the extent that less favourable ground in west Affric was exploited in the same manner as occurred in the Bowmont valley.[58]

In the Outer Hebrides, excavation on sites like Mound 1 at Bornais in South Uist paint a similar picture of arable intensification in the early first millennium AD. There, however, the hulled six-row barley cultivation that dominated the agricultural economy had already been established by at least the Bronze Age, over a thousand years earlier, but new crops – chiefly flax and oats – were added to the repertoire.[59] This trend was not simply accommodated through more land being taken into cultivation, but required changes in both land management practices and cultivation methods. At Bornais, this saw both a well-manured infield and a less well fertilised outfield employed to produce different crops, with the more robust and less demanding oats being grown on the previously unmanured outfield to fill the shortfall in food-cereal production caused by portions of formerly barley-yielding infield being given over to nutrient-hungry flax growing. Although the extent

58 Ibid., p. 2056.

59 Sharples, *A Late Iron Age Farmstead*, p. 339.

of land under cultivation grew, it was probably not increasing the available calorific yield, and here the arable regime was maintained as a predominantly household-level subsistence economy. A more favourable climatic regime may have made it easier to produce crops of flax while maintaining the requisite barley and oat yields without expanding the managed land to any significant extent. The loss of some outfield to regular cultivation, however, perhaps meant that grazing was pushed further into less intensively exploited areas of the Uist blackland or the moorland beyond that. Environmental conditions perhaps encouraged this shift, but it was cultural decision-making that brought the choice of flax, probably chiefly for linen production more than for flax oils, rather than any population- or hunger-related driver. In the midst of this generally positive picture, however, there are hints of contrary movement. We do not know, for example, what factors brought an end to all crop production on some land that had been intensively cropped using spade cultivation in the early centuries AD, such as has been identified on Inner Hebridean sites, such as Ardnave on Islay's north-western coast.[60] Datable to the later fourth or fifth centuries, this apparent abandonment precedes the end of the RWP and the possible population collapse of the sixth century, but could mark the onset and impact of the less predictable conditions suggested by the Uamh an Tartair speleothem record and the Lothian peat moss evidence discussed in the introduction. Wind erosion of sandy soils is a known driver of abandonment elsewhere in the Hebrides in later centuries and on the exposed Ardnave site may have prompted early withdrawal from cultivation.

While the general climatic conditions perhaps encouraged greater investment in arable cultivation, albeit still with extended growing seasons and late harvests, as suggested by Tacitus, soil structures and fertility were not improved by warmth and good weather alone. Within the technological limits of the forms of plough available in the late Iron Age, from the possibly only occasionally cropped fields of the Bowmont valley uplands and through the grain-lands of the south and eastern lowlands to the intensively cropped zones around settlement along the Atlantic edge from Uist to Shetland, fertility remained dependent on manuring. Cattle and sheep dung as well as midden waste provided the bulk of nutrient inputs and would have been essential on areas of more marginal agricultural land brought into cultivation as operations expanded and crop types changed. Manure availability speaks of livestock management, which bring us back to that image of the Celtic cowboys and shepherds, but we should see it in the context of a mixed agricultural economy rather than in extremes of arable or pastoralism, where maintenance of a grazing resource to support herds was as important as the extension of crop-bearing land. Changes in fertilising strategies become clearer in the second half of the first millennium (discussed in Chapter 3), some of it linked to cultural shifts and colonisation but much perhaps associated to a greater extent with declining yields and the difficult environmental conditions encountered as the RWP faded and the LALIA began to make its presence felt.

Creating desolation?

No exploration of the environmental history of the early first millennium AD would be complete without some consideration of the possible imprint of warfare on the record of landscape-scale change in this period. Given the Romans' own recorded history of the deployment of terror tactics that included systematic destruction of settlements and devastation of agricultural resources to undermine their

60 G. Ritchie and H. Welfare, 'Excavations at Ardnave, Islay', *Proceedings of the Society of Antiquaries of Scotland*, 113 (1983), pp. 302–366.

Cill Naoimh, Loch Gruinneard and the headland of Ardnave, Islay, Argyll and Bute. Intensively cultivated into the middle of the first millennium AD, Ardnave slipped into abandonment towards the stuttering close of the Roman Warm Period.

enemies' will to resist, historians and archaeologists have long looked for evidence of the application of similar tactics against the northern British tribes. Amongst the many memorable quotations from the works of Tacitus, the words he ascribed to the Caledonian chieftain Calgacus before the Battle of Mons Graupius are amongst the most repeated. *Prima facie*, the chieftain's speech is a damning indictment of the expansion and imposition of Roman power and of the treatment of conquered peoples: 'To robbery, butchery and rapine, they give the lying name of "government"; they create a desolation and call it peace.'[61] Tacitus probably had in mind well-known episodes in late republican and early imperial Roman history, perhaps most recently the Flavian emperor Titus's wholesale destruction of Jerusalem during the suppression of the First Jewish Revolt in AD 70, but the deliberate wasting of an enemy's land had been used as an instrument of terror and statement of total victory in the ancient Near East since at least the second millennium BC.[62] Rome, however, had practised the destruction of the cities of its defeated enemies during its rise to domination in Italy and the central Mediterranean, best represented by the devastation of Carthage in 146 BC (although the tradition of the ploughing and salting of the site of the razed city by Scipio Aemilianus is now reckoned to be a modern invention).[63] Perhaps more effective in deal-

61 Tacitus, *The Agricola*, p. 81 ('ubi solitudinem faciunt pacem appellant').

62 S. Gevirtz, 'Jericho and Shechem: a religio-literary aspect of city destruction', *Vetus Testamentum*, 13:1 (1963), pp. 52–62.

63 R.T. Ridley, 'To be taken with a pinch of salt: the destruction of Carthage', *Classical Philology*, 81:2 (1986), pp. 140–146; S.T. Stevens, 'A legend of the destruction of Carthage', *Classical Philology*, 83:1 (1988), pp. 39–41; P. Visona, 'On the destruction of Carthage again', *Classical Philology*, 83:1 (1988), pp. 41–42; B.H. Warmington, 'The destruction of Carthage: a retractatio', *Classical Philology*, 83:4 (1988), pp. 308–310.

ing with non- or only proto-urbanised enemies like the Celtic tribes of Britain, however, was the practice of harvesting their crops during the course of a campaign, as is referenced by Tacitus in Calgacus's speech and is illustrated graphically in the commemoration of the Second Dacian War of the early 100s on Trajan's Column in Rome.[64] There, rather than simply burning the unharvested crops, the legionaries are shown cutting the grain with sickles and gathering it in, apparently in the full sight of their Dacian foes. The purpose behind this was as much symbolic as practical, demonstrating the native rulers' inability to protect their land and their people. It deprived them of their food supply for the year ahead and introduced famine and starvation as weapons of war, while at the same time adding to the invaders' food reserves and capability to maintain pressure on their enemies. It is highly likely that similar tactics were employed against the tribes of northern Britain in the later first century and beyond.

Evidence for such tactics, however, is not clearly recognisable in the archaeological record in Scotland. Localised and short-term impacts of warfare might be reflected in the commandeering of conserved resources that was involved in the construction of Roman fortifications. Yet even in the case of the Flavian forts along the southern edge of the Highlands, environmental analysis suggests that native agriculture continued in the surrounding districts and possibly even intensified to meet demand from the new market in the Roman military edifice. It is not that more conclusive evidence has not been sought for, or, in some arguments, found. Between the 1960s and 1990s, for example, it was argued that the rural regime of which the souterrains, discussed above, were central components as stores for agri-

64 Tacitus, *The Agricola*, p. 81; F. Lepper and S.S. Frere, *Trajan's Column: A New Edition of the Cichorius Plates: Introduction, Commentary and Notes* (Stroud, 1988), plate 81: scene 110, 'Roman legionaries foraging'.

Section from Conrad Cichorius's casts of Trajan's Column in Rome (scene 110), showing legionaries foraging, harvesting the grain crop of their defeated Dacian enemies. (Album/DEA/G. Dagli Orti)

cultural surpluses had perhaps been a victim of the brutal war waged by Septimius Severus and his sons against the Maeatae or Maiatai, given that occupation at many excavated sites seems to have ended abruptly in the early third century.[65] Pollen analysis from sites in north-eastern Fife was used to support this hypothesis, with an apparent decline in arable production and regeneration of woodland over an extended period in the early centuries AD being attributed to the genocidal activities of the Roman army.[66] This

65 For a re-evaluation of that war, see Fraser, *From Caledonia to Pictland*, pp. 25–30.

66 G. Whittington and K.J. Edwards, '*Ubi solitudinem faciunt pacem appellant*: the Romans in Scotland, a palaeoenvironmental contribution', *Britannia*, 24 (1993), pp. 13–25. This more assertive interpretation is based on the findings presented in papers such as G. Whittington, K.J. Edwards and P.R. Cundill, 'Late- and post-glacial vegetational change at Black Loch, Fife, Eastern Scotland – a multiple core approach', *New Phytologist*, 18:1 (1991), pp. 147–166. This is the broad thesis followed in Fowler, *Farming in the First Millennium ad*, pp. 62–64.

analysis has been critiqued more recently, and while the apparent collapse of a long-established arable regime in the face of a Roman campaign of terror has its appeal, the tight chronology of the Severan operations sits uncomfortably with the much looser dating offered by the pollen profiles. If collapse there was, it occurred perhaps half a century earlier and seems to have been unrelated to direct military action.[67] As the abandonment of the large souterrains in the country north of the Tay might indicate, intensification of agriculture in response to the stimulus of a new market for produce to 158 might have seen rapid contraction in some districts when those markets were withdrawn. It is, however, still highly contested and conjectural as to what this change represents and whether or not it can be linked to the direct impact of warfare. These are questions that await future palaeoenvironmental and archaeological investigation.

• • •

Across the five centuries explored in this chapter, the generally benign climatic conditions of the RWP contributed to an era of sustained agricultural expansion and social development in Scotland that was further stimulated by engagement with the Roman Empire to the south. The evolving late Iron Age cultures of the southern and eastern regions had perhaps already been advancing along the road to social and political cohesion – first in the form of the tribal confederacies that confronted the legions in the first to third centuries and then in the various kingdoms of the Britons and Picts that we will encounter in the following chapters – but their development was boosted by the stability of the climate and the favourable environmental conditions that characterised the first half millennium AD. These were far more sophisticated cultures than have been imagined by the traditional portrayals of naked barbarians hurling themselves against the armoured might of Rome or of cowed and terrorised tribesmen fleeing into the mist-shrouded safety of the mythical Great Wood of Caledon.

In these first five centuries AD, the foundations were being laid for the kingdoms that would unite into Alba – the Scotland of the later Middle Ages – and for the social and economic structures that sustained them. In that same era, the patterns of land-use and resource exploitation that were largely followed until the Industrial Revolution were also beginning to take shape, defining the zones of intensive and extensive agricultural exploitation and framing our ancestors' perceptions of value and use. In the following chapters, as we enter the Historic era fully, we will see that the people of this land were already viewing the environment in which they lived and toiled as blocks of utility and uselessness. In some senses, they had already formed what could be considered a series of islands of greater or lesser diversity in an expanse of intensively managed and environmentally degraded land.

67 Armit, 'The abandonment of souterrains', p. 593.

Strathcononish, Stirling, Scots pine and birch woodland. Rather than the closed canopy woodland of popular imagination, much of the woodland surviving in Scotland in the Early Historic era was already park-like in character and mixed-species in composition.

CHAPTER 2

After the 'Great Wood': Trees and Woodland Management to AD 1000

> Along the foot of [the] mountains are vast woods, full of stags, roe-deer, and other wild animals and beasts of various kinds; and these forests oftentimes afford a strong and safe protection to the cattle of the inhabitants against the depredations of their enemies.[1]

Although the fantasy of the blanket tree cover of a primeval Great Wood of Caledon, whether in the Highlands or in the Southern Uplands, must be rejected, it is nevertheless probable that around 25 per cent or more of Scotland was still beneath woodland in the first century AD. How much of that woodland could be classed as truly 'natural' even by that date is questionable, for it is likely that human adaptation of and alterations to the composition of forested areas had commenced as early as the Mesolithic era.[2] Fire, it is argued, might already have been employed on a small scale by those early hunters to clear trees and create nutrient-rich swards where game would congregate to feed. Woodland would regenerate in such places if grazing by the large, wild herbivores was managed through hunting, and it is likely that cycles of burning and regeneration had been occurring for millennia before the first farmers entered the northern parts of Britain and began larger-scale clearances to create fields, albeit at first on a short-term basis whereby a period of cultivation was followed by reversion to tree cover.[3] It is unlikely, however, that the first arrivals had seen anything akin to the dense, close-canopy forests encountered by European settlers in North America and that there was immediate and widespread slash-and-burn clearance. Given the naturally prevalent tree species, geology and climatic conditions, the woodland most probably formed a patchwork of zones of different densities rather than an unbroken canopy. These ranged from substantial oak- or pine-dominated forest, expanses of thinner birch-dominated woodland, straggles of scrubby growth and open areas kept clear by large herds of grazing herbi-

1 John of Fordoun, *Chronica Gentis Scotorum*, ed. W.F. Skene, 2 volumes (Edinburgh, 1871–2), volume 2, p. 36.
2 Wickham-Jones, *Scotland's First Settlers*; Smout, MacDonald and Watson, *A History of the Native Woodlands*, p. 28.
3 Smout, MacDonald and Watson, *A History of the Native Woodlands*, pp. 12–13.

vores, to near treeless lowland and upland bogs and montane regions where altitude and weather prevented trees from establishing themselves. Arable farmers moving into this environment would have followed the lines of least resistance, occupying first the best land where least clearance was needed before moving into territory that required a greater investment of time and labour. Rather than some contemporary Amazonian nightmare of blanket clearance that proceeded with breathless rapidity, the erosion of the woodland took millennia and was perhaps as often a consequence of continuing climate change as it was of anthropogenic origin. That observation should not, however, be seen as attempting to diminish the scale of loss. Rather, it is intended to emphasise that the contraction of woodland was a complex process involving both human and wider environmental factors in an interplay that delivered differing outcomes in different parts of the country. That interplay had contrived to reduce the area under woodland in a period of less than five millennia from somewhere around 50 per cent to perhaps only half that level.

Clearance, conservation and coppice

Leaving aside both the montane areas of the Highlands and western Southern Uplands, where most of the land was too high, rocky and exposed to sustain tree cover, and the lower expanses of wetland, whose waterlogged conditions were inimical to anything but scrubby willow and alder, the estimated 25 per cent of land bearing trees would still have included areas of dense woodland. That tree cover was far from static, ebbing and flowing in some areas as human and animal pressure mounted or declined, but it was clearly still reducing in extent at an accelerating rate across the 200 years prior to the Roman invasion. That acceleration was likely primarily a consequence of human action and certainly took some areas of woodland beyond any point of possible recovery through natural regenerative processes. The reduction in cover, however, was uneven across both mainland and island districts and, importantly, there are suggestions of woodland conservation and unexpected survival in some areas alongside other evidence for widespread clearance. As discussed in the introduction, human and natural processes had taken their toll on the Northern and Western Isles, Inner Hebrides and Caithness, central Highlands and Galloway before the end of the first millennium BC, leaving large areas nearly treeless. Palaeoenvironmental analysis of excavation material, however, suggests that woodland remained present in even these seemingly treeless places in probably carefully managed pockets.[4] It was assumed, for example, that Shetland had been devoid of trees since prehistory and dependent on driftwood sources and imports for its wood needs.[5] On the evidence of fragmentary wood remains from alder, hazel and willow recovered in the excavations at Old Scatness at the southwest of Mainland Shetland, however, it seems possible that patches of managed woodland persisted into the middle of the first millennium AD in even these northernmost islands.[6] Excavations at Pool on Sanday in Orkney confirmed the long-established view that significant woodland had been cleared in the smaller islands in the archipelago before the end of the Neolithic. Here too, however, managed scrubland probably provided a limited woodland resource through the Early Historic era for building

4 Armit and Ralston, 'The coming of iron', pp. 42–44.

5 See, for example, A. Fenton, *The Northern Isles: Orkney and Shetland* (Edinburgh, 1978), pp. 111–112, and D. Spence, *Shetland's Living Landscape* (Lerwick, 1979), p. 10.

6 S.J. Dockrill et al., *Excavations at Old Scatness, Shetland*, volume 1: *The Pictish Village and Viking Settlement* (Lerwick, 2010), pp. 196–197.

and equipment uses and as fuel, while large structural beams were obtained from driftwood.[7] The presence on Orcadian sites of wood remains showing clear evidence of long exposure to the ocean – principally the tracks of boring marine worms – possibly indicates the reservation of larch logs carried by ocean currents from North America for the monumental building efforts of the social elites in the islands. Such timbers were used to span the internal spaces of the massive Atlantic roundhouses and brochs and form their main roof members.

Such driftwood also met the large timber needs at the broch and post-broch settlement of Dun Vulan on the western coast of South Uist. Here, however, hazel and willow still grew locally, possibly coppice-managed (with very slow growth as the willow was at the extremity of its natural range), occurring probably as stunted stands in sheltered wetland locations and loch margins.[8] The volumes of similar wood types – both driftwood and probably coppice-managed species – found at the nearby Mound I site at Bornais confirmed the suspicion that while Uist was also a largely treeless area by the late Neolithic or early Bronze Age, small pockets of scrubby woodland survived and were managed carefully as a resource into the first millennium AD.[9] At Bornais, though, the wood remains indicate that small amounts of ash, hawthorn, pine and probably wych elm survived alongside the willow and hazel. In all cases, there is no evidence for wider woodland regeneration during the more benign conditions that prevailed in the North Atlantic from the second century BC, suggesting that land-use and especially grazing pressure prevented regrowth beyond the small, managed patches. The disappearance of these few last traces, or at least their decline to levels where they have left virtually no palynological or charcoal traces in the archaeological record, was perhaps a consequence of the rapidly deteriorating climatic conditions of the later sixth and seventh centuries. The associated changes in human land-use practices that evolved in response to the deteriorating climate combined to place irrecoverable stresses on what had been a fragile resource even during optimal climatic conditions.

Coppice, Inchmahome, Stirling. A few managed coppiced trees could deliver most of the wood needs of small communities, for building materials and wood-turning.

On the Scottish mainland, it is unsurprising that

7 Hunter, Bond and Smith, *Investigations in Sanday*, vol. 1, pp. 5, 120, 170, 200–201.

8 M. Taylor, 'The wood', in M. Parker Pearson and N. Sharples (eds), *Between Land and Sea: Excavations at Dun Vulan, South Uist* (Sheffield, 1999), pp. 188–192.

9 Sharples, *A Late Iron Age Farmstead*, pp. 190–191, 252.

the greatest and most rapid decline occurred where the pre-Roman population was concentrated. Studies undertaken since the 1970s at sites spread between Northumberland and Cumbria and the Scottish central Lowlands have revealed collectively abrupt and near total woodland destruction in the last few centuries BC, maintained across the Roman occupation period.[10] Local palaeoenvironmental studies such as that in the Bowmont Valley in the northern Cheviots shed some light on the rapidity with which such inroads could be made. There, a single episode of clearance removed almost all remaining tree cover in districts where woodland was already scarce at the end of the Bronze Age, 3,000 years ago. In Bowmont, expanses of alder carr in the valley floor that had remained undisturbed throughout prehistory, were cleared in a single event starting around 175 BC, at the transition into the early stages of the RWP.[11] The removal of these wetland-edge trees was probably part of a more general clearance of the valley floor which also eliminated all dryland species and was so systematically thorough that, in Richard Tipping's words, 'the combined effect was the total removal of any plant community that could be seen as a functioning woodland, one in which local seed sources could assure regeneration'.[12] The higher parts of the valley also saw widespread – in most areas probably total – clearance of trees to create grazing land and cultivated fields, but there the survival of localised birchwood is suggestive of managed woodland retained as a source of building materials and fuelwood. The speed and totality of the clearance is indicative of direct human agency with the intention of expanding arable capacity locally, rather than of indirect pressures through increased grazing of flocks and herds. Some of the valley-bottom land was given over to pasture after clearance and the wetter areas seem to have been developed as hay meadows, prob-

10 For example, B.K. Roberts, J. Turner and P.F. Ward, 'Recent forest history and land use in Weardale, northern England', in H.J.B. Birks and R.G. West (eds), *Quaternary Plant Ecology* (Oxford, 1973), pp. 207–221; D.D. Bartley, C. Chambers and B. Hart-Jones, 'The vegetational history of parts of south and east Durham', *New Phytologist*, 77 (1976), pp. 437–468; J. Turner, 'The environment of northeast England during Roman times as shown by pollen analysis', *Journal of Archaeological Science*, 6 (1979), pp. 285–290; L. Dumayne, 'Invader or native? – Vegetation clearance in northern Britain during Romano-British times', *Vegetation History and Archaeobotany*, 2 (1993), pp. 29–36; L. Dumayne, 'Iron Age and Roman vegetation clearance in northern Britain: further evidence', *Botanical Journal of Scotland*, 46 (1993), pp. 385–392; L. Dumayne, 'The effect of the Roman occupation on the environment of Hadrian's Wall: a pollen diagram from Fozy Moss, Northumbria', *Britannia*, 32 (1994), pp. 217–224; Tipping, 'The form and fate of Scotland's woodlands'; R. Tipping, 'Holocene evolution of a lowland Scottish landscape: Kirkpatrick Fleming, II, Regional vegetation and land-use change', *The Holocene*, 5 (1995), pp. 83–96; A. Wilson, 'Roman penetration in Strathclyde south of the Antonine Wall, Part Two: Romanization', *Glasgow Archaeological Journal*, 20 (1996–7), pp. 1–40; R. Tipping, 'Pollen analysis, late Iron Age and Roman agriculture around Hadrian's Wall', in A. Gwilt and C. Haselgrove (eds), *Reconstructing Iron Age Societies* (Oxford, 1997), pp. 239–247; R. Tipping, 'Environment and environmental change in Eastern Dumfriesshire', in Royal Commission on the Ancient and Historical Monuments of Scotland, *Eastern Dumfriesshire: An Archaeological Landscape* (Edinburgh, 1997), pp. 10–25 at 20–21; R. Tipping, 'Vegetation history of southern Scotland', *Botanical Journal of Scotland*, 49 (1997), pp. 151–162; S. Ramsay and J.H. Dickson, 'Vegetation history of central Scotland', *Botanical Journal of Scotland*, 49 (1997), pp. 141–150; L. Dumayne-Peaty, 'Forest clearance in northern Britain during Roman-British times: re-addressing the palynological evidence', *Britannia*, 26 (1998), pp. 315–322; L. Dumayne-Peaty, 'Human impact on the environment during the Iron Age and Romano-British times: palynological evidence from three sites near the Antonine Wall', *Journal of Archaeological Science*, 25 (1998), pp. 203–214; and L. Dumayne-Peaty, 'Continuity or discontinuity? Vegetation change in the Hadrianic-Antonine frontier zone of northern Britain at the end of the Roman occupation', *Journal of Biogeography*, 26 (1999), pp. 643–665.

11 Tipping, *Bowmont*, pp. 182–184.

12 Ibid., p. 183.

Manor Valley from Cademuir Hill, Scottish Borders. Above the managed grassland of the valley bottom and the modern conifer plantations, the open, treeless and *Calluna*-dominated uplands had already become established before the end of the first millennium BC.

ably to provide fodder for the cattle that were so important in the economy of the local Celtic tribes. The chronology of the clearance allows us to be certain of one thing: that it was the work of the indigenous tribal peoples and not, as medieval chroniclers and modern fantasists supposed, of the Roman army.[13] Indeed, by the time Quintus Petillius Cerialis and Cnaeus Julius Agricola were launching their campaigns into what is now northern England and Southern Upland Scotland, much of the hill country stretching south from the Cheviots into the Pennines was already treeless, heather moorland and possibly functioning as the pastoral reserve for lowland communities that had invested heavily in arable cultivation.[14]

The scale and speed of the northern Cheviot clearances together with the pollen evidence for changed land-use make it almost certain that agricultural expansion was the motive. Calculation of the volumes of wood needed for the construction and maintenance of settlements within the district indicates that demand for building materials would have made minimal inroads into local woodland, even at the most generous assumption of timber needs, which leaves agricultural clearance as the most likely cause of its disappearance.[15] It is also clear,

13 W.S. Hanson, 'Forest clearance and the Roman army', *Britannia*, 27 (1996), pp. 354–358.

14 Tipping, *Bowmont*, p. 184.

15 Ibid., p. 187.

however, that the woodland composition had been significantly altered before the start of this major clearance episode, with oak all but vanishing from the valley systems of the northern Cheviots to leave a woodland dominated by birch, hazel, willow and alder. These species are reflected in the wood remains recovered from excavations of settlement sites locally, with no evidence for the importing of larger timber types from outside the valleys to satisfy building needs; communities used what was available locally. The preservation of an apparently birch-dominated woodland in Bowmont when the alder carr was being cleared, growing in places which were least suited to arable development, hints at a conscious decision made at more than single-household level to protect and maintain that timber. A similar trend is present in Glen Affric in the western Highlands, where small clumps of woodland on areas of moderately good ground were preserved when inferior soils were being used for grazing and some arable, indicating that a choice had been made in respect of land-use. Here, it is suggested, the woodland was possibly managed for smallwood and leaf fodder.[16] In these widely separated examples, we might have evidence for the careful reservation of a tree type valued for its fast growth and regeneration, used for building timbers and fencing, wood turning, fuel and cattle fodder, as has been postulated for parts of England and was a practice maintained in parts of Highland Scotland into the nineteenth century.[17] What can be said for sure is that there were individuals or communal decision makers with sufficient authority in later prehistoric Scotland to enforce such set-aside practices and that their role continued into the early centuries AD.[18]

Substantial volumes of heaped earth, turf and especially stone were used in the monumental and domestic architecture of the late Iron Age peoples of much of what is now the Scottish mainland, especially in the Atlantic-facing west and in the northern Highlands. Even in those areas, however, archaeological excavation has shown that wood, ranging from massive oak and pine timbers to small-diameter and probably coppiced oak, birch, elm, hazel, willow and alder was also used at volume. Major settlement sites like Eildon Hill North in the Scottish Borders or Craig Phadrig in Inverness-shire consumed great amounts of timber as palisades for their enclosing circuits and, at the latter, for binding its drystone rampart, while roundhouses like Newmill in Perth and Kinross or Birnie in Moray employed large roundwood for building frame, wicker for walling and smaller poles for roof structures.[19] The volumes and types of timber consumed in these sites are perhaps best understood from examples where material survives, like the late Bronze Age / early Iron Age crannog at Oakbank on the north side of Loch Tay. There, the main supporting piles were alder – favoured for its rot-resisting properties – with oak used for the main beams and planking of the superstructure and hazel for the wall panels and frames. The age of the timber used again points to management of the woodland resource: no oaks were over 100 years old and the alder was mostly around 40

16 Davies, 'Upland agriculture and environmental risk', p. 2058.

17 O. Rackham, *Ancient Woodland: Its History, Vegetation and Uses in England* (London, 1980); O. Rackham, *Trees and Woodland in the British Landscape: The Complete History of Britain's Trees, Woods and Hedgerows*, rev. paperback edition (London, 2001), pp. 87–88; N.A. MacKenzie and R.F. Callander, 'Birchwoods in a Deeside parish', in T.C. Smout (ed.), *Scottish Woodland History* (Edinburgh, 1997), pp. 135–146.

18 Explored in Armit and Ralston, 'The coming of iron'.

19 Watkins and Barclay, 'Excavation of a settlement and souterrain'; F. Hunter, 'Excavations at Birnie, Moray, 2004', National Museums Scotland Research Repository, 2005, https://nms.iro.bl.uk/concern/reports/308939ee-35f4-40bd-9f90-0390237d0426.

Scottish Crannog Centre, Kenmore, Perth and Kinross. These images of the former Crannog Centre, destroyed by fire in 2021, show the structure as reconstructed on the basis of the excavation evidence from Oakbank on the north side of Loch Tay near Fearnan. The excavation and reconstruction provided remarkable insights on the volume of materials, the management of woodland resources and the sophisticated construction techniques required for these supposedly 'primitive' buildings.

years old, pointing to coppice practices already operative early in the first millennium BC in areas viewed traditionally as falling within the coverage of the 'Great Wood'.[20] We lack the hard archaeological and palaeoenvironmental evidence, but it is likely that if settlements along the Highland fringe in less intensively cultivated or grazed zones were already supplying their primary timber needs through coppice management in the late Bronze Age, communities in the more heavily populated and cultivated zones of the country were also satisfying their needs in this way.

Roman inroads?

Did the Roman invasions add to the significant loss of managed or wild woodland? Post-medieval tradition identified trees preserved in peat bogs as the remains of the forests felled by the Romans and spoke frequently of axe marks or partial burning as evidence of the legions' depredations. We have, however, already set aside the myth of Septimius Severus and the attempted felling and burning of the equally mythical Great Wood of Caledon. That is not to say that the advancing armies did not make use of local resources. It is likely that the Romans did utilise the trees they found along the line of their advance. The scale of that opportunistic felling, however, can only be guessed at. Although some have been tempted to see the timber demands of the Roman military for their fort and frontier wall-building efforts as having had a major impact on woodland as far north as the valleys running into the southern edge of the Highlands, this proposition is dismissed by most archaeologists and palaeoenvironmental scientists.[21] The Hadrianic, Flavian and Antonine frontier lines that traversed the country on the Tyne–Solway, Highland Boundary and Midland Valley lines respectively are now understood as running through an already open landscape,[22] and yet the bulk of the timber required for their fortifications and internal buildings, calculated as having been recovered from an area perhaps as small as three times the footprint of their forts, was probably still secured locally. The greatest single construction effort, the legionary fortress at Inchtuthil on the Tay, downriver from Caputh, is reckoned to have required around 30 linear kilometres of wood for its walling needs, estimated as representing a cleared area of over 100ha, which is only around five to six times the area occupied by the fortress.[23] This would have had a tiny impact on the likely surviving woodland area of later first-century Scotland; moreover, given the two-year occupation of Inchtuthil before its abandonment in AD 86, cleared areas would have regenerated quickly provided grazing was kept suppressed. Only some of the larger timbers needed for gate structures or towers might have been transported as seasoned and dressed prefabricated members from military supply bases in northern England, but even this is unlikely given transportation difficulties in advance of road-

20 J.J. Miller, 'An Archaeobotanical Investigation of Oakbank Crannog, a Prehistoric Lake Dwelling in Loch Tay, the Scottish Highlands', unpublished PhD thesis, University of Glasgow (1997).

21 Dumayne, 'The effect of the Roman occupation'; Dumayne-Peaty, 'Forest clearance in northern Britain'; Hanson, 'Forest clearance and the Roman army'; Armit and Ralston, 'The coming of iron', pp. 44–45.

22 For the Antonine Wall's local environment, see W.E. Boyd, 'Environmental change and Iron Age land management in the area of the Antonine Wall, Central Scotland: a summary', *Glasgow Archaeological Journal*, 11 (1984), pp. 75–81; W.E. Boyd, 'Palaeobotanical report', in L.J.F. Keppie et al., 'Excavations at the Roman fort of Bar Hill, 1978–82', *Glasgow Archaeological Journal*, 12 (1985), pp. 79–81.

23 L.F. Pitts and J.K. St Joseph, *Inchtuthil: The Roman Legionary Fortress: Excavations 1952–65* (London, 1985), pp. 45–46; Smout, MacDonald and Watson, *A History of the Native Woodlands*, p. 32.

Strageath Roman fort, Muthill, Perth and Kinross. These images from the 1984 season of excavation at the fort site show the compressed turves of the Flavian-period fort of the 80s AD – a section through the rampart and the pattern of laid turves – which analysis showed to have been stripped from the valley floor. The volume of cut turf needed for the complete circuit of the fort rampart would have scalped a broad area of the low-lying meadow and pasture for several hectares around Strageath.

building work.[24] The Romans' ability to source timber in the open, late Iron Age landscape is suggestive of the presence of managed blocks of the kind postulated by Tipping for Bowmont. They probably commandeered local coppiced woodland for their needs, and it is likely that their military operations caused localised shortages and hardship for native communities dependent on those resources. The impact, however, is likely to have been short-lived and of minor effect on overall woodland cover.

It is also important to reflect on the primary construction material used in Roman fortifications in Scotland, turf. The volume of turf used in building ramparts was significantly greater than the volume of timber, and its local availability to the construction sites again points to a landscape with expanses of grassland and meadow from which turves could be cut. Excavation at first- and second-century forts like Strageath near Crieff produced evidence that is largely supportive of the picture of an open grassland or heath landscape with probably small areas of managed wood. Strageath's first-century Flavian fort's turf-built rampart used material cut from wetter grassland areas, probably on the floodplain of the adjacent River Earn, with corduroy layers of small branches laid at intervals between turf layers to provide lateral stability. That use of smallwood in the rampart tallies with the ash, birch and hazel branches and twigs that predominated in the recovered charcoal remains from the fort's late first-century occupation hearths.[25] The larger wood was probably used in the wattle panels and framing posts of the fort's internal buildings. Pollen analysis of the turves used in the first phase of the rampart identified mainly grass and sedge pollen, with very small amounts of oak, alder and probably hazel, indicative of an open landscape with few trees in central Strathearn at that date. The presence of ribwort plantain, daisies and buttercups supports the idea of mainly grazed grassland, but barley pollen and pollen from the weeds of cultivation indicate that there was also arable activity close to the fort site prior to its construction.[26] The picture produced from this data is one of a well-settled landscape covered by a mosaic of pasture and cultivated ground, probably interspersed on the wetter and poorer areas with stands of coppiced woodland, a picture similar in character to that of the Cheviot valleys at the same date.

Woods and people in Highland Scotland

Although it is likely that the extent of woodland encountered increased as one penetrated the Highlands, with the greatest expanses located in the glens of northern and western Perthshire, Aberdeenshire, Inverness-shire and Argyll, these districts were not covered by a forest blanket in the early centuries AD. Millennia of shifting patterns of climate and weather, further shaped by human settlement and agricultural activity, had created a patchwork of woods interspersed with zones of grassland and cultivation throughout the Highland valley systems. Alder carr and willow fringed the lochs and wetlands, with more extensive cover of larger species on the drier zones that rose from the valley floors. While Scots pine constitutes the romantic view of the ancient forests, oak was dominant in much of this zone, although it was still but one component of a species-diverse woodland that included ash, elm, birch, hazel,

24 The notion of the Roman military machine running a sophisticated stockpiling operation in advance of campaigns is now largely discounted. See Armit and Ralston, 'The coming of iron', pp. 45, 50, and for the contrary view, Pitts and St Joseph, *Inchtuthil*, p. 45.

25 S.S. Frere and J.J. Wilkes, *Strageath: Excavations within the Roman fort, 1973–1986*, Britannia Monograph Series 9 (London, 1989), pp. 271, 272.

26 Ibid., p. 272, 274.

Allt a' Chonais woods, Glen Carron, Highland. The head of the existing semi-natural woodland lies at approximately the same line as its late prehistoric ancestral woodland. This straggle of park-like pines and birch is close in character to its predecessor of the last millennium BC.

willow, alder, aspen, cherry and, in some districts around the Cairngorm fringe, juniper groves. Pine forests there were, but probably extending little beyond their modern distribution and possibly occurring in more of the open, park-like formations still to be seen in, for example, parts of Abernethy, Glen Lui/Glen Derry and Glen Quoich that radiate from the Cairngorm plateau.[27] And, contrary to the vision of the central Highlands as unpeopled wilderness, the archaeological record reveals a succession of landscapes of settlement extending deep into the more mountainous areas.

Excavation and survey in this region have confirmed that widespread clearance for pasture and cultivation had occurred from the Neolithic and continued through the Iron Age into the Early Historic period. In the Dee catchment, for example, the now almost depopulated uplands between the main valley and Glen Gairn was heavily settled by the Bronze Age, with hut circles and clearance cairns pointing to a community practising arable cultivation. On its southern edge, souterrains – subterranean or semi-subterranean passages interpreted since the 1980s as intended for bulk storage of produce[28] – on the low ground of the Dee valley are probably indicative of larger-scale arable cultivation in the early centuries AD that operated at a level beyond basic household subsistence.[29] Further north,

27 Smout, MacDonald and Watson, *A History of the Native Woodlands*, pp. 29, 33–34.

28 The broad thesis is set out in T. Watkins, 'Where were the Picts? An essay in settlement archaeology', in J.G.P. Friell and W.G. Watson (eds), *Pictish Studies: Settlement, Burial and Art in Dark Age Northern Britain*, British Archaeological Reports, British Series 125 (Oxford, 1984), pp. 63–86.

29 Aberdeenshire Historic Environment Record, NO29NE0004, The Strone, https://online.aberdeenshire.gov.uk/smrpub/master/detail.aspx?tab=main&refno=NO29NE0004. For souterrains, see the example at Tullich, 2km east of Ballater: HES Canmore, Braehead Farm, Tullich, Souterrain, https://canmore.org.uk/site/32455/braehead-farm-tullich.

surveys undertaken in Strathdon in the later 1990s revealed a similarly intensively settled late prehistoric landscape, with cultivation and grazing penetrating deep into the mountainous west of the region.[30] Both valley systems bore less woodland than exists there today, with the land broken into zones of pasture and cultivation.

Across the Cairngorms on Speyside, the valley from Inveravon south-west to Newtonmore contains extensive evidence for settlement from at least the Neolithic. Earlier prehistoric settlement was concentrated on the most easily tillable ground, but colonisation spread into less favourable areas as population pressures grew. There is, for example, evidence for the intensification of exploitation leading to a decline in the pinewoods around lochs Garten and Pityoulish in the Bronze Age, and for the expansion of cleared ground into the pinewoods around Rothiemurchus in the Bronze and Iron Ages. By the late Iron Age, along the length of the valley there had been development of important local power centres – fortified sites (including hillforts and duns) and crannogs – and the crystallisation of a clearly hierarchical society whose leaders were buried in extensive cemeteries in the late Iron Age and Early Historic period.[31] Strikingly, the use of stone and absence of evidence for large-scale use of timber in the major fort at Dun da Lamh at the junction of Strathmashie with Strathspey near Laggan, where the two main routes from the Great Glen (via the Corrieyairack Pass and Glen Spean) converge, suggests that trees were perhaps scarcer in this district by the early centuries AD than is popularly assumed.[32] Not all clearance in these regions, however, stemmed from agricultural colonisation; as elsewhere in Scotland, it is likely to have been the result of the interplay of three millennia of anthropogenic and climatic inroads that had presumably taken their toll on the woodlands of these central Highland straths. A combination of continuing agricultural and grazing pressures and demand for specific wood types contrived to prevent widespread regeneration. In such circumstances, it is likely that, in common with more Lowland districts, there was a long-established tradition of coppice management close to the main zones of settlement throughout these valleys to provide essential wood supplies.

The first 'new' forests

Were some species of tree favoured over others? Bowmont's birches and the probably coppiced oak, hazel and alder around Loch Tay indicate that preferential management of tree species was already established in the first millennium BC. That management, however, is unsurprising in that it involved species that were naturally occurring regionally from up to 1,000 years earlier. While the Roman period in southern Britain saw some species introductions, principally chestnut, Romano-British woodland management strategies were basically identical to those of the pre-Roman Iron Age peoples.[33] In northern Scotland, however, there are tantalising traces of something more radical and indicative of forward planning in woodland management over multiple

30 Royal Commission on the Ancient and Historical Monuments of Scotland, *In the Shadow of Bennachie: A Field Archaeology of Donside* (Edinburgh, 2007), chapter 6.

31 Smout, MacDonald and Watson, *A History of the Native Woodlands*, p. 29. For the barrow cemeteries, see, for example, Highland Historic Environment Record, MHG4687, Pictish Barrow Cemetery, West Croftmore, https://her.highland.gov.uk/monument/MHG4687.

32 Highland Historic Environment Record, MHG4666, Fort, Dun da Lamh, Black Craig, Laggan, https://her.highland.gov.uk/Monument/MHG4666.

33 Rackham, *Trees and Woodland*, pp. 40–41.

Dun da Lamh, Laggan, Strathspey, Highland. The massive stone rampart of the Early Historic-era fort on the headland at the junction of Strathmashie with Strathspey shows no evidence for the substantial use of timber in a region seen traditionally as blanketed with 'primeval' woodland.

generations. Pollen analysis at two locations on opposite sides of the country – Morvich in Wester Ross and Reidchalmai in south-east Sutherland – have produced evidence for the apparently deliberate introduction of oak with birch in the early centuries AD at locations where oak had not grown naturally for around two millennia.[34] That chronological gap and the sudden widespread reappearance of oak renders it unlikely that this was a natural regenerative process from a few survivors in remote refugia. Such a sudden appearance of an exotic tree pollen and at a volume indicative of more than a few revenants is strongly suggestive of a deliberate introduction. At Morvich, close to the head of Loch Duich, the oaks were introduced into an area with an old-established grazing regime which was already supporting a birch, hazel and alder woodland. They grew amidst rather than displaced the 'native' species, lasted for perhaps a little more than a single mature generation – 150–200 years – then were clear-felled in a single act towards the end of the sixth century and not replaced. At Reidchalmai, a similar woodland regime was established by around 300 on previously unused ground bordering a small loch and close to cultivated

34 R. Tipping, A.L. Davies and R. McCulloch, 'Introduced oak woodlands in northern Scotland: pollen analytical evidence for early historic plantations', in A. Woolf (ed.), *Landscape and Environment in Dark Age Scotland* (St Andrews, 2006), pp. 29–47.

Morvich and the head of Loch Duich from the west ridge of Beinn Fhada, Kintail, Highland. The sheltered alluvial spread between the foot of the mountain and the sea loch was possibly the location of one of our first examples of anthropogenic woodland planting.

land that continued in use during the 200 years or so of the wood's existence. While the near simultaneous appearance of two areas of oak-rich woodland could reflect some climate event or woodland regeneration following a major local population upheaval, the Morvich and Reidchalmai sites have been postulated as the result of deliberate introductions using seedlings brought from locations further south with the express intention of developing a woodland resource of a species otherwise unknown in northern Scotland at that date.[35] In neither location did the introductions affect local land-use regimes and in neither was there expansion beyond the small wooded areas identified. These were apparently closely confined and managed plantations. We shall never know why these trees were planted, but demand for large hardwood timbers from local elite users might have encouraged such long-term speculation. The clear-felling either marked the abandonment of that strategy or was possibly linked with the changing climate regime at the start of LALIA. At neither location did oaks re-emerge as woodland species until the era of estate planting in the late eighteenth and nineteenth centuries.

If the deliberate planting hypothesis is correct – and it is not without challenge – it is a discovery that has major implications for Scottish woodland history. We have long accepted the very high degree of

35 Ibid., pp. 38–39.

human management and interference in what we oddly insist on calling 'semi-natural' woodland, especially from the sixteenth century onwards. Perhaps because of some lingering belief that the people of a more distant age were incapable of such forward planning, we have barely considered the possibility that such interference occurred centuries earlier and involved species that were not again managed in this way until the modern era. Evidence of preferential planting, 'weeding' and thinning of unwanted species, conscious use of recognisable and well-known practices – in these cases possibly birch coppice with oak standards – all with a view to satisfying an identifiable demand in a manner that we present traditionally as an innovation of the Age of Improvement, suggests that sophisticated woodland management was being undertaken over a millennium earlier. At the very least, it requires us to rethink the ways in which we imagine the people of Early Historic Scotland interacted with woodland.[36]

Managing and maintaining the woodland resource

What this era of climate change, human depopulation and recolonisation meant for Scotland's woodlands is a vexed question. Despite more recent and more granular analysis of pollen evidence from across Scotland, there is little reason to substantially revise Richard Tipping's view that 'much of Scotland had been subject to considerable deforestation' by the early centuries AD.[37] There is also little new evidence to challenge his conclusion that in the northern mainland and the Northern and Western Isles this loss of tree cover arose from a combination of climate change, soil deterioration, blanket peat growth and human impacts spanning millennia. Likewise, his analysis of conditions further south and east, where human agency might have played a greater but still by no means exclusive part in that loss, seems uncontested. As we have already seen, there is some debate over the question of an apparent re-expansion of woodland cover that has been suggested occurred in parts of southern Scotland in the sixth and seventh centuries as a positive environmental consequence of population collapse. Likewise, the cause of an apparently abrupt sixth-century end to areas of oak woodland that seem, as discussed above, to have been deliberately planted and protected in coastal districts of the northern Highlands from the earlier centuries AD remains contested. Was this a felling episode or reflective of a wider failure of oak populations across northern Europe that some have linked to the onset of the LALIA and its initial episodes of climate-cooling volcanic winters?[38] Furthermore, as new data is obtained from archaeological excavations in the Northern and Western Isles, we are slowly revising older perceptions of the likely (non-)survival of native woodland there into the later first millennium AD, as evidence for alder, hazel and willow in Shetland, plus aspen and birch in Orkney, suggests that some stands of such small tree species, possibly coppiced, persisted into the Norse period at least. These, though, are largely nuanced refinements of a picture of continuing decline and intensifying pressure as human land management regimes were reshaped to accommodate the limitations and opportunities of a changing climate.

Most of our evidence for the state of Scotland's woodland in the five centuries after AD 500 comes from archaeological excavations and, in most cases, offers only a general sense of wood use patterns for building and craft-working practices and of wood's

36 For an overview of Celtic culture and society in Scotland to AD 500, see I. Armit, *Celtic Scotland*, new edition (London, 2005).

37 Tipping, 'The form and fate of Scotland's woodlands', p. 37.

38 Tipping, Davies and McCulloch, 'Introduced oak woodlands', pp. 40–42.

Hazel scrub, Tomnadashan, Perth and Kinross. Species like hazel (*Corylus*), which throw multiple new stems when subjected to coppice management, were hugely in demand for the withies and wands produced by this fast new growth. These were used for the wattling and wickerwork employed in buildings and for basketry.

local availability. Thus, to have a contemporary written witness to the functional role of wood in the daily life of a community, to some management practices and to the challenges of obtaining the most suitable timber is priceless. Adomnán's late seventh-century account of the life of the late sixth-century St Columba gives us a wealth of incidental evidence for the condition of woodland in the Inner Hebrides during the saint's lifetime and for how it was managed, obtained and used by the monastic community on Iona. For example, the story of a miraculous crop of grain sown after 15 June and harvested at the start of August reveals that the monks of Iona lacked adequate woodland resources on their island by the last quarter of the sixth century but had rights to supplies from the land of neighbouring laymen.[39] We learn that the monks needed wattle for building a guesthouse at their abbey and that Columba sent the brethren in a freight ship to collect bundles of wattles from the property of the

39 *Adomnan's Life of Columba*, ed. and trans. A.O. Anderson and M.O. Anderson (Edinburgh, 1961), pp. 329, 331.

layman Findchan, which possibly lay in the Ross of Mull. Findchan's own supplies of what was probably willow or hazel coppice seem to have been limited, for it was in recompense for his loss that Columba's prayer delivered his miraculous grain harvest. Here, we seem to have the record of management of small blocks of scrubby woodland as coppice, much as was to be the regional norm in the Inner Hebrides into the eighteenth century and probably not dissimilar to the areas of relict woodland that also survived in parts of the Outer Hebrides until at least the late first millennium AD (see below, pp. 89–91). A second miracle story in the same source, however, narrates Columba's encounter with an especially large wild boar in 'a dense wood' on Skye.[40] This account indicates that larger blocks of woodland still existed on some of the bigger Inner Hebridean islands at the end of the sixth century, again possibly being managed as a source of structural timbers such as were used for shipbuilding and as major constructional beams. Two further conjoined miracle stories appear to confirm the dearth of such woodland close to the monastery. The first concerns large pine and oak timbers that were being transported overland in Adomnán's time to places on the coast for shipment over the sound to Iona, probably from sources on Mull, for use in structural work on a 'great house' and for boat building.[41] The account tells us that the cut timbers were drawn behind boats and currachs 'throughout the long and devious route' by sea from the felling sites to Iona.[42] The second story reveals that wood was being brought to Iona from the district around the head of Loch Moidart and the inland Loch Shiel, again towed behind currachs that the monks had sent to collect it at the mouth of the River Shiel.[43] The distance involved and the dangers of a sea route round the point of Ardnamurchan from Iona to the mouth of the Shiel and back is a good indication of the absence by the last quarter of the seventh century of closer available sources of large structural timbers.

A trend from good availability of wood on or near to Iona in the earlier seventh century to a progressive decline thereafter seems to borne out by evidence from the excavation of parts of the enclosing ditch of the monastery in the 1970s.[44] The excavated wood remains fall into two categories: turned wood, worked wood and their waste, and undressed twigs and branches.[45] The former provide a strong reminder of the likely abundance of treen items used in a region where poor availability of good-quality clay limited the range of ceramics produced. There is reference in the late sixth- or early seventh-century Brythonic poem *Y Gododdin*, set in the south-east of Scotland and north-east of England, to an alder-wood cup,[46] illustrating the ubiquity of turned wood artefacts and utensils at that date and the possible prevalence of alder in its manufacture. The turned wood from Iona was almost entirely alder and possibly obtained locally, for pollen analysis from the excavation was suggestive of a sharp decline in alder in the early phase of occupation and its recovery in later periods.[47] Oak bulked large in the worked wood remains, including what were interpreted as posts that had been set into a sill beam with planking in the spaces between the uprights.[48] Oak, ash, aspen,

40 Ibid., p. 385.

41 Ibid., p. 453.

42 Ibid., p. 455.

43 Ibid.

44 J.W. Barber et al., 'Excavations on Iona, 1979', *Proceedings of the Society of Antiquaries of Scotland*, 111 (1981), pp. 282–380.

45 Ibid., pp. 328–346.

46 K.H. Jackson, *The Gododdin: The Oldest Scottish Poem* (Edinburgh, 1969), p. 153.

47 Barber, 'Excavations on Iona', pp. 335, 337.

48 Ibid., p. 338.

Atlantic oak-dominated woodland, mixed with birch and alder, near Loch Don, Mull, Argyll and Bute. Woodland like this provided Columba and the Iona monks with the large structural timbers they needed for the buildings of their monastery.

birch, cherry, hazel, pine, willow and yew were all used for stakes, pegs or pins,[49] giving some indication either of the diversity of the tree species available close to Iona or, perhaps, of the need to use whatever came to hand in a district where the best woodland had been consumed before the end of the seventh century. Pollen analysis seems to bear out this record of apparent loss, with evidence for oak and ash woodland with more limited birch, hazel and willow on the east side of Iona in the later sixth and earlier seventh centuries but then for a sharp deline in all of these species through the 600s.[50]

The presence of large quantities of worked wood of many tree species in the ditch-fill at Iona but with low levels of pollen for all of them might corroborate Adomnán's tale of wood supplies having to be brought to the island even during Columba's lifetime but certainly by his own day. One species that seems to have thrived for a while was elder, possibly used as hedging along the edges of the ditches. The sudden disappearance of elder pollen is indicative of a felling and clearance event, which might account for the large numbers of twig and branch fragments recovered in higher layers of the ditch-fill. A later phase (possibly tenth century) saw the re-emergence of birch and alder, successive rises and falls in the pollen curve of the former being suggestive of coppice management.[51] Taken together, the wood remains and the pollen illustrate the demands for materials made by a high-status settlement and the mechanisms used to satisfy those demands. Importantly, they reveal coppice practice in operation for species that provided the bulk of everyday wood needs, for wattling, wickerwork and wood-turning, and point to the scarcity of large timber species like oak and pine in the near vicinity of the monastery. The rapid sixth- and seventh-century decline and the non-regeneration of the oak and ash woodland in the eighth and ninth centuries is perhaps also illustrative of initial voracious consumption of local

49 Ibid., pp. 338–339.
50 Ibid., pp. 346–347.
51 Ibid., p. 348.

materials followed by later delivery to Iona of its timber needs from neighbouring island and mainland districts, as suggested by Adomnán's text. More likely, however, we seem to be seeing evidence for the result of long-term declines in substantial woodland areas in the Inner Hebrides as centuries of human exploitation and climate impacts took their toll.

Adomnán's descriptions of the largely wooden built environment of the monastery on Iona and the archaeological evidence for sophisticated beam-and-plank construction, wattle and daub, and ubiquitous turned wood artefacts are reinforced by the excavated remains of the sixth- to eleventh-century monastic site at Whithorn in Galloway.[52] There, successive phases of building, repair and reconstruction made abundant use of wood, from large structural timbers and planks to smaller stakes, poles and enormous quantities of withies and wands for wattling and wickerwork. In all periods, the best timbers appear to have been reserved for high-status structures such as the churches – of which there seem to have been several dispersed around the site – and the large halls and accommodation blocks characteristic of the Anglian occupation of the site from the early eighth to the mid ninth century.[53] In the seventh and eighth centuries, some structures used a mixture of large posts to define their form, with panels of wattling in the interstices,[54] but it cannot be said that this method developed in response to shortages of the large timbers used in the post, plank and stave techniques of the main buildings. Lower-status domestic housing around the monastic core and probable workshop and storage buildings were primarily stake-built or used extensive hazel, willow and alder wattling.[55] Some of the houses of the later tenth or earlier eleventh century were entirely wicker-built, possibly prefabricated inverted like giant baskets.[56] A social hierarchy is clearly reflected in the building materials here, with the bigger timbers from the preferred oaks being reserved for the most prestigious buildings, including the accommodation of the clergy, and with none of the lower-status and non-monastic buildings containing such good-quality structural members.[57] This distinction does not necessarily imply shortages of oak and other slow-growth timber trees locally but more likely indicates control over its use being exercised by a 'lord' such as the monastic community. Certainly, the generous use of beams, posts, planks and staves in the later eighth- and earlier ninth-century buildings at the core of the site indicates that the monastery had access to reliable sources of such timber over an extended period. The profusion of wood from small tree species, principally alder, birch, hazel and willow, however, is suggestive of successful coppice management of species that would have grown profusely as carr in the local environment of fenny hollows and lochans. The continuity in the availability of such materials across half a millennium – and beyond – underscores the point that this form of woodland management was both sustainable and productive.

A similar pattern in the use of major timbers for monumental display purposes and of wattle and wickerwork in other buildings was found in the excavations of the sixth-century phase of the crannog at Buiston in northern Ayrshire.[58] There, a massive

52 P.H. Hill, *Whithorn and St Ninian: The Excavation of a Monastic Town 1984–91* (Stroud, 1997).

53 Ibid., pp. 141, 144, 146–153, 172–178.

54 Ibid., pp. 138–139.

55 Ibid., pp. 183–185.

56 Ibid., p. 185 fig. 5.2.

57 Similar distinctions in the styles of building and materials used were identified in the seventh- and eighth-century Anglian monastery at Hoddom. See C. Lowe (ed.), *Excavations at Hoddom, Dumfriesshire* (Edinburgh, 2006), pp. 38–104.

58 A. Crone, *The History of a Scottish Lowland Crannog: Excavations at Buiston, Ayrshire 1989–90*, Scottish Trust for Archaeological Research Monograph 4 (Edinburgh, 2000), pp. 48–54 for the dendrochronology.

Left. Remains of a wall of a wicker house, excavated in 1986 at Whithorn, Dumfries and Galloway. The white garden tags mark the position of the upright stakes, whilst the sinuous bark band is the remaining horizontal bands of withies, woven through the uprights. Right. The detail is of a modern panel of hazel wattling at Kirkton near Kyle of Lochalsh, Highland, showing the appearance of the woven structure of the Whithorn houses.

outer perimeter using squared oak beams and posts with morticed joints to contain a cut and trimmed alder log palisade enclosed a relatively slightly built house of hazel wattle and alder and hazel post construction but with a fine planked floor of alder nailed to oak joists. Analysis of the timbers and the pollen evidence for the nature of the surrounding landscape suggested that the loch in which the crannog stood was fringed by alder and willow carr, which was carefully managed by the crannog dwellers as a materials source for posts and for the trimmed timbers of the palisade. Further from the loch and probably separated from it by cultivated and livestock-grazed ground was a mixed woodland of oak and hazel. This had been exploited by previous occupants of the crannog and was made up of areas of coppice and young maiden oaks in the clearings they had created, alongside stands of mature old-growth oak standards they had left. Analysis of the wood use in the crannog in the sixth and seventh centuries suggested preferential use of some materials over others – willow, for example, was present but not exploited to any great extent as a building material – and selective felling throughout the whole woodland area that avoided the overexploitation of any one species or stand.[59] As at Whithorn, we seem to be seeing evidence for the long-term sustainable management of the most valuable woodland resources through coppicing and the regeneration of seeded trees in areas of more intensive felling.

How were relict woodlands managed in areas where prehistoric climate change and millennia of human exploitation had massively reduced the availability of timber for even wattle construction? Analysis of the pre-660s pollen evidence from Portmahomack does not present a favourable picture

59 For detailed discussion, see A. Crone and R. Brunning, 'Woodland reconstruction', ibid., pp. 78–82.

Coppiced oak of late medieval date, Dollar Glen, Clackmannan. Although this is of a significantly later date than the woodland around Buiston, the management regime there would have produced a similar structure of coppice-with-standards and a mix of young, new-growth trees and veterans.

of management of woodland resources there: Tarbatness remained a largely treeless landscape through the fifth- to later seventh-century period with occasional patches of scrubby, fen-loving woodland developing as the wetland zones spread in the sixth century and with possible small stands of oak on patches of drier land.[60] Extensive grazing practices probably prevented the regeneration of any significant woodland cover on the peninsula, local demand for wood possibly being met by imports from more distant sources. In Easter Ross, such sources were relatively close and accessible in the upper reaches of the Cromarty and Dornoch firths and the inland valleys feeding into them. In less well-favoured locations, better resource management of patches of relict woodland might have been a socio-economic necessity. Excavation evidence from the Outer Hebrides confirms that those islands already had a largely treeless landscape by the Bronze Age, with trees surviving only in small patches of mainly open birch-hazel woodland and occasionally in more species-rich blocks (oak, elm, ash, alder) in sheltered locations. Peat was already established as the primary thermal energy source by that date in the Western Isles. Where wood other than the ubiquitous heather was consumed as fuel, it is likely to have been waste trimmings and offcuts or wooden artefacts that had reached the end of their usable life; this situation was unchanged in the late first millennium AD. At Dun Vulan in South Uist, tree pollen of any species was barely detectable in the soil samples but the presence of a rare beetle that lives beneath pine bark – a tree species then wholly absent from Uist – is suggestive of timber being imported to the islands from wood-

60 L.S. McHardie, T.M. Mighall and J.E. Schofield, 'The landscape and environmental setting for the Pictish monastery at Portmahomack', in M. Carver, *The Tarbat Discovery Programme*, download 7.4.4, Archaeology Data Service, 2017, https://doi.org/10.5284/1031216.

land elsewhere in the Hebrides or down the mainland Atlantic coast in the early centuries AD.[61] Nevertheless, the survival of turned wood vessels from excavations in the Outer Hebrides suggests either that there was a lively trade in treen from more wooded districts of the mainland's west coast or that the small areas of relict woodland were managed carefully and supplied local producers/consumers of turned wood artefacts into the late first millennium.[62]

Making do with less

Although heather continued alongside peat as the main fuel burned in Hebridean settlements, wood charcoal shows that some tree species were still available. The majority of wood used in all pre-Norse periods of occupation at Dun Vulan and the nearby pre-Norse late Iron Age Mound 1 settlement at Bornais, however, was from various species of larch that could only have reached South Uist as driftwood, as indicated by channels bored into the wood by marine Teredo worms.[63] Such driftwood is likely to have provided the majority of the large structural timbers used in the pre-Norse period. The remainder of the wood identified at Dun Vulan was primarily hazel and alder, pointing to the persistence of some patches of relict but probably managed scrubby open woodland controlled by the site's occupants as coppice to meet their demand for wicker wands, perhaps like Findchan's trees as described by Adomnán. Coppiced willow was also present, but the wood showed the extreme slow-growth traits of trees at the natural limit of their species' distribution and under environmental stress. Such stress could be caused by drought conditions, which would have needed to persist over several years, but is more probably due to the coldness of the general climate in this region at that time.[64] At Bornais Mound 1, wood remains from native species in the pre-Norse settlement included alder, ash, hawthorn, oak, pine and willow, some of which – the oak and pine – must have been imported from sources in the Inner Hebrides or mainland given the absence of pollen evidence for them from the Western Isles by this period. The birch roundwood remains, if from locally sourced coppice, were from trees growing in optimal or relatively stress-free conditions in contrast to the coppiced willow at Dun Vulan. This type of growth is suggestive of a sheltered location and thus probably quite remote from the weather-exposed machair down the west coast of Uist, perhaps in the more protected glens of the island's east coast.[65] Here, we might have evidence for an impact of the climatic cooling from the sixth century.

At Bornais changes in the volume of material and range of species present between the mainly pre-Norse settlement on Mound 1 and the Norse farms and houses of Mound 2 and Mound 3 seem to mark a sharp decline in the availability of native tree species in Uist, perhaps through overexploitation of a precarious resource. The excavated evidence also shows what might be a reduced availability of driftwood.[66] At Mound 3, the presence of juniper, a probably locally sourced wood that was not favoured by the Pictish precursors of the Norse colonists, could signal a shift to a more scavenging approach to whatever growing wood was available. The use of juniper is a

61 M. Parker Pearson and N. Sharples (eds), *Between Land and Sea: Excavations at Dun Vulan, South Uist* (Sheffield, 1999), pp. 54, 55.

62 Sharples, *A Late Iron Age Farmstead*, p. 329.

63 Taylor, 'The wood', pp. 188–189; Sharples, *A Late Iron Age Farmstead*, pp. 228–229.

64 Taylor, 'The wood', pp. 189–190.

65 Sharples, *A Late Iron Age Farmstead*, p. 191.

66 Ibid., pp. 191, 229, 250; N. Sharples (ed.), *The Economy of a Norse Settlement in the Outer Hebrides: Excavations at Mounds 2 and 2A, Bornais, South Uist* (Oxford, 2021), pp. 386–388.

Young birch woodland, Cadzow, South Lanarkshire. These young trees, growing in relatively stress-free conditions, form woodland of the type likely to have supplied the main timber needs of the Mound 1 settlement at Bornais, South Uist.

strong indicator of the consequences of earlier over-exploitation of the more favoured wood species. Through the ninth- and tenth-century occupation at Mound 2, the predominance of heather-wood charcoal in the hearth residues – amounting to 88 per cent of the total assemblage – emphasises the dependence on moorland resources and the overall absence of any accessible fuelwood source in this part of South Uist.[67] The absence of driftwood, while possibly signalling a reduction in the volume or quality of wood being washed up on nearby shores, is perhaps more likely to reflect the exercise of new and rigorous control over such resources by local social leaders. Together with evidence for the management of other coastal resources, this suggests that access was highly regulated and that specific use was allocated to communities by a higher authority.[68]

Similar trends are detectable in the Northern Isles. Extensive woodland had effectively disappeared from the Shetland landscape long before the Historic

67 Sharples, *The Economy of a Norse Settlement*, pp. 267–268, 386.

68 Ibid., p. 317; N. Sharples (ed.), *A Norse Farmstead in the Outer Hebrides: Excavations at Mound 3, Bornais, South Uist* (Oxford, 2005), p. 162.

Jarlshof, Sumburgh, Shetland. The excavated remains of a late Iron Age wheelhouse, showing the stone-built radial spokes that subdivided the outer ring of the internal space and reduced the inner area that had to be roofed using larger timbers. This architecture demonstrates cultural responsiveness to the changing range of building materials available locally.

era, although small patches of willow, hazel and alder might have survived into the Middle Ages.[69] The apparent use of wood on a large scale for house construction at Scalloway in the Pictish period, therefore, is strongly suggestive of timbers being obtained as imports or, as in Uist, from driftwood. The presence also of wood from scrub species raises the possibility that such material came from conserved, sheltered locations. The overall abundance of wood present in the Scalloway house must represent the conspicuous use of a scarce, probably controlled and highly valuable commodity as a statement of power and wealth.[70] Most late Pictish structures that have been excavated in Shetland, however, such as those at Jarlshof and Old Scatness, were primarily stone-built and employed corbelling and radial pier or 'spoke' arrangements in their so-called wheelhouse design to reduce span distances and thus minimise the need for wood in their roof structures. Analysis of charcoal from domestic hearth contexts at Old Scatness shows that a variety of wood species were being consumed there through the Pictish and Norse periods, although peat predominated at all times.[71] Larch and spruce charcoal from Pictish contexts

69 Dockrill et al., *Excavations at Old Scatness*, vol. 1, p. 204.

70 N. Sharples (ed.), *Scalloway: A Broch, Late Iron Age Settlement and Medieval Cemetery in Shetland* (Oxford, 1998), p. 210; Dockrill et al., *Excavations at Old Scatness*, vol. 1, pp. 22–23.

71 Dockrill et al., *Excavations at Old Scatness*, vol. 1, pp. 196–197.

almost certainly represents driftwood, possibly used for firewood after a previous use in house or boat building. In Norse contexts – where pine was also present – the charcoal might represent both driftwood and purposely imported timbers used first for construction and then burned at the end of a structure's lifespan. Unlike the wood from Dun Vulan and Bornais in South Uist, none of the larch, spruce or pine charcoal showed any evidence for the timbers from which they came having spent any significant length of time in the ocean as driftwood, which has led to the suggestion that the wood was imported.[72] The alder, hazel and willow charcoal could have been obtained from managed native sources, but all could also have been imported as finished objects, from alder-stave buckets to willow-wicker goods, burned when they ceased to be usable. Large pieces of charcoal from heather and heath also suggest that these abundant and resilient groundcover plants were exploited for fuel, possibly after previous use as thatch and possibly burned along with the underlying turf. Across both Pictish and Norse occupation periods, the wood remains from Old Scatness point to a commodity in scarce supply, although sources of alder, hazel and willow appear to have persisted. Such wood's consumption as fuel represented a final stage in what may have been a series of uses in construction or as artefacts and tools.

In the late pre-Norse period at Pool on Sanday in the Orkney Islands, environmental analysis has indicated that the settlement stood in the same kind of relatively open landscape that had prevailed in those islands since the Neolithic era. As in earlier periods, this contained significant areas of willow, hazel and birch scrubland but no substantial woodland.[73] Charcoal samples from the site contain traces of all those species, but with a notable drop in the frequency of conifer charcoal between the mid sixth and the eighth century. This might not indicate a lack of availability of conifer, which could only be obtained as driftwood or through import, but rather that it was too valuable to burn and was instead reserved for use as large structural members. The diminishing volume of wood charcoal in this period revealed greater dependence on other fuel types, possibly peat but more probably cattle dung, together with dried seaweed, straw and sandy turves.[74] This is a poor range of fuel resources, and the use of dung in particular would also have reduced the volume of manure available for nutrient enrichment in the heavily cropped infields. How the collection and storage of these materials fared in the colder and wetter phases of the later sixth and seventh centuries, where drying of these naturally wet commodities would have been difficult, remains unanswered. There is, likewise, the question of how increased volumes of such fuel were obtained to support the community through the extreme winters of the era. Here, perhaps, we might see one of the triggers for the shift in domestic architecture from large, open spaces to much smaller and well-enclosed ones, with houses increasingly submerged in insulating mounds of midden waste.

One of the mythic claims made in the thirteenth-century *Orkneyinga Saga* is that the possibly fictitious late ninth- or early tenth-century Jarl (Torf-)Einar 'was the first man to dig peat for fuel, firewood being very scarce on the islands, at Tarbat Ness in Scotland'.[75] While the fictive aspect of this statement about peat use is well understood, it allows us to corroborate two other suppositions. First, that having adopted peat-cutting from their indigenous predecessors, the primary fuel used by the Norse in Orkney to the tenth century was peat. Secondly, that after over a millennium of use by Orkney's pre-Norse inhabitants and

72 Ibid., p. 204.

73 Hunter, Bond and Smith, *Investigations in Sanday*, vol. 1, pp. 118, 200.

74 Ibid., pp. 200–201.

75 *Orkneyinga Saga: The History of the Earls of Orkney*, trans. H. Pálsson and P. Edwards (Harmondsworth, 1981), p. 29.

Willow-dominated carr, Woody Islands, Perth. Wetland-fringing woodland like this, growing in the wet hollow of an abandoned channel of the Tay, was managed for centuries by communities from the Northern Isles to Galloway as a source of wands for wickerwork and larger structural timbers.

their Scandinavian successors, convenient sources of good-quality peat might already have been becoming scarce in the areas of heaviest Norse settlement in the islands.[76] *Orkneyinga Saga*'s thirteenth-century reflection on the scarcity of woodland might also suggest that little had changed from the sixth- and seventh-century picture of surviving patches of scrubby woodland in an otherwise open landscape. Despite a growing difficulty to source peat in some areas, the general shortage of woodland in the Northern Isles meant that this scrubwood continued to be used mainly for structural needs, as were the large driftwood timbers that washed up on island beaches. At Pool, the shortage of other thermal energy sources resulted in dried seaweed probably forming the chief fuel for most domestic needs, with peat, wood and charcoal mixes reserved for more specialist functions that required high temperatures.[77] During the earlier stages of the Norse settlement in the late eighth/early ninth to mid tenth centuries, the occupants of Pool were still burning small quantities of willow, alder and birch, but the predominant fuels included straw, dung and seaweed.[78] What is perhaps most striking in this mix of fuels is the absence of evidence for the

76 B.E. Crawford, *Scandinavian Scotland* (Leicester, 1987), p. 153.

77 Hunter, Bond and Smith, *Investigations in Sanday*, vol. 1, pp. 120, 201–202.

78 Ibid., p. 144.

high-calorific types that would have been necessary for ironworking, despite the increased evidence for metalworking across the Norse era. Clearly, there had been no substantial regeneration of woodland, and what timber resources were available locally were probably reserved principally for building. The longevity of the survival of the areas of scrubby woodland in what was otherwise an intensively cultivated and grazed landscape, however, points to conservation and coppice management being maintained successfully across the transition from Pictish to Norse settlement.[79]

Driftwood is an unreliable source of wood for any purpose and it is likely that Orkney's Norse rulers secured the timbers they needed for building ships and houses, or for making into charcoal for high-calorific fuel, from areas with abundant woodland. Kin connections with western Norway might have offered access to the forests there, but political tensions between the Orkney Norse and the increasingly powerful Norwegian kings rendered that source unreliable. A more secure source, it has been suggested, was northern mainland Scotland, especially the valleys feeding into the head of the Moray Firth.[80] The tenth and early eleventh centuries witnessed clashes for control over this resource-rich region between the Norse jarls based in Orkney and the Gaelic rulers of Moray. This conflict might be visible in the place-name map of Easter Ross, which reveals a scatter of Scandinavian place-names that speaks of Norse successes in asserting more than temporary control over districts south of the Dornoch Firth.[81] These names mark settlements, it is argued, through which the Norse either 'manage(d) this resource for the earl's benefit' or whose occupants 'could compel the native population to fell and log timber according to the requirements for shipbuilding or other domestic needs'. Woods containing substantial timbers also existed down the west coast of the Scottish mainland and on the sheltered east-facing coasts of some of the Inner Hebrides, another zone over which Orcadian lordship was extended between the late ninth and the mid eleventh century. It has long been argued that Norse domination here was expressed principally through tribute gathering, and although usually seen as a silver render, it is possible that material resources like timber also formed part of the tribute being gathered.[82] Such a scenario is not unrealistic, given late twelfth-century chronicle evidence for the timber tribute extracted from the people of Galloway by the late eleventh-century Norwegian king Magnus Barelegs. According to the Manx chronicle, the Galwegians were compelled to deliver the cut timber to the shore of the Irish Sea from where the king transported it to Man to build defensive works.[83] Given the place of the Norse jarldom of Orkney as a maritime superpower in the British Isles into the mid eleventh century, it is clear that the rulers of the effectively treeless Northern Isles were successful in sourcing suitable timber for their galley building through the later ninth and tenth centuries in the regions of northern and western Scotland over which they exerted their authority.

79 Ibid., p. 201.

80 B.E. Crawford, *The Northern Earldoms: Orkney and Caithness from ad 870 to 1470* (Edinburgh, 2013), pp. 153–154, 156.

81 B.E. Crawford and S. Taylor, 'The southern frontier of Norse settlement in north Scotland: place-names and history', *Northern Scotland*, 23 (2003), pp. 1–76; Crawford, *The Northern Earldoms*, pp. 153 fig. 4.2, 154–156.

82 For the historiography of Norse tribute systems in western Scotland and the Isles, see R.D. Oram, 'Ouncelands, quarterlands and pennylands in the Western Isles, Man and Galloway: tribute payments and military levies in the Norse west', in S. Imsen (ed.), *Taxes, Tributes and Tributary Lands in the Making of the Scandinavian Kingdoms in the Middle Ages* (Trondheim, 2011), pp. 57–75.

83 *Chronica Regum Manniae et Insularum: The Chronicle of Man and the Isles: A Facsimile of the Manuscript Codex Julius A. VIII in the British Museum* (Douglas, 1924), s.a. 1098.

A controlled commodity

The political authority of the Orkney jarls that perhaps enabled them to access or control the timber resources of the Easter Ross and Inverness-shire glens was perhaps a manifestation of a recognised trend towards the centralisation of power and of the control of natural resources and human capital within early state formation. It has been suggested that control of woodland resources constituted one of the major demonstrations of such centralisation in the early Scottish state, where, during the tenth century, the kings of Alba were imposing a structure for more rigorous economic organisation and command of manpower for military purposes. The implicit organisational changes that came with the introduction of what we know as the *dabhach* system, discussed in Chapter 4, are likely to have had significant ramifications for woodland management. While the evidence of Irish law codes suggests seventh- or eighth-century recognition throughout Ireland of the impact of unregulated felling, there is no similar evidence that the protection of woodland resources was the subject of laws in the late Pictish state.[84] Dendrochronological data, however, has shown that timbers used in major construction projects in the later medieval period (such as Darnaway great hall roof and Glasgow Cathedral roof), or re-used in urban contexts from Inverness and Aberdeen to Perth, came from trees – mainly oak – that started growing around the middle of the tenth century.[85] Apart from a few rare oak trees that started growing in the later ninth century, it seems that there is a fairly consistent gap in new-growth trees of eighth- or ninth-century date across much of north-eastern Scotland before a mid tenth-century surge. In Ireland, similar gaps have been labelled 'depletion/regeneration periods', where phases of intense construction work and expansion and intensification of agriculture caused the pre-existing woodland to become depleted and then prevented its regeneration.[86] The clustered tree 'birth dates' marked the point at which such pressure was ended for some reason, allowing seedlings to become established successfully. In mainland Scotland, it was perhaps the rapid growth in the reach of kingship and the greater centralisation of overarching political power that occurred in the Pictish state of the early ninth century and its successor, Alba, in the tenth that led to episodes of intensive exploitation of the woodland resources.

We do not need to look far for the kinds of structure that would have consumed substantial amounts of major timbers both during their construction and for ongoing maintenance. At Burghead on the Moray coast, a site where fortification perhaps commenced before the beginning of the fourth century, the headland was occupied by what was perhaps the major royal settlement of Fortriu in the seventh and eighth centuries, which remained a key power centre until its destruction around 900.[87] It was certainly the largest early medieval fortification in northern Britain and consumed an enormous volume of oak in its

84 F. Kelly, 'The old Irish tree-list', *Celtica*, 11 (1976), pp. 107–124. A suggestion of an incipient crisis being recognised in these law tracts is possibly confirmed by Mike Baillie's identification of an absence of evidence for new oaks starting to grow in the period 745–830, which would have meant that the continued depletion of old-growth trees created a shortfall that was not being filled by woodland regeneration: Baillie, 'Bad for trees – bad for humans?', in C.M. Mills and G. Coles (eds), *Life on the Edge: Human Settlement and Marginality* (Oxford, 1998), pp. 16–17.

85 A. Crone, 'A tenth-century tree-ring event in Scotland: proxy evidence for social and economic change?', in A. Woolf (ed.), *Landscape and Environment in Dark Age Scotland* (St Andrews, 2006), pp. 49–56 at 50.

86 M.G.L. Baillie, *Tree-Ring Dating and Archaeology* (London, 1982), p. 211; M.G.L. Baillie, *A Slice through Time* (London, 1995), p. 126; Baillie, 'Bad for trees – bad for humans?', pp. 16–17, 19.

87 K.J. Edwards and I. Ralston, 'New dating and environmental evidence from Burghead fort, Moray', *Proceedings of the Society of Antiquaries of Scotland*, 109 (1980), pp. 202–210.

Burghead, Moray, showing the grass-covered rampart of the 'citadel' area, looking east. Radiocarbon dates from the timbers span the mid fourth to the early seventh century, pointing to construction in the early period of the emergence of the Pictish kingdom of Fortriu.

construction.[88] Transverse squared oak timbers laid in close-set rows were nailed to longitudinally laid plank layers in the defences of the 'citadel' – the 90m x 150m innermost enclosure on the headland summit – and similarly in the series of outer ramparts. The building of these ramparts alone must have involved the felling of hundreds, if not thousands, of mature oaks in the woodlands that clad the southern upslope from the wetlands and lagoons of the Laich of Moray. The squared timbers and planking at Burghead indicate the presence of craftsmen who worked primarily in wood. The sophistication of their work could also be seen in the excavated defences of a smaller coastal promontory fort at Green Castle, Portknockie, a few kilometres east of Burghead.[89] There, a box framework of massive,

88 R.D. Oram, 'Capital tales or Burghead bull?', in S. Arbuthnot and K. Hollo (eds), *Fil súil nglais: A Grey Eye Looks Back: A Festschrift in Honour of Colm Ó Baoill* (Ceann Drochaid, 2007), pp. 241–262 at 253–256; S. Foster, *Picts, Gaels and Scots* (London, 1996), p. 43; I.A.G. Shepherd, 'The Picts in Moray', in W.D.H. Sellar (ed.), *Moray: Province and People* (Edinburgh, 1993), pp. 75–90 at 78–79; A. Small, 'Burghead', *Scottish Archaeological Forum*, 1 (1969), pp. 61–68.

89 I. Ralston, 'The Green Castle and the promontory forts of north-east Scotland', *Scottish Archaeological Forum*, 10 (1980), pp. 27–40; I. Ralston, 'Portknockie: promontory forts and Pictish settlement in the North-East', in A. Small (ed.), *The Picts: A New Look at Old Problems* (Dundee, 1987), pp. 15–26.

Green Castle, Portnockie, Moray, looking north-east. Like Burghead along the coast to the west, this headland site was crowned by a massive rampart, but here the stonework was supported within a box frame of massive, jointed oak timbers.

jointed oak-timber construction enclosed a headland summit some 120m x 30m. Through the seventh and eighth centuries, between the monumental architecture of such centres of power within the kingdom of Fortriu and the equally sophisticated and monumental timber buildings of the major monastic centres that spread across the north with the arrival of Christianity, there was heavy demand for the wood of old-growth oaks at a level where woodland depletion is likely to have been a more than local phenomenon.

Poor conditions for germination and early growth at various points in the post-500 period would have added to pressures at a time when farming communities were apparently seeking to expand cultivation to make up for depressed yield sizes and where wood use in domestic construction was the norm across most of mainland Scotland. Efforts to identify a cause for the end of this pressure in Scotland proposed land abandonment from population relocation as a consequence of Scandinavian raiding as a factor, but the tree birth date event-horizon does not correlate with the main phases of Norse invasion.[90] In light of the political and legal contexts for woodland regeneration in Ireland, similar contexts have been sought for the Pictish/Alban heartland and hypothesised in the new institutional structures of local administration and resource control that were being introduced by the kings of Alba.[91] As discussed in Chapters 3 and 4, these were

90 Crone, 'A tenth-century tree-ring event', p. 52.

91 Ibid., pp. 53–54.

far-reaching developments. They not only entailed changes in social organisation and settlement distribution that would have reconfigured the immediate pressures on resources like woodland, but the local agency of the men through whom these new entities were managed enabled regulation of resource use. It was they who enforced protection of valuable commodities like major timber-yielding oak and pine species and more effective management of the long-established coppicing practices that provided for the building material and treen needs of the bulk of the population. Similar birth dates for the Lennox-sourced oaks used in the roof structure at Glasgow Cathedral in the thirteenth century, however, would require the kings of Cumbria to have adopted the same resource management strategies.[92] If correct, the tree birth date event-horizon of the mid tenth century marks the point at which the first centralised efforts were made to restrain some of the less sustainable aspects of woodland exploitation of earlier centuries, organise and oversee localised access to coppiced woodland resources and conserve at least those parts of Scotland's forests that were most prized by the social elites.

• • •

Looking across the millennium spanned in this chapter, we can see that traditional narratives of continuously intensifying human pressure on woodland require more nuance if they are to reflect the ebb and flow of clearance and regeneration and the irregular pulses of lighter or heavier exploitation that are revealed by the palaeoenvironmental and archaeological records. There is also a need to give greater recognition to the effects of non-anthropogenic impacts on woodland, and to the extent to which these combined with human agency to deliver unpredictable outcomes. Storminess in exposed Atlantic-facing regions, increasing rainfall and waterlogging of the ground, coupled in the sixth and seventh centuries with the coldness of the LALIA, should not be underestimated as contributors to the continuing decline. Unquestionably, however, the increasing rate of inroads into woodland areas that was identified in the introduction to this volume was maintained through the first five centuries of the Historic era, as expansion was driven by population growth and the demands for more managed grassland and arable fields from which to feed more mouths or provide the resources for tribute and trade. But alongside clearance there is also mounting evidence for what would now be labelled sustainable woodland management. In areas where the interplay between human demands and the effects of climate and weather were most pronounced, such as the Northern and Western Isles or the northern parts of mainland Scotland, tree species that in recent centuries have been dismissed as scrub or 'vermin' by commercial foresters provided most of the woodland materials required by the local populations. On the islands, while the large timbers used for spanning wide spaces were obtained mainly from driftwood or were brought, as described by Adomnán, from mainland Scottish sources, the daily needs for smallwood were met from locally available reserves. Coppicing, which stimulates fast growth of shoots that could be harvested to provide the wands used for wickerwork or wattling or the longer poles used in roofing, was already practised in the Bronze Age and would deliver what seems to have been adequate resources in the first millennium AD. It took profound disruption of human cultural systems, coupled with changes in the North Atlantic climatic regime, to

92 The late Alasdair Ross had begun to research the socio-economic structures of pre-eleventh-century Strathclyde / Cumbria but had not yet brought any advanced analysis to publication by the time of his death. His preliminary findings suggested a similar fiscal-administrative structure to that of the *dabhach*.

cause a breakdown in that management method.

What has also become clearer is that the intrusion of Roman power into the northern parts of the British mainland was not the direct cause of widespread woodland clearance there, as was formerly argued. In much of the lowland south and east, and in the Southern Uplands, extensive forested areas had disappeared long before the arrival of the legions. Here too, most timber needs were being met from what appear to have been managed reserves, possibly regulated by the social leaders of the increasingly hierarchically stratified late Iron Age population. The sustainability of that supply is evident in the maintenance of the largely timber-based built environment of the indigenous Celtic people. Intriguingly, there is also some evidence in this period for what appears to be the creation of timber reserves of species that were in demand but were either absent locally or available only in very low numbers. Shortage or absence was being recognised and steps taken to provide remedies that were never going to be quick or easy. The sudden appearance of oak in locations far removed from any other oakwoods, where there would have been no possibility of 'blow-in' from any neighbouring source, speaks of a sophisticated understanding of the creation, protection and control of woodland and of the processes that had led to the absence of the in-demand species. Implicit in the selection of oak for planting is an awareness that this was an investment for the benefit of generations to come, for that species' slow rate of growth would not deliver the desired large, old-growth timbers for centuries into the future. Rather than a relentless cycle of unsustainable, permanent loss, we can now see recognition of human impacts, the application of management regimes as a partial remedy for recurring shortages and the adoption of precocious, forward-thinking measures to ensure future security of supply. The plantation projects of the later seventeenth and eighteenth centuries, it seems, were not another invention of the Improvement era but the rediscovery of more ancient practice. The evidence is fragmentary and contested, but the simplistic vision of continuous, mindless human overexploitation of the woodland resource, much framed by the modern experience of forest clearance, begs some significant revision. We need to rethink traditional notions of the endless slash-and-burn destruction of trees and the simple transference of extractive pressure to previously unexploited areas when local woods were depleted. It might not suit much of the current misanthropic rewilding narratives of the processes of past deforestation, but our ancestors' role in woodland loss is more complex and less black and white than popular understanding suggests.

CHAPTER 3

Riding the Rollercoaster from the Late Antique Little Ice Age to the Medieval Climate Anomaly *c.*500–*c.*1000

The trembling steed,
With this hot impulse seized in every nerve,
Nor heeds the rein, nor hears the sounding thong;
Blows are not felt; but, tossing high his head,
And by the well-known joy to distant plains
Attracted strong, all wild he bursts away;
O'er rocks, and woods, and craggy mountains flies;
And, neighing, on the aerial summit takes
The exciting gale; then, steep-descending, cleaves
The headlong torrents foaming down the hills,
Even where the madness of the straitened stream
Turns in black eddies round: such is the force
With which his frantic heart and sinews swell.[1]

In recent decades there has been a remarkable transformation in our understanding of the history and archaeology of Scotland in the half millennium after AD 500. The fruits of much individual and collective research have enabled us to move far beyond the long-held view of this era as a veritable 'Dark Age' stalked by a succession of often obscure 'warlords and holy men' about whose lives and relationships we knew little. A quick scan of the historiography to the last quarter of the twentieth century reveals the extent to which our historical knowledge of the era was drawn largely from the observations of external commentators in Anglo-Saxon England, Gaelic Ireland or the Scandinavian world, much of it garbled to the point of incoherence and usually hostile in its presentation. Through the nineteenth and earlier twentieth centuries, these limited, external sources were plundered to construct largely one-sided narratives that imagined culturally sophisticated Gaelic Irish Christian missionaries encountering primitive,

1 From James Thomson, 'The Seasons', in *The Scottish Collection of Verse to 1800*, ed. E. Dunlop and A. Kamm (Glasgow, 1985), p. 69.

Rain clouds over southern Mull. It was not the day-to-day changes in the weather that were of interest to chroniclers, but the extreme and spectacular events that departed from the norm. We must be cautious not to read too much into such headline events.

pagan Picts in scenes redolent of contemporary Victorian images of missionaries and explorers in Africa. Scotland's history was usually traced to 'Kenneth McAlpine' and the earlier kings of culturally Gaelic Dál Riata, with the Pictish kingdoms commonly dismissed as a historical cul-de-sac that was swept aside by the vibrant dynamism of the Gaels. Then there were the Vikings, with their glorious saga literature, rich with tales of heroic but humanly flawed warriors who carved out kingdoms in Scotland and the Isles and were infinitely 'sexier' than a people who left only a series of laconic and superficially uninformative king lists. Adding to the difficulties surrounding early medieval Scotland was that our understanding of its culture was based largely on art-historical interpretation and the artefactual evidence from sculpture, chance finds and a handful of excavated sites that focused mainly on the elite end of production.

From the 1960s, a growing body of more forensic analyses of the historical record and a series of paradigm-shifting excavations led to increasingly radical re-evaluations of the early states that emerged, contended, fell and reformed across these centuries. The synthesised fruits of that work now present a compelling, richly textured narrative of state formation in the centuries after the end of Roman influence in northern Britain, of the dynamics of migration and acculturation, dynastic rivalries and the clashes of kingdoms and cultures that led ultimately to the formation of Alba, the future Scottish state.[2] A handful of notable exceptions apart,[3] much of this newer historiography maintained the established narrative focus of the older tradition.

2 See the overarching narratives in Fraser, *From Caledonia to Pictland*, and A. Woolf, *From Pictland to Alba 789–1070* (Edinburgh, 2007). Building on the revision of our understanding of the Pictish kingdoms triggered by Woolf, the essays in G. Noble and N. Evans (eds), *The King in the North: The Pictish Realms of Fortriu and Ce* (Edinburgh, 2019), and the more focused argument in G. Noble and N. Evans, *Picts: Scourge of Rome, Rulers of the North* (Edinburgh, 2022), present a compelling new interpretation of the rise and fall of Fortriu from the sixth to the tenth century. A. Ross, *The Kings of Alba c.1000–c.1130* (Edinburgh, 2011), although concerned primarily with the Alban successor state, provides a detailed exploration of the social and economic foundations of that kingdom that were laid down in the tenth century, delivering a dimension that is not explored in depth by Noble and Evans or their research partners.

3 For example, A. Woolf (ed.), *Landscape and Environment in Dark Age Scotland* (St Andrews, 2006).

Migration and settlement, conversion to Christianity and military expansion maintained a powerful hold as the recognised drivers of political evolution and state formation. Despite a growing body of archaeological and especially palaeoenvironmental evidence, climate, weather and the wider dynamics of environmental change were rarely acknowledged as significant agents in the emergence of new polities throughout the northern parts of the British Isles. This limited engagement with the principal actors in the processes of environmental change persists, seemingly at odds with our contemporary understanding of the mechanisms powering global climate change.

Re-reading the weather record

Scotland's environment runs only as an almost imperceptible undercurrent within most explorations of this period, which reflects the primarily secular and ecclesiastical political nature of the surviving documentary record upon which the narratives are built. Written records detailing climate and weather events are, indeed, scarce for most of Europe before the twelfth century and exceptionally fragmentary for the British Isles generally between the sixth and the early tenth century. But what does exist can be very revealing of the human lived experience of some of the extreme weather and epidemic events that punctuated this era. Leaving aside the question of epidemic disease, which will not be explored in this chapter, analysis of most of the fragmentary surviving early written evidence from mainland Europe has shown that annalists there were concerned most with episodes of intense cold followed by extreme precipitation (rain- or snowfall) and subsequent flooding. No one was interested in reporting 'normality', and while excessive heat or winds might catch the attention of an individual annalist, it was usually spectacular one-off events that they documented rather than any sense of broader trends. This concentration on the spectacular is frustrating for modern scholars seeking to uncover the environmental histories of this era, for while the litany of atrocious weather events might provide a sense of all-pervading near catastrophe, year-on-year incremental changes had a greater impact over time than the usually short-term blow of an episode of extreme weather. But this focus on the reporting of cold is still important in revealing the attitudes and concerns of contemporary observers, for it presumably reflected early writers' anxieties about the consequences of a poor or long-lasting winter for crop growth the following year and the possibility of famine following that. While it also reveals an association of poor or unnatural winters with epidemic and epizootic disease, most often what is recorded is a sense of general dissatisfaction with the physical inconvenience of prolonged cold and wet conditions.[4]

It would transform our understanding of the environmental history of early medieval Scotland if even a handful of written sources produced within the geographical limits of the modern country still – or had ever – existed. The accounts that do survive are chiefly the work of external observers, writing mainly in Northumbrian English or Irish contexts or, for the ninth and tenth centuries, looking back to this period from a Scandinavian perspective in later twelfth- and thirteenth-century Icelandic sagas. For Scotland, the poverty of documentary evidence forces reliance on ice core or tree-ring data and other climate proxies, usually from geographically remote locations like Greenland, Scandinavia and Siberia but with a few, all-important Scottish records, for the reconstruction of broad trends. In this chapter, the fragmentary written record and climate proxy data will be discussed in conjunction with the rich palaeoenvironmental material from archaeological excavations around Scotland. The conclusions are largely deductive and there are substantial parts of the country from which no adequate data has yet been recovered that might allow the formulation of even such tentative deductions. Nevertheless, archaeology, climate proxy data and consideration of the contemporary record evidence from neighbouring territories have been mined to produce a synthesis of some of the broad environmental trends experienced across the half millennium from AD 500.

Changeable times: the Late Antique Little Ice Age in the North

It has long been accepted that the sixth and seventh centuries saw profound and protracted upheaval, with social, economic and political reconfiguration occurring both within the territories of the former Roman Empire and within the emergent polities that developed beyond its borders. Analysis of multiple proxy data sources from across the whole of these European regions and extending west to include Greenland ice core data can add an environmental dimension to that upheaval. Such multiproxy analysis has allowed historians of this 'Late Antique' or early medieval era to formulate a general picture of climatic instability that spanned much of those two centuries, with tree-ring data placing particular emphasis on a relatively short but severe episode that occurred in the decade after *c.*536.[5] Although remote from western Europe, what was happening in the Arctic regions on the other side of the North Atlantic was to have profound ramifications for the lands on the eastern side of the ocean, from northern Scandinavia to the Atlas Mountains of Morocco. Changes in the Greenland data show that sea ice had been increasing around the island in the early decades of the sixth century and continued to expand until *c.*540, indicating a cooling trend that had commenced in the fifth century. Central Greenland ice cores reveal plunging temperatures that started in the early 400s with rapid fluctuations characterising most of the following two centuries.[6] There were periods of greater stability in the temperatures from Greenland's interior to *c.*520 and again from *c.*540 to *c.*560, but also episodes of sudden and deep cold in *c.*526,

4 M. McCormick et al., 'Climate change during and after the Roman Empire: reconstructing the past from scientific and historical evidence', *Journal of Interdisciplinary History*, 43:2 (2012), pp. 169–220 at 191–199; Pfister et al., 'Winter air temperature variations', *The Holocene* 8 (1998), pp. 537, 538.

5 Baillie, 'Bad for trees – bad for humans?', p. 17.

6 McCormick et al., 'Climate change during and after the Roman Empire', p. 191.

535 and 585, apparently coinciding with climate 'forcing' events (the temperature-suppressing effects of volcanic dust-clouds), whose ramifications would have been felt across the wider North Atlantic.

Greenland's cooling events, especially when coincidental with other climate forcing events such as these major volcanic eruptions, affected different parts of Europe in different ways. Northern and western Scandinavia appears to have suffered sharp and extreme cooling in an event that perhaps provided the folk memory for the *Fimbulvetr* or Fimbulwinter, the three-year winter that immediately preceded the world-ending Ragnarök in Norse mythology.[7] In the central European and eastern Alpine region, glacier and tree-ring data reveals more subtle change in the sixth century than was experienced further north, with evidence for a phase of limited glacial advance and an episode of poor tree-ring growth in the late 530s and 540s. Although the local impact of this sixth-century event was probably bad, the worst conditions in this region were not experienced until the second quarter of the ninth century.[8] Further west, the late sixth century and especially its last quarter saw the reversal of a long, dry trend in north-eastern and eastern French weather that had extended through the fifth and earlier sixth centuries. Extremity here took the form of a shift to wetter and cooler conditions, as is revealed in both the tree-ring record and contemporary written accounts.[9] It is also evident in the peat record in upland districts of central Highland Scotland, which suggests a resumption of peat growth consequent on wetter conditions and a rising water table in the years around 600, following a drier episode that commenced in the fourth century.[10]

Meltwater waterfall, Prince Frederick Sound, Greenland. Research programmes like the GRIP and GISP2 projects, using ice core data from the Greenland ice sheet, have transformed our understanding of changing climate regimes and weather patterns over the last twelve millennia. (Image: Anna Ritchie.)

This picture is further corroborated by evidence of an abrupt reversal in the NAO state, which had been positive through the whole of the fifth century, when the more northerly position of the Azores high-pressure atmospheric system had kept the jet

7 H. Lamb, *Weather, Climate and Human Affairs* (London, 1988; reprinted London, 2012), pp. 104–105; P. Holmberg et al., 'The Rök Runestone and the end of the world', *Futhark: International Journal of Runic Studies,* 9:10 (2020), pp. 7–38.

8 McCormick et al., 'Climate change during and after the Roman Empire', p. 192.

9 Ibid., p. 196 and note 21.

10 D. Strachan, D. Sneddon and R. Tipping, *Early Medieval Settlement in Upland Perthshire: Excavations at Lair, Glen Shee 2012–17* (Oxford, 2019), p. 110.

Loch Assynt and Traligill from the summit of Conival, Inchnadamph, Highland. The limestone cave system lies in the river valley, whose winding course can be seen running to left of centre.

stream and the moisture-heavy air moving east across the North Atlantic generally far to the north-west. This NAO state switched to negative in the sixth century, deepening and extending the Iceland low-pressure system southwards and changing accordingly the jet stream's position. This shift brought a widespread increase in precipitation over the northern and western parts of the British Isles and mainland Europe as the associated westerly airflow delivered wave after wave of moisture-laden fronts to Atlantic-facing regions.[11]

Is there direct evidence for volcanic forcing impacts on Scotland? As we have seen, speleothems have been identified as offering a fairly well-nuanced proxy record for winter precipitation with strong regional significance. For north-west mainland Scotland at least, the speleothems from the Uamh an Tartair cave system in the Traligill river basin in Assynt in western Sutherland have begun to yield good evidence for North Atlantic climate variability over the last three millennia. That data allows the construction of some broad models of the winter climate regime experienced in Atlantic-facing northern Scotland.[12] Uamh an Tartair's record has been analysed in most detail for the thirteenth century and later (see below, pp. 200, 202–203, 296–297), but the same kinds of trends of high/low winter precipitation revealing positive and negative NAO states are

11 Baker et al., 'A composite annual-resolution stalagmite record', fig. 4, for a normalised NAO record based on Moroccan speleothem, western Greenland ice core and Scottish speleothem data.

12 Baker et al., 'A composite annual-resolution stalagmite record'.

Tentsmuir, Fife, seen from Drumcarrow Craig to the south. The sandy coastal plain between the estuary of the River Eden and the mouth of the Firth of Tay, much now covered by conifer plantation, was managed as pasture through the Roman era and until the sixth century, when sand-blow overwhelmed the grass and *Calluna* expanses.

identifiable for earlier centuries. The results, compared with tree-ring and speleothem records from Morocco and a West Greenland ice core record, are compelling in the picture they paint of deteriorating conditions. The Uamh an Tartair data points to an abrupt reversal in the north-west Highlands from an era of generally positive conditions that had prevailed since the early third century AD to one of negative conditions from shortly before 600.[13] This major transition point, however, is too late to be consequent on the 535/6 eruption or those immediately following it. Nevertheless, slight dips in the level of precipitation in the first half of the sixth century might relate to these episodes of volcanic forcing. If so, they suggest that the impact was felt less keenly here than in more northerly and southerly European regions, at least in terms of precipitation.

What the trends reflected in this speleothem data meant for human populations, their crops and their livestock in more southerly and easterly districts, we are left largely to conjecture. Some evidence, such as the abrupt transition from a managed grass- and *Calluna*-dominated pastureland on the coastal levels at Tentsmuir in Fife to a sandy waste, reveals how dune deflation and an overwhelming episode of sand-blow inundated the once rich grazing.[14] This sixth-century change coincides with evidence from wetland deposits across Britain and Ireland that indi-

13 Ibid., fig. 4.

14 Whittington and McManus, 'Dark Age agricultural practices'.

cates a general shift along the Atlantic edge into a new era of storminess and precipitation, where well-managed but fragile environments like the Tentsmuir pastures became vulnerable to the more unsettled conditions.[15] This traumatic experience of sudden change is likely to have been an extreme instance of the new instability faced by many communities as the climate patterns across the North Atlantic shifted into a new configuration.

Pollen analysis from excavations at the late seventh- and eighth-century monastic site at Portmahomack in Easter Ross, however, provides some broad-brush evidence for trends that might correlate to the Uamh an Tartair record.[16] This pollen record permits only generalised reading and lacks a tight chronological range, but it does present evidence for a series of significant shifts in land-use and in the nature of the wider landscape character around the site.[17] The record starts in the fifth century, when marshland adjoining the later monastic site appears to have expanded. That spread was possibly as a consequence of an apparent decline in arable and pastoral activity in the vicinity at the end of the fourth century or beginning of the fifth century. A drier episode in the sixth century appears to correlate to the long drought period evident in north-eastern French records and the lower precipitation of the Uamh an Tartair speleothems. The re-emergence of wet-loving plant species points to a wetter phase just before 600, which also ties in with the known record and with the resumption of peat growth identified in the central Highlands. Overall, through this phase of the site's record, there is a trend towards diminishing agricultural activity and re-expansion of wetland that might reflect the combined effects of weather conditions and population decline locally. This depressed picture lasts until the foundation of the monastery shortly after 650, when evidence for locally intensive agriculture resumes. This is where there is some divergence in the record, for at Uamh an Tartair the wettest and generally poorest conditions appear to have occurred in the 680s and early 690s, with a slight recovery commencing in the first half of the eighth century. Although it must be emphasised that this data is obtained from a cave system in a mountainous region of the north-west Highlands and cannot be taken as representative of rain-shadowed and easterly-exposed eastern lowland districts like Tarbatness on which Portmahomack is located, it is illustrative in its generalities of the likely experience of much of Hebridean, north mainland Scotland and the Northern Isles. However, placed alongside evidence for a sudden and increasing frequency of windblown sand events from dune deflation and machair erosion down the west coast of Uist, Barra and their southern satellite islands between *c.*500 and *c.*800, which points to the effects of growing storminess and a strong, westerly airflow off the Atlantic, and the evidence for agricultural contraction and wetland expansion around Portmahomack, this data collectively provides a sense of the increasing precariousness of the agricultural regime that had prevailed across northern and western Scotland throughout the late Iron Age.[18]

Few narrative sources that report climate and weather events anywhere in Europe survive from before the twelfth century. Only fragments exist for the British Isles generally between the sixth and the early tenth century. For Scotland, this scarcity of written accounts requires us to look to climate proxies

15 Whittington and Edwards, 'Climate change', p. 22.

16 Carver, *The Tarbat Discovery Programme*.

17 McHardie, Mighall and Schofield, 'The landscape and environmental setting'.

18 Gilbertson et al., 'Sand-drift and soil formation', p. 465. This precariousness for settlement in marginal zones is emphasised by Baillie, 'Bad for trees – bad for humans?', pp. 18–19, who sees the likely impacts as a reduction in population levels or forced migration.

Looking north over Glen Callater from Carn an Tuirc, Aberdeenshire. Dust-clouds from volcanic eruptions are likely contributors to the darkened and bitterly cold weather that heralded the onset of the Late Antique Little Ice Age.

from far-flung locations around the northern hemisphere to model even broad trends. Where we do have contemporary written references to the weather from regions neighbouring Scotland, however, these permit some insight on wider northern hemisphere climate trends. Some of these references report episodes of summer and winter cooling which modern scholarship has linked to volcanic forcing. Modelling and comparison with better-recorded volcanic eruptions can provide some sense of the likely Scottish experience.[19] The earliest such episode is recorded as a 'failure of bread' which affected the north of Ireland in 536, but with no internal indication of the weather event – if there was one – that had caused it.[20] The first explicit reference to a weather event is a 564 record of a windstorm in the early part of the year.[21] After that, the next explicit record is a reference to 'great snowfall' in the annal for 588.[22] Such terse accounts provide no substantial foundation for the modelling of climatic or weather trends. The synthesis of data for this episode and other data from the Irish annals, however, reinforces a hypothesis that volcanic-induced climatic forcing underpinned the events reported by Irish annalists and their mainland European counterparts.[23] The dimmed and weakened sun that was said in some

19 F. Ludlow et al., 'Medieval Irish chronicles reveal persistent volcanic forcing of severe winter cold events, 431–1649 CE', *Environmental Research Letters*, 8 (2013), pp. 1–10, http://dx.doi.org/10.1088/1748-9326/8/2/024035.

20 Annals of Ulster, https://celt.ucc.ie/published/T100001A/index.html (431–1201) and https://celt.ucc.ie/published/T100001B/index.html (1202–1378), U536.3. A similar failure is recorded in identically laconic terms in 539 (U539.1).

21 Ibid., U564.1: 'A great wind-storm occurred.' The fifth- and early sixth-century portions of the annals were probably retrospectively constructed from other, older and now non-extant records; there appear to be no contemporary records produced in Ireland in a surviving manuscript earlier than the mid sixth century: D.P. McCarthy and A. Breen, 'An evaluation of astronomical observations in the Irish annals', *Vistas in Astronomy*, 41 (1997), pp. 117–138 at 117–118.

22 Annals of Ulster, U588.4. This episode might be the context for the 'very cold winter day' and 'days of rough weather' connected to a miracle of St Columba reported by Adomnán: *Adomnan's Life of Columba*, p. 265.

23 See, for example, P. Peregrine, 'Climate and social change at the start of the Late Antique Little Ice Age', *The Holocene*, 30:11 (2020), pp. 1643–1648, and K.H. Wohletz, 'Were the Dark Ages triggered by volcano-related climate changes in the 6th century?', *EOS Transactions of the American Geophysical Union*, 48:81 (2000), https://www.lanl.gov/orgs/ees/geodynamics/Wohletz/Krakatau.htm.

chronicle accounts to have presaged the cooling event probably stemmed from the effects of volcanic dust-cloud circulating the globe in the high atmosphere, reducing solar irradiance. An impressive range of evidence supports this interpretation. For example, accounts of the 'failure of bread' in 1536 directly correlate with descriptions of a climatic downturn from 536 to 550 evident in Byzantine, Persian and other Eurasian documentary records, northern hemisphere tree-ring data and other proxy records.[24] The eruption and subsequent dust-cloud that mark the beginning of this fourteen-year event is now understood as the first stage in an era of cold summers, sporadically extreme winters, harvest failures and famines spanning the 530s to 660s that has been labelled the Late Antique Little Ice Age (LALIA).[25]

Through crisis to change

Discussion of the apparent impact of the 536 dust-cloud event across northern Europe is nearly unanimous in seeing it as the trigger for a catastrophic period of cooling and subsistence crisis in which hundreds of thousands died, whole populations were possibly displaced, and the social, economic and cultural patterns of large parts of Eurasia began to undergo major reconfiguration.[26] In what became Scotland, this volcanic forcing was perhaps coupled with mortalities arising from the 'Justinianic plague', which ravaged first the Byzantine Empire in the early 540s before spreading west along trade routes to the British Isles (where it was still rampant in the 550s).[27] Depopulation of some areas, linked to plague mortalities, was possibly a significant contributor to some of the woodland regeneration of the sixth and early seventh centuries that was noted in Chapter 2.

A climatic downturn and epidemic were possibly the harbingers of doom for an old order and the catalyst for the concentration of political power and control of economic resources by some overarching authority that are central building blocks in state formation.[28] We have as yet little hard archaeological evidence for this transformation, but it is possible that the dominance by the mid seventh century of a powerful new and apparently territorially extensive kingdom, Fortriu in Moray, might be linked to a wider socio-economic reconfiguration triggered by

24 T.P. Newfield, 'The climate downturn of 536–50', in S. White, C. Pfister and F. Mauelshagen (eds), *The Palgrave Handbook of Climate History* (London, 2018), pp. 447–493; M.G.L. Baillie, 'Dendrochronology raises questions about the nature of the AD 536 dust-veil event', *The Holocene*, 4 (1994), pp. 212–217; Ludlow et al., 'Medieval Irish chronicles'; A. Arjava, 'The mystery cloud of 536 CE in the Mediterranean sources', *Dumbarton Oaks Papers*, 59 (2005), pp. 73–94; U. Büntgen et al., 'Cooling and societal change during the Late Antique Little Ice Age from 536 to around 660 AD', *Nature Geoscience*, 9 (2016), pp. 231–236.

25 Hubert Lamb first highlighted the cold period, which he dated *c.*550–800, in the 1980s (Lamb, *Weather, Climate and Human Affairs*, p. 24), and its potential implications for the Northern Isles were recognised by archaeologists in the later 1990s. It was not, however, until the 2000s that the full scale of the climatic downturn now known as the Late Antique Little Ice Age was recognised.

26 See, for example, Arjava, 'The mystery cloud of 536 CE'; L. Bondeson and T. Bondeson, 'On the mystery cloud of AD 536, a crisis in dispute and epidemic ergotism: a linking hypothesis', *Danish Journal of Archaeology*, 3:1 (2014), pp. 61–67; B. Gräslund and N. Price, 'Twilight of the gods? The "dust veil event" of AD 536 in critical perspective', *Antiquity*, 86:332 (2012), pp. 428–443; Büntgen et al., 'Cooling and societal change'; S. Helama et al., 'Volcanic dust veils from sixth century tree-ring isotopes linked to reduced irradiance, primary production and human health', *Scientific Reports*, 8 (article no.1339) (2018), https://doi.org/10.1038/s41598-018-19760-w; Newfield, 'The climate downturn'; and Holmberg et al., 'The Rök Runestone'.

27 Annals of Ulster, U545.1, U549.3, U554.2, U556.2. It is uncertain if all these references are to the same disease.

28 Fraser, *From Caledonia to Pictland*, pp. 68–69.

The Craw Stane, Rhynie, Aberdeenshire. The symbol stone stands at the entrance to an early Pictish 'power centre' that failed in the mid sixth century. The conical hill in the background is Tap o' Noth, around whose summit was clustered an even larger Pictish settlement that failed in exactly the same period.

the 536–50 episode and the continuation of cooler and wetter conditions into the middle of the next century.[29] Widespread social disturbance linked to the breakdown of the old order is possibly one contributing factor in the apparent proliferation of fortification followed by abandonment or destruction of settlement in northern Britain through the fifth and sixth centuries.[30] The failure of one late fourth- to mid sixth-century high-status centre at Rhynie in Aberdeenshire at some point in the period 510–60 is strongly suggestive of its demise during this era of upheaval, possibly at the hands of an expansionist Fortriu to its north-west. A similar dating for the end of the much larger fortified hilltop site at Tap o' Noth, just 3km north-north-west of the Rhynie site, however, while it could be indicative of a wider conflict, could also signal that both were casualties of the 536–50 episode and the consequent climatic downturn of the LALIA, with its recurrent epidemics that lasted into the mid seventh century.[31] With a population of potentially up to 4,000 at its peak, Tap o' Noth had the volume and density of habitation to

29 For the rise of Fortriu, see N. Evans, 'A historical introduction to the northern Picts', in G. Noble and N. Evans (eds), *The King in the North: The Pictish Realms of Fortriu and Ce* (Edinburgh, 2019), pp. 10–38 at 17–18, and G. Noble, 'Fortified settlement and the emergence of kingdoms in Northern Scotland in the first millennium ad', in N. Christie and H. Herold (eds), *Fortified Settlements in Early Medieval Europe: Defended Communities of the 8th–10th Centuries* (Oxford, 2016), pp. 26–36.

30 G. Noble, 'Fortified settlement in northern Pictland', in G. Noble and N. Evans (eds), *The King in the North: The Pictish Realms of Fortriu and Ce* (Edinburgh, 2019), pp. 39–57 at 50–57.

31 G. Noble et al., 'Rhynie: a powerful place of Pictland', in G. Noble and N. Evans (eds), *The King in the North: The Pictish Realms of Fortriu and Ce* (Edinburgh, 2019), pp. 58–80. For dating, see ibid., p. 74. For Tap o' Noth, see University of Aberdeen News, 14 May 2020, 'Hillfort revealed to be largest Pictish site ever discovered in Scotland', https://www.abdn.ac.uk/news/14019/.

render it an ideal and very vulnerable 'virgin soil' for the spread of previously unknown diseases.[32] Famine and plague deaths in the subsistence crisis and epidemics of the mid sixth century leading to demographic collapse and abandonment is not an unrealistic scenario. Excavation evidence and pollen analysis from sites scattered from Shetland to the Southern Uplands is strongly suggestive of just such a contraction of settlement to address problems of increased marginality. At the very least, there was significant reorientation from Iron Age agricultural systems that had persisted through the preceding Roman Warm Period.[33]

New regimes in fields and pastures

The Northern and Western Isles and the north and west parts of the Highlands were most vulnerable to a prolonged episode of climate cooling and the deteriorating weather conditions attendant on a switch to negative NAO state. Vulnerability increased if epidemic disease had also reduced the labouring population to a level that made previous agricultural and fishing regimes impractical and, in already marginal locations like Shetland, possibly unsustainable. A few harsh winters and poor harvests are unlikely to have triggered agricultural regime change but, in the absence of a cultural shift relating to the arrival of colonists who brought a different suite of cultivated crops, endogenous factors must have underpinned the adoption of new cropping practices. An extended downturn lasting for nearly four generations and poor conditions generally prevailing for over 500 years were probably sufficient to require a shift from old forms of traditional environmental knowledge and social dependencies. Such circumstances possibly brought the introduction or increased sowing of crops better suited to the new weather normal, including wet- and cold-tolerant grains like oats. A situation like that, where land-use and economic structures were 'naturally responsive to climate fluctuations', has long been recognised for the mid sixth to late eighth centuries in the Northern Isles generally but the nature of the impacts has only more recently become clearer.[34]

Recognisable transitions in the agricultural systems of Shetland and Orkney and the appearance of new cultivated plants have traditionally been associated with Norse settlement from the ninth century, but an increasing volume of evidence points to a pre-Norse transition occurring in the sixth- to seventh-century LALIA. Oats appear as a signal of such change. It was not a new crop in the Northern and Western Isles, having been cultivated, as we have already seen, by farmers at Pool on Sanday in Orkney from the third or fourth centuries AD and in Uist from around the same date.[35] Cooler and wetter conditions,

32 T.P. Newfield, 'Mysterious and mortiferous clouds: the climate cooling and disease burden of late antiquity', in A. Izdebski and M. Mulryan (eds), *Environment and Society in the Long Late Antiquity*, Late Antique Archaeology 12 (Leiden, 2018), pp. 89–115; T.P. Newfield with S. Marciniak and B. Cameron-Steinke, '"Verbalist ingenuity" and the evidential basis for virgin-soil smallpox epidemics in the sixth century: from Iona to Ṣanʿāʾ', in R.D. Oram (ed.), *'With Our Backs to the Ocean': Land, Lordship, Climate Change, and Environment in the North-West European Past: Essays in Memory of Alasdair Ross*, Environmental Histories of the North Atlantic World 5 (Turnhout, 2023), pp. 25–82.

33 For an outline consilience and synthesis of data from across Scotland, see Strachan, Sneddon and Tipping, *Early Medieval Settlement*, chapter 1.

34 S. Butler, 'The environmental history of Shetland', in O. Owen and C. Lowe (eds), *Kebister: The Four-Thousand-Year-Old Story of One Shetland Township*, Society of Antiquaries of Scotland Monograph Series 14 (Edinburgh, 1999), pp. 3–7 at 7.

35 J.M. Bond, 'The plant remains', in J. Hunter with J.M. Bond and A.N. Smith, *Investigations in Sanday, Orkney*, volume 1: *Excavations at Pool, Sanday: A Multi-period Settlement from Neolithic to Late Norse Times* (Kirkwall, 2007), pp. 205–207; J.M. Bond, 'Beyond the fringe? Recognising change and adaptation in Pictish and Norse Orkney', in C.M. Mills and G. Coles (eds), *Life on the Edge: Human Settlement and Marginality* (Oxford, 1998), pp. 81–90 at 87 (where a fourth- to fifth-century date is given, subsequently revised back to third- to fourth-century).

coupled with its ability to deliver yields on even poor-quality land, probably encouraged an expansion in oat sowing. The archaeological evidence for such a shift has been found along the whole arc from the Western Isles to Shetland. The early date of excavation at some major sites in Argyll and the Hebrides meant that little or no environmental sampling was undertaken in what were often barely more than glorified treasure hunts, but more recent projects in the Western Isles, and re-excavation in Argyll, have seen extensive palaeoenvironmental analysis. A campaign of excavation in the early 1980s showed that at even relatively southerly and high-status settlements like Dunadd in Argyll, oats appear to have been cultivated locally from the fifth century but became increasingly represented in the palaeoenvironmental record from the seventh century onwards.[36] In Uist, the late Iron Age Mound 1 settlement at Bornais likewise saw an increasing proportion of oat cultivation, but this was occurring at a time when flax was being added to the agricultural repertoire. Here, it has been suggested that the oats were being grown on the poorer-quality soils of previously uncultivated ground that was being broken into use as infield. This adoption of less-demanding and more weather-tolerant oats, it has been hypothesised, was to make up for a shortfall in cereals caused by a concurrent introduction of flax onto the well-manured infield areas where barley had previously dominated.[37] If flax was being grown more for its seed-oil than for the plant fi bres for linen production, the introduction of oats represented further expansion of the calorie sources from cultivated foods available to the community. At Bornais, however, we lack any means of assessing the success of crops and the levels of yields secured, which might mask a cause for diversification that was rooted in deteriorating weather. All that can be said with certainty is that the zone of cultivation was apparently being expanded from a primary concentration on the machair land close to the settlement into the heavier-soil hinterland of the coastal shell-sand strip, a zone otherwise exploited extensively during the summer months for pasture and a source of fuel and other essential materials. In later centuries, such a shift in emphasis can be seen as occurring during peak episodes of deteriorating weather conditions,

Oats, Findogask, Perth and Kinross. Oats were long interpreted as evidence for the depletion of soil fertility, but the transition from an arable regime based on barley to the cultivation of oats throughout Atlantic-facing Scotland is now better understood as a response to the cooler and wetter conditions of the Late Antique Little Ice Age.

36 A. Milles, 'Charred plant remains', in A. Lane and E. Campbell, *Dunadd: An Early Dalriadic Capital* (Oxford, 2000), pp. 221–226 at 226. For the wider environmental context and agricultural production around Dunadd, see A. Lane and E. Campbell, 'Discussion and conclusion', ibid., pp. 233–263 at 255–258.

37 J. Summers and J. Bond, 'The agricultural economy of a late wheelhouse', in N. Sharples (ed.), *A Late Iron Age Farmstead in the Outer Hebrides: Excavations at Mound 1, Bornais, South Uist* (Oxford, 2012), p. 339.

when the machair experienced wind erosion, dune deflation and marine transgressions.

The clearest evidence for a transition in agricultural practices relates to more northerly locations. In addition to Pool in Sanday, the settlement at Howe near Stromness on the Orkney Mainland has yielded evidence for increased oat cultivation alongside the more traditional barley crop in its final settlement phase commencing in the seventh century.[38] In Shetland, the settlements at Old Scatness, where a previously barley-dominated arable regime transitioned between the mid fifth and mid seventh century into one where an increasing volume of oats was grown alongside the still dominant barley, and at Scalloway, where a similar move appears to have occurred in the seventh century, also show wider evidence for a change in processing and storage techniques.[39] At Old Scatness, it has been suggested that the introduction of oats possibly represented an expansion of cultivation into more marginal areas beyond the heavily manured home fields where barley was still grown. These oats were possibly intended as animal fodder, thereby both increasing the amount of the barley yields that was available for human consumption and boosting biomass for animal feed in a location where fodder is crucial for overwintering stock but the growing season for its production is short.[40] Placed alongside evidence from these sites for intensified efforts in cleaning, sorting and especially direct heat-sourced drying of wet or not fully ripe grain for storage, however, as seen at Howe and Pool in Orkney and possibly at Scalloway in Shetland, we might instead be seeing an increasing battle to maintain arable food sources in a time of deteriorating climate.

It is important not to overread the evidence from the Northern Isles and Atlantic-facing west on the basis of just a few high-profile excavated sites in parts of Scotland most exposed to the full force of the North Atlantic and our awareness of the wider European discourse concerning the social, political and cultural impact of the LALIA. There is, however, a growing body of evidence for a contraction in size and a marked decline in material state at multiple, previously important settlement sites in the islands between the mid sixth and early seventh century. For example, the late Iron Age settlement on Mound 1 at Bornais was abandoned and dismantled probably before 600.[41] Pictish-period settlement at Skaill in Deerness on Orkney's East Mainland ended soon after 600.[42] At Pool in Sanday, the early Pictish period had seen expansion of the settlement in the fourth and fifth centuries and the construction of major new stone structures, probably associated with population growth. By the sixth or early seventh century, however, this expansion had gone into reverse with the abandonment, dismantling and covering with midden of large parts of the previously occupied area.[43] The excavator presents a positive view of this late phase of the settlement as being 'a period of

38 B. Ballin Smith (ed.), *Howe: Four Millennia of Orkney Prehistory*, Society of Antiquaries of Scotland Monograph Series 9 (Edinburgh, 1994), pp. 91, 124; C. Dickson, 'Plant remains', ibid., pp. 125–139 at 125, 132.

39 Dockrill et al., *Excavations at Old Scatness*, vol. 1, pp. 192–195; J.M. Bond, E. Guttmann and I.A. Simpson, 'Bringing in the sheaves: farming intensification in the post-broch Iron Age', in R.A. Housley and G. Coles (eds), *Atlantic Connections and Adaptations: Economies, Environments and Subsistence in Lands Bordering the North Atlantic* (Oxford, 2004), pp. 138–145 at 142; T. Holden and S. Boardman, 'Crops', in N. Sharples (ed.), *Scalloway: A Broch, Late Iron Age Settlement and Medieval Cemetery* (Oxford, 1998), pp. 99–106.

40 Bond, Guttmann and Simpson, 'Bringing in the sheaves', p. 142.

41 Sharples, *A Late Iron Age Farmstead*, pp. 223, 348.

42 S. Buteux et al., *Settlements at Skaill, Deerness, Orkney: Excavations by Peter Gelling of the Prehistoric, Pictish, Viking and Later Periods, 1963–1981*, British Archaeological Reports, British Series 260 (Oxford, 1997), pp. vii, 199, 266.

43 Hunter, Bond and Smith, *Investigations in Sanday*, vol. 1, pp. 111–114.

Glen Affric looking east from Beinn an t-Socaich towards the meanders of the River Affric at Carnach Mor. The alluvial spreads on the valley floor here saw the intensification of agriculture in the period from the 300s through to shortly after 725, pointing to a response to the declining climatic conditions of the era.

innovation at Pool: new land appears to have been brought into cultivation according to the presence of lighter soils; cultivated oats may have been introduced; manuring occurred; storage and processing may have been centralised; and dairying appears to have taken place'.[44] These developments, he suggests, relate to the early stages of state formation and the increasing power and influence of the Church. A less positive interpretation might be that a contracting population was cultivating a wider area to maintain yield levels, introducing new grain crops that were better suited to poorer conditions, and had adopted dairying and its associated calf slaughter as winter fodder resources diminished. This is still development, showing responsiveness in the face of mounting challenges, but it might have occurred as a reactive move against a backdrop of dearth and disease.

Agriculture and resource use in Glen Affric in the western Highlands displayed a similar degree of responsiveness that could also reflect adaptation to prevailing conditions. Results from pollen analyses from a series of locations in the upper part of the glen chart pulses of expansion and contraction in the zones of human exploitation across the millennia, possibly relating to shifting opportunities within periods of climatic fluctuation.[45] On the best land of the alluvial floodplain in the valley around Carnach Mor, agricultural exploitation intensified from the fourth century and reached a peak in the second quarter of the eighth century. At the same time, smaller areas of alluvium further west at Camban saw a contrac-

44 Ibid., p. 120.

45 Davies, 'Upland agriculture and environmental risk'.

tion in grazing activities, leading to the re-establishment of a mixed grass and scrubby fen-woodland landscape by the middle of the seventh century.[46] This latter area saw re-expansion of agricultural activity only towards the end of the thirteenth century and is the sole part of the glen investigated to display such a clear hiatus. This gap in exploitation has been seen as a possible signal that the population of Glen Affric did not need to exploit all the areas of soil with good arable potential, and there is no evidence for any commensurate increase in livestock pressures.[47] We might then suggest that here is evidence for a contraction in population rather than a shift in exploitation modes, for the scrub regeneration at Camban does not point to a drift from arable to pastoral practice. Instead, the suspension of intensive use at Camban for around six centuries could imply a lifting of the population pressures that had seen an earlier spread of settlement, cultivation and pasturing in the late Iron Age. The withdrawal from Camban and intensification at Carnach Mor has been hypothesised as consistent with the kind of trend towards a more hierarchical mode of centralisation and redistribution of agricultural produce that interpretation of the broch-building and early post-broch social systems suggests occurred around the north and west of the British Isles in the early centuries AD.[48] Such a system would have provided a degree of buffering against the major social and environmental shocks of the Early Historic period, possibly with only some of the most marginal communities failing entirely during the most extreme phases of the LALIA and its seventh- and eighth-century aftershocks.[49]

In contrast to the Northern Isles and the western Highlands, the northern Cheviot valley of the Bowmont Water offered little evidence for dramatic change from the balance of arable and pasture that had become established in the late Iron Age. Here there was rather a slow evolution through the cooler and wetter conditions from the mid sixth century.[50] After *c.*570 there was a decline in birch woodland on the higher ground around Swindon Hill, possibly due to steadily increasing pressures from grazing that were to continue to the tenth century. The presence of cereal pollen, however, is not supportive of this expansion representing increased dependence on pastoralism in response to climatic deterioration. On the valley-floor mires at Sourhope, peat-cutting that apparently commenced in the early fifth century had possibly changed the hydrology to the point where the ground now supported a *Calluna* heath, but peat exploitation continued. There is also a suggestion that, like their distant fellows in Glen Affric, the local population was making use of areas of poorer soil by allowing trees to grow there, providing them with a timber resource that was otherwise lacking within the wider Cheviot region at this date. Indeed, some of the apparent expansion of woodland cover from the mid sixth century that has been proposed for the region and interpreted as consequent upon an abandonment of agricultural land has instead been hypothesised as evidence of deliberate woodland conservation, like that encountered in Chapter 2.[51] We should certainly not accept at face value the testimony of later medieval written accounts which presented large tranches of upland zones in southern Scotland and northern England as unpeopled wastes prior to the expansion of the Anglian kingdoms

46 Ibid., p. 2055 table 2.

47 Ibid., p. 2056.

48 See, for example, the models proposed in Parker Pearson and Sharples, *Between Land and Sea*, chapter 12.

49 Davies, 'Upland agriculture and environmental risk', p. 2059.

50 Tipping, *Bowmont*, pp. 190, 191.

51 See Turner, 'The environment of northeast England'; K. Dark, *Britain and the End of the Roman Empire* (Stroud, 2000); and Tipping, *Bowmont*, p. 190.

Managed willows, Inveralmond, Perth and Kinross. Stands of willow, alder and birch such as this appear to have been encouraged on areas of poorer, uncultivable ground and conserved for wood needs by the inhabitants of communities spread through the valley systems of the Southern Uplands.

centred on York and Bamburgh. For example, Reginald of Durham's twelfth-century claim that the district between the Tyne and the Tees was in the earlier seventh century 'a desert waste . . . subject to no man's dominion' and the 'the den and dwelling-place of wild and woodland beasts' is a contrivance to explain his monastery's later claim over it.[52] Population may have contracted and exploitation regimes within the region changed in character, but this was hardly an empty waste when settlement began to intensify there once more in the later eleventh century and arable cultivation penetrated deeper into the Pennine and Cheviot hinterland.[53] Instead, we should see them as areas contested by the Anglian incomers and the remnant populations of the older British kingdoms, in which the indigenous population sought to maintain the agricultural and livestock management practices they had pursued in the lowland zones but with progressive adaptation to the different soil and weather conditions they encountered.

The same approach to resource management might also be reflected in the valleys that feed into the head of the Solway in the western Southern Uplands. There, for example, the apparent contraction in grazing and wider agricultural activity and the mid sixth-century and subsequent regeneration of birch and alder woodland at Over Rig in Eskdale, or the woodland clearance and agricultural intensification that occurred around the early Anglian settlement site at Hoddom, both in Dumfriesshire, point to population volatility and agricultural regime shifts.[54] At Hoddom, oats became established as one of the primary crops from the beginning of the settlement around 600 and continued as such for the remainder of the early medieval period. That dominance perhaps signals recognition by its occupants of the particular resilience of oats in poor climatic and soil conditions.[55] The succession of corn-drying kilns built by the community at Hoddom is indicative both of poor ripening and harvesting conditions for the grain crops and of the application of technological solutions to maximise the potential of the harvested yield.[56] Instead of what has been dismissed as 'catastrophist interpretations of population change', this hypothesis sees the population as proactive and responsive to climatic agency.[57] What the palaeoenvironmental analyses cannot do, however, is reveal any demographic changes that might lie behind even subtle shifts in exploitation regimes. Nor does it allow us to identify who constituted that population. It is entirely possible that the apparent intensification of exploitation represented the response of a displaced indigenous British population in the face of the aggressively expansionist Northumbrian Angles from the end of the sixth century, driven

52 Reginald of Durham, *Vita Sancti Oswaldi Regis et Martyris*, in Symeon of Durham, *Symeonis Monachi Opera Omnia*, ed. T. Arnold, 2 volumes (London, 1882–85), volume 1, pp. 326–385 at 339.

53 This process is explored in R.D. Oram, 'Trackless, impenetrable and under-developed? Roads, colonisation and environmental transformation in the Anglo-Scottish border zone *c.*1100 to *c.*1350', in R. Evans and V. Haynes (eds), *Roadworks: Medieval Britain, Medieval Roads* (Manchester, 2015), pp. 303–325.

54 R. Tipping, 'Medieval woodland history from the Scottish Southern Uplands: fine spatial-scale pollen data from a small woodland hollow', in T.C. Smout (ed.), *Scottish Woodland History* (Edinburgh, 1997), pp. 52–75 at 66; R. Tipping, 'Landscape and land-use change in the early historic period on the northern Solway plain', in C. Lowe (ed.), *Excavations at Hoddom, Dumfriesshire* (Edinburgh, 2006), pp. 146–148 at 147.

55 T. Holden, 'The botanical evidence', in C. Lowe (ed.), *Excavations at Hoddom, Dumfriesshire* (Edinburgh, 2006), pp. 150–155 at 150–151.

56 For the kilns, see T. Holden, 'The corn-drying kilns at Hoddom', in Lowe, *Excavations at Hoddom*, pp. 102–113.

57 Tipping, *Bowmont*, p. 191.

into the uplands from their coastal plain and valley-bottom settlements.[58] In the absence of a clearer understanding of settlement distribution and population density across the region in this period, we cannot rule out any permutation of political reconfiguration, climate-driven responses or plague-related mortalities as a driver for systemic change.

Bloody rains and great snows: the later seventh and eighth centuries

While the sixth and earlier seventh centuries present few surviving written records of weather events alongside the limited climate proxy data, a bare handful more exist for trends in the later seventh and eighth centuries. Sadly, these are often as vague and chronologically imprecise in their generalisations as material for the previous century and similarly do not show close geographical alignment with either climate modelling for parts of Scotland or the proxy evidence for north-western Europe generally. While data from the Southern Uplands suggests that the wetter conditions that prevailed there to *c.*650 presaged a longer era of climatic instability that possibly lasted into the tenth century or later, multiproxy syntheses from mainland Europe, the Mediterranean and the Middle East suggest a seventh-century shift from climatic instability towards greater stability that extended to *c.*800 before a further episode of deterioration then a climb into sustained milder conditions.[59] The few written sources from the last decades of the seventh century do provide hints of isolated, exceptional weather events in the north of Ireland, but these offer little sense of the broader prevailing conditions along the Atlantic-facing west. Were these anomalies amidst broadly acceptable conditions or extreme episodes within a declining climatic trend? The speleothem record from the Traligill cave system suggests that high precipitation – rain or snow – was a characteristic of winter weather in the north-west Highlands throughout the seventh century and into the middle of the eighth. Indicative of negative winter NAO, it points to the prevailing weather pattern as moving eastwards across the North Atlantic bringing moisture-laden fronts across Scotland. When that moisture encountered cold Arctic and Siberian air circulating around a Scandinavian atmospheric high-pressure cell, it would have fallen as snow.

It is such an event that is probably described in the first of our surviving later seventh-century Irish accounts. Annals of Ulster and Annals of Tigernach entries, both drawn from the same now lost source, inform us that in the winter of 669/70 'a great snowfall occurred' and that this was followed by 'a great famine'.[60] We must, however, resist the temptation to seize upon this laconic and utterly opaque record as providing evidence for general conditions during the prevailing seventh-century negative NAO state. Greenland ice core and sea ice data, Alpine glacier data and central European tree-rings all indicate that there was an overall shift to slightly warmer conditions in north-western Europe for the century after 650.[61] The wider absence of weather reports in the annals would seem to suggest that this event was exceptional and possibly prolonged. That it is documented in no other Irish sources nor in any mainland British record, however, should also give us pause to reflect on the particularities of weather; as with more recent, better-recorded events, the major snowfall may have been confined to a geographically circumscribed zone, leaving areas further south and east relatively unaffected. The famine with which this

58 Ibid.

59 Tipping, *Bowmont*, pp. 190–191; McCormick et al., 'Climate change during and after the Roman Empire', pp. 199–202.

60 Annals of Ulster, U670.1; Annals of Tigernach, https://celt.ucc.ie/published/T100002A/index.html, T670.1.

61 McCormick et al., 'Climate change during and after the Roman Empire', p. 200.

Looking south from Stob à Choire Dhomhain across the north Affric mountains, Highland. Speleothem data from the Traligill cave system indicates that the north-western Highlands saw increased rain and snowfall through the 600s and into the eighth century.

snowfall is linked in the same annal is equally problematical: there is no note in the 669 annals of a general harvest failure in Ireland and a famine in 670 is unlikely to have arisen solely from one harsh winter, even with a delayed spring and subsequent poor growing season that affected yields, unless the snows persisted into April at least. With no continuation of famine reported in 671 and no reference to one in other sources, if correctly dated this might have been a very localised and short-lived event.[62] We cannot even be sure that areas of Scotland closest to Ulster were affected by it.

Similar issues of particularity attach themselves to what, in terms of survival, is the earliest document of Scottish provenance to provide insight on weather conditions with a more precise locality. Our source is the *Life of Columba*, written in the 690s by Adomnán, abbot of Iona, which, although not an abundant data repository, does contain incidental details that were employed by its author to provide context for events in his subject's life. One such account is a description of a miracle worked through the saint's intervention to end a spring drought in an undated year 'about seventeen years ago', which would fix it in the mid 670s to early 680s. Adomnán called the drought 'persistent and severe, in these lifeless fields', describing how it threatened the coming season's crops.[63] He recounted how the ploughed and sown fields, lying parched and ungerminated beneath a sky that had been 'bare of clouds' through March and April, were well watered by 'great rain, falling by day and by night', which started after his monks had processed with Columba's relics to entreat his saintly intervention. This description seems to demonstrate the effects of a strongly positive NAO, where the westerly airflow over the ocean was shifted far to the north of the British Isles. With no suggestion of cold conditions in the account that would point to a Scandinavian high-pressure cell that would have brought such conditions down from the north and north-east, we seem rather to be dealing with a high-pressure system extending north from the Azores and main-

62 There were major volcanic eruptions through the seventh century that could have precipitated 'volcanic winters', with 639, 643 and 690 suggested as possibly experiencing such extreme conditions.

63 *Adomnan's Life of Columba*, pp. 451, 453.

Lubvan, Ardverikie Forest, Highland. The parched late spring conditions in this district in the early 2010s reflect Adomnán's late seventh-century description of persistent drought lasting through March and April into May, which left fields parched and lifeless, and stunted the spring grass growth.

land Europe rather than a persistent late winter or early spring northerly or north-easterly airflow, keeping the rain-bearing Atlantic fronts from moving east into the Hebrides. Like the 669/70 snows, the drought conditions that resulted, however, appear to have been very localised, for no Irish or Northumbrian source mentions such an episode affecting northern or western districts more generally.

A second late seventh-century event is reported in the *Chronica* section of the later medieval chronicle compilation attributed to the fourteenth-century cleric John of Fordun. This account records that there was 'bloody rain' lasting a week in the late 680s or 690s that affected the whole of Britain.[64] The source for this account might be the reference to bloody rain given for the year 689 in the *Historia Brittonum*, attributed to the ninth-century Welsh monk Nennius,[65] or an unknown source upon which both Nennius and the 'Fordoun' source drew. The same episode may be reflected in the report for that year in the Annals of Ulster and Annals of Tigernach that 'part of the sun was obscured'.[66] *Prima facie*, this could be read as a reference to an eclipse event, but no solar eclipse occurred in 689, presenting the possibility that this was another dust-cloud but perhaps one related to high-altitude Saharan dust rather than a volcanic eruption.[67] High levels of SO_4 in Greenland ice cores, however, are strongly suggestive of a volcanic winter in 690 and it seems from the proxy data that poor conditions prevailed until around 702, during which period it 'may well have been a difficult stretch for food production'.[68]

64 John of Fordoun, *Chronica Gentis Scotorum*, vol. 2, p. 123. This event is reported in similar terms in W. Bower, *Scotichronicon*, ed. D.E.R. Watt et al., 9 volumes (Aberdeen, 1987–98), volume 1, p. 145.

65 Nennius, *British History and the Welsh Annals*, ed. J. Morris (London, 1980), p. 46.

66 Annals of Ulster, U689.8; Annals of Tigernach, T689.5. An earlier Saharan dust episode might be reflected in the Annals of Ulster 684 report of Loch Neagh being 'turned into blood this year' (U684.5).

67 There was a major VEI 6 eruption of Rabaul in 683 and a VEI 5 eruption of nearby Pago in the later seventh century, both in the Bismarck volcanic arc in New Guinea. The six-year lag between the Rabaul eruption and the reports of bloody rain, however, make it unlikely that this was the culprit.

68 McCormick et al., 'Climate change during and after the Roman Empire', p. 200.

None of the Irish sources points to weather events as the cause of a further wave of disease and crop failures at the start of the eighth century. According to the Annals of Tigernach, cattle pestilence struck the island early in 700, while later in the year came the start of 'hunger and pestilence for three years in Ireland so that a man would eat a man'.[69] Poor weather may have played a part in these episodes but there is no explicit causal link made anywhere. We have equally problematical evidence that might provide insight on longer-term climate trends in the eighth century, with just a scattering of references to weather events that do not provide sufficient data to support even a general model of prevailing conditions for western Scotland. 'Great dryness' is reported in summer 714 in Ireland,[70] which is suggestive of an extended episode of positive summer NAO, with high pressure atmospheric systems extending northwards from the Azores bringing hot and dry air north from the Sahara and keeping the rain-bearing westerlies in the northern Atlantic. Conversely, the 'rainy summer' reported in 720 probably saw negative NAO, with the jet stream tracking low across the Atlantic to deliver a succession of rain-laden systems over the west of the British Isles.[71] The chronological separation of these two episodes is not suggestive of any more general climatic disturbance, and even the much-cited reference to the wreck of 150 Pictish ships at the unidentified headland called 'Ross Cussine' in the early months of 729 merely illustrates the dramatic result of a single late winter or early spring storm.[72] With only the Uamh an Tartair proxy evidence pointing to continued positive winter NAO with associated high precipitation levels continuing down to *c.*725 and a move to negative NAO and its drier conditions lasting until *c.*780, the most we can suggest is that such storms were likely to have been brought by northerly or north-easterly winds.

Our speleothem record points to a steady descent into negative winter NAO conditions from *c.*750. This downward trend is reflected in Irish annal accounts that suggest an appreciable worsening in the 760s, with 760 itself being reported as a year of hunger in Ireland.[73] It is not reported if that Irish dearth arose from cold conditions, but a cooling trend consequent on the prevailing positive winter NAO had apparently become established by 762 when great snow was reported in the north-east of Ireland in the late winter.[74] From Ulster to central and south-east Europe, annalists agreed that the winter of 763/4 was exceptionally severe and led into a two-year period of subsistence crises worsened in the Atlantic-facing west by summer drought.[75] That it struck the eastern coast of Scotland with extreme cold and snowstorms is corroborated by a letter from the abbot of the Northumbrian monastery of Monkwearmouth–Jarrow between the estuaries of the Wear and the Tyne in north-eastern England. He wrote of 'the past winter, which very horribly

69 Annals of Tigernach, T700.1, T700.5. Cattle mortality is again recorded in 701 (T701.4).

70 Ibid., T714.6.

71 Ibid., T720.1.

72 Ibid., T729.2.

73 Ibid., T760.2.

74 Ibid., T762.1.

75 Annals of Ulster, U764.1 ('A great snowfall which lasted almost three months'), U764.4 ('A great scarcity and famine'), U764.7 ('An abnormally great drought'); Annals of Tigernach, T764.1. The 'shortage of bread' in Annals of Tigernach T765.10 suggests a subsistence crisis arising from a crop failure after the bad winter of 763/4. See Pfister et al., 'Winter air temperature variations', p. 541. For the causes and impact of this event, see M. McCormick, P.E. Dutton and P. Mayewski, 'Volcanoes and the climate forcing of Carolingian Europe, A.D. 750–950', *Speculum*, 82 (2007), pp. 865–895 at 878–881, and McCormick et al., 'Climate change during and after the Roman Empire', p. 200.

Pitmiddle Wood, Perth and Kinross. Positive NAO and intense winter cold brought snow and ice to much of Scotland in the 760s, delivering extended winters, delayed springs and poor crop growth, which resulted in protracted subsistence crises for farming communities in both Highland and Lowland districts.

oppressed the island of our people by cold and freezing and storms of winds and rain'.[76] The summer weather is indicative of strongly positive NAO, while the winter snows for the Atlantic-facing west were presumably delivered by the continuing positive state bringing intense cold from the north and east to meet moist atmospheric systems coming in from the ocean. Indeed, this winter is classified by Christian Pfister and his colleagues as one of the 'most outstanding cold episodes in the last two thousand years'; a 'great winter' which in some parts of Europe extended from mid December 763 to mid April 764.[77] The speleothem record from Uamh an Tartair likewise reveals a sustained episode of positive winter NAO extending from the 720s to the 790s, indicative of drier but very much colder winter conditions char-

76 Letter from Gutberct of Jarrow and Wearmouth to Lul of Mainz, quoted in McCormick, Dutton and Mayewski, 'Volcanoes and the climate forcing', p. 879.

77 Pfister et al., 'Winter air temperature variations', pp. 541–542; McCormick, Dutton and Mayewski, 'Volcanoes and the climate forcing', pp. 878–881.

acterised by intensely cold easterly to northerly airflow associated with the Siberian high-pressure cell.[78] Characterised by long winters and late, cold springs, which grew in intensity towards the middle of the century, the eighth century likely saw the most marginal communities around Scotland pushed to the brink, or beyond. Persistence of positive NAO in the summer and autumn months is also indicated by records – proxy and written – of recurrent droughts, with 794 being the driest year across most of north-western Europe, including parts of the British Isles.[79]

Continuity or innovation in unsettled times?

Bearing in mind Richard Tipping's strictures against overreading such generalised and usually regional-scale data when set alongside the more granular evidence from closely focused local palaeoenvironmental studies, like Bowmont or Glen Affric, is it possible to detect any broad trends in human responses to later seventh- and eighth-century conditions? Mainland European analyses have led to suggestions that extended episodes of drought or high precipitation there could be identifiable in the archaeological record, as changes of this kind would affect water tables and growing seasons, thus influencing human choices for settlement location or suites of crops grown.[80] In the Bowmont valley, Tipping saw evidence for continuity of established practice rather than significant change; neither upland nor lowland zones within the valley system saw intensification of exploitation for either arable or pastoral needs across this period.[81] Importantly, he found no evidence for the increased erosion that might have been expected through this era of higher precipitation had either of those activities intensified beyond the extensive grazing practices and intermittent cropping of the upland areas that he postulated. What we cannot tell on the basis of the evidence from Bowmont, however, is whether there were changes in grazing load and seasonality, crop mixes and yield levels as farmers strove to maintain agricultural production.

A seventh-century 'bounce back' after the worst extremes of the last decades of the 500s is detectable at a wide range of locations across the zone between the midlands valley and the Great Glen.[82] Within that zone, recent archaeological and palaeoenvironmental work has examined the colonisation and intensified exploitation of an upland district in north-eastern Perthshire. There, it seems that the later seventh century saw a return to intensive land-use across the broad upland district between Glen Isla and Strath Tummel after possibly two to three hundred years of low-intensity exploitation.[83] At the excavated settlement of Lair in Glen Shee, one of a whole group of Pictish farms in the Perthshire uplands of a form named after the type site first excavated at Pitcarmick, south-west of Lair, this transition occurred probably in the middle decades of the 600s and it is likely that similar dating can be extended to

78 Baker et al., 'A composite annual-resolution stalagmite record', fig. 4.

79 The droughts are modelled primarily upon tree-ring evidence, which is lacking for most of Scotland for the whole of the first millennium AD. See E.R. Cook et al., 'Old world megadroughts and pluvials during the Common Era', *Science Advances*, 1 (2015), e1500561. See also T.P. Newfield and I. Labuhn, 'Towards a messy history of crisis and climate in Carolingian Europe', 11 November 2016, https://www.historicalclimatology.com/features/archives/11-2016.

80 McCormick et al., 'Climate change during and after the Roman Empire', p. 206.

81 Tipping, *Bowmont*, pp. 191–192.

82 S. Halliday, 'The wider context of 7th century AD agrarian expansion', in D. Strachan, D. Sneddon and R. Tipping, *Early Medieval Settlement in Upland Perthshire: Excavations at Lair, Glen Shee 2012–17* (Oxford, 2019), pp. 140–143.

83 This landscape is explored in detail in Royal Commission on the Ancient and Historical Monuments of Scotland, *North-East Perth*.

Lair, Glen Shee, Perth and Kinross, looking across the zone of former cultivation and improved pasture of the seventh-century Pictish settlement towards the *Calluna*-dominated upland pastures. Although now classed as very marginal for cultivation, this was a prosperous farming area during a time of climatic instability.

the other sites of this kind.[84] This was not just expansion of existing extensive land-use practices within the region; these seventh-century developments were wholly new farms with apparently high-status occupants. These Pictish farmers were moving into an area that had been heavily grazed but well managed for a long time, possibly with occasional burning to stimulate new growth in the *Calluna* heath that dominated on the grazed land, and with sufficient livestock to suppress scrub woodland regeneration.[85] From the outset of the new settlement at Lair, however, cereal crops were a significant part of the farming economy, with the acid soils supporting cultivation mainly of barley but with some oats and rye also grown.[86] This cultivation should be seen in the context of climatic amelioration at the end of the LALIA and a rapid recolonisation of good-quality land that had ceased to be marginal for crop growing and where hay for winter fodder could be obtained, especially on warmer south- and east-facing slopes such as those around Lair. The better days of the 600s, however, were not necessarily a time of 'barbecue summers' in this region, for although it was more buffered from the still intermittently severe conditions experienced in the more marginal, Atlantic-exposed parts of Scotland, strong seasonality with good summers and harsh winters might still have required indoors overwintering of cattle, hence the need for hay production.[87] Interestingly, the dating of the evidence for oats might indicate that they were a later introduction, possibly added to the repertoire in response to the variable weather of the eighth century, which would tally with the trend towards their cultivation noted in more northerly contexts.

84 Strachan, Sneddon and Tipping, *Early Medieval Settlement*, p. 110.

85 Ibid., pp. 137, 140.

86 Ibid., p. 139–140.

87 The Pitcarmick-type houses have clear internal partitioning into human habitation and storage areas and a byre area. See ibid., pp. 110–119.

Increased marginality or failure are nowhere evident in the pollen record from Portmahomack through the later seventh and eighth centuries. Geography perhaps played an important part in the monastic settlement's apparent success, but other factors also seem to have been at play. From its foundation in *c*.650, the monks of the monastery appear to have broken the surrounding district back into intensive cultivation, clearing and draining ground for cereal crops and also reinstating managed pasture.[88] This concentration on farming is reflected in analysis of the human bone from the site, which revealed little evidence for any marine component in the monks' diet despite the coastal location of their monastery but a high proportion of terrestrial plant and animal protein.[89] This pattern is consistent with the results from analysis of human bone from other Pictish settlements of this time, where there was a similar marked absence of marine protein. While there was strong evidence for pasture in the immediate vicinity of the monastery and pollen from weeds of cultivation was well represented, it is probable that cereals were brought in from fields located more distantly for processing on site.[90] The bone evidence points to a well-nourished community of monks who did not suffer the hardships of hunger or dietary deficiency. The general picture might therefore be presented as one of success amidst the poorer prevailing conditions of the time, but that success appears to have been delivered through intensification and expansion of cultivation and livestock management to maintain yield levels and meat and dairy availability. As illustrated by Adomnán's incidental information on the provisioning of Iona in the late sixth and seventh centuries, foodstuffs were drawn from a widely dispersed monastic estate on which production was managed by the monks' tenants and dependents. Was Portmahomack's apparent success through this long phase of unfavourable conditions delivered through the efforts of its dependent community locally? As suggested for agricultural endeavours scattered from Shetland to the Cheviots in the sixth and earlier seventh centuries, increasing the area under cultivation and investing in manuring to improve the crop was a common response to declining yields in unfavourable weather conditions. It is possible that the monastic household lived comfortably on the produce of an expanded arable zone which their dependent cultivators laboured on to deliver the expected food-rents.

What evidence is there for similar responses elsewhere to adverse conditions in the later seventh and eighth centuries? Although there is nothing immediately obvious in the settlement archaeology or palaeoenvironmental record from south-eastern Scotland to evidence the kinds of climatically induced changes in human behaviours that Michael McCormick and his colleagues suggested might have arisen at Portmahomack,[91] research in the Northern and Western Isles has raised some tantalising possibilities. The steady increase in dependence on a barley and oat mix – in some cases with oats becoming predominant – and the expansion of the cultivated area onto poorer, previously uncultivated ground that was underway in the sixth century continued and possibly accelerated through the seventh and eighth centuries. That physical expansion alone seems to be a clear enough signal

88 McHardie, Mighall and Schofield, 'The landscape and environmental setting'. For the site in overview, see M. Carver, *Portmahomack: Monastery of the Picts* (Edinburgh, 2008).

89 S. Curtis-Summers, J. Montgomery and M. Carver, 'Food for the body, sustenance for the soul: dietary contrast between Pictish and medieval populations at Portmahomack, Scotland, using stable isotope analysis', in M. Carver, *The Tarbat Discovery Programme*, Archaeology Data Service, 2017, https://doi.org/10.5284/1031216.

90 McHardie, Mighall and Schofield, 'The landscape and environmental setting'.

91 McCormick et al., 'Climate change during and after the Roman Empire', pp. 206–207.

of responsiveness to changing conditions and circumstances.[92] Perhaps more interesting, however, is the emergence of new soil-enrichment strategies that created well-manured anthrosols of a distinctive character that can be traced throughout the Hebrides, Orkney and Shetland, as well as in some mainland districts.[93] An Irish monastic link to such modified soils has been postulated since the 1950s, and the discussion of manuring practices in seventh-century Irish law tracts certainly points to a sophisticated understanding of soil-enrichment techniques.[94] In a Scottish context these anthrosols have been associated with Early Christian missionary activity and particularly with the *papar*, those Irish and Scottish monks who established communities throughout the islands from the later sixth to the late eighth century.[95] The palaeoenvironmental signature of their activities has been seen in the anthropogenically deepened soils whose distribution bears a close relationship to locations believed to be the sites of early monasteries. Anthrosols, it has been suggested, are evidence for the arrival with them of 'novel arable land-manuring strategies as part of a "package" of new agricultural methods'.[96] Modifications to soil composition to deepen and enrich cultivated ground were hardly an innovative practice in Orkney or Shetland, where anthrosols were first developed in the Neolithic period.[97] There was, however, a change in manuring practices in the formation of anthrosols that occurred sometime in the pre-Norse late Iron Age, from a technique that used mainly domestic waste including human faecal matter to one that used primarily livestock faecal material mixed with turf and limited quantities of hearth residues.[98] These changes reflect a tight integration between arable cultivation and the keeping of livestock, which analysis of the faecal matter reveals to have been a mix of cattle, sheep and pigs, with the livestock dung being gathered for use as soil-enrichment in the most heavily cropped fields.

This manuring process has been posited as a seventh-century introduction in the Northern Isles, although later eleventh- and twelfth-century contexts for its adoption have also been proposed.[99] The immediate context for a seventh-century introduction has been seen in the presence of Irish monks in

92 See, for example, discussion of the apparent expansion of agriculture in Ireland but the decline in oak regeneration there in the late 700s to early 800s in Baillie, 'Bad for trees – bad for humans?', pp. 16–17.

93 For the first developed theories on chronology of anthrosol formation in the Northern Isles, see I.A. Simpson, 'The chronology of anthropogenic soil formation in Orkney', *Scottish Geographical Magazine*, 109:1 (1993), pp. 4–11. This initial thinking has undergone considerable development, broadening and lengthening the scope of anthropogenic soil formation and the chronology of the practice.

94 L. De Paor, 'A survey of Scelig Mhichil', *Proceedings of the Royal Society of Antiquaries of Ireland*, 85 (1955), pp. 174–187 at 175; I.A. Simpson and E.B. Guttmann, 'Transitions in early arable land management in the Northern Isles – the *papar* as agricultural innovators?', in B.E. Crawford (ed.), *The 'Papar' in the North Atlantic: Environment and History*, St John's House Papers 10 (St Andrews, 2002), pp. 59–67 at 65.

95 Barber et al., 'Excavations on Iona', pp. 359–360; Simpson and Guttmann, 'Transitions in early arable land management'.

96 Simpson and Guttmann, 'Transitions in early arable land management', p. 60.

97 I.A. Simpson, 'Relict properties of anthropogenic deep top soils as indicators of infield management in Marwick, West Mainland, Orkney', *Journal of Archaeological Science*, 24 (1997), pp. 365–380; I.A. Simpson et al., 'Early anthropogenic soil formation at Tofts Ness, Sanday, Orkney', *Journal of Archaeological Science*, 25 (1998), pp. 729–746; I.A. Simpson, S.J. Lancaster and S.J. Dockrill, 'Making arable soils: anthropogenic soil formation in a multi-period landscape', in R.A. Nicholson and S.J. Dockrill (eds), *Old Scatness Broch, Shetland: Retrospect and Prospect* (New York, 1998), pp. 111–126.

98 Simpson and Guttman, 'Transitions in early arable land management', pp. 61, 64.

99 Ibid., p. 66.

Looking north-north-west from near Breacais, Skye, to Pabbay, with Scalpay and Raasay beyond. Pabbay's place-name and the presence of anthrosols near to the chapel on its south side have been linked to the introduction of soil-deepening and -enrichment techniques by Irish monks.

the islands as missionaries from Iona at the end of the sixth century, whose safety was negotiated between St Columba and the Pictish king Bridei, son of Mailcon, who was overlord of the islands.[100] The identification of anthrosols on Iona, enriched through the introduction of turf, shell-sand, animal manure and other organic matter, plus Adomnán's account of the mission to Orkney led by an Iona monk, has been seized upon as evidence of a potential mechanism through which this manuring process was introduced to various island locations.[101] An apparent correlation between areas of anthropogenic soil deepening and enriching and place-names containing the Norse term 'papar' (priest, monk), such as the five Pabay/Pabbay (Priest Island) names in the Hebrides and the two Papa/Papay islands in Shetland and two in Orkney,[102] provides further but still largely circumstantial evidence. Of these places, Papa Stour in the Shetland Islands presents possibly the most extreme example, where the western two-thirds of the island eventually saw the stripping off by the nineteenth century of its peaty turf and its transportation to the eastern third for use as fuel, building material, byre flooring and soil-deepening materials, with fuel residues, waste building remains and the byre materials all subsequently finding their way into the deepened fields.[103] Here and at the other 'papar' place-name locations distributed throughout the Northern and Western Isles and in the northern Highlands, formation of these cultural soils appears

100 *Adomnan's Life of Columba*, pp. 441, 443.

101 Barber et al., 'Excavations on Iona', p. 360.

102 See 'Appendix – *papar* names in the British Isles', in A. MacDonald, 'The *papar* and problems: a brief review', in B.E. Crawford (ed.), *The 'Papar' in the North Atlantic: Environment and History* (St Andrews, 2002), pp. 13–29.

103 D.A. Davidson and I.A. Simpson, 'Soils and landscape history: case studies from the Northern Isles of Scotland', in S. Foster and T.C. Smout (eds), *The History of Soils and Field Systems* (Aberdeen, 1994), pp. 66–74 at 67–68.

to be pre-Norse. Sample assessment of the anthrosols on Pabbay in the Sound of Harris and at Paible on Taransay supported the theory that such names were associated with better-quality or improved land, or with land that could be improved to provide productive yields within the limits of early medieval agricultural technology.[104] The assessment also raised the possibility that some of the 'papar' names in Shetland might be linked to outlying livestock operations associated with a primary monastic centre rather than to primarily arable operations in their own right. Having established the linkage, however, we await the detailed research into the management practices that created their associated anthrosols and how those contributed overall to the shaping of today's environment and the cultural landscape in the Northern and Western Isles.

Outside the primary zone of Norse settlement where the 'papar' names might signpost pre-Norse anthropogenic soil modification, similar anthrosols have been identified on the Tarbatness peninsula in Easter Ross close to Portmahomack. Details of the research and analysis behind this oft-cited identification, however, still await full publication.[105] It was suggested in the 1970s that these soils were associated with the thirteenth-century abbey at Fearn (with a

104 I.A. Simpson, B.E. Crawford and B. Ballin Smith, 'Papar place-names in the Northern and Western Isles of Scotland: a preliminary assessment of their association with agricultural land potential', http://www.paparproject.org.uk/agricultural.html.

105 J.C.C. Romans, the identifier of the Easter Ross anthrosols, is cited as 'pers. comm.' in Barber et al., 'Excavations on Iona', p. 359, then again in almost identical terms thirteen years later in Davidson and Simpson, 'Soils and landscape history', p. 73, and again in 2002 in Simpson and Guttmann, 'Transitions in early arable land management', p. 66. Nearly fifty years from their original identification, the nature and provenance of these anthrosols around Fearn remain unexplored.

tentative allusion to a possibly earlier monastic precursor nearby), which lies central to their area of distribution,[106] but this hypothesis arose before the identification of the monastery at Portmahomack.[107] It is unfortunate that investigation of these anthrosols was not part of the 1990s research programme into the monastic site, for these might be the fields that delivered the grain, manured by the cattle, which the seventh- and eighth-century monks owned and which provided their protein-rich diet. Similar areas of anthropogenic soils were also identified in Strathmore and linked to the activities of twelfth- and thirteenth-century monastic cultivators there on the simple basis that Cistercians, like the occupants of Coupar Angus Abbey, were widely reputed in the Middle Ages as agricultural pioneers and innovators.[108] Like Tarbatness, however, Strathmore was the location of important early monastic centres like Kirriemuir, with which such soil-enrichment strategies might be linked. These soils are a 'document' that potentially records a cultural response to early medieval climate change in northern Scotland and illustrate the mechanisms employed to secure and broaden the arable resource available to communities along the Atlantic edge of these islands. As possible windows onto the ways in which those communities strengthened their social and economic resilience at a time of increasing environmental marginality, they are a record that begs close examination in our own time.

• • •

Although the written evidence for the environmental history of the half millennium after AD 500 is fragmentary and often poetic or literary in form rather than annalistic, when considered together with archaeological and sculptural material it provides a series of rich insights on the changing nature of the climate, landscape and wildlife of the era. We are presented with a record of a land under pressure, from sudden and rapid climate change as much as from human activities. Old ways that had prevailed through much of the earlier first millennium AD were increasingly revealed as untenable in the changing and changed conditions brought by the LALIA, by epidemic disease and by the social and cultural upheavals that were spawned by both. But what we have seen from the material discussed above is that human society in Scotland was neither passive nor helpless in the face of changing conditions but responsive and proactive where necessary and possible. The inhabitants of the agriculturally more marginal districts from Shetland to Barra and through mainland upland zones possibly struggled as the climate cooled and became wetter and growing seasons contracted, but evidence for the adoption of wet- and cold-tolerant crop species underscores the resilience of those communities. Even in less marginal areas, the cultural norms of the Roman era were reconfigured in the face of environmental pressures and as population levels and political factors dictated. What is also clear, however, is that as the impacts of the LALIA faded and new, more centralised social structures emerged in the eighth and ninth centuries, resource exploitation moved into more systematically organised configurations, where access to essential staples was allocated and overseen by a higher, external authority. As the climatic regime of the Medieval Climate Anomaly that prevailed to the thirteenth century began to assert itself towards the end of the tenth century, Scotland's rulers were already well positioned to capitalise on the opportunities it presented. In the next chapter, we will explore the environmental context for the emergence of the medieval Scottish state.

106 Barber et al., 'Excavations on Iona', fig. 45.

107 For discussion of the late sixth- to early ninth-century monastery at Portmahomack, see Carver, *Portmahomack*.

108 Simpson and Guttman, 'Transitions in early arable land management', p. 66.

CHAPTER 4

Raiders, Ranchers, Farmers and Fishers through the Great Winters of the Viking Age

> This counsel then was settled, that they should leave the country, for the sons of Ketill urged it much, and no one spoke against it. Bjorn and Helgi wished to go to Iceland, for they said they had heard many pleasing news thereof. They had been told that there was good land to be had there, and no need to pay money for it; they said there was plenty of whale and salmon and other fishing all the year round there. But Ketill said, 'Into that fishing place I shall never come in my old age.' So Ketill then told his mind, saying his desire was rather to go west over the sea, for there was a chance of getting a good livelihood. He knew lands there wide about, for there he had harried far and wide.[1]

Towards the end of the eighth century, large-scale changes were taking place across much of north-western Europe. The polities that had begun to emerge in the later sixth and seventh centuries, and that had stabilised and matured in the course of the eighth, were soon to face existential challenges from external threats that ultimately brought jarring discontinuities in modes of agricultural production and in the human relationship with the wider environment. Earlier changes had been the responses of long-established indigenous populations to new constraints or opportunities presented by shifting climatic trends and the weather they brought. Now, however, change came from the arrival of people from lands where radically different climatic regimes and physical conditions had delivered significantly different cultural expectations and social responses to resource management. The impact of these newcomers, raiders, traders and settlers, who moved in waves from their homelands in western Scandinavia through the ninth and tenth centuries, has been well explored in terms of the historical development of the British Isles and their role in stimulating a new era of state-building. In this chapter, rather than focusing on the traditional political narratives, we will explore how their arrival and wider influence

1 *Laxdaela Saga*, trans. M.A.C. Press (London, 1899), chapter 2.

The negative NAO conditions of the ninth century brought generally milder conditions to the northern parts of the British Isles, with lower than average rainfall but a steady flow of moist, westerly air moving in from the Atlantic.

led to the adoption of new strategies for exploitation of the resources of the lands and waters of the northern British Isles, changing the pressure on some long-exploited environmental niches and creating new stresses where there had hitherto been only relatively minor impacts.

Blowing hot and cold

After the slow increase in weather observations by monastic annalists through the eighth century, which provided background to the changes discussed in Chapter 3, a spike in such incidental weather records in central European annals in the ninth century provides some insight on the lived human experience of what appears to have been a prolonged episode of warming in the northern North Atlantic. For Scotland, these comments on the weather permit us mere glimpses of the way this event affected communities settled in more southerly segments of the eastern Atlantic seaboard.[2] The written record is terse, often ambiguous and frustratingly imprecise in terms of geographical information. Consequently, it is climate proxies that provide most of our more regionally specific evidence for this sustained northern hemisphere temperature rise. The Greenland ice cores in particular underscore the occurrence of a century-long episode of warming followed by a brief cooling in the first half of the tenth century.[3] Central European glacier records and tree-rings provide much of our proxy corroboration for the effects of this warming in mainland Europe but, again, none of these provide data directly applicable to Scotland. Here, instead, we are fortunate to have the precipitation record of the Traligill cave system. At Uamh an Tartair, the speleothems preserve evidence for sustained negative winter NAO throughout the ninth century but with significantly lower precipitation than at any time during the previous era of sustained negativity in the seventh century.[4] It was not, however, a period of unbroken low rainfall, for there are fluctuations evident within the record, with the first half century and the last quarter century seeing some short bursts of very much wetter weather. Modelling from around the northern hemisphere suggests that this negative NAO, which brings in milder, wetter conditions from the Atlantic over much of western Europe when the Iceland low-pressure cell extends its influence to a lower latitude over the ocean basin, was not accompanied by the cold north-easterly airflow associated with the Siberian

2 Pfister et al., 'Winter air temperature variations', p. 537 fig. 1.
3 McCormick, Dutton and Mayewski, 'Volcanoes and the climate forcing', p. 871.
4 Baker et al., 'A composite annual-resolution stalagmite record', fig. 4.

high-pressure cell and its extension over Scandinavia, which normally results in cold, snowy winters. Instead, the Siberian high was weak, meaning that the milder westerly airflow prevailed, driving up temperatures.[5]

The fluctuating patterns within the speleothem record bear comparison with the evidence for four 'Great Winters' at variance with that warming trend that have been identified for mainland Europe in the ninth century. These winters occurred in 821/2, 845/6, 859/60 and 873/4, with other snowy but less extreme events earlier in the century.[6] Such winters might be the consequence of a brief strengthening of the Siberian high, pushing a ridge of intensely cold air west over the continent, but alternative causes have also been sought in volcanic climate forcing. All four of these Great Winters can be linked to a volcanic eruption, with the Greenland ice cores revealing spikes in SO_4 sulphates, the main signature of aerosol particulates from eruptions. These also correlate with spikes in the chlorides that signify an expansion in sea ice off Greenland, suggestive of phases of rapid cooling potentially delivered by a dust-veil that brought a marked reduction in solar irradiance.[7] The first of these events followed two especially wet summers, in 820 and 821, which affected much of northern mainland Europe, the second also being noted for its coldness.[8] Written witness from Ireland to the Byzantine imperial court, however, tells us that the cold summer was a mere taster for the severity that followed in the winter of 821/2. The worst cold episode hit in the early months of 822, as reported in the second annal for that year in the Annals of Ulster and the *Chronicon Scotorum*, which speak of 'abnormal ice; the seas, lakes and rivers froze and herds of horses and cattle, and loads, were brought across them'.[9] While the following winter seems to have been as severe in much of mainland Europe, there is no record of similar conditions affecting Ireland, suggesting that oceanic buffering might have mitigated the worst impacts there.

The cooling in the 850s, flanked in mainland Europe by the winters of 845/6 and 859/60, was perhaps linked to an unidentified volcanic event

5 McCormick, Dutton and Mayewski, 'Volcanoes and the climate forcing', pp. 874–876.

6 Pfister et al., 'Winter air temperature variations', p. 542.

7 McCormick, Dutton and Mayewski, 'Volcanoes and the climate forcing', fig. 5.

8 Ibid., pp. 881–882.

9 Annals of Ulster, U822.2; *Chronicon Scotorum*, http://research.ucc.ie/celt/document/T100016, CS822.1. The Ulster annals also suggest an especially severe episode in Ulster in January 818, when Lough Neagh froze over for a month: Annals of Ulster, U818.2. This could be a mislocated 822 event or simply a local episode unrelated to the subsequent 821/2 winter.

Looking towards the White Mounth from Carn an Tuirc, March 2019. Amidst the milder conditions of the ninth century, the 'Great Winter' of 821/2 stands out in annals and climate proxy records as one of the most extreme winters to affect north-western Europe, with months of snow, harsh frosts and stormy conditions.

which occurred probably in 853 at a low latitude in the tropics. That eruption possibly triggered the records in Irish sources of a cold spring in 855 and a bitter winter of 855/6, when the Ulster annals reported snow as high as men's belts in late April.[10] Weather disruption may have persisted into 857, when violent windstorms swept across Ireland.[11] With no references at all in the Irish annals, the final ninth-century events, reported on the continent in the mid 870s, seem not to have been experienced on Europe's north-west Atlantic-facing edge; at least, those years pass without comment in the annals, possibly because they coincided with some of the most intense phases of Scandinavian raiding in the British Isles. Extreme though the three episodes that are recorded were, however, it must be remembered that, like the worst episodes preserved in eighth-century annals, these were isolated events driven principally by the extraordinary conditions of major volcanic eruptions. Either side of them, the ninth century saw an extended period of more benign conditions after the poor eighth century.[12]

10 Ludlow et al., 'Medieval Irish chronicles', p. 8; Annals of Ulster, U855.2, U856.1 (lasting from November 855 to January 856).

11 Annals of Ulster, U857.5.

12 Ludlow et al., 'Medieval Irish chronicles', p. 8.

Of volcanoes and fire gorges

At the ninth to tenth century transition, there was a major shift in the North Atlantic climate with an abrupt reversal of the prevailing winter NAO state from negative to positive.[13] This switch meant that the Iceland low-pressure cell was weaker and its influence was felt much further to the north, enabling the Azores high-pressure cell to push the westerlies from the Atlantic into a more northern track and increasing the likelihood of a cold easterly or north-easterly airflow and reduced precipitation. As with the cold spikes in the ninth century, however, there were incidents that disrupted that colder, drier trend, such as the 'dark and rainy year' reported in Irish annals for 912 and/or 913.[14] With no consequences of the dullness and wet reported the following year, it seems that the north of Ireland escaped lightly, for Swiss and French sources report extreme cold into the winter of 913/14 and the Greenland ice cores reveal a spike in sulphates that probably marks a major eruption in the southern hemisphere.[15] This event could account for a dip into higher precipitation in the Traligill cave record in the 910s, but the slightly wetter years passed and the general trend reverted to dry conditions.

There may have been a high degree of volcanic forcing within an especially cold phase towards the close of an already established short cooling episode in the early 900s. The culprit here was almost certainly the dust-cloud arising from a catastrophic volcanic eruption in the 75km-long Eldgjá 'fire gorge' in southern Iceland that has recently been dated to commencing in spring 939 and lasting until at least autumn 940. It has been calculated that this eruption produced in the order of 220 million tonnes of sulphate aerosols which, combined with atmospheric water vapour, created around 450 million tonnes of dilute sulphuric acid circulating the globe in the troposphere.[16] The result was the highest spike of sulphates in the Greenland ice core record in the two centuries after 800.[17] Alongside the evidence from the northern hemisphere tree-ring record of the impact of the eruption on tree growth in areas as widely dispersed as the Canadian Rocky Mountains, Alaska, Central Asia, central Europe and Scandinavia, we can relate this data to the impacts of more recent measured episodes in terms of the effect such an event would have had on global climate. The most pronounced effects were in 940, when tree-rings indicate that there was a cooling of around 1.5°C in summer temperatures, the second largest cooling episode of the last 1,500 years. Unlike the longer-lasting cooling effect produced by eruptions in the tropics, such as the 1991 eruption of Mount Pinatubo in the Philippines (less than one tenth of the scale of the Eldgjá eruption), when aerosols suppressed summer temperatures for between two and five years, the worst of the 939–40 event seems to have passed by 941. Further blows quickly followed, however, with the enormous *c.*946 eruption of Baekdu/Changbai on the modern North Korean/Chinese border, one of the most violent of the last 5,000 years.[18] Chinese and Japanese documentary sources speak of markedly colder weather in the two years following the eruption, affecting a broad swathe

13 Baker et al., 'A composite annual-resolution stalagmite record', fig. 4.

14 Annals of Ulster, U913.7.

15 McCormick, Dutton and Mayewski, 'Volcanoes and the climate forcing', pp. 887–888.

16 C. Oppenheimer et al., 'The Eldgjá eruption: timing, long-range impacts and influence on the Christianisation of Iceland', *Climatic Change*, 147 (2018), pp. 369–381. For the sulphate spike in the Greenland ice core, see McCormick, Dutton and Mayewski, 'Volcanoes and the climate forcing', fig. 5.

17 McCormick, Dutton and Mayewski, 'Volcanoes and the climate forcing', pp. 888–889.

18 J.D. Xu, 'Climatic impact of the Millennium eruption of Changbaishan volcano in China: new insights from high-precision radiocarbon wiggle-match dating', *Geophysical Research Letters*, 40:1 (2013), pp. 54–59.

Eldgjá, Iceland. Looking along the floor of this 75km-long canyon gives a sense of the scale of Eldgjá's 939/40 eruption and the volume of materials expelled into the atmosphere. (Image: Andreas Tille, http://fam-tille.de/island/iceland/high/ 2002_080.html, licensed under the Creative Commons Attribution-Share Alike 4.0 International licence.)

of Manchuria and northern Japan. The dust-cloud from the 100–120 km^3 of tephra ejected into the atmosphere circulated the northern hemisphere in the following years and has been detected in Greenland ice cores.[19] Taken together, amidst the already cooling conditions of the second quarter of the tenth century, the result of these two consecutive eruptions might have been prolonged instability for the whole of the northern hemisphere, marked by colder winters and, probably, by wet summers memorable for acidic rainfall that stunted growing crops.[20]

Visual indicators of the Eldgjá eruption were first noted in annal entries from Ireland, Germany and Italy for 939. The Irish *Chronicon Scotorum* described how the sun turned a blood-red colour,[21] while the Italian *Annales Casinates* and the German *Res Gestae Saxonicae* similarly record the solar colour change and a sense of weakening in solar intensity and loss of heat.[22] The last of these sources also refers to an intensely cold winter in 939/40, while Irish annals speak of floods on major rivers late in 940 and lakes and rivers freezing in early 941.[23] An episode of slightly wetter conditions against a run of very dry years towards the middle of the tenth century in the Uamh an Tartair record probably reflects the north-west Scottish experience of that volcanic forcing.[24] Whilst the degree to which volcanic forcing worsened the human consequences of the episode cannot

19 C.Q. Sun, 'Ash from Changbaishan Millennium eruption recorded in Greenland ice: implications for determining the eruption's timing and impact', *Geophysical Research Letters*, 41:2 (2013), pp. 694–701.

20 Dawson, *So Foul and Fair a Day*, p. 95.

21 *Chronicon Scotorum*, CS939.1; McCarthy and Breen, 'An evaluation of astronomical observations', pp. 125, 135.

22 Oppenheimer et al., 'The Eldgjá eruption', p. 375.

23 For example, Annals of the Four Masters, http://www.ucc.ie/celt/published/T100005B/index.html, M940.10 ('A great flood in this year, so that the lower half of Cluain-mic-Nois was swept away by the water'), and the Annals of Ulster, U941.1 ('Severe frost so that the ice on lakes and rivers was passable'). These effects were also noted in Chinese records across a three-year period to 942: Oppenheimer et al., 'The Eldgjá eruption', p. 376.

24 Baker et al., 'A composite annual-resolution stalagmite record', fig. 4.

Red skies and red-tinged land at sunset over Loch na Sealga in winter 1980/1, following the May 1980 eruption of Mount St Helens in Washington State on the Pacific west coast of the United States. It gives an idea of the 'blood-red sky' effects described by chroniclers in 992.

be established conclusively, 940 and 941 are recorded across a broad swathe from western Europe to China as years of subsistence crisis with poor harvests and widespread food shortages, followed in 942–4 by droughts in northern Europe.[25] Although no Scottish or English source similarly records the aftermath of the 939 eruption, comparison with the effects of the 1783–4 Laki eruption (discussed in the volume *Where Men No More May Reap or Sow*) suggests that it is unlikely that northern parts of mainland Britain escaped unscathed. Yet again, however, despite the severity of this event, it probably constituted a short period of extreme crisis for communities on the environmental margins rather than a phenomenon that forced wider changes in land management and resource exploitation in a direction contrary to that encouraged by ninth- and tenth-century climatic conditions.

As with the ninth century's Great Winters, the 912 or 913 and 939/40 events went against a broader trend to warmer and drier conditions across the tenth century. Certainly, the Traligill speleothem record suggests that even north-western Scotland was sharing the same extended episode of general dryness that afflicted mainland Europe. Reconstruction of drought conditions from European tree-ring sequences suggests that the mid tenth century in particular witnessed an extended period of below-average rainfall across much of northern Europe.[26] Short, headline episodes of extreme wet or cold weather continued through this period, despite the proxy data indication of an intensification of drought in the last quarter of the century in the north-west Highlands. In central and western Europe, a short phase of harsh winter weather is recorded in 974/5, with a final such episode in 992/3.[27] The first of these periods, marked by an observation in Irish annals that there was 'very bad weather' in this year, seems to have seen poor conditions in both summer and winter.[28] There is, however, no suggestion of harvest

25 Oppenheimer et al., 'The Eldgjá eruption', pp. 376–377.

26 Cook et al., 'Old world megadroughts and pluvials', fig. 3.

27 Pfister et al., 'Winter air temperature variations', p. 542.

28 Annals of Ulster, U975.6.

failure or food shortages in the following year, which gives us pause to consider what reports of 'very bad weather' mean in terms of crop or livestock success or failure in any one year.

Blood-red skies reported in the Annals of Ulster on 26 December 992 probably signify a volcanic dust-cloud event,[29] given the wider impact of extreme winter weather reported across Europe. Coming so late in the year, it seems to have had no appreciable impact on crop growing or harvests, but an annal entry in the second half of 993 records 'a great mortality of people, cattle, and bees throughout Ireland this year'.[30] We cannot, however, prove any direct correlation between these events. As the memory of the mortalities of 933 faded, the weather along Europe's north-west Atlantic-facing seaboard began to improve towards the beginning of the long era of generally milder conditions – what is often referred to as the Medieval Warm Period or Medieval Climate Anomaly (MCA) – that spanned the eleventh and twelfth centuries.[31] The rollercoaster had been ridden and the north-west of Europe was about to enter on one of the longest, most sustained episodes of generally benign conditions of the medieval period, one that allowed a great expansion of agriculture and population but that brought new pressure to bear on the wider environment.

Resilience, adaptation and adoption

How did human culture and society across Scotland adapt to the changing circumstances of the ninth and tenth centuries? Until recently, our models have been constructed mainly on the basis of data from archaeological excavations in the Northern and Western Isles. The focus there has been primarily on evidence for the implantation of Norse culture by western Scandinavians who came first as raiders in the eighth and early ninth centuries and then began to settle in increasing numbers as the ninth century progressed. Not only did these colonists bring a different cultural portfolio in terms of art, architecture and artefacts but they also brought different agricultural and livestock management traditions, different trade networks and new shipbuilding and sea-faring techniques that allowed a quite different approach to the exploitation of marine resources. Even those parts of Scotland that experienced the raiding but saw little or no lasting settlement of Scandinavians were transformed by the encounter, adopting the techniques of the newcomers or remodelling their own social and economic organisation to enable them to better counter the military threat.

It was Shetland and Orkney that probably experienced the earliest raids, although the first recorded attacks were much further south on the Dorset coast (*c.*789) and on Lindisfarne (793).[32] It is likely that the Northern Isles also saw the first permanent settlements early in the ninth century, before the *Orkneyinga Saga*'s traditional *c.*870 date for their creation as a jarldom for Rognvald of Møre and his planting of his own dependents in the islands, though how much earlier remains an open question.[33] There is also a continuing debate over what the colonists' arrival meant for the indigenous Pictish population, with the process of transition from native to Norse culture that is evident in the archaeological record being read by some as an indication of elimination of the previous occupants by the invaders and by others as pointing towards assimilation and the absorption of the one into the other. Even that

29 Ibid., U992.4.

30 Ibid., U993.7.

31 Pfister et al., 'Winter air temperature variations', p. 542.

32 Crawford, *Scandinavian Scotland*, pp. 39–40.

33 Dockrill et al., *Excavations at Old Scatness*, vol. 1, p. 78.

Jarlshof, Sumburgh, Shetland, looking across the remains of the Norse farmstead towards Scatness. At Old Scatness and Jarslhof, pre-existing Pictish estate centres were taken over and adapted by incoming Scandinavian settlers.

second proposition, which has become the more favoured view in recent years as DNA-based projects have opened new lines of research, is fraught with difficulties, for the apparent takeover of the important Pictish estate centres by new Norse owners could be indicative of processes of enslavement, intermarriage or elite social replacement. Continuity of settlement, furthermore, is not conclusive evidence for continuity of the land management and resource exploitation practices which went with that settlement, especially where traditional regimes might already have been under pressure as a consequence of the changing climatic conditions of the eighth century.

At Old Scatness in the south of the Shetland Mainland, the earliest indications of the Norse takeover of the Pictish settlement are suggestive of occupation and reordering of the pre-existing Pictish buildings and their slow replacement over time. For the excavator, this was evidence of a direct appropriation of a clearly economically successful high-status estate centre.[34] What is less clear is whether there was a similar continuity in the agricultural system that underpinned it. In one key area, that of manuring strategies, there was a definite shift away from inherited practice, whereby an anthrosol formation that had used a rich mix of domestic waste, animal faecal matter, byre materials and turves through the seventh and eighth centuries gave way to one in which domestic waste alone and especially domestic hearth residues predominated. As has been observed for that site, this is suggestive not of Norse adoption of the strategies that had maintained the arable output of their predecessors but of a cashing-in and steady depletion of the stored-up soil-fertility capital of the Pictish farmers.[35] The change in manuring strategy found at Old Scatness is at odds with evidence from elsewhere in the Northern Isles for the maintenance of the late Iron Age/Pictish approach. It perhaps illustrates how communities in Shetland and Orkney responded differently to local environmental conditions. At Old Scatness, analysis of excavated bone suggests that there was a local decline in the previous emphasis on a balance of arable and pastoral farming, from a farming model in which the livestock had provided the bulk of the faecal matter used in soil-enrichment, to one that supported fewer sheep and cattle. A less successful manuring strategy would have progressively reduced the soil fertility necessary for success with barley and also for the flax which the Norse colonists also invested in on the better-quality land. But this shift may have been compensated for by further expansion of the cultivation of less demanding and more resilient oats. It may also represent a conscious response on this site to the changed weather conditions of the ninth century, perhaps involving greater reliance on trade to satisfy grain needs. Such a strategy was certainly important for Shetland in later centuries. On the other hand, a second very significant shift in resource exploitation that was evident in the Old Scatness data, from shore-based and inshore to deep, open-water fishing, suggests that it might have been pursued locally from much earlier.[36] Better ship technology, better seamanship in the rough waters and strong currents off Shetland and a long-established cultural tradition of deep-water line-fishing, primarily for cod and saithe, perhaps provided a source of wealth that made up for the progressive deterioration of the old arable economy of this site.[37]

More generally in Shetland, the excavated

34 Ibid., pp. 78–82, 96–97, 361–363.

35 Ibid., p. 202.

36 Ibid., pp. 156–167, 202–203, 362.

37 J.H. Barrett, R.A. Nicholson and R. Céron-Carrasco, 'Archaeo-ichthyological evidence for long-term socioeconomic trends in Northern Scotland: 3500 BC to AD 1500', *Journal of Archaeological Science*, 26 (1999), pp. 353–388.

Deerness, east Mainland Orkney, looking south towards Skaill, Orkney. Animal burrowing and weather erosion of the thin soil over the underlying bedrock that was left after the stripping of turf for soil-deepening on neighbouring farmland has created an impoverished and degraded zone along the clifftops and high ground.

evidence does not support any suggestion that the possible decline in animal husbandry at Old Scatness in the Norse period was a widespread phenomenon. The expansion in oat cultivation that started in the pre-Norse period but continued in the ninth and tenth centuries could have been in part an effort to increase fodder availability, especially for the overwintering livestock. Importantly, however, there does seem to have been a switch in livestock strategies from meat production to dairying, a trend that intensified across the medieval period and that would have kept the overwinter herd sizes small in number. This switch is revealed through a marked increase in evidence for calf mortalities, with neonates and young calves being slaughtered to allow the cows' milk to be taken for human consumption in cheese, butter and the plethora of dairy preparations for which the Northern and Western Isles were later famed, while also minimising winter fodder demands.[38] While the colder and drier conditions of the tenth century might have retarded grass growth and made winter fodder production from the traditional Norse hay crops more difficult, it is unlikely that the move towards a dairying economy was driven solely by environmental factors. It is more likely to represent the local adaptation of Scandinavian cultural preferences.

In Orkney, some indications of continuing challenging conditions in the Northern Isles during the mild but wet years of the ninth century might also be read in the evidence from Pool for the processing of barley and oats which on occasion were being

38 J. Mulville, J.M. Bond and O. Craig, 'The white stuff: milking in the outer islands of Scotland', in J. Mulville and A.K. Outram (eds), *The Zooarchaeology of Fats, Oils, Milk and Dairying: Proceedings of the 9th ICAZ Conference, Durham 2002* (Oxford, 2005), pp. 167–182; D. Serjeantson, *Farming and Fishing in the Outer Hebrides ad 600 to 1700: The Udal, North Uist*, Southampton Monographs in Archaeology, New Series 2 (Chandlers Ford, 2013), pp. 53–56.

harvested unripe and then dried artificially.[39] Here, however, the integrated livestock and arable economies of the seventh and eighth centuries were maintained by the incoming Norse, with no apparent break in the soil-enriching processes that saw a substantial input of organic material to maintain fertility. The difference might have been in a higher level of survival of and integration with the indigenous Pictish population, allowing for a more effective exchange of cultural information around the anthrosol-forming strategies that had helped to secure the arable basis of Orcadian society through the major climate downturn of the sixth to eighth centuries. Indeed, this process continued throughout the Norse period in Orkney and was maintained into the eighteenth and nineteenth centuries in some parts of the islands. The soil-deepening and manuring of arable land in Orkney, which accelerated during the MCA in the twelfth and thirteenth centuries, has left a legacy of significantly improved soils in arable zones but a degraded upland from which much of the turf material was stripped. The stripped turves were most often applied directly to the cultivated ground but also provided animal bedding, building materials and fuel. What needs to be stressed in the case of Pool, however, and probably for most later ninth- and earlier tenth-century Norse farms in Orkney, is that this settlement was successful in cultivating arable crops and possibly in producing surplus grain amidst growing conditions that further south would have been regarded as far from optimal. Norse maintenance of Pictish practice kept land that may not have been naturally viable for cereal cultivation at least marginally viable across the long era of climatic disturbance before 1000.

Despite the warming that occurred in the ninth century, the wetter weather of that century and the long-term deterioration in climate conditions across the previous half millennium meant that by the later 900s oats were displacing barley as the dominant crop in Orkney, even in the traditional 'bread-basket' and grain-exporting islands like Sanday.[40] This does not indicate that barley was abandoned but rather that its cultivation was concentrated increasingly on the best, heavily manured ground to maximise yields. Differently from Old Scatness in Shetland, however, there was no move towards greater dependence on deep-water fishing to offset any deterioration in the grain yields; the cultivation of cereals continued to be a mainstay of Norse Orkney's economy. Fishing formed a distinct economy in its own right and would evolve later in the Middle Ages into a commercial fish trade with at least regional if not international significance.[41] What we do see, however, is an increase in efforts to ensure the viability of the harvested grain in less than desirable weather conditions, such as prevailed through the ninth century and which the intermittent bad years of the tenth century continued to deliver. At Pool and elsewhere in Orkney, it is clear that variations in weather from year to year meant that grain was on occasion harvested not just wet but unripe and required drying artificially before it could be milled, which used more of another commodity scarce in Norse Orkney, fuel.

Maximising opportunity in the Western Isles

Norse settlement of the Western Isles, as revealed through the excavation of sites in Lewis, North Uist and South Uist especially, again represents a process of intrusion into a long-exploited landscape. Here

39 Hunter, Bond and Smith, *Investigations in Sanday*, vol. 1, p. 143.

40 Ibid., p. 519.

41 See, for example, the excavation of the apparently important fish-producing centre at Freswick in Caithness: C.D. Morris, C.E. Batey and D.J. Rackham, *Freswick Links, Caithness: Excavation and Survey of a Norse Settlement*, North Atlantic Biocultural Organisation and Highland Archaeology Monograph 1 (Inverness, 1995).

The Udal, North Uist, Western Isles, looking across the excavated area of the Neolithic to seventeenth-century settlement site towards the machair and shallow coastal waters. Analysis of the cattle bone from the site suggests intensification of dairying practice through the late Pictish and into the Norse period, which would have increased grazing pressure on the pasture resources locally.

too we can see patterns of cultural introduction and adaptation of indigenous practices that we have already encountered in the Northern Isles. What is most striking, however, are the discontinuities, even where the Norse were occupying previously Pictish settlement sites. As in Shetland, the biggest transition in exploitation regimes was in respect of marine resources. At every Norse site so far examined in the Western Isles, there was a marked shift in the profiles of fish caught, away from the inshore species that were more favoured in the late Iron Age and Pictish period to deep-water, offshore species.[42] The trend in the later Norse period through the eleventh to thirteenth centuries was towards emphasis first on mature cod and saithe and later on herring catches, possibly linked to a large-scale fishery operation that supplied the important island communities and was likely involved in a trade network that extended to the Scandinavian colonies in Ireland.[43] The fish were caught using boats and nets and were processed probably at coastal locations. Currently available evidence for the presence of processed remains on settlement sites, but with limited evidence for that processing, points to the fisheries as being seasonal ventures organised at a community level and representing the introduction to the Western Isles of practices from the Norse homelands. Norse shipbuilding and seafaring technology no doubt provided the knowledge base upon which this fishery was developed, but its effective operation was probably also eased by the more benign sea conditions of the era and the settlers' continued links to both Norway and other

42 M. Parker Pearson et al. (eds), *Cille Pheadair: A Norse Farmstead and Pictish Burial Cairn in South Uist* (Oxford, 2018), p. 508.

43 Ibid., pp. 508–509; Sharples, *A Late Iron Age Farmstead*, p. 347; Sharples, *The Economy of a Norse Settlement*, pp. 421–422.

Feral goats, Craignaw, Dumfries and Galloway. Modern feral goat herds are largely descended from domesticated ancestors. Goats formed an important component of the suite of domesticates kept by the Norse settlers of the Western Isles. Their grazing behaviour probably accounted for the rapid decline of the remaining woodland areas in the islands in the tenth century.

areas of Scandinavian settlement for the materials they needed in order to maintain their maritime culture, especially the wood for shipbuilding that was unavailable in the Outer Hebrides.

In the terrestrial economy, livestock practices seem largely to have continued from the late Pictish to Norse periods in Uist, with perhaps only a slight drop in cattle numbers by comparison with sheep and pigs early in the settlement phase in the ninth and tenth centuries.[44] There do seem to be demonstrable changes in approaches to the management of cattle, however, with maintenance of the traditional Outer Hebridean dairying practices at some sites and a shift towards beef production at others. For the former, analysis of calf mortality at the Udal in North Uist suggests that most were neonates that died within a week or two of birth, with the numbers indicating slaughter rather than natural perinatal mortality.[45] This slaughter profile is strongly indicative of continuation of the dairying culture that had prevailed through the earlier Historic period at this site, something that is also indicated by the slaughter ages of the adult cows, which suggests that they were killed once they had ceased calf-bearing. Collectively, this emphasis on dairying at the Udal points to a sustained pressure on available grazing and fodder resources in this district, for the protein return to the herder from grass consumption for milk is five times greater than for meat.[46] As in Shetland, however, some Norse farms chose to buck the general trend,

44 Serjeantson, *Farming and Fishing*, p. 43. Pigs seem to have been entirely domesticates and to have been kept at the farming settlements, where they foraged in middens and were probably fed domestic food waste.

45 Ibid., pp. 50, 52.

46 Ibid., p. 52.

as at the Mound 1 site at Bornais in South Uist, where the decline in evidence for neonatal slaughter is matched by an increase in the age profile of the cattle generally.[47] Here, during the late ninth and tenth centuries, meat production was perhaps a specialism for this farm within a wider estate economy. This seems to have been even more pronounced in the tenth-century Norse occupation at nearby Cille Pheadair, where a higher percentage of older animals seems to have been maintained in the early settlement phase, probably for beef and hides and possibly even for traction power.[48]

It is unclear what prompted this shift towards meat production, but it could be linked to communal cultural practices such as feasting.[49] While climatic conditions had improved through the ninth century, this would not have increased the availability or quality of grazing to such an extent as to encourage a large-scale shift away from dairying. We are instead probably looking at a conscious decision linked to Norse displays of conspicuous consumption as a reflection of social status and leadership. Specialism in the production of beef for elite consumption and chieftainly feasting continued throughout the Middle Ages in the Hebrides.[50] Maintenance of a beef herd would have required a different approach to grazing management and fodder provision, although the warming in the ninth century would probably have delivered more days over winter on which the critical 6°C daytime temperature threshold for grass growth was crossed, perhaps providing more opportunities for outside grazing on winter grass on the machair.[51] The Great Winters of the ninth and tenth centuries might have presented grave challenges to such a regime but single-year failures are unlikely to have made Norse farmers rethink their livestock strategy.

Further change can be seen with regard to sheep and goats. Comparison of data from Bornais, Cille Pheadair and the Udal points to a shift in the management of sheep flocks from a primarily lamb / mutton production regime to one where ewes' milk and more probably wool and fleeces were the main outputs.[52] That shift might also have resulted in a change in grazing practices, away from intensive management close to the farm settlements and towards a more extensive mode with the flocks probably being kept on the better-quality and higher-altitude summer pastures of the islands' interior and east coast.[53] This was part of a broader transition in the exploitation of the moorland resources in Uist, of which the introduction of goats in number during the early Norse period was a major part. At Cille Pheadair and Bornais, goats were present from the early tenth century and continued as important elements of the livestock range through the later Norse periods. Probably valued more for their milk and skins than for their meat, it was the ability of goats to browse for their fodder and to metabolise woody twigs and tough leaves that perhaps brought them to the fore during the colonisation.[54] That kind of browsing, however, had consequences for the relict woodland and coppice areas that had been curated to the early ninth century, perhaps bringing available woodland quickly down to the levels

47 Sharples, *A Late Iron Age Farmstead*, p. 345.

48 Parker Pearson et al., *Cille Pheadair*, p. 445.

49 Sharples, *The Economy of a Norse Settlement*, p. 429.

50 See, for example, R.D. Oram and P. Adderley, 'Lordship, land and environmental change in West Highland and Hebridean Scotland c.1300–c.1450', in S. Cavaciocchi (ed.), *Economic and Biological Interactions in Pre-industrial Europe from the 13th to the 18th Centuries* (Florence, 2010), pp. 257–268.

51 Serjeantson, *Farming and Fishing*, p. 30.

52 Parker Pearson et al., *Cille Pheadair*, p. 445.

53 Sharples, *The Economy of a Norse Settlement*, pp. 422–423.

54 Sharples, *A Late Iron Age Farmstead*, p. 346; Parker Pearson et al., *Cille Pheadair*, p. 446.

recorded in the post-medieval period and forcing increased reliance on peat and heather for fuel.[55] Together, the grazing habits of sheep and goats would have been a significant check on woodland regeneration and consequently on timber and fuel-wood availability for the island communities.

Cultivation practices saw general continuity in terms of the composition of the crops grown but agriculture seems to have intensified and the cultivated area to have been expanded by the newcomers. There were also some new species introductions or changes in emphasis that marked a cultural preference rather than being a response to climatic or environmental conditions.[56] At all excavated sites, barley still dominated the cereals sown and would have continued to thrive on the stabler, well-manured parts of the machair where the Pictish inhabitants of the islands had previously grown it in the slightly improving weather of the period. There was, however, a shift in the focus of agriculture, with expansion starting in the eleventh century onto blackland areas east of the machair, whose heavier soil was better suited for supporting demanding crops like barley. Oats were perhaps of increasing significance due to their ability to thrive on both the sandy machair and the peatier soils of the blacklands, their steadily rising presence perhaps reflecting the physical expansion of the cultivated area rather than a cultural preference. But there was also the introduction of rye, which like oats is tolerant of poor soils and was to become of major significance later in the Middle Ages. It was probably grown on the freer-draining portions of machair that other crops – apart from flax – did not favour, with the possibility that new areas were brought into cultivation for this purpose.[57] There were possibly attempts to grow wheat at Cille Pheadair,[58] but the fact of its very specific soil and weather requirements perhaps implies unfamiliarity with the weather regime likely to be experienced in the islands; alternatively, the few grains found may represent a contaminant introduced with barley or oat seed-corn by the colonists.[59] The oats reflect a greater realism, for they are more tolerant of poor and calcareous soils like those of the Uist blackland and machair and of the wet and windy island weather. Flax was the final major introduction, going from a crop that was merely 'present' in the late Iron Age to one that was being cultivated at scale in the Norse period, probably for both its oils and its fi bres. Intolerant of extreme cold or heavy rains and favouring free-draining soils, the flax would have been ideal for cultivation on the machair, so long as its fertility levels were maintained by regular application of seaweed or other soil enrichers.[60] Taken together, the extension of the cultivated range in the excavated sites in Uist points to both intensification of agriculture and expansion of the arable zone, which might reflect a positive process in response to greater agricultural opportunity, especially in the second half of the tenth century as the climate began to improve. The switch to more extensive pastoral practice and the importance of dairying, however, serves as a check to this rosier image, signifying continuing pressure on grazing quality and the limited availability of winter fodder sources for livestock. But these pressures and limitations were not entirely driven by climate. Most, indeed, likely arose from conscious decisions by the occupants of Bornais, Cille Pheadair or the Udal about what they wanted from their land. The outcomes, whether in terms of fluctuating

55 Sharples, *The Economy of a Norse Settlement*, p. 387.

56 Sharples, *A Late Iron Age Farmstead*, p. 348; Sharples, *The Economy of a Norse Settlement*, p. 428.

57 Parker Pearson et al., *Cille Pheadair*, pp. 530–531.

58 Ibid., pp. 516–517, 528–529.

59 Ibid., p. 523.

60 Ibid., p. 529.

Upper Strathspey, looking west from Carn Dearg towards Garva and the Corrieyairack Pass. Record evidence shows that the strath west of Laggan was divided into a series of *dabhaichean*, with arable blocks in the valley floor and detached areas of summer pasture stretching all the way to the watershed with Glen Roy. This was an intensely exploited and intensively settled landscape.

yields, acceptance of poorer-quality cereal types to accommodate other in-demand crops like flax, changes in the protein basis of their diet, the breaking-in of new land, accelerated machair erosion or the loss of residual woodland, are unlikely to have been what the Norse farmers had envisaged for their new life in the west.

Sketching out a bigger picture

The discussion so far has focused on micro-level changes in exploitation regimes at individual communities and what that can tell us of broader processes affecting Scotland's environment in regions extending from Shetland to Uist. For much of mainland Scotland in this era, however, we are still lacking the depth of published palaeoenvironmental data from individual sites that might support such sharply focused exploration of human impacts on the environment or enable analysis of underlying trends in cropping patterns or livestock management systems. The rich environmental record from Portmahomack, for example, changes abruptly following the destruction of the monastery in *c.*800, probably in a Norse raid.[61] Here, a sharp decline in pollen associated with cultivation and a rise in that of plants associated with fenny and wetland areas is most likely the consequence of an extreme anthropogenic event – the end of the monastery and collapse of its estate – rather than the impact of any climate event. Illustrative of the local environmental changes arising from the disruption of a highly organised agricultural system, and perhaps suggestive of how the progress of Viking raiders around the country might be charted by similar abrupt transitions in the palaeoenviron-

61 McHardie, Mighall and Schofield, 'The landscape and environmental setting'. The abrupt change in the environmental record is represented by the transition from Phase 2 into Phase 3 in the report's division of the periods of activity at the site.

mental record, this kind of data is less informative in respect of wider non-anthropogenic trends.

Macro-scale research on the territory of the former kingdom of Fortriu and the heartlands of its successor kingdom of Alba cannot come close to offering the finely nuanced evidence of the kind obtained at Old Scatness, Bornais or Cille Pheadair. It can, however, give us some sense of the way socio-economic organisation changed in the tenth century and of the environmental conditions implicit in that change. Debate has raged since the mid nineteenth century about a key element of the pre-twelfth-century social organisation of much of north-eastern Scotland, the *davoch* / *davach*, in Gaelic 'dabhach' (pl. 'dabhaichean').[62] Various interpretations of what constituted a *dabhach* had mostly focused on its origins, function and status as a base unit in a system of land and resource organisation, concentrating on the likelihood that it was a measure of the arable production of a piece of land of a fixed extent. These older theories have more recently been overturned and it is now generally accepted that the *dabhach* was an entity that dated from at least *c*.900 and probably had its origins much earlier still in north-eastern Scotland's Pictish culture.[63] More importantly, from the perspectives of socio-economic and environmental history, it remained in some form of use as an identifiable social unit in parts of central Highland Scotland into the 1930s. This longevity enables long-term analysis of land-use patterns, availability of key resources and, of greatest value, environmental change across a millennium. For early medieval historians familiar with the identically purposed Carolingian *mansus* or the Anglo-Saxon *hide*, it may seem unremarkable that these *dabhaichean* were components in an apparently planned or systematised division of the land. Their function was to give agricultural communities apportioned access to a suite of key resources: arable land, summer and winter grazing, water, fuel and building materials. The introduction of such a system underscores the now widely recognised level of political, social and fiscal sophistication in Pictish Fortriu and its successor state Alba. Through the *dabhach*, we can see an over-arching authority exercising effective political and legal domination over populations distributed from the fertile and heavily settled lowlands to the thinly peopled remote upland districts of central and north-west Highland Scotland.

Research into the suites of resources intrinsic to the *dabhach* exposed commonalities in the way each was structured to ensure provision of adequate access to the different environmental niches that supplied those resources. Earlier discussion had dwelt heavily on the arable element within the unit, regarding it as the *sine qua non* of a *dabhach*'s function. What such discussion overlooked, however, was the 'competent quantity of mountain and grazing ground' that later medieval charters reveal was also attached to them.[64] Recognition of that broader physical and environmental range overturned the traditional arable prescription: since it had no exclusive link to grain – whether as seed, yield or render – arguments which focused on measured extents were insupportable, being founded on the devolving premise of a yield of a specific volume being the produce from cultivated ground of a fixed area, an idea rooted in a grain-based hypothesis. Systematic mapping of *dabhaichean* in northern Scotland by the late Alasdair Ross confirmed that they were of widely varying extent, partly determined by topography, altitude and predominant landscape character, that they frequently comprised non-contiguous parts so as to ensure access to the resources necessary to

62 A. Ross, 'The *dabhach* in Moray: a new look at an old tub', in A. Woolf (ed.), *Landscape and Environment in Dark Age Scotland* (St Andrews, 2006), pp. 57–74.

63 The argument is set out in detail in A. Ross, *Land Assessment and Lordship in Medieval Northern Scotland* (Turnhout, 2015).

64 Ibid., pp. 14–15.

The Sneck, Cairngorm Mountains, looking north into Garbh Choire and the Slochd Mor, Aberdeenshire. These tributary glens of Stratha'an, in the south-western part of the Forest of Stratha'an, provided summer grazing for the *dabhaichean* of the Moray coastlands.

sustain a community expressed as a number of households, and that the component elements had roughly equivalent carrying capacities for livestock.[65] This broad suite of equivalent capacities convinced him that the *dabhaichean* originated as the planned base unit in a common assessment system that served both fiscal and military purposes, generating revenues in kind for a superior authority and providing a fixed number of appropriately equipped men in times of war.

How access to and exploitation of key resources was allocated and managed within the *dabhaichean* displays great planning sophistication and an intimate understanding of variations in local environmental conditions. It was a mechanism that provided adequate resources for both summer and winter grazing for agreed numbers of livestock and, contrary to a deeply entrenched modern historiographical convention according to which hay production was absent from much of pre-Improvement Scotland, access also to meadowland for provision of a hay crop for supplementary winter fodder. Few

65 Ibid., chapters 2–3.

dabhaichean were compact and consolidated, containing a full suite of necessary resources within a single block. Instead – and as best illustrated in the *dabhaichean* located in the upper reaches of Strathspey and Stratha'an – they comprised scattered blocks of land in different environmental zones, from valley bottom to mountain plateau, dispersed through a landscape of similar blocks allocated to other *dabhaichean*. While this subdivision of the land could have arisen from countless intercommunal negotiations, that it was a system established and overseen by an overarching authority is evident in arrangements made to provide summer grazing for lowland, primarily coastal *dabhaichean* in distant upland areas remote from the core *dabhach* elements in the lowlands.[66] For example, the people of the *dabhaichean* of the Moray coastal district – the so-called Laich, rich in arable, fishing rights and winter grazing but poor in summer grass resources – took their livestock over thirty miles inland to the Forest of Stratha'an in the northern valleys of the Cairngorm Mountains. While they no doubt built on older foundations, perhaps similar to the Pictish estates that seem to be present at sites like Old Scatness, Pool or the Udal and Bornais, the distribution of a pattern of *dabhaichean* across the whole of the kingdom of Alba from the Tay to Easter Ross suggests a planned restructuring organised by the powerful early tenth-century rulers of that kingdom. These rulers and their provincial leaders were at the forefront of the conflict with the Norse that had seen the demise of the old kingdom of Fortriu and needed to direct the resources of their realm – human and natural – towards ending that threat and asserting the independence of their kingship in the north of Britain. Coming towards the end of an era of profound social, political and especially environmental upheaval,the *dabhach* system also presented them with an opportunity to reconfigure the fundamental organisation of their territories at the local level and distribute access to its resources that would give local communities the future resilience and sustainability needed to actualise royal power.

Can we see evidence of such external direction of settlement patterns and resource allocation in action? At present, the only large-scale study to have provided some possible evidence for organisational direction and external decision-making concerning land-use choices is that at Lair in Glen Shee in northern Perthshire. There, the substantial Pitcarmick-type farms that had been established in the later seventh century seem to have been abandoned in the later tenth century, and although cultivation continued on a reduced scale and livestock were still present to an extent, regeneration of scrubby woodland suggests that a major change in the local land-use model had taken place.[67] The intensive use of the upland zone that had been a characteristic of the later seventh-century recolonisation was effectively finished before 1000 and the residual areas of arable cultivation were gone before the end of the eleventh century. Climatic deterioration was not an issue in this period. Indeed, despite greater levels of precipitation in the eleventh century the climate was warming as the MCA developed after the Oort Solar Minimum (see Chapter 6), and, in any case, the arable zones at Lair were on good, free-draining soils and so unlikely to have suffered from waterlogging. Socio-economic trends linked to the creation of hunting reserves are inherently improbable, given the accepted twelfth-century chronology of that cultural introduction.[68] There is, however, greater mileage in the possible reconfiguration of local settlement, with a shift from dispersed to nucleated being a recognised characteristic of the late first millennium AD, and place-names analysis around Lair has highlighted the

66 Ibid., pp. 68, 101, 148.

67 Strachan, Sneddon and Tipping, *Early Medieval Settlement*, pp. 70, 76, 110, 147.

68 Ibid., p. 147.

presence of a number of post-900 'Bal-' names (from Gaelic 'baile', a township), which might reflect movement from the individual Pitcarmick-type farms to a clustered community setting.[69] A trend towards *pro rata* contribution to the expensive and resource-intensive new agricultural technology of the eleventh century, especially the carruca-type plough and its oxen traction team (see below, p. 264), which accelerated nucleation in more southerly districts, does not seem to have been a driver behind this trend. Indeed, the abandonment of the arable land at Lair points rather in a different direction. Here we are perhaps seeing the consequence of the socio-economic reordering of which the *dabhach* was the expression. Shared access to allocated resources would certainly be an encouragement towards nucleation. Reordering of resource use and possible reassignment of rights to land and its resources was not cost-free. It is likely that cessation of arable activity in some zones, regardless of its viability, might have been the price of creating the summer pastures and hill-resource districts for settlements that were deficient in those things when the notional *dabhach* map was drawn up. In Lair, we may be seeing the external imposition of a new order in the landscape. Importantly, it serves as a reminder, at the rise out of the last dip of the pre-MCA climate rollercoaster, that some dramatic changes in land-use which had profound effects on Scotland's environment were made, even in the tenth century, for purely human political and economic reasons.

• • •

In looking at the degree and nature of change that occurred in the two centuries to AD 1000, we find echoes of what modern Scottish society has experienced since the eighteenth century. As discussed in the volume *Where Men No More May Reap or Sow*, what until the comparatively recent past was viewed solely as a human and largely social process driven by economic motives founded on new political philosophies can now be recognised as a response both impelled by and capitalising on changing environmental conditions. Although our evidence base for the ninth and tenth centuries is considerably less substantial than that for the eighteenth and nineteenth, the material reviewed here has underscored the need to look beyond the simple, traditional narratives of colonial arrivals bringing new cultural and technological practices. In almost every case from Shetland to Galloway in this period we see displays of resilience and adaptation by both newcomer and native, with the former overlaying their practices onto the pattern of inherited indigenous regimes and the latter opportunistically reframing exploitation regimes to deliver the resources necessary to meet new political challenges. Certainly, the colonists brought new methods, especially for the more intensive exploitation of marine resources, and integrated their colonies into an expanding network of North Atlantic and north-western European trade and exchange. But they also adapted their expectations to what they encountered in the lands they took in the Northern and Western Isles, responding to the limits and opportunities presented by the significantly different environmental contexts in which they found themselves. Importantly, they continued to adapt and adopt as the shifting climate brought new weather patterns that favoured alternative agricultural and fishing strategies, retaining those aspects of the indigenous regimes that they could see worked in the particular contexts of their new homes. Colonisation here did not mean a rupture with the past and unthinking efforts to replicate their homeland experience in the Scottish islands.

69 Ibid., p. 16. For the trend towards nucleation, see P. Dixon, 'Rural settlement patterns on the ground', in K.J. Stringer and A.J.L. Winchester (eds), *Northern England and Southern Scotland in the Central Middle Ages* (Woodbridge, 2017), pp. 237–271.

What is also clear is that indigenous Scottish cultures did not stand still in the face of changing times. Traditional environmental knowledge is only as good as the stability of the climate and weather regime in which established practices first delivered the desired results. While some of the key organisational changes, like the development of the *dabhach* as a fiscal unit, were top-down processes designed to provide the manpower and material that the rulers of the emerging Scottish state required both to consolidate and then expand their personal authority and to meet the intensified external threats of that turbulent era, they were at root reactions to the changing availability of key resources that climate, weather and past human exploitation patterns had shaped. Fundamentally, the *dabhach* was a response to the environmental conditions of the time, one that accommodated the different spatial availability of certain resources into its planning and assigned access to essential elements for those in areas where past overexploitation or prevalent natural conditions presented no opportunities to acquire them. The success of both the hybridised regimes that developed in the areas of most intensive Scandinavian colonisation and the new structures for the better control of the human and natural resources of the kingdom of Alba ensured that they constituted the bedrock of social organisation and environmental exploitation across most of Scotland for centuries to come.

CHAPTER 5

Beasts of the Wild Wood, Mountain and Sea Shore

I saw the Hurcheon, and the Hair,
Quha fed amangis the flowris fair,
Wer happing to and fro:
I saw the Cunning, and the Cat,
Quhais downis with the dew was wat,
With monie beistis mo:
The Hart, the Hynd, the Dae, the Rae,
The Fowmart, and the Foxe
War skowping all fra brae to brae,
Amang the water broxe.
Sum feiding, sum dreiding,
In cais of suddain snairis,
With skipping, and tripping,
Thay hantit all in pairis.[1]

To most people nowadays, Scotland appears alive with wildlife and is still rightly celebrated as being home to a wide diversity of wild species. It is seen especially as a home to animals and birds found in the wild nowhere else in the British Isles. Many are surprised when told that what we see today is heavily degraded and lacks the kind of biodiversity that could be expected in a country at this latitude and in the type of environmental range it occupies at the interface between oceanic and continental regimes. Losses have been huge, principally in the varieties of invertebrates and plants that have disappeared or been pushed to the brink of oblivion by intensive land management regimes that originated in relatively recent times, but also in the higher-order mammalian species that have been driven to extinction over the last 10,000 to 12,000 years. Humans have played a major part in most of those extinctions or near-extinctions, through over-hunting, deliberate policies of extermination, habitat destruction and,

1 From Alexander Montgomerie, 'The cherie and the slae', in *The Scottish Collection of Verse*, p. 63.

more recently, chemical pollution and pesticide use. But losses are not solely a consequence of past anthropogenic (in)action, and declines to critical levels and beyond have also been caused by long cycles of climate change. These cycles have brought wider environmental change which has eliminated, and is still transforming, the distinctive ecological niches occupied by some highly specialist species, progressively narrowing the habitat ranges required for maintaining stable populations, or reducing genetic diversity to the point where reproductive viability has been threatened. In other cases, Scotland's unique location and patterns of environmental development have created isolated populations whose exposure and consequently resistance to exotic pathogens has been fatally compromised. These trends are explored in depth in the companion volumes *Where Men No More May Reap or Sow* and *Standing on the Edge of Being*, where they are discussed against the backdrops of systemic change in landscape organisation and management and wider patterns of environmental degradation through industrialisation and pollution as well as overexploitation for grazing and hunting, and are set against the accelerating pace of global heating. Here, it is the evidence for greater species diversity and large mammalian predators especially that will be explored, alongside the record of exploitation of a broad range of avian and aquatic species.

Wild grazers and browsers

Notwithstanding those thousands of years of human and environmental inroads on biodiversity, Scotland in the early Middle Ages still abounded with wildlife. Some of the larger herbivore species that had been present in the Mesolithic and in some northern districts into the Neolithic, like giant elk (*Megaloceros giganteus*), elk (*Alces alces*) or reindeer (*Rangifer talandus*), were already long extinct through a combination of post-glacial climate-change-related habitat loss and human hunting pressures.[2] Nevertheless, sufficient numbers of large wild herbivores remained to contribute to the continuous reshaping of the woodlands through their grazing habits. Between dietary preference for some plant and tree species and avoidance of others, and the disturbance of the ground by hooves, antlers or horns, their presence was a central contributor to the shaping of the whole structure of woodland biota, from invertebrates and small mammals to the constitution of woodland floors, understorey vegetation and even the high canopy of the trees.

Of the remaining large herbivores that survived into the Early Historic era, red deer (*Cervus elaphus*), roe deer (*Capreolus capreolus*) and wild pigs (*Sus scrofa*) were the most common. They were widely hunted for their meat, skins, antlers or ivory, but were possibly already seen as the preserve of a social leadership group for the sport they gave. As the Pictish sculpture of the early seventh and eighth centuries illustrates for their descendants, red deer were probably prized for hunting by men of high status in the late Iron Age. Late fourth-century references to large hunting hounds from Ireland owned by the Roman quaestor Quintus Aurelius Symmachus, who used them for fighting lions and bears, might identify ancestral dogs of the type later shown on the Pictish stones in mounted chases aided by hounds.[3] Rather than being the quarry of huntsmen on horseback, boar are shown being hunted on foot by men with bows or

2 The species diversity and dates of extinction are explored in D. Yalden, *The History of British Mammals* (London, 1999).

3 *Arrian on Coursing: The Cynegeticus of the Younger Xenophon, Translated from the Greek, with Classical and Practical Annotations, and a Brief Sketch of the Life and Writings of the Author: To Which Is Added an Appendix, Containing Some Account of the Canes venatici of Classical Antiquity* (London, 1831), pp. 11, 296–297; F. McCormick, 'The dog in prehistoric and early Christian Ireland', *Archaeology Ireland*, 5:4 (1991), pp. 7–9.

Reindeer on Braeriach, Cairngorm Mountains, Highland. Descendants of the first reindeer to be reintroduced to Scotland in 1952 after around 8,000 years of absence, the Cairngorm herd offers a tiny glimpse of a species that roamed early post-glacial Scotland in their thousands.

crossbows. Red deer remained common across all of mainland Scotland, the Western Islands, the Inner Hebrides and Orkney[4] – and remain in very high numbers in most of these areas today, with roe deer almost ubiquitous in mainland districts – while wild pigs survived as a game species certainly into the later twelfth century in southern Scotland and possibly into the later thirteenth century in some northern areas, but were perhaps effectively extinct before the fourteenth century. The decline of wild pigs, as discussed in Chapter 7, was perhaps linked most closely to the contraction of the oak woodland they preferred, where they would already have encountered competition from human swineherds who wished to fatten their animals on acorns in autumn and had no wish for their herds to be tainted by interbreeding with non-domesticated pigs. Red deer, by way of contrast, made a transition from woodland to open-country animals, as seen especially from later prehistory in the Western Isles, where they survived in apparently sustainable numbers despite the effective disappearance of extensive woodland. Roes also

4 For the survival of red deer into the medieval period in Orkney, see J.M. Bond, 'The mammal bone', in J. Hunter with J.M. Bond and A.N. Smith, *Investigations in Sanday, Orkney*, volume 1: *Excavations at Pool, Sanday: A Multi-period Settlement from Neolithic to Late Norse Times* (Kirkwall, 2007), pp. 207–242 at 214.

Aberlemno roadside cross-slab, Aberlemno, Angus. The central panel on the reverse face shows a hunting scene with the riders' hounds bringing down a stag. Dogs like these may have been the prized animals brought to Rome to fight lions and bears in the Coliseum in the fourth century.

made that transition, perhaps aided in more lowland mainland areas by the retention of managed woodland for timber, which provided them with shelter.

Although there are suggestions of herds of giant wild cattle surviving in forested areas of Scotland into the later Middle Ages, and John Bellenden's translation of Hector Boece's 1527 description of the denizens of the Great Wood of Caledon referred to white bulls with 'crisp and curland mane, like feirs lionis', the giant prehistoric aurochs (*Bos primigenius*) is likely to have become extinct in the early Bronze Age, some two millennia before the arrival of the Romans.[5] Wild cattle, smaller than the aurochs but larger than the domesticated breeds that existed in the late Iron Age, survived and perhaps persisted in semi-domesticated herds to become the ancient

5 H. Boece, *The Chronicles of Scotland by Hector Boece, Translated into Scots by John Bellenden, 1531*, Oxford Text Archive, https://ota.bodleian.ox.ac.uk/repository/xmlui/bitstream/handle/20.500.12024/A16292/A16292.html; A.C. Kitchener and J. Doune, 'A record of the aurochs, *Bos primigenius*, from Morayshire', *Glasgow Naturalist*, 25:4 (2012), https://www.glasgownaturalhistory.org.uk/gn25_4/kitchener_1.pdf. For discussion of dating, see D.W. Yalden and A.C. Kitchener, 'History of the fauna', in S. Harris and D.W. Yalden (eds), *Mammals of the British Isles: Handbook*, 4th edition (Southampton, 2008), pp. 17–31. There are possible archaeological finds of aurochs bones from fourth-century-ad contexts in North Wales and from Roman forts along the Hadrianic frontier line, but these are incomplete skeletons and more probably represent portions of large, domesticated bull carcases rather than of wild cattle.

Bull sculpture from Burghead, Moray, now in the National Museum of Scotland, Edinburgh. The bull sculptures from various locations in the coastlands of the Moray Firth are perhaps representations of those white bulls with 'crisp and curland mane, like feirs lionis' that the sixteenth-century historian Hector Boece identified as denizens of the mythical 'Great Wood'.

white cattle variety that once existed in the wooded parkland at Cadzow in Lanarkshire and can still be seen at Chillingham in Northumberland. It is probably these animals that Boece referred to in the 1520s and that Bishop John Leslie claimed had been widespread in Scottish forests throughout the medieval period but, through over-hunting, had been reduced by the later sixteenth century to three herds, one in the royal park at Stirling, a second in the woods around Cumbernauld and a third at Kincardine.[6] A Privy Council case of 1570–1 refers to the final extermination of the white cattle at Cumbernauld during the ravaging of the hunting forest of Cumbernauld in the recent civil war. As woodland browsers, such animals are likely to have helped to shape a structure of open glades and denser stands of trees, akin to the parkland of eighteenth- and nineteenth-century mansions. The persistent early post-medieval traditions regarding their former ubiquity could indicate that they were numerous

6 J. Leslie, *The Historie of Scotland, Written First in Latin by the Most Reuerend and Worthy Jhone Leslie Bishop of Rosse and Translated in Scottish by Father James Dalrymple*, ed. E.G. Cody and W. Murison, 2 volumes (Edinburgh, 1888–95), volume 1, p. 30. A herd of ten white cattle (nine cows and one bull) are recorded as being bought for the King's Park at Stirling in 1509: G. Burnett and A.J.G. Mackay (eds), *The Exchequer Rolls of Scotland*, volume 13: *1508–1513* (Edinburgh, 1891), p. 60.

until well into the Historic era, yet they remain a species whose presence has to date barely registered in archaeological or historical discussion, or in the proposals of restoration ecologists.

A further herbivore with a likely widespread distribution across Scotland in the early centuries AD is the beaver (*Castor fiber*).[7] Before they were hunted to extinction between the twelfth century (when the Welsh population disappeared) and somewhere from the sixteenth to the eighteenth century in Scotland, beavers were established in most of the major river catchments, where abundant river-edge woodland on alluvial silts provided an ideal habitat for these medium-sized mammals. Beaver bones of Mesolithic date have been found at sites from Berwickshire to central Perthshire, and while bone evidence from later than around the third century AD is lacking, they are described as still quite common in the waterways of the Great Glen into the early 1500s. Their dam building and preferential dietary selection of tree species will have been most evident on slower-moving rivers and streams, where their activities will have stimulated fast-growing shoots of the trees whose bark, twigs and leaves were most favoured. Some populations might have become extinct earliest in areas where anthropogenic woodland clearance was most advanced by the Iron Age, but in the birch-, hazel-, willow- and alder-rich Highlands they are likely to have remained numerous through the first millennium AD.

Predatory rivals

Before the numbers and diversity of large herbivores began to be affected by human activities, herbivore populations were to a degree regulated by the presence of large numbers of apex animal predators. The truly large predatory carnivore species had probably been reduced to wolves (*Canis lupus*) during prehistory, with the smaller carnivores headed by foxes (*Vulpes vulpes*) and wildcats (*Felis silvestris*). Although lynx (*Lynx lynx*) had been present since the Mesolithic period, when they probably crossed the land bridge that still linked what became southern England to the European mainland in pursuit of their favoured prey, roe deer, they appear never to have been common and might just have survived in some remoter districts into the early centuries AD.[8] Another marginal survivor might have been the brown bear (*Ursus arctos arctos*), which had also been present in the post-glacial fauna population but which seems to have been hunted to near-extinction before the end of the pre-Roman Iron Age and is commonly said to have disappeared shortly after the Roman invasion.[9] In the Roman poet Martial's epigrams written in AD 80 to celebrate the opening games at the emperor Titus's newly completed Coliseum, reference is made to a man being tied to a rock to be slain by a Caledonian bear.[10] It is possible that this animal had been amongst the exotic plunder sent back to Rome by the advancing armies of the Flavian campaigns in the north. If so, however, it might already have been a great rarity, for there is no subsequent reference to bears from the wilds of Caledonia in the Roman period (see below). Prized for their fur and to a lesser extent their meat, and possibly for the trophy value of their teeth and claws, it is unlikely that Scotland's native bear population was ever large, and it possibly dropped below sustainable levels quickly once the human population began to make inroads into its habitat. More recently, however, that

7 A.C. Kitchener and J.W.H. Conroy, 'The history of the Eurasian Beaver *Castor fiber* in Scotland', *Mammal Review*, 27:2 (1997), pp. 95–108.

8 Yalden, *The History of British Mammals*, p. 165.

9 Ibid., pp. 111–112.

10 A. Amos (ed.), *Martial and the Moderns* (Cambridge, 1858), p. 9.

Lone wolf on a slab from Ardross, Easter Ross, now in Inverness Museum.

traditional view of early disappearance has been questioned, and with a neck bone of a brown bear found in a mid fifth- to late sixth-century context in Yorkshire triggering discussion of an early medieval extinction in England, it has been suggested that the species survived much later in Scotland and possibly into the sixteenth century in parts of the Highlands.[11] The Pictish sculptural depictions discussed later in this chapter suggest that the animal was not unknown in northern Scotland into the eighth century, but the absence of any reference to it in any surviving medieval written source is surely indicative of its effective extinction before the end of the first millennium.

While it is likely that humans and wolves came into competition for resources early in prehistory, the targeted hunting of wolves probably became normal practice when episodes of climatic deterioration such as that of the middle Bronze Age brought extremes of weather that perhaps saw increased wolf predation on domestic flocks and herds. One consequence of the Roman Warm Period might have been that wild game resources remained sufficiently abundant and extreme winters became sufficiently rare

11 H. O'Regan, 'The presence of the brown bear *Ursus arctos* in Holocene Britain: a review of the evidence', *Mammal Review*, 48:4 (2018), pp. 229–244; D. Hetherington, 'A short history of Scotland's lost species 2: the bear', https://www.linkedin.com/pulse/short-history-scotlands-lost-species-2-bear-david-hetherington/.

Wolf on a Pictish sculpted stone of probably sixth-century date, found at Newbigging Leslie and now at Leith Hall, Aberdeenshire.

for wolf depredation on domestic livestock to be minimal, as wolves are likely to shun areas of human settlement unless driven by desperation. Large litters helped to keep pack sizes large – hence the later medieval deliberate targeting of females – and in upland areas extensive (as opposed to intensive) human exploitation of the land left large areas of woodland and open grass or heathland sufficiently empty to encourage the wide-ranging movement of wolf packs. The intensification of land-use that occurred during the late Iron Age and Roman period saw a contraction of that free zone in many areas outside the Highlands and especially in the Southern Uplands, presumably leading to greater levels of collision between human and lupine populations as they competed for the same territory. Even there, however, despite centuries of persecution, wolves were still a present threat to pastoralists in the thirteenth century.[12]

In later centuries, another form of human–animal competition for a sought-after resource – fish – led to dramatic declines in aquatic mammals.[13] In earlier centuries, otters were hunted principally for their skins and it is in respect of these that they generally appear in the surviving medieval record. There is little evidence for them before the 1420s, when Inverness appears as an important centre for a trade in the skins of otters, pine marten, polecats, foxes and deer (as well as cattle hides).[14] The custom payable on the valuable skins exported from the kingdom was set in 1424 at the considerable price of sixpence per bundle of ten,[15] which is probably indicative of the level of demand for what was an increasingly rare commodity. The domination of Inverness in the otterskin trade – Ayr also emerges as an exporting burgh but at a much lower number of skins[16] – is probably a reflection of the greater survival of otters in the coastal and inland waters of less intensively exploited regions of the Highlands and the Galloway uplands by the later Middle Ages, but archaeology confirms their much wider prehistoric and early medieval distribution.[17] The collapse in their numbers after centuries of hunting for their skins, however, came in the later eighteenth and nineteenth centuries when they joined a long list of species branded as 'pests' on sporting estates.

Also on nineteenth- and twentieth-century lists of pests was another aquatic mammal hunted previously for its skin, blubber and meat, the seal. Growing populations of grey (*Halichoerus grypus*) and harbour (*Phoca vitulina*) seals around Scotland's coasts since the late twentieth century have tended to blur memories of earlier low numbers attributable to millennia of hunting for their food and material value and later to culling in response to their perceived competition for prized fish resources and the damage they inflicted on fixed fishing nets and creels. Both species were hunted by Mesolithic people in the Hebrides some 8,000 years ago (as illustrated by butchered bones found in the mainly shell middens on Oronsay), were exploited by the late Iron Age people of South Uist, and were represented in midden material ranging from the Bronze Age to the tenth century AD at Jarlshof in Shetland.[18] The very low presence of seal bones at the apparently high-status Pictish-period site of Old Scatness, just north-west of Jarlshof, might either reflect a social differentiation in the consumption of seal meat – cattle, sheep and pig predominated here – or provide evidence of earlier overexploitation.[19] By the Norse

12 Wolves are explored in more depth in the volume *Where Men No More May Reap or Sow*, since our main surviving evidence for their presence and impact in Scotland dates from later in the medieval and early post-medieval periods.

13 Yalden, *The History of British Mammals*, p. 174.

14 G. Burnett (ed.), *The Exchequer Rolls of Scotland*, volume 4: *1406–1436* (Edinburgh, 1880), pp. 452, 497, 538, 577.

15 *The Records of the Parliaments of Scotland to 1707*, ed. K.M. Brown et al. (St Andrews, 2007–24), http://www.rps.ac.uk/ (hereafter *RPS*), 1424/12.

16 Burnett, *The Exchequer Rolls of Scotland*, vol. 4, p. 451.

17 Yalden, *The History of British Mammals*, pp. 69, 72–73.

18 C. Grigson and P.A. Mellars, 'The mammalian remains from the middens', in P.A. Mellars (ed.), *Excavations on Oronsay* (Edinburgh, 1987), pp. 243–289; J. Cartledge, C. Ingrem, J. Mulville and A. Powell, 'Animal bone', in N. Sharples (ed.), *A Late Iron Age Farmstead in the Outer Hebrides: Excavations at Mound 1, Bornais, South Uist* (Oxford, 2012), pp. 119–121; M.I. Platt, 'The animal bones', in J.R.C. Hamilton (ed.), *Excavations at Jarlshof, Shetland* (Edinburgh, 1956), pp. 212–215.

19 J.E. Cussans and J.M. Bond, 'Mammal bone', in S.J. Dockrill et al., *Excavations at Old Scatness, Shetland*, volume 1: *The Pictish Village and Viking Settlement* (Lerwick, 2010), pp. 132–156 at 135–137.

period in Shetland – and probably representing continuity from the Pictish past – and continuing through the Middle Ages, exploitation of seal herds was controlled by the more important landholders and managed in a manner akin to the management of domestic cattle, their bones occurring with a frequency comparable to those of cattle on some sites.[20] The late seventh-century *Life of St Columba* composed by Adomnán, abbot of Iona, provides evidence for such management by his monastery during Columba's lifetime, which establishes it as probably common practice by the second half of the sixth century and likely of much earlier origin. Adomnán reported a vision of the saint's in which he foresaw a raid on the breeding ground of 'the sea-calves that pertain to us' by a thief named Erc, locating the seals and their pups on Erraid or one of the other small islets off the coast of the Ross of Mull.[21] The abbot's language indicates that the seals were regarded as a managed economic resource in the property portfolio of the estate and their value is further underscored by Erc being given sheep instead of the seals on his capture and release on Columba's orders.

Similar regulation of seal hunting might also have been established around the Scottish coasts, possibly as a response to declining numbers as much as to control access to an in-demand resource. On the east coast, significant seal populations still existed in the Firth of Forth in the twelfth century, with charters granting every seventh seal taken on the shore of the royal estate at Kinghorn in Fife to the monks of Dunfermline suggesting that they were still sufficiently abundant to slaughter in some numbers and were considered a valuable resource whose gift to the abbey would be recognised for its worth and appreciated for the meat, blubber and skins that it yielded.[22] The disappearance of this gift from the list of Dunfermline's rights before the end of the thirteenth century is perhaps an indication of the scale of overexploitation of seals and its consequences, although it must also be recognised that changes in fish migration consequent on climate change might also have affected the local presence of these fish-eating predators. The absence of seal bones from midden assemblages in eastern Scottish burghs, however, suggests either that numbers declined rapidly in the twelfth century when most burghs were founded or that the dietary preferences and cultural traditions of the colonists moving into them did not recognise or demand seal meat. Instead, it was amongst populations like that of Shetland, where the availability of seal ensured that it constituted an important part of the diet of the human population, that its consumption continued longest.

The fate of all of these animal species is linked by a common denominator: how they were perceived or valued by the human population. Sharp declines in number, if not extinctions, were due to the intensity of human exploitation or persecution. Wild mammal species regarded as of high – if not prized – food value, or desired for their skins and furs, were subject to mounting pressure that in many cases led to collapses in the wild population. Where they were not conserved as the exclusive hunting preserve of men of high status, unregulated capture, especially where accompanied by habitat loss as human populations expanded their arable land and managed pasture, brought swift decline. Those animals seen as competitors for resources or who on occasion preyed on domesticated livestock were subject to similarly increasing pressure. Early medieval populations did not recognise that the competition most commonly arose when clearances for agriculture expanded into the hunting territories

20 J. Marttila, 'Resources, production, and trade in the Norse Shetland', *Journal of the North Atlantic*, 29 (2016), pp. 1–20 at 13.

21 *Adomnan's Life of Columba*, pp. 295, 297.

22 C. Innes (ed.), *Registrum de Dunfermelyn* (Edinburgh, 1842), p. 4.

Common seals, Howmore, South Uist. Seals of this species had been hunted in the shallow inshore waters and beaches around the Hebrides since the Mesolithic period. The right to hunt seals was, by the later sixth century AD, a jealously guarded right of territorial lordship.

of carnivorous mammals. Quite simply, these were beasts that were preying on the livestock they needed for their own subsistence. Although the skins of wolves, foxes, martens, polecats and bears had a trade value or social importance, these animals were perceived primarily as a threat and their elimination was a deliberate action aimed at the preservation of some other, more valued commodity. Put simply, if a species could be profitably exploited its survival was desirable, but by no means guaranteed. If it was perceived as having little or no intrinsic value, it was afforded no space within land that was being worked increasingly to support a growing human population.

'A boar of remarkable size': wild predators, hunting and prey in early medieval Scotland

Amongst the various kinds of evidence that can inform an environmental history of later first-millennium Scotland, sculptural representations are perhaps the most neglected. There is perhaps too great a reticence over using this form of evidence, for there must always be a lingering doubt as to whether the depiction represented an experienced contemporary reality, folk memory, or an imagined and perhaps semi-mythical past. Placing it alongside other forms of evidence, mainly archaeological but with some literary and poetic material, allows some

Above. Bear stalking a doe and fawn, from the reverse side of the Drosten Stone, St Vigeans, Arbroath, Angus.

Opposite top. Bear-like animal devouring a human corpse, from the end panel of a recumbent slab of eighth-century date from Meigle, Perth and Kinross.

Opposite bottom. Bear on a slab from Old Scatness, Sumburgh, Shetland.

degree of sifting of the fantastic from the real. The biggest corpus of data comes from Pictish sculpture of the sixth to early ninth centuries, which portrays a rich repertoire of wildlife and elite hunting activity, as well as fantastic and probably mythological beasts. Bear, boar, deer and wolves appear in sculptural depictions, as do crows, eagles, geese, hawks, ravens, salmon and snakes. Perhaps significantly for current calls for their reintroduction as relatively recent losses from the population of native species, lynx seem to be absent, which might suggest that they had already been hunted to extinction before the second half of the first millennium. But also absent are beavers, which we know from other evidence were one of the larger mammals still present in Scottish fresh waters in the Middle Ages.[23] The value of such sculptural representations for the environmental history of Scotland is the evidence they provide for the survival of some of the large apex predators into the medieval period – these are observed animals rather

23 For lynx extinction, see Yalden, *The History of British Mammals*, p. 75. Gerald of Wales, writing in the late 1180s, observed that in his day beavers were entirely absent from Ireland, occurred only in one river in southern Wales, and were already very rare in Scotland: Gerald of Wales, *The History and Topography of Ireland*, trans. J. O'Meara (Harmondsworth, 1982), book 1, chapter 26. Bryony Coles has argued for the possible late eighteenth- or nineteenth-century extinction of beavers as opposed to the traditional medieval date for England and Wales and the sixteenth- or seventeenth-century date for Scotland: B. Coles, *Beavers in Britain's Past* (Oxford, 2006); B. Coles, 'The European beaver', in B. Sykes and T. O'Connor (eds), *Extinctions and Invasions: A Social History of British Fauna* (Oxford, 2010), pp. 104–115. More recent research, however, has supported the early extinction dates. See L. Raye, 'The early extinction date of the beaver (*Castor fiber*) in Britain', *Historical Biology*, 27:8 (2015), pp. 1029–1041.

than artistic impressions gleaned second-hand from manuscript sources – and for human interactions with them, principally as hunters. They do not, however, necessarily record the presence of such animals in the districts where the sculptures are found, either at the date of the carving or at some unknown point prior to that.

Pictish carvings of bears are a case in point, as it is unclear if any survived in northern Britain after the fourth century, although, as mentioned above, recent discovery of part of a bear skeleton from a fifth- to sixth-century context in Yorkshire has sparked new debate over the date of their extinction in the wild.[24] A bear appears amongst a group of animals on the so-called 'Drosten Stone' from St Vigeans in Angus, where it is shown in the distinctive flat-footed walking posture of the species, moving in towards a doe suckling a fawn.[25] The stone in question is from the second quarter of the ninth century and the realistic representation of the bear has been taken to suggest that the sculptor might have seen one in the flesh. It is, however, unlikely that bears were living in the wild in a heavily settled and agriculturally developed area like eastern lowland Scotland at that date. A manuscript source seems more likely to have provided the model for animal representations such as this. Certainly, there is no question that the maned lions that surround Daniel on the well-known panel from Meigle in Perthshire were inspired by anything other than a depiction in a book or some such source. A bear-like creature also appears on the side of a recumbent slab from the late eighth- or early ninth-century church site at Meigle, devouring the dismembered remains of a human corpse.[26] Perhaps the finest bear depiction is that found in the excavations at Old Scatness in Shetland, which, although lacking its upper parts, is one of the most beautifully executed pieces of incised Pictish animal sculpture to survive anywhere in Scotland.[27] Its quality is seen in the well-understood anatomical details of the bear and its walking posture, aspects which the excavator thought might indicate that the sculptor was perhaps somehow acquainted with such animals. It is carved on Shetland sandstone from the vicinity of Old Scatness, but, regardless of whether or not bears were yet extinct in the British mainland, it is highly unlikely that such animals had ever been present in these northern islands and certainly not by the seventh century.[28] Stylistically, the sculptural affinities of the Old Scatness carving are with animal representations in mainland Scottish examples from around the inner reaches of the Moray Firth rather than with the southern Pictish artistic tradition exemplified by the Meigle and St Vigeans examples, rendering it likely that a carver from the heartland of Fortriu had taken his expertise to Shetland. Why a bear was the chosen subject of his work, however, will forever remain unknown and we can only guess at the symbolic meaning that it may have held for his patrons.[29]

Carvings of boar are perhaps best known from the hilltop of Dunadd and from Knocknagael near Inverness, but they also occur on stones from Shandwick in Easter Ross, Meigle in Perthshire and, in a scene apparently showing a hunting Pict using a crossbow to shoot at one, on the reverse of the

24 Yalden, *The History of British Mammals*, pp. 112, 129; O'Regan, 'The presence of the brown bear'.

25 J.R. Allen and J. Anderson, *The Early Christian Monuments of Scotland: A Classified Illustrated Descriptive List of the Monuments with an Analysis of Their Symbolism and Ornamentation* (Edinburgh, 1903), p. 236.

26 Ibid., p. 304.

27 Dockrill et al., *Excavations at Old Scatness*, vol. 1, p. 366.

28 For discussion of bear extinction in Britain, see Yalden, *The History of British Mammals*, p. 112, and the contrary view in O'Regan, 'The presence of the brown bear'.

29 Dockrill et al., *Excavations at Old Scatness*, vol. 1, pp. 366–367.

Pictish huntsman shooting a wild boar with a crossbow, from the reverse panel of the Drosten Stone, St Vigeans, Arbroath.

Drosten Stone at St Vigeans.[30] Literary references to boar in this period are also a sign of the ubiquity of boar distribution across the mainland and larger islands. The poem *Y Gododdin*, attributed to the possibly sixth-century Lothian-based Brythonic poet Aneirin, speaks of one warrior charging like 'the tearing onset of the woodland boar' and of another who rushed impetuously like a boar to mount for battle, and elsewhere applies the name 'the Boar' to a despised and 'haughty' messenger of the Britons' Anglian enemies.[31] Set in the British kingdoms of what is now Lothian, the Scottish Borders and Northumbria, the poem uses language that suggests the poet's personal familiarity with the aggression of the hunted boar in the woodland of the Southern Uplands. Adomnán's account of Columba's encounter with 'a boar of remarkable size' in a dense wood on Skye confirms both the existence of these woodland denizens even on the inner Hebridean islands towards the close of the sixth century and the hunting of them for sport by men with teams of dogs.[32] The mid twelfth-century B version of the St Andrews foundation legend and the account of the foundation of the Augustinian priory there, composed by its first prior, Robert, references material found 'written in old books of the Picts'. These 'old books' detailed the gift of the royal demesne bounded by 'Cursus Apri' (the Boar's Raik or Chase) made by the monastery's

30 A. Lane and E. Campbell, *Dunadd: An Early Dalriadic Capital* (Oxford, 2000), pp. 19–22; J.M. Gilbert, 'Crossbows on Pictish stones', *Proceedings of the Society of Antiquaries of Scotland*, 107 (1978), pp. 316–317; Allen and Anderson, *The Early Christian Monuments*, pp. 73 fig. 69, 103–104.

31 *Y Gododin*, trans. J. Williams (London, 1852), verses 30, 39, 69, 89.

32 *Adomnan's Life of Columba*, p. 385.

The missing lynx

Our known corpus of Pictish sculpture includes a wealth of non-human animal and bird representations, as well as a remarkable array of fantastic beasts. The animals comprise those likely to have been encountered in domestic contexts and in the wild in Early Historic Scotland but also some which can only have been known from manuscript illustrations or representations on portable objects. There are horses, both wild and domesticated, ponies and what seem to be mules (given their donkey-like tails), bulls, cows and oxen, boars and sows. Dogs are regularly shown, often as part of a hunt scene and usually depicted as long-limbed hounds. Fish, clearly salmon from the form of the fins, and serpents are also fairly common. The wild animals, depicted singly and in groups, shown either as part of hunting scenes or as isolated images, include 'game' such as wild boar, red deer stags, does and fawns. Wild bulls should probably be included in the game category, although there has to be a question mark against the single identification of a bull in a hunt scene (from St Donnan's church on Eigg). Birds, game or domesticated, are less readily identifiable, with geese being the only unmistakable species and probably representing migratory varieties. It is a rich repertoire.

Large mammals and birds, mainly predators, form another important group of the non-human representations. There are what may be otters, finned or flippered 'Pictish beasts' or 'swimming elephants' that could be porpoises or seals, if they are not purely mythical, crows and eagles. The most striking carvings, however, are of bears and wolves, like the magnificent creatures at Old Scatness, Meigle, Ardross and Leith Hall. These animals are all likely to have been known to the inhabitants of fifth- to ninth-century northern Scotland, but others were clearly not denizens of the British Isles in the Historic (or any other) era. Lions feature prominently, usually as part of a biblical scene (Daniel in the lions' den or David slaying the lion), and are clearly not creatures with which the sculptors were familiar. The known examples are probably derived from Late Antique manuscript images, including Gospel Books.

Lions, however, are the only large cats present in the corpus of animal representations. There is nothing that looks remotely like a wild cat and, perhaps significantly, there are none that can be identified as a lynx. This absence might be an important piece of evidence for when this largest of our native wild cats either ceased to be present in numbers that made them visible to the people of Pictland or became altogether extinct. With our earliest recognisable animal sculptures dated currently to the sixth century, the non-presence of the lynx in any recognisable form is strongly suggestive that it was already absent in the wild and unfamiliar even in popular tradition, from which sculptors might have drawn inspiration. That absence is also evident in our early literature and in orally transmitted tales, which together with its invisibility in the visual arts makes it likely that lynx had disappeared from what we now know as Scotland by the fifth century AD at the latest, and possibly by much earlier still. Their likely extinction by between 1,500 and 2,000 years ago should not be used as an argument against their reintroduction, but if any large predator is to be considered it should perhaps be the wolf, which was finally hunted to extinction only at the end of the seventeenth century.

Meigle, Perth and Kinross. Section of Pictish sculpture showing in its middle section Daniel in the lions' den, from the Book of Daniel, chapter 6. The standing figure of Daniel is flanked by four lions with manes and curling tails in a highly stylised representation derived from manuscript illuminations.

eighth-century Pictish founder, King Onuist, son of Urguist.[33] This extensive district, approximately 17.5 x 9.6km and encompassing much of what became the medieval parishes of St Andrews and St Leonards, Dunino and Kemback, possibly took its name from boar hunts based on the royal residence at Cennrigmonaid which Onuist gave as the founding endowment of the new monastery.[34] It is perhaps in this context that we should understand a twelfth-century gloss in the version B text, which tells us that another former name for the monastery site was 'Mucros', which is explained as the 'grove of the pigs'.[35] From the eighth century at least, it seems that the St Andrews district was prized by Pictish and Scottish rulers as prime boar hunting territory.

Deer hunting scenes are more common on sculptures from right across the former Pictish territories, and, through their depictions of mounted hunters accompanied by hounds pursuing stags and does, emphasise that this was already an elite activity in Pictish society.[36] The size of the animals depicted suggests that these are red deer as opposed to roe, although it is impossible to be categorically certain of this except where the stags are shown with the multi-branched antlers of the adult male *Cervus elaphus*.[37] The famous Hilton of Cadboll stone from Easter Ross shows two riders with spears pursuing an already wounded doe that is being brought down by two long, rangy-looking dogs of probably deer hound type.[38] The Aberlemno roadside cross-slab presents a cavalcade of rivers accompanied by hounds and pursuing three female deer, one of which is being brought down by two dogs while a third dog is leaping at the neck of another.[39] On the Kirriemuir No. 2 stone, a stag is being pursued by a spear-wielding rider while his hound has its jaws fixed on the stag's haunch.[40] A complex hunting scene on the granite slab now in Elgin Cathedral in Moray shows a rider with what appears to be a falcon on his wrist and accompanied by two hounds along with a further rider whose hound is felling a stag.[41] Together with the fine carvings of unaccompanied deer, like the especially graceful example of a stag from Grantown-on-Spey and the family group from St Vigeans,[42] the ubiquity of these creatures in Pictish

Opposite. Hunting scene from the reverse side of the Hilton of Cadboll cross-slab, now in the National Museum of Scotland, Edinburgh. Here, we see what later in the Middle Ages would become the reserved sport of royalty and their favourites already depicted as an elite sport in the eighth century.

33 S. Taylor, 'The coming of the Augustinians to St Andrews and version B of the St Andrews foundation legend', in S. Taylor (ed.), *Kings, Clerics and Chronicles in Scotland, 500–1297* (Dublin, 2000), pp. 115–123 at 118 map 8.2, 119, 122–123; Fraser, *From Caledonia to Pictland*, pp. 361–362; Woolf, *From Pictland to Alba*, pp. 89–90.

34 M.O. Anderson, 'The Celtic church in Kinrimund', in D. McRoberts (ed.), *The Medieval Church of St Andrews* (Glasgow, 1976), pp. 1–10 at 7. For the refinement of the boundary, see Taylor, 'The coming of the Augustinians', p. 122 note 13.

35 Taylor, 'The coming of the Augustinians', p. 115.

36 Allen and Anderson, *The Early Christian Monuments*, p. 73 fig. 69, shows a kneeling crossbowman firing at a stag.

37 As on the fragment from Burghead, where two large hounds are attacking a stag with seven tine antlers: ibid., pp. 137–138.

38 Ibid., pp. 61–62.

39 Ibid., pp. 214–215.

40 Royal Commission on the Ancient and Historical Monuments of Scotland, *Early Medieval Sculpture in Angus Council*, RCAHMS Broadsheet 11 (Edinburgh, 2003). There is a similar hunting scene on the No. 3 stone from Inchbrayock near Montrose: Allen and Anderson, *The Early Christian Monuments*, p. 255.

41 Allen and Anderson, *The Early Christian Monuments*, pp. 134–135.

42 Ibid., pp. 126, 269.

carving underscores their widespread presence throughout even the more heavily settled, lowland districts of north-eastern Scotland and not simply the upland zone traditionally associated with red deer.

Totems and symbols

While the Old Scatness bear tells us little about the fauna of seventh-century Shetland and probably more about pre-Christian Pictish cosmology, other pieces of Pictish animal sculpture are perhaps more valuable in illustrating important aspects of the local wildlife and its environmental setting. The symbolic and mythological significance of particular animals in most Celtic cultures is well known, like that of the Picts, and represented concepts such as universal knowledge (the salmon) or death and rebirth (the skin-shedding snake). Although these creatures appear amidst a repertoire of clearly fantastic beasts, the detail of their representation points to the familiarity of the sculptors with these commonly occurring animals. It is difficult not to see the beautifully carved slab from Easterton of Roseisle in Moray depicting a backward-looking goose and a salmon, for example, as a nod towards the migratory wildfowl in the now drained Loch of Roseisle that once occupied the low ground beneath the stone's findspot and the abundance of salmon in the nearby rivers that debouched into the Moray Firth.[43] A backward-looking goose also occurs on a standing stone near Blackford in Perthshire, in another district still noted for the flocks of overwintering migratory geese to be found in its wetlands. Eagles on the slabs from Brough of Birsay and Knowe of Burrian in Orkney are probably sea eagles of the kind once common in the Northern Isles and whose remains were found in the Neolithic chambered tomb at Isbister on South Ronaldsay.[44] At more inland and upland Strathpeffer in Easter Ross, or Fyvie and Tyrie in Aberdeenshire, the carved eagles are perhaps more likely to be golden eagles.[45] Non-food species like these should perhaps be considered as totemic, but, again, the detail in their many representations points to a personal familiarity with the physicality of the birds and their plumage.

Opposite. Red deer stag, Pictish sculpture from Grantown-on-Spey, now in the National Museum of Scotland, Edinburgh.

The appearance of wolves in sculptures across most regions is one of our surest indicators of the ubiquity of this apex predator in the Pictish parts of Scotland. They are carved most commonly as single animals, as on the slabs from Ardross near Alness in Easter Ross, Newbigging Leslie (now at Leith Hall) in Aberdeenshire and Hill of Keillor in Angus,[46] never appearing in any known representation as a pack. Although the sculptures do not depict wolf hunts, poetic evidence exists for this practice elsewhere in early medieval Scotland. In the *Gododdin*, for example, there is a warning against tackling an enemy unprepared which uses a wolf hunt metaphor: 'He who grasps the mane of a wolf, without a club in his hand, will have it gorgeously emblazoned on his robe', meaning that the man would be mauled.[47] Here, we perhaps have a brief glimpse of the close physical nature of the final confrontation on foot between the two chief predators of the early Middle

43 Allen and Anderson, *The Early Christian Monuments*, pp. 124–126 fig. 130.

44 J.N.G. Ritchie, 'Two new Pictish symbol stones from Orkney', *Proceedings of the Society of Antiquaries of Scotland*, 101 (1968–9), pp. 130–133 no. 5.

45 Allen and Anderson, *The Early Christian Monuments*, pp. 59–60, 165–166, 187–188.

46 Ibid., pp. 55–56, 177–178, 207–208.

47 *Y Gododdin*, verse 41.

Opposite. Backward-looking goose and salmon, Pictish sculpture from Easterton of Roseisle, Moray, now in the National Museum of Scotland, Edinburgh.

Above. The so-called 'Craw Stane' at Strathpeffer, Easter Ross, Highland, depicts what is probably a golden eagle.

Ages. The use of a club is also perhaps illustrative of the hunting of the animal for its pelt, as it would inflict less damage than a bladed weapon. The *Gododdin* also records the use of wolf skins in clothing in a reference to the garb of a warrior called Tinogad, 'which was of divers colours, made of the speckled skins of young wolves'.[48] Pelts from wolves continued to be listed amongst customable Scottish exports into the seventeenth century, speaking to their value.

• • •

Taken together, our sculptural and literary sources provide a rich record of the wildlife favoured especially as game that was present in Scotland in the second half of the first millennium AD and of human conflict and interaction with those species. We can never know for certain if some of the animals represented, such as the bears, were drawn from life or were part of folk memory and self-identity, preserved in sculptural form long after the species had become extinct in Scotland. Were such images culled from foreign-produced visual sources? But most of the carvings are of larger game animals or birds and the predators with which the human populations were in competition through both the grim days of the LALIA and the improving weather of the ninth century. These were usually woodland denizens or animals of the upland moors, but wetland wildfowl also figure prominently and the falcon-carrying rider from Shandwick speaks to the pursuit of game birds over heather moorland or open grassland. We have evidence of many different species being hunted for sport, usually as the preserve of the aristocratic warrior elites, favoured for their meat which would be consumed as part of early medieval feasting culture. Solitary huntsmen with bows or crossbows, however, suggest that not all huntsmen conformed to the spear-wielding horseman motif and perhaps point to a difference between hunting for pleasure and hunting for subsistence. Skins and pelts were also prized and we should perhaps consider how the killing of a wolf, stag or boar might have been regarded as a rite of passage in which such trophies might have been perceived as essential tokens of the transition into adulthood sought by young men as they competed for status amongst their peers and for recognition of their worth by their lord. These social and cultural themes become more prominent in our richer documentary record of more recent centuries, but the sculptural and poetic renderings of this earlier age provide a window onto a little-explored dimension of environmental interactions, challenges and opportunities across an era of profound transformation.

Opposite. Serpent, salmon and mirror symbol on the reverse side of the cross-slab in the Manse Garden, Glamis, Angus.

48 Ibid., verse 90.

sita est sup ripam fluminis twede. in loco

CHAPTER 6

An Era of Increase and Abundance? Scotland *c.*1000–1256

He always produced abundance of the ripe fruit of good deeds.[1]

A persistent theme in the history of medieval Scotland is that of a lost golden age. Most traditional narratives describe this as an era of peace, prosperity and abundance, usually constructing the image as a stark contrast to the troubled times that lay ahead, times of political instability, war, famine and disease. It is a striking image of a land transitioning from light into darkness, pivoting around one tragic moment, the untimely accidental death of its ruler, King Alexander III, which removed a strong, guiding hand and left Scotland facing a void of uncertainty and a downward spiral into chaos and near oblivion. As we shall explore in Chapter 9, that traditional view is open to challenge in many fundamental respects, for the later medieval presentation of Alexander's reign as a time of fruitful summers and bountiful harvests is not borne out by any record, either written accounts or climate proxies. What these sources do reveal, however, is that the amelioration of climatic conditions that had begun to become established in the tenth century settled into a long episode in which more benign weather prevailed, a period once referred to as the Medieval Warm Period and now more commonly as the Medieval Climate Anomaly (MCA). Although all forms of record evidence show this era to have been punctuated by episodic extreme events, both severe winters and poor, rain-drenched summers that resulted in harvest failures and dearth, the generally propitious weather was sustained from the later eleventh through to the mid thirteenth century and led to favourable conditions that supported population growth and economic prosperity. It is the record for this climatic shift and the weather it delivered that is the focus of this present

Opposite. Miniature portraits of King David I (1124–53) and his grandson King Máel Coluim IV (1153–65), depicted enthroned under the arches of the capital 'M' of 'Malcolmus', from the charter issued to Kelso Abbey in 1159. Throughout the earlier Middle Ages, the prosperity, fertility and peacefulness of a kingdom was often seen as God's judgement on its ruler. In the case of David I, a godly king presided over an era of unprecedented economic growth, in which agricultural expansion was aided by the spread of new agricultural technologies, rising population levels and intensifying pressure on Scotland's natural resources. (Image reproduced from the facsimile in C. Innes (ed.), *Liber S. Marie de Calchou*, volume 1 (Edinburgh, 1846).)

1 From a description of the virtues of King David I, in Bower, *Scotichronicon*, vol. 3, p. 127.

chapter. It ends in 1256, however, rather than at the traditional 1286 date, for the MCA was already in decline at the start of Alexander III's reign. Rather than representing the apogee of the 'golden age', Alexander's Scotland was basking in the afterglow, spending the golden inheritance and sliding towards a more uncertain future.

Framing the narrative

Modern accounts of Scotland's history across the two and a half centuries after 1000 that span the MCA have been dominated not by discussions of the environmental trends that enabled growth but by a focus on political and cultural themes, woven into the idea of the 'making of the kingdom'.[2] From the early nineteenth century, the central narrative was one of state-building and transformation, of violent but visionary kings who welded together the petty kingdoms that jostled for supremacy within what became Scotland to create a realm that ultimately emerged as one of only two fully fledged states in the British Isles.[3] And it is a remarkable story, in which the colour and excitement of warfare and dynastic challenge vies with the often drabber but equally compelling processes of administrative and legal development, religious reform, colonial settlement and economic growth to take centre stage as the key agency of transformative change. Within much of the historiography there is a cleavage into a black-and-white presentation of engagement with these processes, which projected Scotland as a land of binary opposites: Celtic versus English or anglicised, Highland versus Lowland, native versus newcomer, conservative or traditionalist versus moderniser or innovator. The shifting emphasis of these deceptively simple opposites in fact has more to do with internal questions of culture and national or even regional identity current in the nineteenth and twentieth centuries than with the complex realities of medieval Scotland. Through the kaleidoscope of historiographical trends, however, there is one unifying strand, a certainty that this was an era of development, betterment and economic growth in which conflict was short and led to greater unity and stability, and where the lives of ordinary people gained security as they shared in the new prosperity.

Historians rarely agree about key pivot moments, the events that led to decisive change, all the more so in a discipline that has largely abandoned 'great man' narratives and in which political decisions are no longer accepted as the sole or even primary arbiters of change. Politics do not occur in a vacuum and even the unwonted success and endurance of King Máel Coluim II mac Cinaeda with whose twenty-nine-year reign (1005–34) this period commences can no longer be seen only as a consequence of his ruthless elimination of dynastic rivals to secure his grip on the throne of Alba.[4] In an age where harvest failures and epidemics could unseat a king seen as 'unlucky' or judged against by God, the effects of climate and weather on perceptions of good and effective rule could be just as devastating as defeat in battle. We can look at the climate proxies and surviving written records and see how Máel Coluim benefited from a long period of favourable weather conditions and probably successful harvest, with only a few poor years to cause unease. Full bellies helped to quell discontent at home and reduce

2 Exemplified by the title of Archibald Duncan's opening volume of the four-volume Edinburgh History of Scotland series, published in the mid 1970s: A.A.M. Duncan, *Scotland: The Making of the Kingdom*, paperback edition (Edinburgh, 1978).

3 There is still no better exploration of that process across all component parts of the British Isles than R.R. Davies, *Domination and Conquest: The Experience of Ireland, Scotland and Wales 1100–1300* (Cambridge, 1990).

4 For a summary overview of the man and his reign, see Woolf, *From Pictland to Alba*, pp. 232–248.

the need for risky predatory raids into foreign territory to win plunder and take slaves, when check or defeat could signal the end for an unsuccessful war leader. Although there are only hints of poorer harvests in the reign of his hapless grandson and successor Donnchad mac Crinain (1034–40), the 'King Duncan' of Shakespearian tradition, the warfare and raids that contributed to his defeat and death at the hands of his Moray-based rival, Macbethad, were perhaps a response to demands from his kingdom's chief men for plunder to replenish depleted resources and satisfy their own followers.[5]

Our richer historical records from this period allow us to see the role of human agency in facing the challenges of Scotland's ever-shifting and multiply diverse environment in greater detail than in earlier periods, as well as illustrating the nature of those challenges. The nuance in these records shows how climate and weather affected perceptions of a ruler and could reinforce or undermine his reputation and ability to rule effectively. This insight is perhaps nowhere clearer than in respect of Donnchad mac Crinain's killer and supplanter, Macbethad, whose historical reputation still languishes in the shadow of later negative portrayals constructed by the supporters of Donnchad's ultimately victorious heirs. During Macbethad's seventeen-year rule (1040–57), we are told by the fifteenth-century chronicler Walter Bower, who drew on older sources like the eleventh-century *Duan Albanach* and the positive view of the king presented in the 'prose and verse' chronicle section of the Melrose Chronicle, Scotland saw a time of 'great fertility'.[6] As in the case of Máel Coluim mac Cinaeda, we can look at the ice core and speleothem proxies to see how Macbethad benefited from a long period of relative stability in which generally benign weather conditions delivered sufficient produce to ensure that his rule passed unchallenged other than by the close kin of his predecessor. It was only the intervention of the English king in the 1050s in support of the exiled son of King Donnchad that began to destabilise Alba and led ultimately to the deaths of Macbethad and his briefly reigning stepson, Lulach, at the hands of supporters of Máel Coluim III mac Donnchada (1058–93).[7]

Traditional narratives of the new king's reign usually present him as either a benevolent, visionary leader who 'modernised' his kingdom or a bloody-handed warlord in the Dark Age tradition, ultimately meeting his deservedly violent end during a raid into Northumberland in November 1093. More recent analysis, however, suggests that he was a far more complex character whose reign may have witnessed significant innovation within Scotland's political and religious culture and society and increased exchanges of trade and ideas with his kingdom's English, Scandinavian and continental neighbours.[8] Amongst these innovations, we can see responses to changing environmental conditions that encouraged population growth and the expansion of agriculture to sustain it, delivering surpluses that could be exchanged with external partners. Most discussion still focuses on the controversial role of the king's second wife, St

5 For Donnchad's brief reign and the politics of his overthrow, see ibid., pp. 252–256.

6 Bower, *Scotichronicon*, vol. 3, p. 23. Bower almost certainly drew this claim from an earlier source that was more sympathetic towards the kings of Moray than those accounts he used to construct his more detailed narrative of Macbethad's reign.

7 For a reappraisal of these events, see Woolf, *From Pictland to Alba*, pp. 260–266.

8 Máel Coluim III mac Donnchada has been the subject of a major historical re-evaluation. See N. McGuigan, *Máel Coluim III 'Canmore': An Eleventh-Century Scottish King* (Edinburgh, 2021). For an alternative summary of the last years of his reign, see R.D. Oram, *Domination and Lordship: Scotland 1070–1230*, New Edinburgh History of Scotland 3 (Edinburgh, 2011), chapter 1.

Oakwoods in winter, Birnam, Perth and Kinross. The lower slopes of King's Seat, rising above the village of Birnam and the location of the famous 'Birnam Oak', are again cloaked with an oak-dominated woodland. Birnam Wood first emerged into record in the sixteenth-century chronicle of Hector Boece, but it is unlikely that much dense woodland still covered the sides of the Tay gorge at Dunkeld by the mid eleventh century, when according to Boece and Shakespearian tradition Máel Coluim mac Donnchada's men found cover here for their mythical advance on Dunsinnan Hill.

Margaret, in the process of religious reform in Scotland. That limited view, however, neglects the part she played – and the roles of her kin and followers who accompanied her – in fostering wider European connections.[9] In addition to the first hints of regular trade in luxury items from Europe, perhaps in exchange for exotic and highly valued Scottish commodities like freshwater pearls and fur pelts, it is likely that semi-processed bulk staples, like salted fish, began to be exported from Scotland in this period.[10] But it was probably the knowledge exchange that came with the exiled nobles, clerics and their followers who accompanied Margaret and her brother to Scotland that contributed most to the rapid development of an agricultural 'revolution' that saw new inroads into previously lightly exploited parts of Scotland's environment. Most important amongst the knowledge reaching Máel Coluim's kingdom was the introduction of new agricultural technologies, including the carruca type of plough that would transform the face of the land.[11]

Scotland's turbulent internal politics in the years that followed Máel Coluim's death have generally been presented as a brief outburst of conservatism and tradition against the forces of innovation, brought to an end when his son Edgar was established on the throne with Norman aid.[12] Through the reigns of Edgar (1097–1107) and his brother Alexander I (1107–24) we gain glimpses of what innovation meant, beyond the conventional historiographical representations of the introduction of foreign clerics and knights and a slow growth in a market economy with the formal establishment of planned towns like Stirling. From Edgar's time, we have recorded accounts of the expansion of permanent settlement and agriculture into land that had been lightly exploited during the preceding half millennium, providing our first historical evidence for a process that was already well underway across the whole of northern Europe from Poland to the British Isles.[13] It is, however, the reign of their youngest brother, David I (1124–53), that is often seen as marking a decisive transition in Scotland's development.[14] This perspective is largely a product of the marked increase in record evidence from his reign – as will be seen in the following pages – and of a historiographical tradition that presented this burgeoning bureaucratic parchment output as a symptom of the transformation he wrought. Change, we see, was driven by his introduction of foreign knights and clerics, the adoption of English styles of government, law and administration, and the economic advancement of his kingdom through the foundation of privileged market and trade centres in his burghs and his introduction of a native silver coinage.[15] There is no denying the economic – and environmental – impact of the changes his reign witnessed, from the material demands in stone, lime, wood and lead for the cathedrals, monasteries, churches and castles that were built from the second quarter of the twelfth century,

9 Oram, *Domination and Lordship*, chapter 1; R.D. Oram, *David I, King of Scots 1124–1153* (Edinburgh, 2020), chapters 1, 12 and 13.

10 For the earliest known record of salt production in Scotland, dating to Máel Coluim's reign, see R.D. Oram and C.A. Whatley, 'A brief history of Scottish salt from the eleventh century to the late twentieth century', in S.A. Whatley and J. Hambly (eds), *Salt: Scotland's Newest Oldest Industry* (Edinburgh, 2023), pp. 16–33 at 16–17.

11 For trade, see A. Stevenson, 'Trade with the south, 1070–1513', in M. Lynch et al. (eds), *The Scottish Medieval Town* (Edinburgh, 1988), pp. 180–206 at 180. For discussion of the impact of the carruca, see below, p. 264.

12 Duncan, *Scotland*, chapter 6.

13 R. Bartlett, *The Making of Europe: Colonisation, Conquest and Cultural Change 950–1350* (London, 1993), chapters 1, 2, 5 and 6.

14 The most recent major treatment of the reign is Oram, *David I, King of Scots*.

15 Ibid., chapter 14.

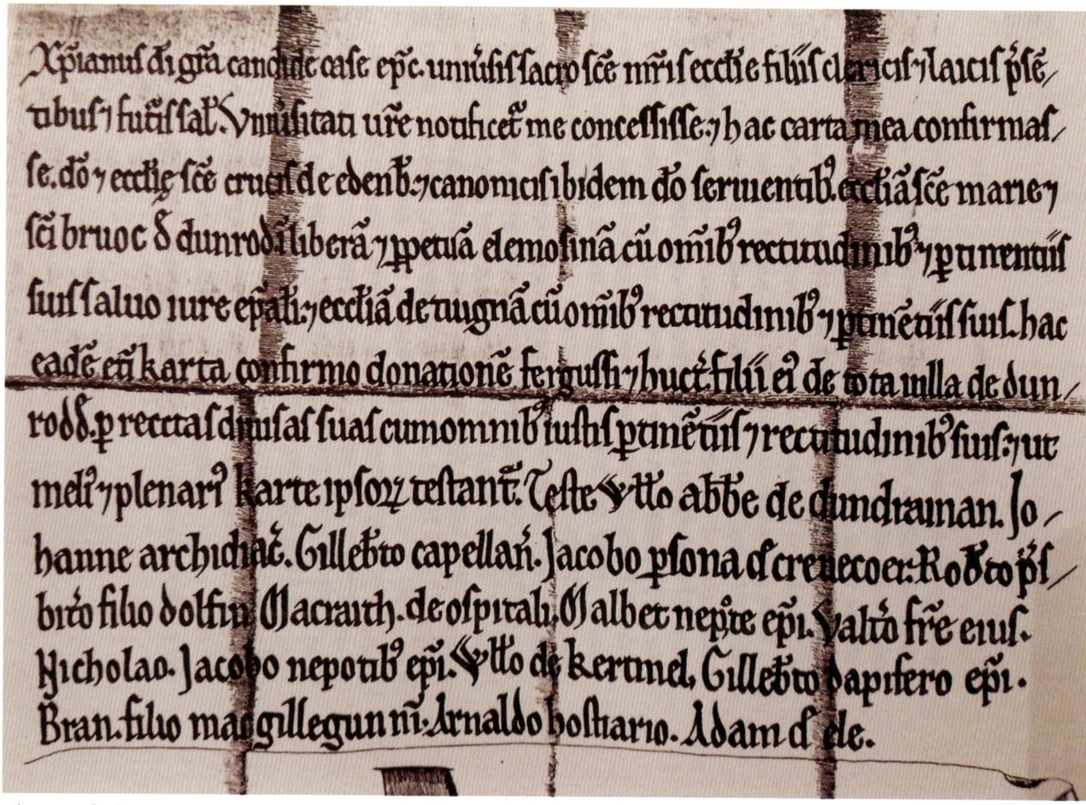

Charter of Christian, bishop of Whithorn, confirming the grant of the churches of Dunrod and Twynholm, and the whole toun of Dunrod, to the canons of Holyrood. Parchment records such as this start to shed light on the human organisation of the land and its resources, on exploitation practices and on intensifying pressure through human population growth. (Image reproduced from the facsimile in C. Innes (ed.), *Liber Cartarum Sancte Crucis* (Edinburgh, 1840).)

to the reordering of the land to accommodate the cultural expectations of incomers and in response to the economic opportunity of booming commercial links with mainland Europe. The novelty of much of what he 'achieved' stems from its first surviving notice in his parchment records, but most often these reveal that the institutions, economic systems and land management practices they document were already in place and in many cases were likely to have existed for generations past.[16] Beneath the formulaic language of the charters, in the detail of their subject matter, we start to see how people viewed, used and negotiated the land, waters and seas of Scotland and their rich and varied resources. We also see competition, conflict and compromise in action, where individuals and communities competed for access to and control of specific resources. Here, in black and white, we have the first recorded evidence for concerns over the future availability of commodities and resources and for attempts to regulate their use

16 The re-evaluation of the development of the state in Scotland presented in Alice Taylor's *The Shape of the State in Medieval Scotland 1124–1290* (Oxford, 2016) is a powerful reminder of the continuities in practice that run through David's reign and into the later thirteenth century.

through legislation and the use of legal violence.

Our knowledge of Scotland expands gradually through the reigns of David, his grandsons and successors Máel Coluim IV (1153–65) and William (1165–1214) and his great-grandson Alexander II (1214–49) as they were able to push their authority more effectively to the geographical limits of the Scottish mainland.[17] Although our understanding of how this was achieved in some areas of the country remains patchy, it is in this period that we gain our first detailed written insights on the social organisation, resource base and economic systems of regions as widely dispersed as Caithness, the Hebrides and Galloway but also of zones within the heartland of the kingdom like Angus and the Lennox. We remain, however, as deeply dependent on archaeological evidence for an understanding of how life operated for the bulk of Scotland's population. The people themselves remain but shadows in the record, for it must be emphasised that the 'voices' we have for this supposedly golden era of Scotland's historical development and its apogee of pre-modern environmental status, preserved in our few chronicle accounts, are almost exclusively those of the socially elevated and male. The main contemporary Scottish material survives in the Melrose Chronicle, compiled at the Cistercian abbey in the Scottish Borders, the north-west English Lanercost Chronicle, the so-called *Gesta Annalia I* which covers events in Scotland to 1285 and is embedded within the later fourteenth-century chronicle attributed to John of Fordoun, and the two early fifteenth-century chronicles of Andrew of Wyntoun and Walter Bower. All were compiled from the later twelfth century onwards but contain material gathered from earlier, now lost annals and from English and continental sources. At best, they are a snapshot record of mainly what we might call headline news. Providing a start to our more detailed exploration of this era, however, they also contain scattered references to weather and weather-related events, often reported second-hand and with strong regional as well as social biases. If we were reliant on surviving Scottish chronicle accounts for these two and a half centuries spanned by the MCA, we would be left fumbling in the dark for much of the time. Fortunately, they can be supplemented from Irish annals and English records to create a synthetic – but still far from complete – weather narrative. These sources, however, are all the product of privileged male ecclesiastical and often monastic writers who were reliant on others for the data they presented and who viewed the experiences of others largely through the lens of their own perceptions of hardship and loss influenced by biblical exemplars. What they do not tell us is often more important than what they do, for lack of comment on weather, harvests and yields speaks of relative success and 'normality'. It is the exceptional that is given space in their narrative and, as what follows will demonstrate, the exceptions were rare. The silence that is normality suggests that for many the 'good times' of this era meant at least adequate food upon which to subsist and potentially an opportunity to profit from surplus production. That, after all, is the benefit of peace and stability, and it is perhaps in this respect that the sources reveal the reality of the 'golden age'.

Climate and weather I: a century of contrasts, 1000–1099

Returning now to the years immediately after 1000, it has been said that there is little written record of Scotland's weather in this early phase of the MCA.[18] Indeed, few documentary sources survive from the eleventh century until its final decade, but there is more information for the earlier part of the century than such statements imply. Scanning across the

17 Oram, *Domination and Lordship*, chapters 4 and 5; R.D. Oram, *Alexander II, King of Scots 1214–1249* (Edinburgh, 2012).

18 Dawson, *So Foul and Fair a Day*, p. 96.

An era of contrasts, Clackmannanshire and the Ochil escarpment, looking west from near Saline, Fife. The 1000s witnessed wide variation in weather which delivered bounty and dearth in equal measure.

century, Scottish, English and Irish sources can be most easily mined to form a picture of several years of adverse weather, harvest failures, famine-related mortalities and probably associated pestilence in the 1080s and 1090s. Looking deeper, there are further hints of earlier poor conditions at the beginning and middle of the century. Unfortunately, apart from an uncorroborated early sixteenth-century claim by the notoriously unreliable Hector Boece that in the time of King Máel Coluim mac Cinaeda there was one unspecified year when frosts at midsummer killed the growing corn and caused great dearth throughout the country as a result,[19] none otherwise relate directly to Scottish events. Was the event noted by Boece 'real' or does it represent his interpretation of the famine in 1006 mentioned by Walter Bower in the early fifteenth century, which one of Bower's still earlier sources drew from eleventh-century Flemish accounts of events on the near continent?[20] That particular north-west European famine followed poor weather in 1005, which had delivered a meagre harvest and may be related to the 'great famine, the most severe in living memory' that struck southern England in 1005.[21] There is no mention of such an event in sources from the north or west of the British Isles. Irish annals, usually meticulous recorders of bounty and dearth, describe only a severe late winter in 1008/9, with snow and ice persisting from early January to late March.[22] If there was an episode of poor weather in the 1000s it passed quickly. A noteworthy good but very hot summer in 1010 brought 'a fruitful autumn' in Ireland.[23] From then until the 1030s the silence of the annalists implies a long period of relatively stable, productive but unremarkable conditions for the Atlantic west of the British Isles. In this unremarkable phase, we might suggest, the MCA was beginning to make its mark.

A high point was reached in the early 1030s. In north-east Ireland, 1032 was a year of abundance, celebrated in rhyme by the Armagh-based Ulster

19 H. Boece, *The Chronicles of Scotland by Hector Boece, Translated into Scots by John Bellenden, 1531*, ed. E.C. Batho and H.W. Husbands, volume 2 (Edinburgh, 1941), p. 142.

20 Bower, *Scotichronicon*, vol. 4, p. 67. The source for this event seems to be the chronicle of the Flemish monk Sigebert of Gembloux (1030–1112): ibid., p. 207.

21 *The Anglo-Saxon Chronicle*, ed. and trans. G.N. Garmonsway, new edition (London, 1972), pp. 136, 137; Henry of Huntingdon, *The History of the English People 1000–1154*, ed. D. Greenway (Oxford, 1996), p. 8.

22 Annals of Ulster, U1008.4.

23 Ibid., U1010.5.

annalist.[24] This peak, examined below, transitioned into the half century of the episode of low sunspot activity known as the Oort Solar Minimum, which lasted from around 1040 for a period of roughly four decades. Certainly, 1037 was observed as a wet and stormy year in Ulster but the absence of reference to harvest failure or to food shortages the following year suggests no noteworthy damage to growing crops.[25] What we see from this point is a marked cleavage in the record evidence which provides confirmation of the prevalence of the potentially broad divergences in weather experience across the British Isles that more modern records chart, where the extension of northern European atmospheric conditions into south-eastern England, usually in the form of a high pressure cell, draws colder air around its northern edge and forms a blocker to the eastwards movement of warmer, moister air coming in from the Atlantic. There is, for example, no evidence that southern England's 'most disastrous year' of adverse weather, crop failure and cattle mortality in 1042, which brought accounts of widespread suffering in southern and eastern annals,[26] was shared by the peoples of northern and western parts of the British Isles. Deteriorating conditions persisted in the south, with the winters of 1046/7 and 1047/8 being especially severe. In the first of these, England experienced a 'cruel winter', with frost, snow and stormy conditions striking in early February 1047 and resulting in 'great mortality over all England' of men and livestock.[27] These years were not especially severe on mainland Europe, being generally snowy but apparently not causing a delayed spring and harm to the following year's crops.[28] It is possible that the severe winter of 1047/8 reported in the Annals of Ulster is a misdating of this English experience. That event saw heavy snowfall between early December and mid March which caused widespread cattle mortalities, coupled with storms at sea that brought numerous deaths amongst 'sea-beasts' and sea birds. Rather than being a north-westward extension of the conditions recorded in England, however, it seems to relate to a separate episode of extreme weather coming eastwards across the Atlantic.[29] A lull in weather reporting in Irish annals after 1047/8 suggests that better – or at least, unnoteworthy – conditions returned and continued into the 1050s. Perhaps significantly, however, the Durham-based *Historia Regum* records a severe winter storm on 21 December 1052 which caused widespread damage to buildings and devastated woodland throughout north-eastern England.[30] What is significant was its exceptionalism, being the only such storm recorded in this period in that source, suggesting an intensity without precedent in the chronicler's experience. This single, possibly seasonal storm apart, other English sources also give little notice of weather-related events in these years, which suggests a shared run of the unremarkable years experienced in the north of Ireland. It may be to these more stable conditions that the first weather-related observation in this period that was associated directly with Scotland referred, when it was claimed that 'in the time of [Macbethad] there was great fertility'.[31]

No significant events are recorded for the 1060s and 1070s, which on the continent saw hard winters

24 Ibid., U1032.8.

25 Ibid., U1037.7; Annals of Loch Cé, http://www.ucc.ie/celt/published/T100010A/index.html, LC1037.6.

26 *The Anglo-Saxon Chronicle*, p. 163.

27 Ibid., pp. 164, 165, 166.

28 Pfister et al., 'Winter air temperature variations', p. 542.

29 Annals of Ulster, U1047.1.

30 Symeon of Durham, *Historia Regum*, in *Symeonis Monachi Opera Omnia*, ed. T. Arnold, 2 volumes (London, 1882–85), volume 2, p. 170.

31 Bower, *Scotichronicon*, vol. 3, p. 23.

Storm-thrown trees, Panmure estate, Angus. The winter storm of December 1052 recorded by Symeon of Durham reportedly devasted woodland throughout the north-eastern English coastal plain. As an isolated event, it is unrepresentative of general trends, but it was noteworthy to the chronicler as an episode of hitherto unprecedented severity in his own experience.

in 1059/60 and 1068/9, suggesting that for the northern and western British Isles there was instead an extended period of relative weather 'normality'. This was the time of William the Conqueror's 'Harrying of the North' in the winter campaign of 1069/70, when he ravaged the country between the Humber and the Tees. Although famine is recorded in 1070, this was attributed to the king's actions, not to adverse weather conditions.[32] Surprisingly, there is no specific record in British and Irish sources of the exceptionally cold winter of 1076/7 recorded in continental accounts,[33] an absence which underscores the ameliorating effects of NAO in limiting the influence of the Siberian and Scandinavian atmospheric high-pressure systems that deliver these episodes of intense cold. There is instead a gap of four decades in references to weather events before the Annals of Ulster record a year of exceptional acorn production in 1087, pointing to the warm, damp conditions favoured by oak trees.[34] These good conditions persisted in Ireland into the early 1090s, with 1091 being described as 'a fruitful year with good

32 The fullest account is in Symeon of Durham, *Historia Regum*, p. 188. See also *The Anglo-Saxon Chronicle*, pp. 203–205.

33 Pfister et al., 'Winter air temperature variations', pp. 538, 542–543.

34 Annals of Ulster, U1087.8.

Winter waterlogging on Orchardhead Carse, Airth, Stirlingshire. The saturated conditions of the 1090s delivered a run of poor sowing, growing and harvesting years, leading to widespread dearth throughout the British Isles and the near continent.

weather'.[35] In England, however, 1086 was afflicted by a murrain amongst cattle and poor fruit and cereal harvests, all of which the chronicler attributed to adverse weather conditions. As a consequence, 1087 saw famine, pestilence and high human mortality.[36] Conditions were variable along the north and west sides of Ireland in the following years, with the Munster-based Annals of Innisfallen recording severe frosts and snow over the winter of 1092/3, the destruction of crops by rain and epidemic disease afflicting the human population throughout 1093. In that same period, however, northern Irish annals record only 1093's abundant acorn crop.[37] Bower's earlier sources reported a sequence of weather events, crop failures and subsequent famines in southern England and northern parts of France throughout the 1090s, after the Irish reports of a good year. The sequence commenced with descriptions of incessant rains and flooding affecting England in autumn 1093 leading into a harsh, icy winter in which

35 Ibid., U1091.6.

36 *The Anglo-Saxon Chronicle*, pp. 217–218; Henry of Huntingdon, *The History of the English*, p. 31.

37 Annals of Ulster, U1093.7; Annals of Innisfallen, https://celt.ucc.ie/published/T100004/index.html, AI1092.11, AI1093.9, AI1093.10.

rivers froze, and then a destructive thaw. Poor conditions continued in 1094 and 1095. A harvest failure and famine followed, worsened, allegedly, by King William II Rufus's ruthless taxation of his kingdom.[38] The result was apparently great mortality, but the account forms part of a litany of complaints against the 'ungodly' William Rufus composed by its clerical chronicler, rather than being a reliable report of famine.[39]

Parts of Ireland also experienced a poor year. There, 1094 was described as seeing extremely bad weather throughout the island and widespread food shortages.[40] A subsistence crisis continued in 1095, with Irish sources again reporting a hard winter.[41] In England, apparently on the back of a further harvest failure, simmering social stresses erupted into violent attacks by the poor, hungry and dispossessed against those seen as richer and better provided for. Again, these conditions were presented in England in the light of King William's unjust and ungodly actions, with gales and an earthquake reported in September 1095 being viewed as further evidence of Divine displeasure.[42] In Ireland, famine effects might be reflected in the great sickness and high mortalities reported from August 1095 through to May 1096.[43] A further year of deluge followed in 1096/7, when the rains disrupted the autumn sowing and caused another poor harvest with much low-lying cultivation waterlogged by constant rain,[44] but the wet conditions in Ireland yielded an abundant nut harvest in autumn 1097.[45] Far worse was to come, however, with the west-central Irish Annals of Loch Cé reporting that 1099 was a year of general famine.[46] Concluding the poor decade, November 1099 saw storms down Britain's east coast which brought destructive sea-floods.[47]

The sole suggestion that Scotland shared in some of these poor experiences comes from the early sixteenth-century chronicle of Hector Boece. Speaking in general terms of undated events that occurred during the reign of King Máel Coluim III, he described the widespread devastation of the eastern coasts of Britain caused by sea-floods and made specific mention of inundation in Moray. Those floods, we are further told, were accompanied by violent storms with lightning strikes that destroyed trees and cornfields.[48] These events may be a misdated reference to the 1099 eastern English floods mentioned above, which occurred six years after Máel Coluim's death. But sea-floods down the English North Sea coast occurred at various times across the eleventh century, such as the devastating September 1014 event in Kent and East Anglia,[49] without any indication of a similar experience afflicting the Scottish coast at those times. Boece's source, we can suggest, made specific reference to an event that affected north-eastern Scotland alone, unmentioned in any southern British record. This chronological imprecision combines with geographical vagueness to make it difficult to construct a coherent

38 *The Anglo-Saxon Chronicle*, pp. 232, 233.
39 Bower, *Scotichronicon*, vol. 4, p. 89.
40 Annals of Ulster, U1094.8.
41 Ibid., U1095.1; Annals of Innisfallen, AI1095.2.
42 Bower, *Scotichronicon*, vol. 4, p. 115.
43 Annals of Ulster, U1096.8; Annals of Innisfallen, AI1095.13.
44 *The Anglo-Saxon Chronicle*, p. 234; Bower, *Scotichronicon*, vol. 4, p. 143.
45 Annals of Ulster, U1097.8.
46 Annals of Loch Cé, LC1099.1.
47 Bower, *Scotichronicon*, vol. 3, p. 101.
48 Boece, *The Chronicles of Scotland*, vol. 2, p. 175.
49 *The Anglo-Saxon Chronicle*, p. 145.

Frost and ice at Loch Monzievaird, Crieff, Perth and Kinross. Across the middle decades of the eleventh century, the prevailing NAO state saw a succession of intensely cold but dry winters, with severe frosts and prolonged spells of ice.

weather narrative for Scotland in the eleventh century from these few, disjointed fragments. Furthermore, the contrary evidence from Ireland underscores the danger of extrapolating Scottish experience from southern English record sources alone and affirms the marked regional variations in weather experience encountered across these islands.

If the written record is deficient, can climate proxy data offer a clearer framework on which to hang these fragmentary historical records? Proxies – mainly Greenland ice core data and Irish, Scandinavian and Ural–Siberian tree-ring records – all indicate a continuation through the eleventh century of a strongly cyclonic weather system across the whole North Atlantic region, in which a westerly air stream over the ocean brought predominantly mild and damp conditions for winter and warm and moist summers.[50] In general, the dominance of this pattern meant few winters that saw long spells of anticyclonic airflow bringing cold, dry Arctic air to deliver severe frosts or collide with moist Atlantic fronts and deposit significant snowfall over Scotland. There were exceptional episodes that punctuated this trend, the worst being a reversal in the prevailing winter conditions from the 1030s to 1070s, which saw a succession of dry and probably very cold winters. This disruption correlates with the Oort Solar Minimum (*c.*1040–*c.*1080), where the reduced solar activity was perhaps sufficient to cool ocean surface temperatures in the North Atlantic and interfere with the relative strengths of the prevailing atmospheric pressure cells that sit over Iceland and the Azores.[51]

50 Campbell, *The Great Transition*, pp. 45–46.

51 It was this disruption that produced the severe winter of 1076/7 in southern Europe, while the north saw low precipitation: Pfister et al., 'Winter air temperature variations', pp. 542–543.

Dune systems at the southern end of Lunan Bay, Angus. These towering dunes are suffering erosion on their seaward sides as a consequence of the increasing frequency and intensity of easterly-driven storms from the North Sea. Destabilisation through turf-stripping and cultivation on their landward side in the later eleventh or twelfth century led to catastrophic deflation and sand inundation of arable land.

This episode saw fluctuations in sea ice in the North Atlantic and temperatures in Greenland which suggest that there was prolonged disruption of the NAO.[52] As we shall see, further but shorter such episodes occurred in the later twelfth and earlier thirteenth centuries.

The three main sets of currently available proxy data from Scotland – the Traligill speleothem record (mainly winter precipitation), the Loch Sunart bottom water temperature record and the Cairngorm tree-ring sequence – are all most closely linked to weather patterns originating in the NAO, and some inferences can be drawn from these and correlated with northern and western Irish annals for general trends affecting western Scotland. Of these, only the Traligill and Loch Sunart records extend currently into the tenth and eleventh centuries; the Cairngorm data starts only in the early thirteenth century.[53] The Traligill sequence commences with three decades of high winter precipitation to around 1030, associated with a high winter NAO that is characterised by strong westerly airflow and cyclonic rainfall. It then plunges to the lowest precipitation – signalling a low winter NAO – recorded for the north-west Highlands until the early seventeenth century, with the dry conditions lasting into the 1080s.[54] Low winter NAO is associated more with cold, dry winters and a northerly or easterly airflow. Here, we see the possible influence of the Oort Solar Minimum. Loch Sunart shows a similar trend, with generally warmer conditions maintained for over a century from 916 until 1035, when there was an abrupt and deep cooling episode that persisted into the mid twelfth century.[55] Unfortunately, neither of these proxies can be extended with confidence to the eastern Lowlands of Scotland, making it impossible to understand the weather

52 Ibid., p. 207 fig. 3.22.

53 For the potential of the north-east Scottish pine sequences, see R. Wilson et al., 'Reconstructing Holocene climate from tree rings: the potential for a long chronology from the Scottish Highlands', *The Holocene*, 22:1 (2011), pp. 3–11, and Mills, 'Historic pine and dendrochronology'.

54 Proctor et al., 'A thousand year speleothem proxy record', p. 818 fig.4; Campbell, *The Great Transition*, p. 46 fig. 2.6 (A).

55 Cage and Austin, 'Marine climate variability', pp. 1641 fig. 7, 1643.

regime in the midst of the solar minimum that supposedly made Macbethad's reign a time of great fertility in Scotland, when England and Ireland were experiencing some extreme winters and disastrously poor summers for crops.

One piece of archaeological evidence that casts some light on conditions in east coast districts relates to a catastrophic but undocumented sand-blow event affecting settlement and agriculture just inland from Lunan Bay near Inverkeilor in Angus.[56] Sampling in the area extending west from the modern dune systems revealed that arable cultivation here had been submerged beneath 'a natural sudden deposit' of sand at depths of up to nearly 2m.[57] There was nothing to suggest that this accumulation was gradual or occurred over an extended period. Layering visible in the sand is the result of changes in wind direction during a single and brief episode of sand-blow. Darker sand included field silt and represented deposits by westerly winds while the cleaner dune sand was the deposition of easterlies. The excavator concluded that 'there is no reason to extend the accumulation over more than a single season'.[58] It was clear, however, that the suddenness of the major deposition had been presaged by a more limited episode that had blown a thin layer over the fields, which had been ploughed in for a single season and produced a distinctly sandier layer at the top of the plough-soil. Here, it seems, peasant farmers had recovered arable land of very high quality from a major environmental event in one season, only for a catastrophic dune deflation the following year to obliterate their fields and lead to permanent abandonment of their cultivated land. Exactly when this happened is, sadly, unclear. It is argued only to predate the later twelfth century, by which time the centre of local settlement lay inland at Inverkeilor rather than nearer to the coast.[59] If any period in

56 D. Pollock, 'The Lunan Valley Project: medieval rural settlement in Angus', *Proceedings of the Society of Antiquaries of Scotland*, 115 (1985), pp. 357–399.

57 Ibid., p. 389, 390, 393–395.

58 Ibid., p. 391.

59 Ibid., pp. 392–393, 396–397.

which such an event is likely to have occurred is to be inferred from the wider historical and climate proxy evidence, however, the turn from the eleventh to the twelfth century seems the most likely.

Climate and weather II: the best and worst of times

Written records for Scottish weather trends start the new century as patchily and vaguely as for much of the previous one. Once again, we are forced to turn mainly to English and Irish records for descriptions of the general effects of weather on human and animal populations throughout the British Isles. Frustratingly, these are largely expressed in simple terms of 'good' and 'bad' seasons, years or harvests. They do, however, indicate that the new century saw little change to the unpredictability of the previous decade. Summer 1103 was noted for adverse weather and a poor harvest for grain and fruit across southern England, with an especially severe windstorm on 10 August reported as flattening cereal crops.[60] A second extraordinarily poor harvest followed in southern England in 1105, which went unreported in northern chronicles, but it was 1110 that was marked as exceptionally bad, with both southern English fruit and cereal harvests lost to the weather.[61] This poor year extended into what was noted as a remarkably long winter, with bitter and severe weather that affected the spring growth of the crops and threatened another poor harvest in 1111.[62] The run of poor summers and bad winters for southern England broke that year, however, and 1112 delivered an abundant harvest.[63] A hot, dry summer affected the south alone in 1113, with reports of the Thames being so low east of London that people could cross it on foot, while the autumn was remarkable for violent windstorms through October and November.[64] The winds were a forerunner of bad times ahead and by winter 1114/15 there was a return to previous extremes: late summer 1115 saw incessant rains that ruined the harvest and there was also a failure of the beech and oak mast on which pigs were fattened for winter.[65] In much of Europe, winter 1115/16 was reported as severe, but there is no specific comment in English or Irish sources.[66] After a respite in 1116, a year in which Flemish accounts report the availability of ripe strawberries locally at Christmas,[67] the poor summer weather returned to England in 1117 and devastated the East Anglian grain harvest.[68] The winters of 1119/20 and 1120/1 were notable only for their violent and destructive winds, but that of 1121/2 seems to have been marked by calm, clear and cold conditions in which the aurora borealis was visible in southern parts.[69] Those of 1123/4 and 1124/5 (and

60 *The Anglo-Saxon Chronicle*, p. 238; Symeon of Durham, *Historia Regum*, p. 236.

61 *The Anglo-Saxon Chronicle*, pp. 240, 243.

62 Ibid., p. 243.

63 Ibid.

64 Ibid., p. 244. This event is assigned to 1114 in Bower, revealing his source to be Version E of the Anglo-Saxon Chronicle, whose annual entries are misdated by one year at this point: Bower, *Scotichronicon*, vol. 3, p. 115. It is described in great detail, again dated to 1114, in Symeon of Durham, *Historia Regum*, p. 248, which indicates that the wider south-east of England was afflicted by the drought.

65 *The Anglo-Saxon Chronicle*, pp. 246–247. Symeon of Durham, *Historia Regum*, p. 248, has the winter of 1114/15 as notable for extreme cold and frosts.

66 Pfister et al., 'Winter air temperature variations', p. 543.

67 Ibid., p. 538.

68 *The Anglo-Saxon Chronicle*, p. 247.

69 Ibid., pp. 249–250.

Winter snow lying late over post-medieval rig-and-furrow cultivation near Foulford, Perth and Kinross. Extended winters delayed preparation of the ground for sowing the new year's crops and slowed grass growth, leading to fodder shortages for flocks and herds.

possibly 1125/6) were especially bad throughout much of northern Europe, with long periods of extreme cold in early winter, then more disturbed conditions with spells of snow, rain and ice extending into early spring.[70] The summers of both 1124 and 1125 were also poor, with heavy rain that 'almost drowned the crops in the fields' causing poor harvests and widespread famine after successive failures. The dreadful year of 1125, however, represented the final sting in the tail of an exceptionally poor quarter century, with conditions improving markedly before the end of the decade. In the fifteenth century, Walter Bower, our one Scottish source for these events, again drew on an earlier southern English source that otherwise finds no corroboration in any northern accounts.[71] For a probably more reliable source of written information to illustrate the Scottish experience, we must again look to Ireland.

In sharp contrast to those southern English reports of recurrent poor conditions to 1125, Irish sources have much less to say. Here, we again seem to be seeing the recorded effects of the oceanic systems along the north and west sides of the British Isles. A single night of blizzard conditions in March 1107 that killed numerous cattle, followed by an unsettled summer and a poor harvest, is our only indication of any exceptionally poor weather in the first decade of the new century.[72] The following year,

70 Pfister et al. 'Winter air temperature variations', pp. 543–544.

71 Bower, *Scotichronicon*, vol. 4, p. 161. See also *The Anglo-Saxon Chronicle*, pp. 254, 255–256.

72 Annals of Ulster, U1107.1, U1107.6.

moreover, saw benign weather, good harvests and an abundance of oak mast on which to fatten up the pigs for slaughter.[73] North-eastern Ireland shared in the extreme winter of 1110/11 noted in English sources, with numerous livestock deaths in the snow and ice.[74] That concurrence underlines the probable severity of this episode for the whole of the British Isles and points to the likely disruption, albeit temporarily, of the more established seasonal weather patterns. If both north-east Ireland and northern England were affected by this event, it is likely that southern Scotland also saw extreme conditions. North-eastern Ireland also suffered in the poor winter of 1114/15, with significant livestock losses and a delayed spring sowing which threatened future food shortages.[75] Other than these headline events, however, the period had recurrently acceptable if not benign conditions. There is certainly no indication of the run of wet and then extreme winters that affected England and much of Europe in the early and mid 1120s,[76] suggestive of a sustained shift into the prevailing NAO positivity of the MCA.

Some of our climate proxy data from the north-west Highlands supports this impression from northern parts of Ireland of a return to the pre-Oort Solar Minimum high NAO conditions. At Traligill, increasingly thick annual speleothem rings indicate that winters were again wet and (probably) for the most part mild.[77] In Loch Sunart, bottom water temperatures had recovered from the lows of the 1080s and early 1090s but plunged again in the later 1110s, possibly reflecting the protracted disturbances recorded in England and the shorter episodes that afflicted Ireland at this time.[78] Taken together, the data is suggestive of milder but wetter conditions in northern and western Scotland through to the early 1110s. There then seems to have been a slackening of the NAO mid decade, contributing to a series of colder but still high-precipitation winters for the Atlantic-facing parts of Scotland. For the east and south-east, that probably meant snowy winters and possibly delayed springs, but there is no indication that these harsher and protracted winters were harbingers of dearth later in the year, as occurred later in the Middle Ages.

There are few narrative written records from any part of England for weather conditions through the second quarter of the twelfth century. Most English accounts are preoccupied with the political turbulence and civil wars that followed the death of King Henry I in 1135. Surviving general weather reports derive mainly from the remote witness of Robert of Torigni, abbot of Mont-Saint-Michel in Normandy, and say little for southern England, let alone for the northern parts of Britain far beyond the information networks through which the abbot derived his news.[79] His reference to a severe drought in northern France in summer 1137,[80] however, accords well with a strongly positive NAO suggesting good summers and milder – but very wet – winters in the north and north-west British Isles. Indeed, the Traligill speleothems show the 1130s to have seen particularly wet winters.[81] A notably harsh winter is to be inferred for southern England from descriptions of the empress Matilda's December 1142 escape from the

73 Ibid., U1108.8, U1108.9.

74 Ibid., U1111.1.

75 Ibid., U1115.1.

76 Pfister et al., 'Winter air temperature variations', p. 546.

77 Proctor et al., 'A thousand year speleothem proxy record', p. 818 fig. 4; Campbell, *The Great Transition*, p. 46 fig. 2.6 (A).

78 Cage and Austin, 'Marine climate variability', p. 1641 fig.7.

79 Robert of Torigni, *Chronicle of Robert of Torigni, Abbot of the Monastery of St Michael-in-Peril-of-the-Sea*, in R. Howlett (ed.), *Chronicles of the Reigns of Stephen, Henry II and Richard I*, volume 4 (London, 1889).

80 Ibid., p. 133.

81 Proctor et al., 'A thousand year speleothem proxy record', p. 818 fig. 4; Campbell, *The Great Transition*, p. 46 fig. 2.6 (A).

besieged town of Oxford, dressed in white to blend in with the snow and able to cross the frozen Thames on foot.[82] That account is strongly suggestive of an intensely cold easterly airflow over southern England in the winter months, colliding with a moist, westerly airflow from the Atlantic. This same winter was recorded as especially severe on both sides of the English Channel, with the Meuse estuary in Belgium becoming ice-bound.[83] It is unlikely that eastern Scotland was unaffected by the strong flow of icy air from Scandinavia and Siberia which delivered such conditions. Much worse, however, came in winter 1149/50, recorded as one of the 'Great Winters' for much of Europe, where an extremely cold phase that lasted between three and six months in different parts of northern France, England and Germany saw sea ice form off the Dutch coast and snow cover for most of the time.[84] For Robert of Torigni, the bitter and lasting cold brought what he described as hardship to men and sterility to the land, probably resulting in a famine the following year.[85] We can attribute this episode to strong anticyclonic circulation around a high-pressure cell that had formed over the southern Baltic, or to a high-pressure ridge extending from the Azores high to the Baltic. As in earlier cases of similar phenomena, that high-pressure system kept the eastwards flow of moist air much further to the north, where it would have collided with the cold air drawn westwards from Siberia and south from the Arctic in the prevailing anticyclonic conditions. With high precipitation dominant in the Traligill speleothem record, such atmospheric conditions likely brought heavy snowfall to at least northern parts of Scotland, but possibly drier weather with only extreme cold to more eastern districts.

It is unfortunate that neither contemporary Scottish nor Irish sources survive for these years. In Ireland, the Annals of Ulster record an abundant fruit harvest in 1130, after which its account is interrupted for twenty-five years.[86] The Annals of Loch Cé, which continue across part of this gap, note only a windstorm in early 1137 which caused devastation of woodland and unexplained 'scarcity' in Connacht later in the year, with high human mortalities.[87] Unfortunately, these too fail entirely for several years, leaving us with no contemporary written record of central-western Irish conditions to the 1170s from which to gain a sense of the weather in the Atlantic-facing west during the Great Winters of 1142/3 and 1149/50. Other than a retrospective reference from 1156 to a 'great crop' nine years previously,[88] there are no accounts that might illustrate the general trends for the northern and western British Isles throughout this period.

Across much of northern Europe, the next five decades were characterised by good if unexceptional conditions, with few references to adverse weather, failing harvests or episodes of extreme heat or cold. Scottish written accounts remain sparse, with only a single reference in the contemporary Chronicle of Holyrood to 'great famine' in Scotland in 1154 giving any suggestion of a failed harvest during this period.[89]

82 William of Newburgh, *Historia Rerum Anglicarum*, in R. Howlett (ed.), Chronicles *of the Reigns of Stephen, Henry II and Richard I*, volume 1: *Containing the First Four Books of the Historia Rerum Anglicarum of William of Newburgh* (London, 1881), pp. 42–43.

83 Pfister et al., 'Winter air temperature variations', p. 544.

84 Ibid.

85 Robert of Torigni, *Chronicle*, pp. 160, 163.

86 Annals of Ulster, U1130.6.

87 Annals of Loch Cé, LC1137.1, LC1137.8.

88 Annals of Ulster, U1156.5.

89 *A Scottish Chronicle Known as the Chronicle of Holyrood*, ed. M.O. Anderson with notes by A.O. Anderson (Edinburgh, 1938), pp. 125–126.

Snow lying late on Ben Wyvis, seen here from Cullicudden in the Black Isle, Highland, keeps spring temperatures suppressed in the lowland districts to its east. The Traligill speleothem sequences suggest that the middle of the twelfth century saw some of the bitterest winters of this era, with heavy snowfall over northern and western mountains, but colder and drier conditions further east.

The limited evidence for sporadic poor winters points to generally favourable or unremarkable atmospheric conditions. Meltwater spates on the Tweed recorded at the end of 1178, for example, suggest that the Southern Uplands saw a hard early winter with high precipitation falling as snow and then a prolonged period of intense cold, sufficient to cause this major river to freeze over.[90] Irish sources are equally unforthcoming, with the Ulster annals reporting only a 'great crop' in 1156 throughout the whole island.[91] The 'snow of the destruction' recorded at the end of the annals for 1179 is perhaps a mislocated reference to the harsh winter suggested for southern Scotland the previous year.[92] Otherwise, weather-related references to the end of the century are limited to the note of a good acorn crop in 1185, which points to a warm, wet summer.[93] A warming of bottom water temperatures in Loch Sunart and fairly steady evidence for high winter NAO from Traligill suggest that this reticence in the historical sources is indicative of unremarkable and largely stable conditions for much of the 1150–90 period, interrupted only by the single severe winter of 1178/9.[94]

A more disturbed period of weather commenced in the 1190s, according to reports of widespread famine across northern Europe from around 1192

90 Bower, *Scotichronicon*, vol. 4, p. 333.

91 Annals of Ulster, U1156.5.

92 Annals of Ulster, U1179.11. Robert of Torigni's great frost at the start of 1179 appears to corroborate the Scottish record for a harsh winter in 1178/9: Robert of Torigni, *Chronicle*, p. 281.

93 Annals of Ulster, U1185.1.

94 Cage and Austin, 'Marine climate variability', p. 1641 fig.7; Proctor et al., 'A thousand year speleothem proxy record', p. 818 fig.4.

The Tay in spate (9 October 2023).

The Tay untamed

With the largest catchment area of any river in Scotland (5,200km^2) and the greatest water volume discharge in Britain (averaging 220 cubic metres per second), the Tay is one of the most powerful natural forces in these islands. Its basin extends from the mountains of Argyll in the west, bounded to the north by Rannoch and the southern Cairngorms and to the south by the Ochils, encompassing terrain that sees significant precipitation from weather systems moving eastwards from the Atlantic or south-westwards from Scandinavia. As a consequence, the river and its many tributaries are prone to spates, whilst the long funnel of the Firth of Tay is subject to storm surges and backing up behind the saltwater barrier of high spring tides. Combinations of meltwater and rainfall spates with strong easterly winds and high tides have caused catastrophic floods of the river throughout history, especially in the vicinity of Perth, with some of the most devastating on record occurring in the 1980s and 1990s.

Perth's low-lying location at the highest tidal reach of the Tay and just below the confluence of the main channel with its major tributary, the Almond, made the town highly susceptible to flooding. Contour mapping of the historic core of the burgh shows that almost the entire footprint of the twelfth- and thirteenth-century burgh was vulnerable, with only small islands of higher ground – such as that occupied by the medieval church of St John – escaping inundation during extreme events. Our first detailed record of such an occurrence is an account of the flood of September 1209, which saw the destruction of the bridge over the river and the old chapel of St Mary that was described as standing upon it:

> This year [1209], or as some would have it around Michaelmas [29 September], there was such a downpour of rain that the torrents and rivers burst beyond their former channels and swept the crops out of the fields. The waters of the Tay and Almond flowed right through the greater part of the town which used to be called Bertha and is now called Perth in Scotland. A mound which fell into the river not only destroyed several houses, but also completely swept away its bridge with the old chapel on it. The lord king, with his noble son Alexander and his brother the earl of Huntingdon left the town in a very small boat and looked for dry land, accompanied by a few of the magnates who happened to be there at that time. Some others among the nobility of Scotland who had likewise been in the same town only just escaped, saving themselves as best they could in small boats or in upper rooms. For the level of the sea had suddenly risen, swollen by a high tide and the great increase of excess rain water, so that not only cobles and skiffs but even huge ships were able to pass freely through the royal town without encountering any obstacle, at any rate through the streets and open spaces, owing to the high level of the water (Bower, *Scotichronicon*, vol. 4, p. 457).

The nineteenth-century tradition whereby this event also saw the destruction of the old royal castle is based on an inaccurate reading of the medieval narrative of events, but the burgh proper suffered badly through direct flooding from the Tay itself and indirectly from the Almond, perhaps via the town's mill lade from that river. The Tay rose very rapidly, taking by surprise the king and his household who were then in residence and forcing them to take to boats or seek refuge on the upper floors of buildings to escape the water. For several years after the event, once valuable burgage plots in the districts closest to the river remained derelict, suggesting a lingering fear of a repeat episode, perhaps made real by the steadily deteriorating weather from the later 1250s onwards. The river level was unpredictable at the best of times, for rapidly melting snow or heavy rainfall over the distant mountains could see the river surge in spate sixty or more miles distant, and the transition from the MCA into the unstable regime of the late thirteenth century delivered a new source of anxiety to those who lived along the banks of the Tay.

Late-lying spring snow east of Milquhanzie Hill, Perth and Kinross, reveals a palimpsest of past and present upland land-use. As in the middle of the twelfth century, the first decade of the thirteenth century saw a succession of extreme weather events, including two severe winters that are recorded as devastating livestock in the south-east of Scotland and delayed spring planting.

and continuing into 1196.[95] In this last year, untimely rain was blamed for the famine continuing and also for the pestilence that afflicted the starving people.[96] Reference to these events affecting Scotland occurs only in 1196, possibly following a poor harvest there the previous year.[97] Until this final episode, Scotland may have been buffered from some of the worst effects of the year-on-year harvest failures affecting England and northern Europe by its more mixed agricultural economy. Destructive storm spates and river-floods in Scotia (Scotland north of the Forth) in late summer 1199 brought what had otherwise been a century of relatively benign weather to a miserable end.[98]

Climate and weather III: passing the peak

A slight dip in the precipitation record of the Traligill speleothems in the 1200s might not, *prima facie*, appear significant.[99] Weather models, however, indicate that it marks an episode of slightly weaker NAO sufficient to cause successive poor winters in eastern Scotland, dealt by a more northerly or easterly airflow encountering moist Atlantic air over the country. The first bad winter came in 1204/5, with a dry, intense cold indicative of a northerly airflow delivering sharp frosts in which large numbers of livestock perished. The weather was recorded in the

95 William of Newburgh, *Historia Rerum Anglicarum*, book 5, in R. Howlett (ed.), *Chronicles of the Reigns of Stephen, Henry II and Richard I*, volume 2 (London, 1885), pp. 460, 484.

96 Ibid., pp. 484–485.

97 Bower, *Scotichronicon*, vol. 4, p. 419.

98 Ibid., p. 423.

99 Proctor et al., 'A thousand year speleothem proxy record', p. 818 fig. 4.

chronicle of Melrose Abbey, which sustained substantial losses to sheep flocks and cattle herds that year.[100] In west-central Ireland, the Loch Cé annalist reported this episode as bringing snow that lasted into mid March.[101] This was the first European Great Winter of the thirteenth century, lasting for up to six months and bringing an intensity of cold that was justifiably said to have been unexperienced within living memory.[102] Three centuries later, in his Scots translation of Hector Boece's Latin chronicle, John Bellenden described this memorably as a winter of bitter frost that lasted through into March and delayed the spring ploughing, with heavy snowfall at the winter's end causing many deaths amongst cattle and sheep. It was so cold, Bellenden claimed, that 'þe ayill was sa frosin þat it was sald be pund wechtis' (the ale was so frozen that it was sold by pound weights).[103] A second bad winter possibly followed in 1205/6, according to the Melrose chronicler, and again it was in the early spring that the worst snows arrived, with a major fall reported in late April, a time when significant damage to budding fruit trees and growing crops was likely to occur.[104] After a brief lull, early autumn 1209 brought intense rainfall in time to destroy crops throughout the Tay basin, where flooding caused by a combination of spate conditions, high tides and a storm surge up the river's tidal reach devasted Perth, swept away its bridge and forced the king to escape with his brother and only son in a rowing boat.[105] A final very harsh winter spanned November to April 1210/11, with a return to extreme cold as in 1204/5 and the consequent loss of large numbers of livestock 'in a disaster of incredible proportions'.[106] Then, as abruptly as it had come, it was over, as the NAO strengthened and the flow of cold air from the north and east passed.

Chronicle silences on weather matters for the next decade probably represent a run of mild and

100 *The Chronicle of Melrose* (facsimile edition), ed. A.O. Anderson et al. (London, 1936), sub anno 1205. This event is repeated in Bower, *Scotichronicon*, vol. 4, p. 437, with additional reference to the deaths of wild creatures.

101 Annals of Loch Cé, LC1205.3.

102 Pfister et al., 'Winter air temperature variations', p. 545. A sense of its impact across England can be gauged from the geographical spread of reports of the winter in monastic chronicles: the chronicle kept at the West Country abbey of Tewkesbury reported frost as late as midsummer 1205: *Annales de Theokesberia*, in *Annales Monastici*, ed. H.R. Luard, 5 volumes (London, 1864–69), volume 1, pp. 41–190 at 57; at Winchester in Hampshire, the annalist reported twenty weeks of frost over this winter: *Annales Monasterii de Wintonia*, in *Annales Monastici*, vol. 2, pp. 1–126 at 79; at Bermondsey, east of London, the frost lasted for nine weeks from early January until early March: *Annales Monasterii de Bermundeseia*, in *Annales Monastici*, vol. 3, pp. 423–488 at 449; and at Worcester, it was reported that the frost lasted from the Feast of St Thomas (21 December) until the Feast of the Annunciation of the Virgin (25 March): *Annales Prioratus de Wigornia*, in *Annales Monastici*, vol. 4, pp. 353–562 at 393. For these and other reports, see also A.O. Anderson (ed.), *Early Sources of Scottish History to a.d. 1286*, 2 volumes (Edinburgh, 1922), volume 2, p. 367 note 2.

103 Boece, *The Chronicles of Scotland*, vol. 2, p. 214.

104 *The Chronicle of Melrose*, sub anno 1206; Bower, *Scotichronicon*, vol. 4, p. 437. The Melrose chronicler(s) possibly used sources with different dating systems, with 1204/5 aligning with a year that started on 25 March and 1205/6 possibly aligning with a 1 January new year. The winter of 1205/6 appears to have been mild and dry in northern France and Germany, which might support this suggestion of a duplicated incident in Scottish sources: Pfister et al., 'Winter air temperature variations', p. 546.

105 Bower, *Scotichronicon*, vol. 4, p. 457.

106 Ibid., p. 461. Although extreme frost from December 1210 to February 1211 is reported by the Tewkesbury chronicler (*Annales de Theokesberia*, in *Annales Monastici*, vol. 1, p. 59) and at Worcester priory, whose chronicler reported that the Severn froze over for four miles below Gloucester during this period (*Annales Prioratus de Wigornia*, in *Annales Monastici*, vol. 4, p. 398), this winter is not described in these terms in mainland Europe, suggesting that the cold was centred further north over Scandinavia than in 1204/5.

productive years that extended into the early 1220s. The first clear indications of poorer conditions setting in occur in southern English sources. At Bermondsey and Waverley in Surrey, the monastic chroniclers reported dreadful weather in April 1222, with windstorms flattening trees followed by a prolonged frost that withered leaf buds and blighted blossom, ruining the fruit crop.[107] To make matters worse, this cold was followed by a long drought which delayed the grain crop. The Bermondsey chronicler added that gale-force westerlies in mid December 1222 further devastated woodland. There are indications of a similar deterioration affecting western Scotland from late summer 1221, when King Alexander II aborted his naval expedition into the Hebrides on account of a storm, but it was autumn 1222 that saw the most significant deterioration at the end of his expedition into Caithness, when his army was hit by 'a very great storm, with floods of rain' and progress was obstructed by rivers in spate and widespread flooding.[108] Exceptionally windy conditions were also reported by the Loch Cé annalist for 1222, but it was gales starting on 22 September 1223 that caused the worst damage, flattening much of the oat crop shortly before harvest started.[109] In Scotland, the east of the country was struck on 20 December 1223 by blizzard conditions 'the like of which had not been seen before [lasting] for so long, as far as those alive on the land at the time could remember'.[110] The depth of the snow killed many cattle outright and, when it froze, left roads impassable, grazing inaccessible and mills deprived of power when their lades and ponds froze solid. There is no explicit reference to similar weather in English or Irish accounts, but in Ireland 1224 was very poor generally. In Ulster, it was reported that the '1224' harvest was brought in only at the start of February 1225 in a year that saw 'great destruction of people'.[111]

Our climate proxies mark this as a period of weakening winter NAO, when the prevailing Atlantic westerlies encountered more pervasive cold conditions brought by anticyclonic winds from the north and east. Although there was an appreciable drop in signals for precipitation in the Traligill speleothem record, winters were still wet in the north and west of the country, but probably more often with snow than with rain.[112] We are fortunate that the third current suite of climate proxy data, the Cairngorm tree-ring sequence, commences around 1200 and gives us insight on summer conditions in the central and north-east Highland uplands in this same period. Summers in this region were generally cooler from the start of the century, declining slightly through the 1210s but becoming significantly cooler through the 1220s, with 1227 being the fifth coldest across the entire 800-year sequence.[113] All told, the decade seems to have been marred by cold and wet summer and winter weather.

The extreme European winter of 1233/4, which has been described as 'the most outstanding

107 *Annales Monasterii de Waverleia* and *Annales Monasterii de Bermundeseia*, in *Annales Monastici*, vol. 2, pp. 296, 455.

108 Both these events are reported in the probably mid thirteenth-century *Gesta Annalia I* section of John of Fordoun's later fourteenth-century chronicle (*Chronica Gentis Scotorum*, vol. 2, pp. 284, 285), but not in Melrose or Lanercost.

109 Annals of Loch Cé, LC1222.3, LC1223.6.

110 Bower, *Scotichronicon*, vol. 5, p. 117. The source of Bower's information is unknown, occurring in neither the Melrose chronicle nor in Fordoun's *Gesta Annalia I*. None of the weather events of the 1204/5 to 1225 period are reported in Lanercost.

111 Annals of Ulster, U1225.8, U1225.9. Heavy snowfall and thunderstorms were reported in Hampshire at midsummer 1225 (*Annales Monasterii de Wintonia*, in *Annales Monastici*, vol. 2, p. 84), but not in any other account.

112 Proctor et al., 'A thousand year speleothem proxy record', p. 818 fig. 4; Campbell, *The Great Transition*, p. 46 fig 2.6 (A).

113 Rydval et al., 'Reconstructing 800 years of summer temperatures', pp. 2959 fig. 5 (for the northern Cairngorm temperature sequence), 2960.

Scots pine woodland, Glen Derry, southern Cairngorms, Aberdeenshire. The processed dendrochronological data from the Cairngorm pines kicks in from the early thirteenth century and provides a vital source of information for annual variations in temperature and precipitation in the north-eastern Highlands.

meteorological episode of the thirteenth century',[114] received little reporting in England and none in contemporary Scottish records, despite proxy data which suggests that this was indeed a time of great cold. A single reference from western England, made in the Worcester chronicle, to severe frost between Christmas 1233 and mid January 1234 states that the autumn-sown grain was largely lost, but otherwise describes this event in less graphic terms than the severe winter of 1210/11.[115] In Ulster, it was likewise reported that the cold spell started at Christmas with snowfall that lasted through to 6 January, then continued with an extended period of bitter cold that saw loughs and rivers freeze.[116] Unlike the reports of previous events, the annals say nothing about longer-term impacts on the harvest.

It is the proxy record that again indicates Scotland's experience of these events through the decade of the 1230s. At Traligill, the speleothem growth rings indicate that these years were the driest decade of the thirteenth century, albeit with still relatively high levels of precipitation in comparison to the drier eleventh century.[117] What this points to is a further marked weakening of winter NAO which brought a much colder anticyclonic airflow pulling Arctic air south and across Scotland. This cooling is present

114 Pfister et al., 'Winter air temperature variations', p. 545.

115 *Annales Prioratus de Wigornia*, in *Annales Monastici*, vol. 4, p. 425. See note 106 above.

116 Annals of Ulster, U1234.4.

117 Proctor et al., 'A thousand year speleothem proxy record', p. 818 fig. 4; Campbell, *The Great Transition*, p. 46 fig 2.6 (A).

Post-medieval shieling hut, Barvick Burn, Crieff, Perth and Kinross. The cool and unsettled conditions of the mid thirteenth century may have affected dairying at the upland, seasonal pastures used in short-range transhumance more than the arable lowland districts.

also as a sharp dip in the Loch Sunart deep water temperatures – followed by an equally sharp positive movement – perhaps pointing to a combination of cooler ocean surface temperatures and cold run-off from the surrounding land.[118] In the Cairngorms, summer 1232 was the coldest of the previous 800 years.[119] Collectively, these fragments of evidence point to the first half of the 1230s as a period of sharply deteriorating summer and winter weather conditions, but only from south-western England is there a suggestion of direct impacts on growing crops, in contrast to the previous cold episodes of the early 1210s and mid 1220s. Scottish proxy data suggests that high-altitude summer grazing might have been affected temporarily, but there is no indication of wide-ranging or lasting consequences for Scotland's people.

Although the 1240s and early 1250s saw a return to the dominant weather trends that had prevailed since the end of the eleventh century, the successive shocks delivered by the recurrent cold and wet episodes signalled that the generally milder conditions of the MCA were becoming less stable. Few at the time might have noticed, as the middle years of the thirteenth century were marked by a run of good years with warm summers and good harvests for

118 Cage and Austin, 'Marine climate variability', p. 1641 fig. 7.
119 Rydval et al., 'Reconstructing 800 years of summer temperatures', p. 2960.

some or, at worst, unremarkable but acceptable conditions.[120] There were extreme episodes, but of heat rather than of cold and wet. In north-east and west-central Ireland, summer 1252 was especially hot, with drought affecting the Shannon, the harvest being brought in by early August and trees withering.[121] It was, however, the summer of 1253 that brought out the superlatives, delivering an abundance of acorns, fruit, corn and healthy livestock, while reports for 1254 were only slightly less exuberant.[122] But these were the last summers of plenty for decades to come. From 1256 and its awful weather and through the 1260s, it becomes clear to us modern observers that the halcyon days of the MCA were over. While it had persisted, the lot of most of Scotland's people is likely to have improved to the extent that even the relatively poor years with which the era is punctuated nevertheless delivered adequate enough conditions and sufficient crop yields to avert major crises. Times might still have been hard and hunger never far from the door of many peasant families, but the generally more benign conditions provided opportunities to expand production, deliver surpluses of staples and permit diversification into the production of commodities that could be traded for cash that could buy luxuries in times of abundance or provide a buffer for subsistence in times of dearth.

• • •

Weather, as reviewed in this chapter, is but one of the key influences to exert agency over the broader developmental trends that were occurring across the MCA. It created opportunity, presented challenges and stimulated responses from the human population – and from plant and animal communities – that helped to shape Scotland's environmental trajectory. But weather was not a determinant that only constrained responses: it worked in conjunction with human agency and other factors to stimulate a range of decisions and deliver outcomes, good and bad, that moved us further down particular paths. Perhaps one of the most enduring images of this era of benign weather and infrequent harvest failures is that of the rapid expansion of the area under permanent cultivation to feed Europe's burgeoning population. Peasant land-hunger and pressures to produce more dietary staples and raw materials to satisfy rising market demand for finished products, especially woollen cloth and leather, were met through expansion into zones that had seen no intensive exploitation possibly for centuries. Scotland's experience of this era of growth has often been viewed in isolation, but it was part of a pan-European phenomenon that saw the expansion of what is known as Latin Christian culture from its centres in what are now Italy, Germany and France. Physical expansion saw this culture spread to Norse Greenland in the far northwest, through the movement of German peasant settlers into the Slav lands east of the Elbe, through the English and Norman invasions and colonisation of Ireland, and through the planting of northern European knights and their dependents into the eastern Mediterranean via the Crusades and into the Iberian lands at the spearhead of the Reconquista.[123]

120 The only significant adverse event is a report of two weeks of snow, ice and strong winds in Surrey in December 1241, reported in the Annals of Waverley: *Annales Monasterii de Waverleia*, in *Annales Monastici*, vol. 2, p. 329. In Ulster, 1249 brought a good acorn crop, signalling warm, moist conditions throughout the summer in which oak trees flourished: Annals of Ulster, U1249.5. In the Irish mid-west, the Annals of Loch Cé reported violent thunder and lightning in summer 1251 and violent downpours with localised flooding: LC1251.9, LC1251.2. This was followed by a cold snap in early winter in which the loughs and bogs of the region were frozen: LC1251.12.

121 Annals of Ulster, U1252.1; Annals of Loch Cé, LC1252.9.

122 Annals of Loch Cé, LC1253.12, LC1254.13.

123 This pan-European process is described in Bartlett, *The Making of Europe*.

Scotland experienced its share of these inflows of colonists, many from central and eastern England but others from mainland Europe, Flanders in particular.[124] In common with other European regions, however, it also witnessed internal colonisation, where peasants were relocated to provide muscle for pioneering ventures that brought more land into cultivation.[125] There is a long tradition of seeing the Cistercian monks introduced to Scotland in the 1130s as active in these processes of colonisation and clearance. Cistercian lay brothers are regularly depicted in English and European manuscript illustrations as clearing woodland, digging drainage ditches and harvesting grain. Much of that imagery, however, is recognised as Christian allegory (making the wilderness blossom = finding Christ in a spiritual desert) or part of a polemical debate that advanced the superiority of Cistercian monasticism over that of other orders.[126] Secular landowners, however, were as attuned to the new market opportunities for the produce of their lands, with settlement and clearance directed by such lords evident in the valleys of the eastern Borders, upper Clydesdale, the Carse of Gowrie and the districts enclosing the inner Moray Firth.[127] Indeed, all over Scotland settlement was advancing and the exploitation of natural and farmed resources intensifying. Historical discussion of such colonisation has viewed it traditionally in purely human terms, tracing colonists' origins and analysing their social, economic and, especially, cultural impact on the receiving territory.[128] How the colonists transformed the face of the land onto which they were settled, and how they exploited it to deliver the commodities they wanted, have been less well explored, but some closely focused studies provide insights on the nature and scale of their impact.[129] It is to the wider ramifications of these processes that we turn in the following chapters.

124 G.W.S. Barrow, *The Anglo-Norman Era in Scottish History* (Oxford, 1980); L. Toorians, 'Twelfth-century Flemish settlement in Scotland', in G.G. Simpson (ed.), *Scotland and the Low Countries 1124–1994* (East Linton, 1996), pp. 1–14; R.D. Oram, 'A family business? Colonisation and settlement in twelfth- and thirteenth-century Galloway', *Scottish Historical Review*, 72 (1993), pp. 111–145; R.D. Oram, 'Castles and colonists in twelfth- and thirteenth-century Scotland: the case of Moray', in *Chateau Gaillard*, 22 (2006), pp. 289–298; Oram, *Domination and Lordship*, chapters 7 and 9; Oram, *David I, King of Scots*, chapters 9–10.

125 Duncan, *Scotland*, p. 327.

126 C. Berman, *The Cistercian Evolution: The Invention of a Religious Order in Twelfth-Century Europe*, illustrated edition (Pennsylvania, 2010); R.D. Oram, 'Landscape, land use and environment', in B.M. Kaczynski (ed.), *The Oxford Handbook of Christian Monasticism* (Oxford, 2020), pp. 418–434.

127 Oram, 'Castles and colonists'; R.D. Oram, 'Estuarine environments and resource exploitation in eastern Scotland *c.*1125–*c.*1400: a comparative study of the Forth and Tay estuaries', in E. Thoen et al. (eds), *Landscapes or Seascapes? The History of the Coastal Environment in the North Sea Area Reconsidered*, Comparative Rural History Network 13 (Turnhout, 2013), pp. 353–377.

128 See especially Barrow, *The Anglo-Norman Era*, chapters 1–4.

129 For example, J. Gilbert, 'The monastic record of a Border landscape 1136 to 1236', *Scottish Geographical Magazine*, 109 (1983), pp. 4–15. The surviving charters of Dryburgh and Melrose abbeys on which this study is based reveal lay landlord and peasant tenant clearance efforts.

CHAPTER 7

Forests and Hunting Reserves

> William, by God's grace King of Scots, to all his worthy men of his whole land, clerical and lay, greetings. Let them know, present and future, that I have given, granted and by this my charter confirmed to God and to St Mary of Coupar Angus and to the monks there serving and to serve God, a half carucate of land as a site for making their abbey, and also Campsie, namely my chase, with all my waste which pertains to that land. To hold and have by them in pure and perpetual alms, as freely, quietly, fully and honourably as any alms of all my land is held and possessed more freely, quietly, fully and honourably. I wish also and command that the aforesaid monks have and hold the aforesaid land, and chase, and waste in free forest, and no one is to cut timber or hunt deer within the same land without licence of the said monks, on pain of the king's full forfeiture of £10.[1]

In the context of debates over the state of Scotland's environment at the beginning of the MCA and the impacts of the expansion of agriculture it witnessed, it may appear counterintuitive to begin a deeper examination of changing environmental conditions in this era with an exploration of the interlinked issues of woodland and hunting. As seen in earlier chapters, however, the extent of woodland and the management of reserves – for timber conservation or for specialised activities like hunting – has been used as a proxy for assessment of wider environmental health. By the end of our discussion of woodland management at the close of the first millennium in Chapter 2, the area of Scotland's land surface under trees was probably already similar to our present coverage and falling, as a combination of increasing human pressures and the vagaries of climate and weather had an impact on woodland clearance and regeneration. In the present chapter and in Chapter 8 we will examine the processes that contributed to what appears to have been one of the sharpest declines in the extent of the surviving woodland areas across much of mainland Scotland since the rapid losses of the Neolithic and Bronze Age periods. It is one of the ironies of this era that the more intensive management that is documented by the growing volume of parchment records

1 *Regesta Regum Scottorum*, volume 2: *The Acts of William I 1165–1214*, ed. G.W.S. Barrow (Edinburgh, 1971), no. 154.

from eastern, central and southern Scotland also saw the most intensive exploitation of the remaining wooded areas in these regions. Such records offer fascinating insights on the conflicting demands and needs of different users and, more fascinating still, on the degree to which peasant land-hunger ultimately forced lords into slackening controls over expansion into woodland and progressively eroded the already shrunken extent of Scotland's forests.

Creating reserves

When record evidence for woodland begins to appear in the twelfth century, it seems *prima facie* to be already concerned with its protection from clearance by land-hungry peasants. Much of it charts efforts to impose limitations on the range and duration of peasant exploitation of wooded areas, in almost every instance involving measures linked to the safeguarding of deer hunting privileges reserved to the king or favoured lay magnates. The preservation of trees was certainly not just about the protection of game species habitat, but in most acts for the reservation of hunting rights the two issues are inseparable. According to most traditional historical narratives, the legal establishment of large game hunting as a reserved activity was an innovation introduced by Scotland's supposedly anglicised and Francophile rulers after 1124, as an import from England where it had itself been an introduction from Normandy in the later eleventh century. That view is well founded in the surviving parchment record, but it ignores the Pictish sculptural evidence discussed in Chapter 5. There is no rational explanation for this hard-edged exclusion of the Pictish representations of apparently noble horsemen engaged in chasing deer with hounds, other than that the twelfth-century and later privilege is traceable in the law codes of the time as a royal right alienated to chosen nobles through an act of jurisdictional enfranchisement, while the Pictish activities have no such association with a jurisdictional right to control access to hunting. It seems perverse, however, to dismiss depictions of what was clearly an elite social activity in the Pictish period simply because they are not supported by a similar representation of social exclusivity in some early written record. Apart from their supposed personal passion for hunting and their enthusiastic embracing of the seventh-century Merovingian Frankish innovation of deer hunting as an elite preserve, there has been no satisfactory explanation for why Scotland's kings imported the cultural package of forest law – regulations that established limitations both on who could hunt what and where and on the wider economic exploitation of land reserved for hunting – and hunting forests into their kingdom.[2] Indeed, its appearance as a topic for discussion in general histories of medieval Scotland gives a sense of something that just 'happened'. Contrary to this popular mythology, however, the woodland in question was already perceived as 'private' property before the 1100s,[3] being a possession of the king or of one of the great aristocratic kindreds who dominated the provinces of the kingdom. It is uncertain if hunting within that woodland had also been 'priva-

Opposite. Aberlemno roadside cross-slab, Aberlemno, Angus. Section from the reverse face of the eighth-century cross-slab showing high-status mounted hunters with hounds, pursuing deer.

2 The most detailed discussion of the introduction of hunting reserves in the twelfth century remains J.M. Gilbert, *Hunting and Hunting Reserves in Medieval Scotland* (Edinburgh, 1979), chapters 1 and 2.

3 Earl (later King) David's *c.*1120 'foundation' charter of Selkirk Abbey gave the monks wood from 'boscos meos' (my woods) for building and firewood needs, indicating private ownership of some woodland and suggesting that there were other woods that were not his: G.W.S. Barrow (ed.), *The Charters of David I: The Written Acts of David I King of Scots, 1124–53, and of His Son Henry, Earl of Northumberland, 1139–52* (Woodbridge, 1999), no. 14.

tised' as a reserved activity for the region's social leadership,[4] but what was reserved for twelfth-century royal hunters is stated clearly. In Annandale in 1194 x 1214 it was 'stags and hinds, wild pigs and roe deer' and in 1230 in Moray it was 'stags and hinds, roebucks and does, wild boar and sows'.[5] Here, we see the same repertoire of wild beasts as was represented in sixth- to ninth-century sculptures.

Parallel to the notion of twelfth-century novelty runs a deeply ingrained popular assumption that prior to King David I's apparent introduction of Norman-French forest law, all game was regarded in Scotland in terms similar to the principle of *res nullius* in Roman law, according to which it 'belonged' to no one and became the property of the hunter regardless of the ownership status of the land on which it was killed.[6] As explored in Chapter 5, however, hunting of particular game – deer and wild pigs especially – gained an elite association probably in later prehistory, as illustrated by the competition for the award of the 'hero's portion' of the catch recorded in early Irish poetry. Whether this was in any sense formally expressed as a reservation of hunting rights to certain people is unknown, but the depictions of hunting deer with hounds in Pictish sculpture are strongly suggestive of its being an activity already restricted by the eighth century to those at the apex of the social hierarchy. Also unknown is when that limitation was extended to a reservation of the right to hunt to the owner of the land on which the hunting took place, but the territorialisation of power that was in train in the eleventh and twelfth centuries suggests that it is again likely to have predated the introduction of Norman-French legal practices.[7] And what of the assignation of their introduction to David I? There is no surviving manuscript gathering of 'forest laws' from earlier than the fourteenth century and those that do exist are clearly composites comprising enactments instituted for different purposes at different times rather than one code enacted in a single process.[8] The attribution of the codification of the laws to King David is a modern assumption based on the traditional assignation to him of the first substantial codification of Scots laws.[9] Although the language of forest law was clearly employed in Scotland from David's reign, it seems likely that the formalisation of a suite of forest laws belonged to the reign of his grandson William (1165–1214) or of his great-grandson Alexander II (1214–49), who used attribution to David to lend antiquity and authority to legal innovation in their day.[10]

4 Ibid., where Earl David also granted the monks the teind of the skins of all stags/bucks and hinds/does caught by 'veltrarii mei' (my fewterers or greyhound keepers), pointing to specialist deer hunting with hounds.

5 Sir William Fraser (ed.), *The Annandale Family Book of the Johnstones, Earls and Marquises of Annandale*, 2 volumes (Edinburgh, 1894), volume 1, pp. 1–2 no. 2 ('cervo et bissa, porco et capreola'); S.R. MacPhail (ed.), *History of the Religious House of Pluscardyn* (Edinburgh, 1881), pp. 69 (where the editor translates 'caprosum' as 'goat' rather than roe deer), 199–201 ('cervum et cervam caprosum et caprosam, aprum et apram').

6 This view was promoted stridently in the early nineteenth century by the Celtophile James Logan in his *The Scottish Gaël, or Celtic Manners as Preserved among the Highlanders*, 2 volumes (London, 1831), volume 1, chapter 9, in an extended rant against the territorially and socially exclusive and oppressive hunting practices of Highland landowners and the impact of the introduction of large-scale sheep farming on former hunting land. For discussion of the nineteenth-century development of deer forest, see the final companion volume to the present one, *Standing on the Edge of Being*.

7 Taylor, *The Shape of the State*, pp. 18–19, 26, 176–179, 184–187.

8 Gilbert, *Hunting and Hunting Reserves*, chapter 13.

9 Ibid., p. 244.

10 The case for the later attribution of laws to David I by kings from William through to Robert I in the first quarter of the fourteenth century is made in A. Taylor, '*Leges Scocie* and the lawcodes of David I, William the Lion and Alexander II', *Scottish Historical Review*, 88:2, no. 226 (October 2009), pp. 207–288, and A. Taylor, 'The assizes of David I, king of Scots, 1124–53', *Scottish Historical Review*, 91:2, no. 232 (October 2012), pp. 197–238.

Having been raised from 1093 in the household of England's Norman kings, imbued with the elite culture of England's rulers and from 1115 admitted to the topmost ranks of that society as earl of Huntingdon, David I was intimately familiar with Norman-French hunting practices and forest law. The men he brought into Scotland in the 1110s, and then in increasing numbers after he became king in 1124, were also deeply familiar with England's hunting law and its language. The appearance of that language in parchment records from David's reign does not necessarily imply the conscious importation of innovative hunting practice and land management for hunting, but rather the employment of familiar terminology by the incomers to describe what they encountered in Scotland.[11] For example, when the king confirmed the privileges of Dunfermline Abbey in *c.*1128, they included the teind of his *venacio* in Fife, using the Latin term for hunting which already carried a specific association in Anglo-Norman territories with deer (and from which the modern name for deer meat, venison, is derived).[12] The implication in this grant is that this *venacio* was reserved to the king and that the reservation was something that predated David's reign. Around a decade later, the term 'foresta' – perhaps already meaning both forest in the woodland sense and land that was reserved for hunting – had also established itself in legal terminology within royal charters, being applied to a wooded component of royal land.[13] Again, there is no sense of innovation in the distinction between land that was largely under trees and used for hunting and land used for agricultural purposes. Such documents are concerned with long-established woodland economic activities like seasonal pig and horse grazing or firewood gathering as well as hunting. The novelty is in the vocabulary and a progressively formalised compartmentalisation of the land into zones of hunting and non-hunting function.

During David's reign, specific named royal 'forests' are mentioned at Selkirk, Traquair, Stirling and Clackmannan and 'between Gala and Leader', and soon after 1150 he gave Robert de Brus II the right to have a forest in his lordship of Annandale. Within that district, Robert received exclusive rights to hunt – presumably empowering him to define reserved hunting zones – and to control movement of people, thereby preventing free access to its other resources.[14] This apparent southerly distribution of forests is perhaps an accident of the survival of written records, for named forests at Drimmie and Clunie in eastern Perthshire connected with David's castle at Clunie are recorded by *c.*1161, and rights to have a forest at Dunlappie in Angus were given to the native nobleman Aodh son of Orm before 1165.[15] These references suggest that such royal reserves were associated with most royal castles throughout southern, central and north-eastern Scotland by the mid twelfth century. It would seem, then, that when David reserved to himself the hunting of large game

11 Gilbert, *Hunting and Hunting Reserves*, p. 13, suggests an incremental development but concludes that David established forests as a conscious act at some point in the 1130s.

12 Barrow, *The Charters of David I*, no. 33.

13 Ibid., no. 98. The king's 'forests' of Selkirk and Traquair were named as such between 1136 and 1142 in the 'foundation' charter of Melrose Abbey. These had economic functions different to and distinct from the 'wood' ('boscus') of Gattonside assigned to them in the same charter, principally pasture for horses and oxen, seasonal acorn grazing for pigs, firewood and building materials in the forests but with no defined activities in the *boscus*: ibid., no. 120. John Gilbert links the term 'boscus' with managed, coppiced woods: J.M. Gilbert, 'Place-names and managed woods in medieval Scotland', *Journal of Scottish Names Studies*, 5 (2011), pp. 35–56 at 38.

14 Barrow, *The Charters of David I*, nos 120, 147, 210.

15 *Regesta Regum Scottorum*, volume 1: *The Acts of Malcolm IV 1153–1165*, ed. G.W.S. Barrow (Edinburgh, 1960), nos 226, 311.

Looking west across the Forest of Clunie from the park at Buzzart Dykes, Perth and Kinross. One of a series of mid twelfth-century royal hunting reserves strung along the southern edge of the Mounth from the Stormont district north of Perth to Kincardine, Clunie covered a swathe of upland between the valley of the Lunan Burn west of Blairgowrie on the south and Strathardle on the north. Parts of this upland zone had been settled and cultivated in prehistory and were still exploited as extensive pasture in the twelfth century.

within the bigger woodland areas on his own properties, giving certain favoured nobles – both native and incomer – and senior clerics and monastic communities those same rights over some of their land, he was continuing established practice. In all such cases, the primary concern was the reservation of deer for hunting by the king or by a chosen few amongst the nobility. Lesser game, principally hares and, once they had become feral rather than purely domesticated, rabbits, lay outside forest constraints, but from at least the later 1150s grants of what were referred to as warren rights reserved to their recipients the hunting of smaller game animals on their property.[16] In this development, we might finally be seeing the importation of a novel concept, extending the exclusivity of hunting rights from deer and wild boar to lesser and smaller game animals.

Controlling and managing the forests

Before examining the exploitation of these 'forest' environments for pasture and cultivation, let us consider how they were used in practice for hunting. The office of royal forester, who had oversight of all activities within a forest, existed by the 1140s and possibly before the 1120s.[17] Men who had custody of royal forests, styled 'keepers' in the later Middle Ages, and sought to preserve their hunting quality are recorded from the reign of King William. A succession of royal writs, brieves and records of their actions to protect royal hunting interests reveals the nature and extent of their roles and authority.[18] A principal concern was the maintenance of the *vert*, literally the green vegetation, within the forest area. The *vert* provided both cover and fodder for the deer,

16 Gilbert, *Hunting and Hunting Reserves*, p. 208.

17 Barrow, *The Charters of David I*, nos 147, 265.

18 The most detailed discussion of the protective role of forester concerns the Morville family and the royal forest between Gala and Leader through the later twelfth century: Gilbert, 'The monastic record of a Border landscape'.

The 'forest between Gala and Leader', looking north from the Eildon Hills over Melrose towards the Ale Valley and the upland heart of the afforested area that was the focus of competing interests between the Morville lords of Lauderdale (the king's foresters there), who were charged with protecting the king's hunting, the abbeys of Dryburgh and Melrose, who wanted to expand their grazing within the forest, and the peasant farmers of Lauderdale and Wedale who sought to develop new arable assarts there.

but it was also attractive to browsing cattle and to peasants for smallwood cutting.[19] Many mandates to foresters concerned supervision of timber extraction by third parties to whom the possessor of the forest had granted felling rights for building or fuel needs,[20] reflecting the foresters' duty to ensure that woodland cover was not so badly damaged by irregular or unregulated felling as to affect the hunting, but also charging them not to interfere with the fellers' activities unless they threatened the overall forest quality. Such a threat to the extent of *vert* through clearance or encroachment for cattle grazing underlay a highly

19 Gilbert, *Hunting and Hunting Reserves*, pp. 103–104

20 See, for example, the grants of timber cutting rights and instructions to royal officers in the forest or wood of Clackmannan before 1150: Barrow, *The Charters of David I*, nos 126, 135, 147, 135. In the early 1200s, Walter II son of Alan granted the monks of Paisley access to building timber and fuelwood in his Renfrewshire forest, giving them access to 'the green' (live trees) for the timber and 'dry or dead wood' for the fuel, without restriction, but their men were to cut timber and gather fuel under his forester's supervision: C. Innes (ed.), *Registrum Monasterii de Passelet 1163–1529* (Edinburgh, 1832), p. 17 (hereafter *Paisley Registrum*).

The south-eastern edge of the Forest of Atholl, Perth and Kinross, looking towards the Beinn a' Ghlo massif, with Airgiod Bheinn in the centre of the image. The low hill in the valley to its right is Meall na h-Eilrig (the Lumpy Hill of the Deer Trap), where the deep folds between the main summits and the Meall formed natural traps for deer driven over the open ground.

informative dispute between the Morville family and Melrose Abbey in the forest between Gala and Leader in the Scottish Borders. A negotiated settlement gave both parties the right to have a forester to oversee activities there and protect their lords' interests.[21] A second but associated concern was the monitoring of activities that affected the *fermeisun* or close season, to prevent the disturbance of hinds or does with fawns, for which the *vert* provided protective cover. In the early 1200s, on the Steward of Scotland's Renfrewshire lands, differential fines for grazing infringements by the herds of Paisley Abbey in the Steward's forest in what are now known as the Fereneze Hills were levied for open and close seasons, with the higher fines imposed for the close season.[22] How the Avenel lords of Eskdale were able to police the activities of the monks of Melrose in western Ettrick, where they had granted the abbey pasture and wood-cutting rights in their baronial forest, but prohibited them from hunting anything other than wolves, without a forester to supervise their activities, remains a mystery.[23] These penalties and the

21 The beginning of Melrose's encroachment can be seen in a grant to the abbey of access to cow pasture in the district in 1162 x 1165: *Regesta Regum Scottorum*, vol. 1, no. 235. The escalation of conflict between the Morville foresters and the monks is evident in the 1180 settlement between Richard de Morville and Melrose: *Regesta Regum Scottorum*, vol. 2, no. 236. By 1185, however, the abbey had again received a further extension of its rights over those of the Morvilles: *Regesta Regum Scottorum*, vol. 2, no. 265. See Gilbert, 'The monastic record of a Border landscape', and Duncan, *Scotland*, pp. 420–421.

22 Gilbert, *Hunting and Hunting Reserves*, pp. 192–194; R.D. Oram, 'Paisley Abbey and its estates in the Middle Ages', *Renfrewshire Local History Forum*, 18 (2016), pp. 5–23. The name 'Fereneze' arises from a misreading of the Latin 'in fermeisun' (in the close season) as a place-name.

23 C. Innes (ed.), *Liber Sancte Marie de Melros*, 2 volumes (Edinburgh, 1837), volume 1, no. 39 (hereafter *Melrose Liber*). The monks infringement of their privileges resulted in a later Avenel lord of Eskdale taking his dispute with Melrose to the king: ibid., no. 198.

prolonged dispute between the Morville keepers of the royal forest between Gala and Leader and the monks of Melrose underscore the lengths to which possessors of hunting rights went to protect game and the nature and level of the challenges they faced.

A particular issue for foresters was the monitoring of livestock grazing in upland areas. The presence of flocks and herds with their keepers not only risked disturbance of game but could also interfere with styles of hunting which ranged over extensive districts.[24] Drives, where dogs and beaters drove the quarry towards hunters waiting at *trysts* or into *eileircg* (Gaelic, pronounced 'eilirig', an elrick or deer trap), or chases, where mounted hunters with hounds pursued deer for miles across country, were practised for centuries before 1100 and remained popular into the eighteenth century, when stalking re-established itself as the chief method of hunting. Numerous place-names containing the term 'eileircg' spread from Galloway to the northern Highlands chart the presence of these funnel-shaped traps where game animals were driven towards huntsmen waiting at the apex of the V for the kill.[25] The pursuit or stalking of individual animals is a low-impact form of hunting, but the potential mass-kill event and associated feasting culture represented by the *eileircg* had the potential to alter ecological balances within a hunting landscape.[26] The presence of domestic flocks and herds and their attendants, plus the construction of shelters, folds and fences for them, were unques-

24 For the types of hunt practised, see Gilbert, *Hunting and Hunting Reserves*, chapter 4.

25 A fine example of a surviving structure can be seen at Sròn an T-saighdeir on the island of Rùm, which was reserved hunting land for the Lords of the Isles in the later Middle Ages: https://canmore.org.uk/site/21933/rum-sron-an-t-saighdeir.

26 Mass-kill hunting and its ecological consequences have not been researched in Scottish contexts, but recent work in the Arabian peninsula on sites spanning at least four millennia has highlighted their potentially transformative impacts: H.S. Groucutt and W.C. Carleton, 'Mass-kill hunting and Late Quaternary ecology: new insights into the "desert kite" phenomenon in Arabia', *Journal of Archaeological Science*, 37 (2021), https://doi.org/10.1016/j.jasrep.2021.102995.

tionably impediments to these styles of hunting and underlay the Morville/Melrose dispute. The distribution of the related place-names and physical evidence for traps, however, suggest that huntsmen were not everywhere the losers, or at least not until the eighteenth-century development of intensive livestock management and new hunting methods in Highland districts.

One response to the incompatibility of hunting and other modes of exploitation within a forest was the creation of zones of greater restriction, principally for exclusive hunting or deer management purposes but also for better protection of the most valuable woodland resources. The first clear reference to the construction of one of these 'parks', as they were known, is dated 1165 x 1174 and pertains to King William's enclosure of what is still known as King's Park in Stirling.[27] The key features were physical demarcation and exclusive possession of the area within the enclosure; at Stirling, this required an exchange of property with Dunfermline Abbey.[28] The early park at Stirling formed a triangle roughly 1.6km in length on every side enclosing 1.1km^2 of land. While this is already a substantial enclosure, surviving park boundaries at Kincardine in the Mearns district of southern Aberdeenshire run for over 7.5km across what is now largely heather moorland at an elevation of up to 315m; a further 1.6km of surviving earthwork encloses a second area to the north of the main enclosure.[29] The Kincardine Park dyke is a massive structure that survives mainly as an earthwork bank formed from the upcast from an internal ditch, the form of which was established through excavation at two other park enclosures, Buzzart Dykes near Blairgowrie in Perth and Kinross and Durward's Dyke at Lintrathen in Angus. In these cases, evidence was found for posts to support a timber pale on the crest of the earthwork.[30] Together, pale, earthwork and ditch constituted a formidable boundary, high enough to prevent deer within the enclosure from escaping and so guaranteeing the king game to hunt when he visited Kincardine. Here, over 8km^2 of land was enclosed for the king's exclusive hunting, its trees and *vert* managed carefully, vermin and raptors eliminated and peasant graziers excluded. The park's boundary, moreover, required thousands of small diameter timbers to form the pale, with larger ones for the posts and stakes, underscoring the intensive coppice management of forest resources to provide the materials necessary to build and then maintain the enclosure. Kincardine Park is a remarkable monument to the long-term environmental legacy of this medieval landscape and land-use policy.

27 *Regesta Regum Scottorum*, vol. 2, no. 130.

28 Gilbert, *Hunting and Hunting Reserves*, pp. 82–83 and map 3. A second park at Stirling, the 'New Park', was established further south in the reign of Alexander III. A detailed account of land surrendered by a tenant for enclosure within his lord's new park relates to Yester in East Lothian, where in *c.*1250 Adam de Morham surrendered to Sir Hugh Giffard an extensive block of woodland for inclusion within the park 'wall', with a further strip three feet wide for digging a drainage gully to prevent erosion of the boundary: C.C. Harvey (ed.), *Calendar of Writs Preserved at Yester House 1166–1503* (Edinburgh, 1916), no. 16.

29 Gilbert, *Hunting and Hunting Reserves*, pp. 83–85; D. Hall, R. Oram and K. Malloy, 'A hunting we will go? Stirling University's medieval deer parks project', *Tayside and Fife Archaeological Journal*, 17 (2011), pp. 59–67; D. Hall and K. Malloy, '"Always chasing deer" – further excavations at the deer parks of Buzzart Dykes and Kincardine Park and new excavations at Kincardine Castle', *Tayside and Fife Archaeological Journal*, 21/22 (2015–16), pp. 25–34; K. Malloy and D. Hall, 'Archaeological excavations of the medieval royal Kincardine landscape, Aberdeenshire, Scotland,' *Medieval Archaeology*, 61:1 (2018), pp. 157–176.

30 K. Malloy and D. Hall, 'Medieval hunting and wood management in the Buzzart Dykes landscape', *Environment and History*, 25 (2019), pp. 365–390.

Bank and ditch of the northern perimeter of the King's Deer Park at Kincardine, Aberdeenshire, lying just to the north of the crest of Arnbarrow Hill. The snow is filling the ditch that lay on the inside of the bank which, when capped by a fence of tall palings, would have presented a formidable obstacle for any deer attempting to escape the emparked area. The scale of the earthwork gives a vivid impression of the control over human and natural resources that was exercised by Scotland's medieval kings.

Woodland: used, abused and protected

The widespread distribution of land designated as 'forest' for hunting has helped to embed the popular misconception that medieval Scotland was still heavily wooded and that the decline in woodland cover to modern levels was consequent on later medieval and post-medieval human action. As we have already seen in the introduction and Chapters 1 and 2, however, pollen evidence from sites spread across the country from Berwickshire to Sutherland and from the Northern Isles to Galloway indicates that tree cover was already at around its present extent in the late first millennium BC and that this situation was probably driven at least as much by climate change in later prehistory as by human interventions. During the MCA, inroads into woodland, including in areas where it had regenerated to a degree since the early post-Roman period, further reduced cover. Long before the MCA, moreover, even the greatest expanses of woodland, such as those in the central Borders region, the central Highland and Atlantic

A land bereft of trees? Upper Clydesdale in the twelfth and thirteenth centuries

It is possible to catch glimpses of the character of parts of the Scottish landscape in some of our earliest surviving written records of those areas. Charters and the place-names fossilised within them, references to modes of exploitation and jurisdictional rights, especially when coupled with archaeological and palaeoenvironmental datasets, modern place-name studies and sixteenth-century and later map evidence can throw otherwise poorly documented, tightly defined areas briefly into the spotlight. One such district is the highest part of the Clyde basin, which archaeology shows to have been well populated from the later Bronze Age through into the Roman era and until the early post-Roman period, before experiencing a long episode of relative emptiness that stretched into the 1100s. Resettlement of Upper Clydesdale commenced in the middle decades of the twelfth century, and by the 1190s a series of lordships had been created that stretched from Biggar southwards to the watershed with the basins of the Annan and Nith. Land which had been only seasonally exploited for summer pasture and hunting through the second half of the first millennium ad had been repeopled, mainly by Flemish settlers and their dependents. The biggest landholding at the valley head, however, which centred on Crawford and encompassed the headwaters of the Clyde and catchments of the Daer, Potrail, Elvan and Gonnar, was in the possession of the eastern English Lindsay family. Shortly before 1200, David de Lindsay awarded a large block of his property around Glengonnar to the west of Crawford itself to the monks of Newbattle Abbey in Midlothian (C. Innes (ed.), *Registrum S. Marie de Neubotle* (Edinburgh, 1849), no. 135), with David's sons and grandsons extending the award and granting wider rights within it to the middle of the thirteenth century.

Even as a remote landlord, Newbattle was active in exploiting its Clydesdale property as a grange for livestock grazing and as a source of lead from the mines it worked in the Lowther Hills. Those activities helped to form the modern appearance of the South Lanarkshire landscape, but they did not create it. Before the later twentieth-century introduction of extensive commercial conifer plantations over the hills south of Crawford, much of this district was grass and *Calluna* moorland that had supported sheep flocks since the Middle Ages. Earlier still, in later prehistory, the Celtic Iron Age cattle herders whose settlements clustered thickly through the valleys and hills of the central Southern Uplands worked what was an already largely treeless land as extensive cattle ranges. In common with other areas of human depopulation from the sixth century, however, whilst deteriorating climatic conditions had kept the higher districts open, there had been an expansion of birch-dominated scrubby woodland in the valley bottoms. English place-names like 'Byrkebanke' (birch bank), recorded in the 1220s (Innes, *Registrum de Neubotle*, no. 138), indicate the character and existence of such woodland at the time of the colonisation of the district around Crawford by English-speaking settlers. It is clear from such records and from palaeoenvironmental analyses that at the time of resettlement this was a landscape of scrubby woodland patches amidst open expanses of heather moorland and grass, rather than one swathed by trees.

By the 1230s, the birch woodland within the lordship of Crawford was under pressure. The Lindsays enjoyed free forest jurisdiction over the district by that date – the modern place-name 'Parkend' at the bottom of Glengonnar is a possible relict name related to an enclosed reserve there – this designation as forest reflecting the hunting value of the land rather than signalling any abundance of woodland. In the 1240s, Gerard de Lindsay resigned to the monks the forest and hunting rights that his predecessors had retained when they granted the land to Newbattle, reserving only the easement enjoyed by his tenants in Crawford of taking building materials in the wood of Glengonnar, albeit under the supervision of the monks' foresters (Innes, *Registrum de Neubotle*, no. 144). There is no evidence to suggest that Newbattle received royal confirmation of Gerard's alienation to it of his free forest jurisdiction, and the monks concentrated instead on maximising the extraction of lead from their properties in Glengonnar and Glencaple. Any significant remaining birch woodland had disappeared by the early fourteenth century, by when all reference to forest rights, foresters, wood easements and building timber had ceased in the parchment records of Newbattle's lands around what became known as Crawfordmuir. When Timothy Pont visited the district to research for his manuscript map of southern Lanarkshire at some point in the period 1583–96, he recorded no woodland of any type anywhere in the upper reaches of the Clyde basin (NLS, Timothy Pont, [Glasgow and the county of Lanark] – Pont 34, https://maps.nls.uk/view/00002331).

Lowther Hills, South Lanarkshire. This part of the Southern Uplands was largely devoid of trees by some 4,000 years ago, with the grassy uplands supporting seasonal use for livestock grazing. Birch woodland repopulated some of the lower valley districts in the later first millennium ad before disappearing again as the human settlement of Upper Clydesdale intensified after 1200.

oakwoods and the north-east Highland pinewoods, were mosaics of mainly valley-bottom and lower slopes tree cover interspersed with grass 'lawns', patches of arable or expanses of heathland. Generations of cattle wood pasture and pig foraging practices had also inhibited tree renewal. Annual demands for woodland materials and access to woodland grazing resources required the regulation and management of what remained.

Contrary to the long-established popular view of the Middle Ages as witnessing unbridled destruction of the remaining 'natural' woodland across Scotland, there is good evidence that protection and management of growing timber was important to landholders and their tenants.[31] Management, it is also clear, extended into woodland that has long been held to be 'wild' or 'natural', to protect it from irregular use and unnecessary damage or loss, and to reserve its resources for the use of specific individuals or communities. For landholders, however, private possession of woodland for their exclusive use was the central concern; notions of planned management for sustainable use were secondary and much later developments.[32] In legal terms, woodland was a component of the land on which it stood.[33] Most protection for trees, therefore, related to their status as property and damage such as barking or peeling, illegal fuel gathering or unlicensed clearance were to become crimes against items of real, privately owned property. Protections afforded to trees were incidental to any deterrent effect of penalties for criminal damage and property theft. On common land, including territory that had been assigned for the common use of the inhabitants of the new towns that were being established by royal edict, all men with rights in the commons could exploit it subject to regulations governing common property. Misuse was again managed through fines. On both private and common land, the level of fine reflected the trees' present commodity value as timber and not the potential current and future loss of wider benefits from the woodland.

In the 'assizes' attributed to Scotland's twelfth-century kings, property rights were expressed in general terms.[34] These codes name neither specific property types nor commodities, but it is likely that assumed widespread popular awareness of what was encompassed, set out in traditional oral laws, probably underlies the generic labelling. The generality of property types is best illustrated in the ritualistic jingle of subject matter in charter formulae for property transactions used from the middle of the twelfth century onwards, expressed in minor variants of 'in woods and plain ground, in waters, in meadows and pastures, in moors and marshes, in pools and mills and in all other easements pertaining to [the land in question]'. Such broad categories were intentionally comprehensive, making it clear to all that everything that could be associated with these generalities was encompassed by an award of possession under broad headings. A grant of woods, therefore, unless qualified by a burden or servitude due to others, meant exclusive possession of all living (or 'green') and dead wood (the bark of the trees as much as the wood), the fruit of the trees (both on the branch and on the ground) and the understorey of young trees, scrubby species and herbs, as well as all rights to charcoal making and wood pasture, within the land specified

31 For the longevity of managed native timber resources, see Mills and Crone, 'Dendrochronological evidence'.

32 For an overview, see Smout, MacDonald and Watson, *A History of the Native Woodlands*, pp. 37–44.

33 For the development of the legal protection of trees and woodland and emergent 'policy', see R.D. Oram, 'Public policy and private practice: protection, management and development of woodland resources from the 12th to 17th centuries', in M. Stewart (ed.), *Policies and Products through the Ages: The Role of the State and Private Interests*, Scottish Woodland History Conference Notes 21 (2016) (Native Woodland Discussion Group, 2022), pp. 1–7.

34 For discussion of these laws and their dating, see Taylor, '*Leges Scocie*', and Taylor, 'The assizes of David I'.

by the charter. How the recipient chose to exercise such rights over these items of property was, again subject to any burden, servitude or restriction, entirely at his discretion.

Wood pasture and pannage

As already encountered in respect of forests and hunting, the importance of woodland as pasture gave rise to conflicts over its exploitation. Although its use as pasture was largely discontinued with the introduction of 'Improved' animal husbandry techniques in the eighteenth century, its economic and even more important ecological value has seen its recent experimental reintroduction.[35] Post-medieval accounts record wood pasture throughout much of the Highlands, but it is mainly from the Southern Uplands and the eastern interface of Highland/Lowland zones that medieval records of this practice survive. The progressive intrusion of the herds of Melrose and Dryburgh abbeys into the forest between Gala and Leader is perhaps the best-known example of grazing pressures on woodland, but it is paralleled across the country.[36] When sometime between 1131 and 1145 David I gave Dunfermline Abbey Penick near Auldearn in Nairnshire, for example, he included grazing rights in its woodland and open ground whilst reserving for himself all hunting in the same lands.[37] In the glens of north-east Perthshire, in *c.*1161 Coupar Angus Abbey received from King Máel Coluim IV the right to graze livestock in his forest of Drimmie, north of Blairgowrie, provided that the animals – with the interesting exception of wild (or unbroken?) horses ('equos silvestres') – returned to the monks' own property nightly.[38] In later centuries, the abbey attempted tight regulation of its own tenants' access to its woodland to graze their animals (see below). In eastern Teviotdale in the late 1140s, David I confirmed Jedburgh Abbey's rights to grazing anywhere in his woods, reserving just one area called 'Quikege'.[39] Walter son of Alan, in his 'foundation' charter for Paisley Abbey, gave it pasture rights in his Renfrewshire forest for cattle and pigs alongside his and his tenants' own, with the additional provision that should he or any of his heirs wish to have their own cattle graze in the forest, he would provide the abbey with another part of it sufficient for its animals' needs.[40] This last case highlights the multiple overlapping and conflicting rights in woodland pasture, beyond the potential clash with hunting interests, that were produced by the interplay of ownership, tenancies and gifts of easement rights in the same resource.

A recurring feature of several twelfth-century forest pasture awards was pannage rights relating to the grazing of horses and more especially pigs on acorns in woodland.[41] Some of Scotland's earliest surviving medieval records preserve evidence of that practice which indicates that it was well established in the later eleventh century and probably had been for centuries earlier. It is likely, for example, that the name 'Swinewde' (*sic* = Swinewood) for a property listed in a charter of King Edgar (r.1097–1107) to the

35 BBC Scotland News, 22 January 2022, 'Grazing cattle boost capercaillie habitat', https://www.bbc.co.uk/news/uk-scotland-highlands-islands-60084433.

36 Gilbert, 'The monastic record of a Border landscape'.

37 Barrow, *The Charters of David I*, no. 131.

38 *Regesta Regum Scottorum*, vol. 1, no. 226. The grant also conferred easements in the forests of Drimmie and Clunie (west of Blairgowrie) including building timber, firewood, bark, pannage and charcoal burning.

39 Barrow, *The Charters of David I*, no. 174.

40 Innes, *Paisley Registrum*, p. 6.

41 Duncan, *Scotland*, p. 364.

Pigs feeding on acorns in the New Forest, Hampshire. Although there is little evidence for it being practised in Scotland after *c.*1300, pannage rights are still enjoyed in some areas of ancient oak woodland in England, as shown in this photograph, taken in 2010. (Image: Tim Woodcock / Alamy Stock Photo.)

Benedictine monks of Coldingham, dated *c.*1100, derives from its use as a pig-grazing area through at least the previous century and possibly since the arrival of the first English-speaking settlers in this district in the seventh century.[42] Swinewood continued to be managed through the twelfth century, possession being confirmed to the monks by David I in the 1120s, and the priory's exclusive game and material resource rights in it were explicitly confirmed by King William in the later 1160s;[43] these confirmations speak of a still functioning resource. By the second quarter of the twelfth century, multiple references to pannage – from the Latin 'pannagium', the name of the payment charged for autumn mast grazing by pigs, later extended to the process of grazing itself – reveal that mature oak woodland throughout eastern and south-eastern Scotland was exploited for the fattening of pigs on the calorie-rich nuts (one acorn contains on average 144 kcals) prior to slaughtering in November. That this was an important and large-scale resource throughout lowland oakwoods from Berwickshire to the inner reaches of the Moray Firth is shown by grants of pannage made before 1153 to Newbattle Abbey in royal forests

42 A.C. Lawrie (ed.), *Early Scottish Charters prior to 1153* (Glasgow, 1905), p. 17.

43 Barrow, *The Charters of David I*, nos 31, 32; *Regesta Regum Scottorum*, vol. 2, no. 46.

in Lothian and to Melrose in the forests of Traquair and Selkirk and between Gala and Leader,[44] by *c.*1161 to Coupar Angus in the eastern Perthshire forests of Drimmie and Clunie,[45] by the early 1190s to the bishop of Moray and his men in the king's forests near Elgin, Forres and Inverness,[46] and soon after 1200 to Paisley in Walter son of Alan's baronial forest there.[47] By the 1200s, however, references to pannage were often simply fossilised relics in royal confirmation charters.[48] By the second half of the thirteenth century, on royal properties, it seems to have been collected only in the Forest of Selkirk, appearing in the accounts of the sheriffdom which encompassed the main surviving forest district in the Southern Uplands.[49] The decline of pig numbers in Scottish oakwoods in the 1200s is so ubiquitous as to be strongly suggestive of recognition of the negative impact of these destructive foragers on a resource which was coming under intense pressure in the most densely populated areas of the kingdom and of a consequent sharp decline in available areas of pannage foraging by that date. With only mature oaks of fifty years and older producing acorns, the felling of young-growth trees for wood needs and the destruction of seeds and seedlings by animal grazing would have left a dwindling core of old-growth veterans in increasingly park-like settings.[50]

A brief episode of intensive woodland pig grazing occurred in eastern Galloway from the later 1160s into the early 1200s between the rivers Urr and Nith. When Uhtred of Galloway granted the canons of Holyrood the church of Urr, probably shortly after 1160, it was hunting rights that were considered one of the most important economic assets locally.[51] A few years later, when Uhtred leased Kirkgunzeon in the centre of the district to Holm Cultram Abbey in Cumbria, reference to hunting had gone, the abbey evidently having other plans for the 5,000ha it had acquired.[52] Under its lease, the abbey received free pannage for its pigs anywhere in Uhtred's demesne woods in the area, later reduced in an agreement with the monks of Dundrennan to pannage for both abbeys outside the demesne woods during acorn season, but with Holm Cultram retaining general access to the woods at other times.[53] Although Galloway was noted primarily as cattle country, pigs were abundant in the region in the later twelfth century and not just on these expanding monastic

44 Barrow, *The Charters of David I*, nos 98, 120, 164.

45 *Regesta Regum Scottorum*, vol. 1, no. 226.

46 *Regesta Regum Scottorum*, vol. 2, no. 362.

47 Innes, *Paisley Registrum*, p. 17, where the monks were given grazing rights for 100 pigs in the forest for a month in 'tempore de pessun' (the time of pasture).

48 For example, King Alexander II's June 1224 confirmation of the grants of David I and Máel Coluim IV to the monks of Newbattle: NRS GD40/1/15. It was also in the wooded uplands west of this forest in Eskdale that an agreement between Roger Avenel, lord of Eskdale, and the monks of Melrose saw Roger agree not to put his pigs – or any other domestic animal – on the monks' Eskdale properties to their injury: *RPS* 1235/1.

49 J. Stuart and G. Burnett (eds), *The Exchequer Rolls of Scotland*, volume 1: *1264–1359* (Edinburgh, 1878), pp. 29, 30, 35. In 1288, twenty pigs taken in default of pannage payments in Selkirk forest were salted for the king's use.

50 J.M. Gilbert, 'Woodland management in medieval Scotland', *Proceedings of the Society of Antiquaries of Scotland*, 146 (2017), pp. 215–252 at 219.

51 C. Innes (ed.), *Liber Cartarum Sancte Crucis* (Edinburgh, 1840), p. 19 (hereafter *Holyrood Liber*).

52 F. Grainger and W.G. Collingwood (eds), *The Register and Records of Holm Cultram* (Kendal, 1929), pp. 48–49 (hereafter *Holm Cultram Register*).

53 Ibid., p. 52. For development of the Kirkgunzeon property as a grange, see K.J. Stringer, 'Reform monasticism and Celtic Scotland: Galloway, *c.*1140–*c.*1240', in E.J. Cowan and R.A. McDonald (eds), *Alba: Celtic Scotland in the Medieval Era* (East Linton, 2000), pp. 127–165 at 147–151.

Mixed oak and birch woodland near Whithorn, Dumfries and Galloway. Similar woodland in the district between the rivers Urr and Nith in the east of Galloway was possibly cleared in the later twelfth century using the grubbing and rooting habits of pigs to prevent regeneration.

estates: in 1174 an annual tribute of 500 pigs was part of the settlement offered to King Henry II of England by Gillebrigte, lord of Galloway, for the murder of his half brother and Henry's cousin, Uhtred.[54]

Pigs played an important part in the medieval rural economy, being valued for their skins and fat as much as for their meat,[55] which was regarded as a high-status food. The effective disappearance of pigs from the grazing livestock record by the early thirteenth century thus suggests that something had changed to make large-scale pig management less attractive.[56] The decline was perhaps related to a shift from a wooded to a more open landscape that favoured sheep or cattle rather than woodland-grazing pigs, or had been broken into cultivation. A transition from woodland grazers in large numbers to sty-dwelling farmyard animals is evident generally across Europe in this period,[57] perhaps reflecting a reduction in habitat and pannage availability as the MCA and attendant expansion of arable reached its medieval climax. The possible contribution of pigs to the clearance of the pre-*landnám* birchwoods of Iceland has been recognised, and pig farming saw a brief tenth-century efflorescence there before they disappeared from large-scale management.[58] In

54 R.D. Oram, *The Lordship of Galloway* (Edinburgh, 2000), p. 96.

55 Pigs would, for example, have been amongst the animals slaughtered for the king's use north of Lammermuir and north of the Tay of whose skins, tallow and grease a half was given to the canons of Scone by King William: *Regesta Regum Scottorum*, vol. 2, no. 18.

56 Gilbert, 'Woodland management in medieval Scotland', p. 219.

57 For the transition from forest-grazed to farmyard pigs, see A. Ervynk et al., 'An investigation into the transition from forest dwelling pigs to farm animals in medieval Flanders', in U. Albarella et al. (eds), *Pigs and Humans: 10,000 Years of Interaction* (Oxford, 2007), pp. 170–193.

58 McGovern et al., 'Landscapes of settlement', pp. 29, 30, 45.

Iceland, there seems to have an awareness of the role of pigs – and goats – in landscape clearance and degradation. The question remains whether they were used as initial agents for clearance of new land before their more destructive impact was understood. Elsewhere in Europe, there is a suggestion of pigs' involvement in deliberate deforestation, where their grazing and rooting habits were valued for preventing regeneration after clearance.[59] Whether or not pigs were the instruments of clearance in eastern Galloway, the woods in Kirkgunzeon were evidently much reduced even before 1200. They survived mainly in the higher ground north of Dundrennan Abbey's grange,[60] and in the lord of Galloway's demesne properties on the west side of the tidal estuary of the Nith, from Troqueer round the Solway coast to Southwick.[61]

Reserving and conserving timber

In addition to grazing pressures, clearance for agriculture and demands for building materials and firewood saw fresh and increasing inroads being made into remaining areas of woodland across much of lowland Scotland in the twelfth century. Assarting, the process of clearing land from woodland and scrub and converting it to agricultural purposes, was perhaps already in progress across parts of the Forth basin from the edge of the Trossachs to the southern fringes of the Ochils by the eleventh century and certainly by the early twelfth, perhaps mapped by the spread of 'Gart-' place-names (from Gaelic 'gart'/'gort', a field (of standing corn), garth, cultivated plot) in the districts around Drymen and Aberfoyle in the west and Clackmannan in the east.[62] In Scots, there was an understanding of 'garth' as a cultivated plot surrounded by 'waste'. Those in Clackmannanshire may have included sites that were the locus for tree felling operations by the servants of the various monasteries granted free access to the timber and firewood of the royal forest there in the twelfth century.[63] Elsewhere, pollen evidence illustrates fresh intakes from scrubland that had regenerated in previous centuries and containment of remaining woodland zones. Along the northern rim of the Cheviot Hills, for example, such evidence points to the reduction of scrub woodland cover, mainly of willow, hazel, birch and alder, into widely dispersed and heavily managed copses in the valley bottoms and isolated upland stands that were possibly enclosed to protect them from grazing animals.[64] In the western Highlands, the land-use history of Glen Affric suggests a similar approach, with wood-

59 An example might be Le Tchèté d'la Rotche in the Belgian Ardennes, where pigs seem to have been prominent in the diet of the castle inhabitants during the initial phases of settlement but declined as the surrounding landscape was cleared of woodland. See A. Ervynk and J. Woollett, 'Top-predator or survivor? The Castle of Sugny as seen through its animal remains', *Les cahiers de l'urbanisme*, September 2006, pp. 78–89. For current issues with wild boar and their more destructive traits, see BBC Scotland News, 'Thriving wild boar population creating havoc in the Highlands', [11 March 2022], https://www.bbc.co.uk/programmes/articles/2gLYpzPmBtbH4qTt3VQR9bF/the-thriving-wild-boar-population-creating-havoc-in-the-highlands.

60 For the character of the landscape in the hilly district north of the grange, bounded by a roadway described as cutting across a wooded valley, see Grainger and Collingwood, *Holm Cultram Register*, p. 50 nos 122, 123.

61 K.J. Stringer, 'Acts of lordship: the records of the lords of Galloway to 1234', in T. Brotherstone and D. Ditchburn (eds), *Freedom and Authority: Historical and Historiographical Essays Presented to Grant G. Simpson* (East Linton, 2000), pp. 203–224 at 217–222 nos 15, 17, 21, 23, 27, for easements in the demesne woods for building timber and firewood.

62 P. McNiven, 'The *Gart*-names of Clackmannanshire', *Journal of Scottish Name Studies*, 1 (2007), pp. 61–76.

63 Ibid., p. 66. For the forest of Clackmannan, see below, p. 213 note 20.

64 Tipping, *Bowmont*, p. 45.

Looking west up the valley of the Forth from Stirling Castle towards the Trossachs and Ben Lomond. The later eighteenth- and nineteenth-century drainage and clearance of the mosses in the central part of the valley overshadows the thirteenth-century and later medieval assarting around the Lake of Menteith and, further west, the probably later tenth- to twelfth-century agricultural colonisation reflected in the region's 'Gart-' place-names.

land patches persisting amidst a landscape of dispersed arable cultivation, valley-bottom meadow, rough grazing and peaty wetland, even at the peak of the MCA. This unnatural disposition of the woodland possibly reflects a conscious retention of some informally managed areas of trees for the various wood needs of the local peasant communities.[65] In both of these upland contexts, a degree of management or protection of remaining woodland resources seems to have been established between at least the tenth and the thirteenth century.

Small patches of woodland rather than extensive woods or forests appear to have become well-established landscape features throughout the eastern Borders over the course of the twelfth century. This picture of small, scattered stands is exemplified by the nine named blocks of woodland within Coldingham parish identified in a late 1160s royal charter to the priory there.[66] A similar proliferation of smaller, named woods can be seen in other zones of intensified agricultural activity, such as the Laich of Moray, where the wood of 'Inverdamin' (lost, near Nairn) and the 'great wood of Kelbuthac' (Kilbuiack in Alves parish) were named in late twelfth-century grants to Kinloss Abbey, along with an unnamed wood (*nemus*) between Forres and Burgie.[67] We can see a similar landscape of small woods in a largely cleared expanse of managed, open land in the district around Renfrew, where Paisley Abbey's property at 'Penuld' contained three pieces of woodland – two unnamed *colles* and the 'great wood that is called Barpenuld'.[68] In both Moray and on the Paisley estate, these surviv-

65 Davies, 'Upland agriculture and environmental risk'.

66 *Regesta Regum Scottorum*, vol. 2, no. 46. Five of the nine woods can be identified precisely: Reston wood, Brockholeswood, Aikieside wood, Swinewood and Houndwood. For a discussion of the locations of the woods, see ibid., p. 154.

67 C. Innes (ed.), *Registrum Episcopatus Moraviensis* (Edinburgh, 1837), Carte Originales, nos 2 and 3 (hereafter *Moray Registrum*).

68 Innes, *Paisley Registrum*, pp. 48–49. Penuld was known in the later Middle Ages as Foulton, now represented by East and West Fulton, near Linwood: ibid, p. 412.

ing woods were under pressure before 1200 and most had disappeared from the documentary record before 1250, except where managed carefully in the royal hunting forests of Darnaway and Longmorn in Moray and the baronial forest of Renfrew.

Accelerating inroads

In Moray, Kinloss and its tenants had made significant inroads into Kilbuiack wood by the 1220s, probably clearing the wooded district north of the Forres–Elgin road. In a 1221 confirmation of a perambulation of the monks' lands of Burgie, one boundary marker was 'the great oak' in Malevin which had been marked with a cross cut into its bark.[69] It seems that such isolated veterans as this solitary oak remained in a largely cleared landscape, preserved for their landmark status in fixing boundaries. The same fate is likely to have befallen the one 'old elm' that served as a boundary marker for a portion of property given to Melrose Abbey, having survived from a once more extensive woodland at Greenan in Ayrshire that had been cleared for fuel and for cultivation.[70] By the late 1180s at Kirkgunzeon in Galloway, references to single cross-marked oaks already serving as boundary markers between the monks' property and the lands of neighbouring lay tenants suggest that a few substantial trees survived in a landscape of scrubby wood and recent clearance.[71] Similar clearance was taking place east of Kirkgunzeon, where cross-marked oaks survived as boundary features at Mabie.[72] Although these marked oaks were still being noted towards the end of the thirteenth century,[73] significant woodland had otherwise disappeared in the wider territory between the Nith and the Urr, replaced by a mix of arable and sheep- or cattle-grazed pasture interspersed with scrubby wetland hollows and upland blanket peat bog.

Steadily increasing inroads into woodland can be traced throughout southern Scotland in both charter evidence and pollen records through the later twelfth and thirteenth centuries. These inroads appear to be related to the more benign conditions of the MCA. Combined with population growth, those conditions encouraged efforts to bring more land into intensive agricultural use and more wood and wood-related products were consumed. Hunger for pasture and for arable land and its products spurred further encroachment into wood and scrubland, with woodland gradually reduced to ribbons of trees separating blocks of cleared ground. Such was the situation by 1164 between Gala and Leader, where Dryburgh Abbey yielded Colmslie near Lauder and 'the wood which divides the clearing of Colmslie from a certain other clearing' to Melrose Abbey.[74] Here, a dispute over woodland property was resolved in a negotiated settlement which suited the needs of both monasteries, but in this rapidly clearing landscape possession of dwindling woodland resources could be more fiercely contested between parties with conflicting interests or rights. The 1164 agreement, rather than permanently resolving issues relating to the use of these woods, simply spawned further disputes, as the king's forester attempted to prevent Melrose from extending its encroachment on one of the dwindling areas of woodland cover for deer under his oversight in the Lauderdale uplands. Melrose, however, saw this as desirable wood pasture and continued to claim

69 J. Stuart (ed.), *Records of the Monastery of Kinloss* (Edinburgh, 1872), pp. 112–113.

70 Innes, *Melrose Liber*, vol. 1, no. 34.

71 Grainger and Collingwood, *Holm Cultram Register*, p. 50 nos 122, 123.

72 Ibid., p. 56 no. 142.

73 Ibid., p. 59 no. 154.

74 C. Innes (ed.), *Liber S. Marie de Dryburgh* (Edinburgh, 1847), no. 80 (hereafter *Dryburgh Liber*).

Rig-and-furrow cultivation between Hillslap and Colmslie, Scottish Borders. Although these areas of rig are post-medieval in date, they lie in the area of the former forest between Gala and Leader that the monks of Melrose were bringing back into arable cultivation from the later twelfth century.

access to the disputed woodland, which by the late 1170s was appropriately named Threpwude, nowadays Threepwood/Threipwood (from Scots 'threap', an argument, contention, dispute). In 1180, the monks secured what they wanted and the grazing of the disputed woodland was reserved to them as pasture for 100 cows.[75] A similar case occurred across Lauderdale in *c.*1220 at Bemersyde, whose lord contested Dryburgh's claims to another Threipwood, which was separated from a larger area of woodland called Flatwood by the high road through his property.[76] Such disputes were not restricted to Lauderdale, as a case from Coldingham shortly after 1124 shows. There, King David I used a dispute between the priory and a local landholder over possession of a wood to have his *rogus* (logpile) at Berwick replenished with tree trunks from the contested property.[77]

Rather than a free-for-all of destructive woodland exploitation, such protracted disputes and agreements point to a far more complex picture. Analysis of place-names and the charter terminology for woodland of different types and purposes suggest that there was a sophisticated management system behind the legal conflicts.[78] What we are seeing in the records is that system breaking down under the increasing pressures of rising population and overlapping layers of often conflicting rights to use the woodland. Amongst the clearest evidence is management for the production of underwood within more mature woodland. Underwood is the spindly, rapid-growth shoots that are produced when young trees of timber-producing species – like oak or elm – are cut, the resulting slender stems being ideal for poles and rods for use in building or for making equipment.

75 *Regesta Regum Scottorum*, vol. 2, no. 236.

76 Innes, *Dryburgh Liber*, nos 133–135; Gilbert, 'Place-names and managed woods', pp. 41–43.

77 Barrow, *The Charters of David I*, no. 43.

78 Gilbert, 'Place-names and managed woods'.

Ancient managed alder, Craignavar, Glen Almond, Perth and Kinross. Alders line the river for most of the middle reach of the glen from Newton Bridge to Connachan Lodge, showing evidence of coppice management spanning many centuries. Trees such as this provided the types of materials that were gifted to the monks of Lindores from northern Perthshire.

This coppicing process created a woodland of tall trees reserved for timber, often controlled by the lord of the property, with an extensive underwood that was made available either through easement rights – where tenants have assured access to the wood for fuel and materials – or through leases. The underwood, however, was also managed carefully, with coppicing occurring on a rolling cycle of five to twenty years to allow recovery and new growth. That new growth required protection in woods where there was grazing livestock, and references to portions of wood 'in defence' suggest that sections were enclosed by banks, hedges or fences to keep grazing animals out.[79] We have no certain physical evidence for such woodland enclosures from this period but we do have record evidence for them in abundance from the twelfth century through to the end of the Middle Ages. In Highland areas such practices continued into the eighteenth and nineteenth centuries in some districts where wood pasture remained common. It is probably such a piece of coppice management that David I referred to as his 'virgultum' (a spinney of rod-producing small trees) at Roxburgh in the early 1130s,[80] from which the smallwood needs of the castle household were perhaps met. And those needs could be substantial in a built environment where wattling, light fencing and pole-supported thatch were ubiquitous and where smallwood still predominated as a fuel type.

Coppice and its products

Although the bulk of our surviving record evidence for intensive woodland management relates to the Southern Uplands, it is clear from charters in districts from Inverness to the Clyde estuary that identical practices were employed throughout more northerly

79 Ibid., p. 39.

80 Barrow, *The Charters of David I*, no. 42.

districts. As the Scottish crown's control of Moray strengthened in the later twelfth century, for example, we see its bishops and their tenants receiving confirmation of easement rights for building materials and fuel from underwood in the king's forests throughout the region, with identical rights also confirmed to the burgesses of Elgin, Forres and Inverness.[81] In the Mearns, Humphrey de Berkeley granted the monks of Arbroath and their tenants free access to his wood for building and fuel needs, albeit supplemented by rights to cut peat for fuel.[82] Evidence for what appears to be specialist management of the underwood for particular purposes can be seen in a grant by Conan, a son of the earl of Atholl, who gave the monks of Lindores rights in his wood of 'Tulyhen' (near Blair Atholl). In addition to dry logs of dead wood for firewood, he gave them annually 'the logs that are called *Wrawes* of *bule* [birch] and of *auhne* [alder]', 100 loads of hazel rods for making their sleds and 100 long rods for making hoops.[83] These two latter items are indicative of coppice. Management of woodland in the Lennox to produce poles, rods and withies is implicit in records from the 1230s of 'a great house made from withies [*fabricata de virgis*]' standing immediately east of the parish church at Old Kilpatrick.[84] This structure was probably a particularly large example of the 'creel house' type, such as has been excavated at Whithorn, Dublin and elsewhere in northern and eastern Ireland.[85] The distribution of evidence for such management is so widespread through upland and lowland zones as to suggest that well before 1200 there was little if anything in most of at least lowland Scotland that could be labelled truly 'natural' woodland.

How sustainable were these practices? Scotland's medieval population would not have understood modern concepts of conscious sustainability, but it would have understood the need to ensure that there were accessible and adequate future supplies of the right types of wood for its varying needs. It is probably in this context that we should see the long-term survival of small areas of woodland in the Highlands, as suggested for Glen Affric, where trees were left on some areas of better-quality land through the MCA and only disappeared in the later Middle Ages.[86] In such areas, however, population pressures might never have reached the levels experienced in some Lowland districts, where lords were attempting to accommodate multiple demands for the different resources or different land-use potential of the same land. Furthermore, what might seem to be an absence of careful management evident in the inroads made into woodlands for fuelwood purposes might actually be evidence for the capacity of regulated access to deliver unintended and undesirable outcomes. The coppicing of young trees to promote underwood growth alongside the reservation of a small number of large, mature trees appears to

81 *Regesta Regum Scottorum*, vol. 2, no. 362.

82 C. Innes (ed.), *Liber S. Thome de Aberbrothoc*, 2 volumes (Edinburgh, 1848–56), volume 1, no. 89 (hereafter *Arbroath Liber*).

83 J. Dowden (ed.), *Chartulary of the Abbey of Lindores 1195–1479* (Edinburgh, 1903), no. 73 (hereafter *Lindores Charters*), my translation.

84 Innes, *Paisley Registrum*, p. 166, my translation.

85 Hill, *Whithorn*, pp. 55–56, 216–217; C.J. Lynn, 'Ballywee', *Excavations*, 5 (1974), pp. 4–6; H. Murray, *Viking and Early Medieval Buildings in Dublin*, British Archaeological Reports, British Series 119 (Oxford, 1983); D.M. Waterman, 'The excavation of a house and souterrain at Whitefort, Drumarood, Co. Down', *Ulster Journal of Archaeology*, 19 (1956), pp. 73–86; D.M. Waterman, 'The excavation of a house and souterrain at Craig Hill, Co. Antrim', *Ulster Journal of Archaeology*, 19 (1956), pp. 87–91. For discussion of these building traditions in the medieval and early post-medieval Highlands, see H. Cheape, '"Every timber in the forest for MacRae's house": creel houses in the Highlands', *Vernacular Building*, 37 (2013–14), pp. 31–50.

86 Davies, 'Upland agriculture and environmental risk'.

deliver 'sustainability'. But the young-growth wood is not producing seed and the diminishing numbers of seed-bearing old-growth trees would, long-term, have led to a decline in woodland regeneration. Eventually, with the loss of the mature trees through a combination of felling, storm damage and age, the balance would have tipped decisively towards young-growth underwood and the development of scrubby woodland.[87]

Fuelwood

This trend towards enclosed coppice was probably evident first in districts where there was intensive exploitation of the trees for fuel. Specialist charcoal production for craft processes that required high temperatures was well established in Scottish oakwoods by the 1100s. At Edzell in Angus, for example, Arbroath Abbey had access to the woods around the mouth of Glen Esk for its charcoal burners to work there on a seasonal basis.[88] In most cases, however, it was straightforward fuelwood rather than charcoal that was being extracted, for use in domestic and industrial hearths. Record evidence for the highest volumes of fuelwood consumption in the twelfth and thirteenth centuries relates to salt manufacture. On the carses flanking the Forth between Kincardine and Bo'ness in the east and at Stirling in the west, this industry was in large-scale operation throughout the twelfth century,[89] but a gift to the monks of Lochleven in the third quarter of the eleventh century reveals the existence of operative pans on the Forth estuary by the second half of the eleventh century.[90] It is in the reign of David I, however, that record evidence for widespread salt production on the Forth carses proliferates, with references to royal pans at Stirling and others granted to the monasteries of Cambuskenneth, Dunfermline, Holyrood, Jedburgh, Kelso and Newbattle on either side of the estuary downriver from there.[91] Beyond the Forth, we have record of large-scale salt production in regions as dispersed as the Solway coast in the south-west and East Lothian, the sea lochs of the Lennox, the Angus and Kincardineshire coast and the Laich of Moray.[92] As the sole bulk preservative for perishable meat, fish and dairy products, salt was required in abundant quantities by all large-scale producers of such commodities, so most landowners with suitable locations had pans on their properties or sought access to such places where they did not.

How saltpans were fired is not made explicit before the second half of the twelfth century. To the 1150s, royal grants of places at which to manufacture salt from the boiling of brine were often just a small part of a bundle of rights that included general access to woods, which was presumably for gathering fuel close to the saltpans' locations. From the third quar-

87 Gilbert, 'Woodland management in medieval Scotland'.

88 Innes, *Abroath Liber*, vol. 1, no. 72.

89 R.D. Oram, 'The salt industry in medieval Scotland', *Studies in Medieval and Renaissance History*, 3rd series, 9 (2012), pp. 209–232 at 215–217, 221–222; Oram and Whatley, 'A brief history of Scottish salt'.

90 Lawrie, *Early Scottish Charters*, p. 5; Oram and Whatley, 'A brief history of Scottish salt', pp. 16–17.

91 Barrow, *The Charters of David I*, nos 67, 98, 114, 147, 151, 159, 164, 174.

92 *Regesta Regum Scottorum*, vol. 1, nos 109, 195; *Regesta Regum Scottorum*, vol. 2, nos 62, 83, 88, 226, 545; D.E. Easson (ed.), *Charters of the Abbey of Coupar Angus*, 2 volumes (Edinburgh, 1947), volume 1, no. 148 (hereafter *Coupar Angus Charters*); W. Fraser (ed.), *Registrum Monasterii S. Marie de Cambuskenneth* (London, 1872), no. 51; Grainger and Collingwood, *Holm Cultram Register*, nos 120, 121; Innes, *Arbroath Liber*, vol. 1, no. 142; Innes, *Holyrood Liber*, nos 1, 32; C. Innes (ed.), *Liber S. Marie de Calchou*, 2 volumes (Edinburgh, 1846), volume 1, nos 2, 151 (hereafter *Kelso Liber*); Innes, *Melrose Liber*, nos 34, 37, 65; Innes, *Paisley Registrum*, pp. 210–211; Innes, *Moray Registrum*, no. 211; R. Fawcett and R. Oram, *Elgin Cathedral and the Diocese of Moray* (Edinburgh, 2014), pp. 145–146.

ter of the twelfth century the imprecision was removed, with royal grants and confirmations now being explicit that access to woodland was for the fuel required for the saltpan, as in a confirmation to Newbattle giving fuel for its pan on the carse near Falkirk in the local Callendar woods.[93] The need for such specificity suggests that the pan operators were finding it difficult to secure their fuel needs and were possibly being challenged or obstructed by royal foresters, by other holders of fuel easement rights and by local communities. Why pan operators might have faced supply difficulties can be illustrated by the case of Melrose Abbey's saltpan at Greenan on the coast of Carrick, south of Ayr. The monks received possession from the local lord, Roger of Skelbrooke, who shortly after 1185 gave them 'a fishery upon the sands at the mouth of the river Doon, a saltpit and salt pan wherever they wished upon his lands of Greenan, and fuel from [his] nearest wood for the aforesaid saltpit and pan'.[94] The charter also reveals that there was an older pan site in the vicinity, suggesting that an established manufactory was already drawing on the local woodland fuel resources, hence Roger's careful stipulation of the monks' fuel source. The Melrose pan functioned through the 1190s, but shortly after 1200 Roger's overlord, Duncan of Carrick, granted the monks two alternative pan sites near Turnberry 'because the wood which they had for the pan at Greenan had been completely burned and destroyed'.[95] In less than twenty years, it seems, the monks' fuel needs for their Carrick saltpan had seen the overuse of the locally available underwood, forcing them to seek a new site with a fresh supply. It is hardly an edifying picture of much-vaunted Cistercian advanced estate management practices, let alone of the sustainable management of a renewable resource.

Conservation versus consumption

Despite an increasing volume of twelfth-century litigation involving woodland and woodland rights, it was not thought necessary to frame laws specifically to deal with trees. The lack of evidence for any emergent 'policy' towards woodland, however, does not imply a lack of political interest in matters directly related to it. By the early 1200s, shortages of building materials and firewood surviving on the common lands of Scotland's burghs had become a cause for concern. Where burghs had not been planted on already agriculturally developed land, the establishment of a new community brought extensive clearance. We can see such clearance in action at Ayr in the six acres to be cleared from the brush (*boscus*) for each burgess's toft granted to the new burgh by King William,[96] with further inroads made into local woodland to meet their fuel and materials needs. Details of arrangements for the regulation and policing of the exploitation of these common burgh resources are lacking from this time, but perceptions of a free-for-all 'tragedy of the commons' situation should be counterbalanced by indications of coppice management on burgh land.[97] Urban expansion, however, saw demand outstrip the capacity of the accessible resources to supply community needs. The pressure for remedies came from burgesses to whom the shortage of such materials was potentially devastating economically. Every aspect of burgh life was affected, from commercial processing activities and

93 *Regesta Regum Scottorum*, vol. 1, no. 109.

94 Innes, *Melrose Liber*, vol. 1, no. 34.

95 Ibid., no. 37.

96 *Regesta Regum Scottorum*, vol. 2, no. 462; Cooper, *Charters of the Royal Burgh of Ayr*, no. 1.

97 A. Crone and F. Watson, 'Sufficiency to scarcity, 500–1600', in T.C. Smout (ed.), *People and Woods in Scotland* (Edinburgh, 2003), pp. 60–81 at 66–67.

domestic routines to the built environment in which these functions were performed. Around 1209–12, the leaders of several east-central Scottish royal burghs secured new royal charters that expanded their communities' legal and economic privileges. The best-known of these charters, that granted to Perth in 1209, contained a new clause designed to encourage traders to bring timber and firewood into the burgh for sale.[98] These were the only commodities identified so specifically in any burgh charter of this period, suggesting that urban communities saw the protection of timber suppliers as an essential extension of their burgh's economic privileges; they well understood the threat to their economic well-being posed by wood shortages.

We lack contemporary price records for the economic context of these acts, but wood's high cost is unquestionable. Its value is illustrated by gifts of oak building timbers from royal or lordly demesne woodland to ecclesiastical institutions and noble favourites throughout the thirteenth century and beyond. Grants such as that made by Walter the Steward to Paisley Abbey from the forest of Sanquhar around St Quivox in Ayrshire in *c.*1213, or that made by Gille Chriosd, lord of Tarbet on Loch Lomond, of unrestricted access to his woods for mature timbers for building the monastery, reveal that such wood was still plentiful but in controlled contexts.[99] As one-off noble gifts, they show recognition of the high value of old-growth oaks. But AD hoc gifts were inadequate for burgess needs; their concern was simply maintenance of access to an in-demand commodity, rather than safeguarding – let alone regenerating and extending – the sources of that commodity. King William's concessions deregulated trade and encouraged owners of woodland assets at a distance from burghs to increase the exploitation of their property through the commercially liberalised supply of materials to burgh markets. The perception of unregulated exploitation of woodland to which this development gives rise is difficult to counteract, but some early thirteenth-century charter evidence does exist for continued royal efforts to protect such resources. In April 1236, King Alexander II feued the five pennylands of Alloway, Corton and Carcluie to the burgesses of Ayr but prohibited them from burning, alienating, selling or in any way destroying the growing timber there unless to take what was needed for the burgh's buildings and ships.[100] Again, this was no conservationist measure directed at addressing wood shortage, but the straightforward protection of a lord's asset from abuse by his tenants. Assertive measures to protect and from the fifteenth century to plant and extend the woodland reserves in Scotland lay far into the future. Until that time, the remaining woods saw only continuing loss in the face of growing demand from consumers.

Woodland beyond the lowlands

How general was the pressure on woodland described in the previous sections? Modern percep-

98 *Regesta Regum Scottorum*, vol. 2, no. 467. Perth's privilege has attracted much discussion, but it was not unique, for a similar clause was contained in Alexander III's 1285 reconfirmation of the burgh privileges of Lanark and Robert I's 1318 charter to Haddington, both apparently verbatim rehearsals of earlier William I charters: *Regesta Regum Scottorum*, volume 4 part 1: *The Acts of Alexander III 1249–1286*, ed. C.J. Neville and G. Simpson (Edinburgh, 2013), no. 163; *Regesta Regum Scottorum*, volume 5: *The Acts of Robert I 1306–1329*, ed. A.A.M. Duncan (Edinburgh, 1988), no. 142.

99 *Regesta Regum Scottorum*, vol. 2, no. 518; Innes, *Paisley Registrum*, p. 157. The Sanquhar woods might be the source of the massive old-growth oak timbers of early thirteenth-century date found in the Hunterston Sands in 2014: Wessex Archaeology Coastal Archaeological Landscapes: Intertidal & Estuarine, https://blogs.wessex.co.uk/coalie/2014/10/30/coalie-project-report-online/.

100 Cooper, *Charters of the Royal Burgh of Ayr*, no. 5.

tions of where the most extensive afforested areas survived to the eighteenth century have focused on the Highlands, seeing in the pinewoods there the relics of that 'Great Wood' of Graeco-Roman tradition, whose imaginary extent was discussed in Chapter 1. The expanses of woodland whose existence can be traced in the earliest surviving detailed maps of Scotland – the extant portions of the manuscript maps drawn in the 1590s by Timothy Pont – and which remain as major landscape features today have often been presented as fragments of that mythical forest.[101] Their persistent presence in the Scottish landscape has been used to argue that, in Highland areas, the drive to clear woodland to make room for arable expansion was largely absent before the later eighteenth century and that there was some nascent form of environmental consciousness that maintained a kind of balance between nature and culture.[102] Proponents of such views have pointed to the documentary sources, where evidence of the mounting pressure on Scotland's wooded areas and the accelerating rate of clearance that occurred across the MCA is weighted heavily towards regions outside the Highlands and Islands. This unquestionable bias, however, is almost purely an artefact of the nature of the surviving written records, most of which are drawn from the muniments of the major monastic foundations in the southern and eastern parts of the country. The relative dearth of similar material specific to other parts of Scotland does not mean that equivalent pressure did not exist in those places.

In the central and north-east Highlands, the *dabhach* system established by the tenth century, discussed in Chapter 4, presents us with a record of the land and its key resources being systematically divided between communities. The emphasis here was on access for the settled population of each *dabhach* to adequate summer and winter grazing alongside an adequate extent of arable land for its grain needs. Surviving descriptions of a *dabhach*'s component elements show them to have been grouped into these basic pastoral and arable zones. Woodland figures rarely in most accounts of *dabhaichean* and we must be cautious about associating the major blocks of forested land that were established through this region from the later twelfth century with extensive woodland. These were areas under forest law restrictions, reserved primarily for hunting purposes, rather than the great woodland areas to which we apply the name 'forest' today. Importantly, as the late Alasdair Ross identified, most of these hunting forests were also designated as zones of summer grazing, possibly including wood pasture, and contained extensive shieling grounds for seasonal use.[103] The forests of Glenmore, Stratha'an and Glenfiddich had ceased to be expanses of close canopy woodland by the time they entered the documentary record in the thirteenth century, and the division of those districts into parishes paying regular teinds in grain, cheese, wool and hides to support the Church in the later twelfth and early thirteenth centuries speaks of substantial agricultural settlements there. As the formation of the parishes of Moray diocese underscores, the units of settlement through which parochial economies were sustained had already spread deep into the catchment of the Spey and the other rivers flowing out of the central Highlands.[104]

Extensive woodland, however, certainly existed and survives still, after episodes of intensive felling

101 For discussion of the woodland, see T.C. Smout, 'Woodland in the maps of Pont', in I.C. Cunningham (ed.), *The Nation Survey'd: Timothy Pont's Maps of Scotland* (Edinburgh, 2006), pp. 77–92.

102 The articulation of this sentiment is exemplified by J. Hunter, *On the Other Side of Sorrow: Nature and People in the Scottish Highlands* (Edinburgh, 1995).

103 Ross, *Land Assessment and Lordship*, pp. 76–80, 99–101, 148.

104 Ibid., pp. 86–92.

Scots pine, Guay, Perth and Kinross. This tree is most likely a seedling from the extensive larch and pine plantations created for the 4th duke of Atholl on the east side of Strath Tay in the early 1800s. Tall, straight specimens of this sort were carefully preserved by medieval landowners for their personal shipbuilding needs or for long-span beams and joists.

Oak woodland, lower Glen Kinglass, Argyll and Bute. The oak woods on the shores of Loch Etive and in the glens emptying into the loch on its eastern shore have been exploited for timber, bark and charcoal for centuries, most recently from the 1720s to the 1870s by ironmasters first at Glen Kinglass then at Bonawe. These woods probably met the needs of the Campbells of Glenorchy, their MacDougall of Lorne predecessors and the monks of Ardchattan Priory.

and subsequent regeneration, in the present age.[105] But place-names like 'Kingussie' (Gaelic 'Ceann a' Ghiuthsaich', head of the pine forest), which is first recorded around 1200,[106] suggest that the extent of the semi-natural woods in the valley of the River Spey that we see today is probably not that different from what it was at the end of the twelfth century. Upstream of Kingussie into the headwaters of the Spey beyond Laggan, and in tributary valleys like Glen Banchor and Glen Tromie, there is little evidence for anything other than the expanses of mainly birch and alder woodland that still characterise this area. Here, the valley is divided into a succession of *dabhaichean*, those at the upper end of the catchment consisting of core settled areas with detached portions of grazing.[107] Centuries, indeed millennia, of intensive human exploitation, coupled with the extensive waterlogging of much of the valley bottom from Rothiemurchus southwards and westwards to Laggan, had created an open landscape of grazed grassland and heather, patches of arable on the freer-draining ground, seasonal water-meadow and wetland, and scrub carr and woodland.

As the Kingussie place-name implies and the surviving semi-natural woodland in Glen Feshie, Rothiemurchus, Glen More and Abernethy attests, the dominant tree species of the upper Spey catchment was the Scots pine. In the lower Spey catchment and in at least the lower parts of the glens that feed into the head of the Moray Firth, later thirteenth-century evidence indicates that oak predominated. We have, for example, royal writs that show woodland in the royal forests of Longmorn south of Elgin and Darnaway south-west of Forres to have been the source of 200 oaks allocated as a gift or compensation to Sir Ralph Cheyne, lord of Duffus, by King Edward I of England.[108] Darnaway oaks remained an important resource in royal hands into the later Middle Ages.[109] Oaks also seem to have been readily available in Inverness in the first half of the thirteenth century, where the oft-cited reference to the preparation of a ship there by the Flemish count Hugh of St Pol might reflect the continuation of the wood management for shipbuilding by the Scandinavian rulers of Orkney and Caithness that we have seen postulated for this region in the earlier medieval period.[110] The most important centres of shipbuilding, however, were located down the Atlantic coast south of Skye, presumably using the oak-dominated woodland that still forms the so-called 'rain forests' that fringe Kintail and Glenelg, the sea lochs from Morar, Moidart and Arisaig (where Iona's monks obtained their wood in the seventh century) and areas down the coast of Argyll. Here was the locus of the galley-building tradition on which the heirs of the mid twelfth-century Gaelic ruler Somhairle mac Gillebrigta founded their power.[111] Argyll and Hebridean galley-based lordship was a reality into the later Middle Ages and speaks of careful conservation of the material resource

105 For discussion of the experience of the woods around the southern and western peripheries of the Cairngorms, see R.D. Oram, *Where Men No More May Reap or Sow: The Little Ice Age: Scotland 1400–1850* (Edinburgh, 2024), chapters 2, 4 and 7.

106 *Regesta Regum Scottorum*, vol. 2, no. 476.

107 For these 'scattered' units, see Ross, *Land Assessment and Lordship*, pp. 71–74 and diagram 2.

108 J. Bain (ed.), *Calendar of Documents Relating to Scotland*, volume 2: *1272–1307* (Edinburgh, 1884), no. 1736.

109 Oram, *Where Men No More May Reap or Sow*, p. 67.

110 For Hugh of St Pol and his ship, see Matthew Paris, *Chronica Majora*, ed. H. Luard, volume 5 (London, 1880), p. 93. The broad context is discussed in Smout, MacDonald and Watson, *A History of the Native Woodlands*, p. 42. For earlier timber management, see above, pp. 81–2, 89.

111 D. McWhannell, 'Ship service and indigenous sea power', *West Highland Notes and Queries*, 3rd series, 1 (August 2000), pp. 3–18; D.C. McWhannell, 'The galleys of Argyll', *Mariner's Mirror*, 88:1 (2002), pp. 14–32.

necessary for the major timbers used in shipbuilding at the scale described in our medieval sources. That conservation clearly persisted into the post-medieval era, when galleys finally ceased to have significance in the warrior culture of the region.

It is questionable whether chieftains of the major Gaelic kindreds of the Hebrides had direct access to such timbers on their own land at any time in the Middle Ages. All of our written and archaeological record evidence points to an absence of large expanses of woodland and large tree species through most of the island areas. For the Western and Northern Isles, the apparent breakdown in the ninth and tenth centuries of arrangements for the conservation of small blocks of managed woodland to supply possibly coppiced smallwood saw no recovery after 1000. Some woodland survived in remoter eastern enclaves in Uist and perhaps in Harris, but most of the timber and general wood needs of the Outer Hebrides were met by driftwood sources and by the importation of materials from the Inner Hebrides and the west coast of mainland Scotland. The situation in Uist in particular is discussed in greater detail in Chapter 8. Our earliest systematic description of the Hebridean islands, a record probably composed as a 'briefing note' for King Alexander III around the time of negotiations for the Scottish takeover of lordship of the region in the mid 1260s and surviving as an account embedded into the later fourteenth-century *Gesta Annalia I*, confirms this picture of an effectively treeless Atlantic rim of islands.[112] The situation in the Inner Hebrides is more complex, with woodland noted in only one island location in the record, Rùm, but with later record evidence and palaeoenvironmental sources suggesting more widespread wooded areas on some of the larger islands, like Islay, Mull and Skye.[113] The usual translation of the description of Rùm as 'wooded and hilly with excellent sport, but few inhabitants'[114] has led to suggestions that it was reserved as a woodland hunting forest for the enjoyment of the earlier Norse rulers of the Kingdom of the Isles.[115] This view has been reinforced by what is presented as the traditional Gaelic name for Rùm, Rìoghachd na Forraiste Fiadhaich, translated as 'the kingdom of the wild forest', although the adjective 'fiadhaich' also means 'full of deer'.[116] The use of such names to support the notion of extensive woodland on the island in the Middle Ages is problematic, however, particularly as the Latin adjective 'silvestris', translated in respect of Rùm to mean 'wooded', more commonly means 'wild', while the noun 'forraist' is not common in Gaelic usage for any form of woodland (the common nouns are 'frìdh', 'coille' and 'fàsach') and represents a modern adoption into Gaelic of the English noun 'forest'. That word was only just entering Scots common usage in the later fourteenth century, adopted from the Latin legal term 'foresta', itself an import of late eleventh-century creation from the earlier eleventh-century Norman-French term for land under forest law, i.e. the laws governing hunting use. There is, furthermore, no implication in the use of the noun 'foresta' of the land in question carrying much or indeed any woodland. In light of this late

112 John of Fordoun, *Chronica Gentis Scotorum*, vol. 2, pp. 39–40. The compiler of the description was possibly the royal clerk, Richard Vairment or Veremundus, identified by Dauvit Broun as the composer of much of the historical account of Scotland to the time of Máel Coluim III that he dubbed 'proto-Fordun': D. Broun, *Scottish Independence and the Idea of Britain: From the Picts to Alexander III* (Edinburgh, 2007), chapter 9.

113 For coppicing on Islay in the still surviving small woodland areas of oak and hazel thickets into the modern era, see D. Caldwell, *Islay: The Land of the Lordship*, paperback edition (Edinburgh, 2017), p. 175.

114 John of Fordoun, *Chronica Gentis Scotorum*, vol. 2, p. 40 ('silvestris et montana, venacionibus apta, paucos habens incolas').

115 D. Rixson, *The Small Isles: Canna, Rum, Eigg and Muck* (Edinburgh, 2001), p. 99.

116 Ibid., p. 110.

Rùm from Coire Ghrunnda, Cuillin Mountains, Skye. Recent tradition based on mistranslated Latin has seen Rùm presented as, literally and figuratively, an island of woodland in an almost treeless zone throughout the Middle Ages. Evidence for open-landscape deer hunting methods – using the type of trap known as an 'eileircg' in Gaelic – suggests that this presentation is unlikely.

importation of the term, it is likely that it is a late medieval coining at the earliest and that we therefore need to decouple the reference in *Gesta Annalia I* to woodland from the Gaelic reference to forest. More importantly, we need to reflect on the shared meaning of 'wild' in the Latin 'silvestris' and the Gaelic 'fiadhaich'. It looks very much as though we have combined a mistranslation of the one with the 'forraist' of the other to create a woodland as fictitious as the 'Great Wood of Caledon'.

We lack any further topographical descriptions of the Hebrides that might allow us to test this idea further until the mid sixteenth-century account written by Donald Monro, archdeacon of the Isles.[117] Although composed significantly later than the period under review here, Monro's account is of great value as it presents the general socio-economic and environmental condition of the Inner and Outer Hebrides following both the long era of population growth and intensified agricultural exploitation to

117 The text of Archdeacon Monro's account is printed in M. Martin, *A Description of the Western Islands of Scotland circa 1695, by Martin Martin, Gent., Including A Voyage to St. Kilda by the Same Author and A Description of the Western Isles of Scotland by Sir Donald Monro*, ed. D.J. Macleod (Stirling, 1934), pp. 477–526.

the mid fourteenth century and the subsequent era of increased storminess, cold and precipitation in the first phase of the so-called 'little ice age' (discussed in the volume *Where Men No More May Reap or Sow*). His use of language (he wrote in Scots) is very specific and draws a clear distinction between what he referred to as 'forrests' and what was labelled as woodland. This distinction is illustrated by his description of Arran, which is 'ane grate ile, full of grate mountains and forrests, good for hunting, with pairt of woods'.[118] The 'forrests' are associated with hunting, while the woods are not specifically identified in that way. This differentiation is underscored in respect of Jura, the whole of which is labelled 'ane ather fyne forrest for deir', with no mention of any woodland, while neighbouring Scarba was 'ane high rough yle, inhabit and manurit, with some woods in it'.[119] Monro's forests, it is clear, are hunting grounds in the formal, legal sense, while woodland is a topographical feature with connotations of being a resource of value. That separation into land under a forest jurisdiction and land that is not is illustrated in respect of Islay, which he described as 'fertill, fruitful and full of natural grassing, with maney grate diere, maney woods, faire games of hunting beside everey toune', and Skye, which had 'maney woods, maney forests, maney deire, fair hunting games'.[120] On Islay, the woods and hunting are not explicitly linked and there is no suggestion that any of the island was reserved as forest. That lack of a reference to forest on Islay is striking, for Monro was otherwise very careful to reference explicitly the presence of forest jurisdictions. On Skye, there is an equally clear distinction made between woodland and forest.

What, then, does Archdeacon Monro have to say about Rùm? His description contains more detail in respect of both topography and nature observation than most of his other island accounts. He labels it 'ane forest of heigh mountains' that contained an 'abundance of litle deire in it'.[121] After a description of the type of deer hunting undertaken there, which used the drive or 'tynchell' (Gaelic 'timchioll') whereby deer were forced towards the huntsmen waiting in an *eileircg*,[122] he added that 'in this ile will be gotten about Britane als many wild nests upon the plane mure as men pleases to gadder, and yet by resson the fowls hes few to start them except deir'. There is no mention of woodland, only of 'plane mure' (open moorland) rich in wildfowl. The description of the forest on Rùm as consisting of 'heigh mountains' also brings us back to the reporting of the island in *Gesta Annalia I* as 'wooded and hilly', with its mountains and value for sport being the common denominators. The wooded island of *Gesta Annalia I* contrasts sharply with Monro's 'plane mure', presenting us with the possibility that Rùm's trees had been destroyed in the two centuries between these two accounts. Alternatively, it is possible that, if 'silvestris' was indeed used in *Gesta Annalia I* to mean 'wooded' rather than 'wild', the earlier description was overstated. Given that both accounts speak of the absence of any significant settled population on the island, and therefore no one to make inroads into woodland to extend agriculture or obtain materials, the former explanation would have required an unregulated free-for-all of felling to have occurred. Whilst overzealous timber extraction followed by failure to control the grazing impacts of the island's abundant 'litle deire' (perhaps mainly roes) could explain the absence of regeneration, that position contrasts sharply with the evident careful conservation of woodland through coppice on other

118 Ibid., pp. 485–486.

119 Ibid., pp. 488–489.

120 Ibid., pp. 493, 503.

121 Ibid., pp. 501–502.

122 For the presence of these forms of deer traps on Rùm, see above, note 25.

islands and Monro's otherwise meticulous noting of those instances where woodland of any type existed in his time. Bringing all these points together, it seems most likely that 'silvestris' in the Latin of *Gesta Annalia I* carried the meaning of 'wild' rather than 'wooded' and that modern translators, historians and environmental scientists have repeated the mistranslation and created the modern factoid that Rùm is another example of the creation of an anthropogenic wasteland through centuries of mismanagement.

• • •

The general perception gained from the material reviewed in this chapter might be that there was a long and ultimately unsuccessful rearguard action to prevent further loss of the woodland areas that still survived at the end of the first millennium AD. That perception is founded on two main pillars. First, there is the very clear record from most of the Lowland and Southern Upland districts of the accelerating erosion of the remaining major blocks of woodland from the eleventh to the thirteenth century as human demand for more arable and pasture land increased. Second, there is archaeological interpretation of a seeming breakdown in long-established wood management regimes in the Hebrides and Northern Isles, which reduced these areas to dependence on driftwood or import for large timber and on scavenging from relict wood patches for other smallwood needs. The result has been a picture of a lowland zone already largely devoid of significant woodland cover by around 1200 (which we shall examine more closely in Chapter 8), island areas reduced to a few significant preserves – with Rùm presented traditionally as the chief source of major timbers in the Inner Hebrides – and the main pine and oak areas being concentrated in the central and north-east Highlands and down the Atlantic coast from Loch Duich to Knapdale. All of these traditional assumptions can be questioned.

Perhaps the greatest obstacle to gaining a clearer understanding of the state of Scotland's medieval woodland remains the deeply embedded misconceptions that surround the use of the word 'forest' in medieval records. To many nineteenth- and earlier twentieth-century writers, the forests encountered in these records were dense woods that some saw as 'primeval' places, true relict woodland of the mythical 'Great Wood of Caledon', haunted by wild beasts and equally wild and free-roaming Highlanders. Scratch at the surface of that interpretation and we find that the sources upon which those views were based are referring to legal entities, not to real woods. Often, we are witnessing the projection into the past of the later medieval – and nineteenth-century – hunting forests that had been created as the exclusive preserves of the nobility. Again, there is a blurring between the fictive or notional and the real, for it was what we might term the noble hunting interest that had first attempted to assert its priority for woods as hunting reserves and had ensured that the forest law which protected that function survived as the overarching legal status of areas named as forest, even where no woodland existed. It was not just in the northern Highlands or the Hebrides that the forests in question were often nothing more than legal notions rather than real expanses of woodland. We must bear in mind that the forests that emerged in the royal charters in the twelfth century contained land that was as often open moor and wetland as it was treed, and that any woodland was as likely to consist of birch, alder and willow scrub as of large broadleaf or conifer varieties.

Having set aside the red herring of 'forest', however, we are still left with clear evidence for the survival of woods in many parts of the Highlands that are likely to have been at least as extensive as the semi-natural woods that survive today. This was richly biodiverse woodland. It constituted habitat for a broad range of wildlife, mammals, birds, and insects, and supported a diversity of plant and fungal life that was unique to different types of higher-order trees. While the foregoing discussion has concentrated on a series of heavy-demand uses for that woodland – as wood pasture, as a source of pannage and for timber and fuel needs – we must also remember its far broader role as a place for forage. Here was a source of forage foods, from nuts and berries to wild bee honey, of edible fungi and of inedible varieties and lichens for dyes and medicinal uses. The loss of woodland impoverished everyone through the loss of these multiple functions, services and resources that it contained and sustained.

On the other hand, whilst we lament the dramatic decline in woodland cover that occurred across the era, most at the time celebrated the gain it represented. In an era where the environment in general was viewed in adversarial terms as something that threatened and challenged human society, as a wild thing that biblical authority insisted was there for God's greatest creation to bend to its will and exploit, the conversion of such visibly wild land to uses deemed by humans as more useful and productive was to be pursued to total victory. The contraction of the woods was a necessary corollary of the expansion of agriculture, the spread of human settlement and the replacement of trees with more valuable – at least for a time – fields or livestock. As illustrated by the many medieval European depictions of the contrast between the culturally inferior inhabitants of mountains and forests on the one hand and those of cities and well-ordered fields on the other, the mastery of woodland and its conversion to what were regarded as more productive uses by those who did not depend on its bounty were worthy objectives. It is those supposedly more productive uses that form the subject matter of the next chapter.

CHAPTER 8

Feeding and Fuelling Growth

> William, by God's grace King of Scots, to all his worthy men of his whole land, greetings. Let them know, those present and those to come afterwards, that I have granted and this my charter confirms, to God and the church of St Mary of Newbattle and the monks there serving God, in perpetual alms, that peatary of Loquhariot, which is called *Wulvestrother*, which David son of Robert gave to them and his charter confirmed. I have granted also to them that easement in the land of the same David for drying their fuel, which the said David gave to them, and free entry in the land of the said David and egress for carrying it.[1]

In the previous two chapters, increasing pressure on the resources of the land and the primary factors contributing to that pressure were the main themes explored. We have seen how the ameliorating climate and (generally) more benign weather conditions of the MCA provided opportunities to expand economic activities that supported a steady upward trajectory in population levels everywhere across Scotland. Those same opportunities, of course, were being experienced throughout the western parts of mainland Europe, creating demand for the raw materials necessary to sustain that increased population in food and clothing and adding to the push and pull factors that encouraged migration into new regions, where access to those resources was easier and where there was active encouragement of new settlement by the indigenous rulers. From Poland to Ireland and Scotland, this was the great era of the medieval expansion of Latin Christendom, with colonists from already densely populated areas like Flanders, the Rhineland-Westphalia and some eastern parts of England moving into remoter and, in both their own eyes and those of their ambitious new lords, under-developed regions. There, they began the process of transforming the patterns of resource exploitation, intensifying and extending demands and introducing mechanisms of environmental change both to facilitate those processes and to make the desired resources more abundant and more accessible. It was not just people who moved, however, for they

1 *Regesta Regum Scottorum*, vol. 2, no. 191.

brought with them new technologies and, perhaps more important still, different cultural expectations from those of the native peoples of Scotland. They had different ways of working the land, different dietary preferences and, for some, different, more commercial mindsets that affected the way they looked at and valued land and the exploitable resources it could yield. In this chapter, we will explore how those different perspectives began to deliver widespread environmental change.

Rising demand and dwindling supplies

Despite the evidence considered in Chapter 7 for increasing difficulties in sourcing fuelwood and building timber as the thirteenth century progressed, Scottish historians have rarely discussed fuel supply. Transitions between dominant fuel types have been seen as smooth, technologically driven processes that arose from conscious choices rather than being adopted by necessity.[2] More recent research suggests that change was also driven by fuel shortages, especially in urban domestic contexts but also as affecting proto-industrial processes like salt manufacture. There is abundant evidence for shifting strategies in the quest for fuel security in regions where access to wood was already difficult in the later Neolithic and the Bronze Age, and we have noted in earlier chapters of this volume that the quest for adequate sources of thermal energy had led to the adoption of alternatives, ranging from peat to dried animal dung. Looking at just the era of the MCA, we can see that similar transitions occurred in successive waves from at least the twelfth century, depending on location, population density and the nature of available alternatives.[3] A new transition from wood to peat in areas where wood had formerly been available is the first significant shift that can be observed. Early twelfth-century record evidence survives for this shift in areas like the eastern lowlands of the Scottish Borders, where we have already seen woodland reduced to a few, tightly controlled blocks, as occurred in the grant to Kelso Abbey in the late 1120s of the right to cut peat or turf on nearby Ednam Moor 'for making fire'.[4] Elsewhere, the earliest transition episode – from increasingly scarce firewood to cheaper and more widely available peat – is evident in charters after *c.*1175 and affected urban and rural populations alike. Ready access to alternative forms of fuel drove the transition but cultural accommodation and environmental change were key underlying factors. For rural populations, as encountered in previous chapters, mixes of fuel of varying composition dependent on local availability had been the norm for centuries. Dead wood gleaning or windfall from woodland, heather and small scrub species like broom, gorse or whin, dried seaweed, peat and dried animal dung provided the main thermal energy sources – as indicated by excavated hearth residues – with fuelwood supplies maintained through natural regeneration and the stimulated growth of coppicing. As illustrated by Humphrey de Berkeley's grant of land in Balfieth in the Mearns to Arbroath Abbey, wood easements were accompanied by access to peateries locally.[5] In such cases, fuelwood may have

2 Duncan, *Scotland*, pp. 359–363; D. Ditchburn and A.J. Macdonald, 'Medieval Scotland, 1100–1560', in R.A. Houston and W.W.J. Knox (eds), *The New Penguin History of Scotland: From the Earliest Times to the Present Day* (London, 2001), pp. 96–181.

3 Oram, *Domination and Lordship*, pp. 248–250; R.D. Oram, 'Environmental change, resource conflicts and social change in Scotland c.1250–1500', in C-F. Mathis and G. Massard-Gilbaud (eds), *Sous le soleil: Systèmes et transitions énergétiques du Moyen Âge à nos jours* (Paris, 2019), pp. 211–224. These transitions are explored in depth in Oram, *Where Men No More May Reap or Sow*, especially chapters 2, 8 and 12.

4 Innes, *Kelso Liber*, vol. 1, no. 2.

5 Innes, *Arbroath Liber*, vol. 1, no. 89.

As references to the king's log pile at Berwick in the reign of David I (1124–53) might suggest, stacks of cut firewood such as this might have become the preserve of the wealthy and powerful before the end of the twelfth century.

been reserved for particular uses that did not require fires of very high temperatures.

For the inhabitants of burghs and monasteries, many of whom were colonists from regions where wood was the sole fuel type, anything other than fuelwood was culturally alien. All of the first townsmen gained shares of arable and pastoral resources in the burghs' common lands for production of their dietary staples. All had equal shares in building materials and fuel resources from those lands, with which to erect and heat their homes, cook and provide the thermal energy for craft processes.[6] Pollen evidence suggests, however, that the hinterlands of the new settlements were already largely treeless before *c.*1100, with woodland reduced to islands of large standards and saplings but with extensive tracts of scrub in valley floors and marginal areas.[7] While urbanisation was new in the early twelfth century, rural settlement was already nearly five millennia old in most parts of Scotland and the lowland zones into which most burghs were implanted had, after the contraction of the post-Roman period, seen renewed acceleration of intensive agricultural exploitation from the mid ninth century AD. As disputes over rights to wood indicate, by the early 1100s there was mounting competition for overexploited fuel resources.

In urban contexts, the evidence for rising demand

6 Oram, *Domination and Lordship*, pp. 269–283.

7 Tipping, 'Medieval woodland history'; R. Tipping, 'Palaeoecology and political history: evaluating driving forces in historic landscape change in southern Scotland', in I.D. Whyte and A.J.L. Winchester (eds), *Society, Landscape and Environment in Upland Britain*, Society for Landscape Studies, supplementary series, 2 (2004), pp. 11–20; Oram, *Domination and Lordship*, pp. 238–250.

Flood-borne flotsam trees, River Almond, Inveralmond, Perth and Kinross. Trees carried downstream in spate events on the River Tay and its major tributaries have been seen as a possible source of fuelwood. While such serendipitous deliveries were likely seized upon by Perth's burgesses, the random nature of this supply makes it unlikely that it satisfied their daily fuel or materials needs.

and diminishing local availability is clear in the recorded experiences of Edinburgh and Perth. At the former, woodland was already scarce locally before its foundation as a trading burgh in the reign of David I, as illustrated between 1128 and 1150 in a royal grant of fuel made to Holyrood Abbey, which spoke solely of wood or brushwood.[8] Given that the territory awarded to the early burghs for arable and grazing land and as a source of building materials and fuel was relatively small, the burgesses' demands might quickly have outstripped the capacity of their common property to supply, forcing a reliance on external suppliers or the sufferance of neighbours like the canons of Holyrood. The same charter forbade neighbouring rural communities from cutting peat in lands granted to the monastery, despite peat not being listed amongst the resources available on its property. This omission implies that it was the sources of wood locally that were most valued by the canons of Holyrood and that were already the object of competition from other users, while peat was abundant but clearly not considered to be of economic significance. Holyrood's rights in the woods therefore required explicit definition to ensure they were enforceable, whereas its control of the peateries did not.[9] Further charters indicate that

8 *Regesta Regum Scottorum*, vol. 2, no. 39.

9 Barrow, *The Charters of David I*, no. 148.

Wild apple tree, North Muirton, Perth and Kinross. Archaeological evidence from nearby Perth indicates that tree species like apple and blackthorn and scrubby cover such as broom or gorse were cut for use as building materials or for burning as kindling, or were bundled into faggots for fuel. The use of such species is strongly suggestive of a lack of larger timber species locally.

woodland around Edinburgh was being cleared rapidly, underscoring the need for explicit reservation of the resource. [10] By *c.*1150 peat was already the most common fuel named in Edinburgh records,[11] with few grants involving fuel thereafter mentioning firewood.[12] Before 1200, the burgh was evidently a primarily peat-consuming town in an almost treeless district. The wetlands which extended west from the hilltop burgh site probably provided most of the community's fuel needs throughout the thirteenth century, but the complex landholding pattern in Edinburgh's hinterland beyond the burgh commons meant that much of the available fuel lay in private lands outside the reach of the burgesses. As Edinburgh grew and its mossland became located further and further from the urban area proper, townsfolk sought alternative and more convenient sources of thermal energy.

Perth's rapid twelfth-century expansion placed similar stresses on local wood supplies. The position there worsened when clearance for agriculture accelerated after 1150, as neighbouring peasant communities and lords expanded arable and pasture land to capitalise on Perth's demand for foodstuffs, tradable

10 J.B. Paul (ed.) *Registrum Magni Sigilli Regum Scotorum*, volume 2: *1424–1513* (Edinburgh, 1882), no. 610.

11 Ibid.

12 *Regesta Regum Scottorum*, vol. 1, nos 75, 109.

wool and hides.[13] Recent discussion has suggested that seasonal flood deposits of trees washed down the River Tay from Highland Perthshire provided a serendipitous source of fuelwood to make up for woodland lost to agricultural clearance, but this would hardly have been a reliable method of supply.[14] While wood imports could make up for deficiencies in the availability of large, structural timbers, they could not satisfy Perth's requirements for the high-calorific, thermally efficient charcoal necessary for the specialised craft processes upon which its prosperity was founded. As the Edinburgh case illustrates, Perth's wood supply problem was by no means unique, but in burghs throughout east-central Scotland it had become so severe by the early 1200s that specific provision was made in 1209–12 to encourage traders to bring fuelwood and building timber to market in Perth, Haddington and Lanark.[15] That urban communities saw it as desirable to enshrine the protection of timber suppliers as an essential extension of their burgh's economic privileges in a revised royal charter shows an understanding of the threat to their economic wellbeing posed by wood shortages. We lack price records to provide a market context for these acts, but as noted in Chapter 7 in respect of oak's importance as a high-status gift, the high monetary value of wood is unquestionable. In short, wood became too expensive a commodity for most townsfolk simply to burn. Although charred remains from domestic hearth residues indicate that some wood was still used for cooking and heating purposes after 1300 and charcoal remained essential in specialist metalworking,[16] most evidence from Perth indicates that, as in Edinburgh, a transition from firewood to peat as the principal source of thermal energy was already underway in the late 1100s.

What that meant for the local environment is not difficult to reconstruct. At Perth, the increasing use of species that would not be regarded as normal building timbers or fuelwoods, such as apple, and the prevalence of smaller branches and twigs in hearth residues points to a more scavenging approach to obtaining necessary materials. Within the burgh's lands, the standards of the big deciduous species (oak and elm especially) had long gone. Coppice might only have started to provide supplies of withies and wands from willow and hazel and of poles from regenerating oak stools that had not been rooted out for agriculture. Woodland birds and mammals would have fled the devastation; even the riverside beavers would have moved on as the wood-hungry burgesses took the river-edge trees for their needs. Along the banks of the Tay, the townsmen's hunt for wood is likely to have brought instability and erosion, starting an increased flow of sediment in the river waters that would ultimately lead to silting of the deep channel downriver on which the burgesses relied for foreign trade, and affecting the fishery that provided food and employment. The marshy areas that fringed the burgh's two 'inches', the islands of drier land to its north and south that were the main locations of its cultivation and common grazing, were probably still edged with carr, but inroads were already being made into these. Across the River Tay on the Kinnoull side and up the slopes to the west, woodland was probably still

13 D.E. Easson and A. MacDonald (eds), *Charters of the Abbey of Inchcolm* (Edinburgh, 1938), no. 37; Oram, 'Estuarine environments and resource exploitation', pp. 362, 363–364.

14 *Perth High Street Archaeological Excavation 1975–1977*, fascicule 4: *Living and Working in a Medieval Scottish Burgh: Environmental Remains and Miscellaneous Finds* (Perth, 2010), p. 84.

15 *Regesta Regum Scottorum*, vol. 2, no. 467. This clause occurs in Alexander III's 1285 confirmation to Lanark (*Regesta Regum Scottorum*, vol. 4 pt 1, no. 163) and Robert I's 1318 confirmation to Haddington (*Regesta Regum Scottorum*, vol. 5, no. 142), pointing to widespread supply problems before 1214 when the grants had first been made.

16 *Perth High Street*, fasc. 4, p. 84.

Methven Moss, looking east across the formerly peat-filled basin towards Tibbermore, Perth and Kinross. Finally cleared and converted to arable use in the nineteenth century, the mosses that stretched from here to Blackruthven were a vital source of fuel for Perth and the parishes to its west into the eighteenth century.

present but also in shrinking stands of isolated mature trees around which the rural peasant communities were steadily expanding their fields. The burgesses of Perth might have turned their eyes towards those trees, encouraging the local landowners to consider transporting the wood to the burgh market for sale. They had to secure a royal charter giving them powers to protect the suppliers of fuel and building materials and thereby encourage them to come, but the cost of obtaining it would have been worthwhile for the respite it afforded them. Flotsam hauled from the river after spates was not sufficient to supply Perth's long-term needs, and, following the example of the local peasantry, it was to the blanket of peat that covered many parts of the burgh's hinterland that they now looked. That alone offered fuel security into the distant future, but it opened a new chapter in anthropogenic impacts on the face of the land.

An age of peat?

In contrast to wood, peat was abundant and had been exploited since at least the second millennium BC. The implanting of colonial settlements cut across established fuel-gathering rights enjoyed by pre-existing rural communities in lowland moss areas, but the parchment alienation of access rights to peateries granted by the crown and recorded in burgh charters gives no hint of conflict – yet. Stricter enforcement of exclusive peat-cutting rights came in the later 1100s as firewood pressures mounted and what many townsfolk had regarded as culturally inferior fuel became the only affordable option. The shift is clearest first in records relating to industrial processes which consumed high volumes of fuel, like horn-working, cloth-dyeing and particularly salt production,[17] which at this date involved the boiling of sleeched brine. As discussed in Chapter 7, earlier

17 For horn-working and cloth-dyeing, see A.I. Cameron and J.A. Stones (eds), *Aberdeen: An In-Depth View of the City's Past*, Society of Antiquaries of Scotland Monograph Series 19 (Aberdeen, 2001), pp. 105–106, 112–113.

Looking north from Coupar Grange, Blairgowrie, Perth and Kinross. The site of one of Coupar Angus Abbey's arable granges, the lands of Coupar Grange were taken from mossland in the later twelfth and thirteenth centuries as part of a programme that, over the next three centuries, saw the eventual draining and cutting of the blanket of peat that once extended north from the River Isla towards Blairgowrie.

twelfth-century grants of saltpans had included rights to firewood, but from the 1170s onwards the fuel rights granted were exclusively of peat.[18] At Altens on the coast just south of Aberdeen, for example, Coupar Angus Abbey acquired a saltpan in the early 1230s with access to a 'sufficient peatery for the making of salt'.[19] Aided by the generally warmer and drier conditions of the MCA, which increased the likelihood of securing an adequate supply of cut peat in May and its drying by September for burning over the next twelve-month cycle, the dependability of a fuel type whose main limitation was in terms of the volume that could be cut with the available labour ensured its widespread adoption. Whatever burgesses' cultural perception of peat's inferiority, its higher calorific value and greater thermal efficiency relative to volume, coupled with its greater abundance and easier accessibility, saw its rapid displacement of firewood as the bulk fuel of choice.

We should not underestimate the significance of this transition or its lasting environmental legacies. As conflicts over control of peat sources from the fifteenth century onwards illustrate,[20] the voracious consumption of this fuel by both urban and rural populations led to future shortages as easily accessible areas of moss were stripped. The eighteenth-century removal of the raised peat bog at Flanders Moss, discussed in the volume *Where Men No More May Reap or Sow*, is often cited as the classic example of the destruction of a wetland that had long served as a fuel resource, but while clearance of peat for fuel in places like the Moss of Blair between Coupar Angus and Blairgowrie took several centuries, it likewise saw peat-stripping that exposed the boulder clays and gravels below, and their subsequent exploitation as arable land.[21] No less than woodland clearance, the

18 Oram, 'The salt industry in medieval Scotland', pp. 215–219; Oram and Whatley, 'A brief history of Scottish salt'.

19 Easson, *Coupar Angus Charters*, no. 48.

20 Oram, *Where Men No More May Reap or Sow*, chapter 2.

21 For more recent discussion of the Modernity of large-scale peat removal, see F. Watson and P. Dixon, *A History of Scotland's Landscapes* (Edinburgh, 2018), p. 14.

inroads into Scotland's lowland peat mosses delivered a profound transformation of the land, removing distinctive environmental zones with complex ecologies that had supported species-diverse ecosystems and satisfied other resource needs. As the moss areas contracted, so too did supplies of smallwood from the scrubby woodland supported on the peat and cattle grazing on the poor grass and richer scrubland *vert*. Seasonal exploitation for material needs and hunting gradually declined as the mossland shrank, scrub disappeared and small game and wildfowl were driven elsewhere.

To pastures new

Urban transitions to peat for fuel had significant environmental ramifications for their hinterlands, but at this early period the impacts were still localised. The far more important mechanism for large-scale changes, affecting swathes of upland moor, moss and woods, was increasing levels of grazing by domesticated livestock during the MCA. It is not that the value of upland was suddenly newly appreciated at this time but rather that the longer growing season for grass and the stabler weather conditions across much of the era created opportunities for renewed or intensified exploitation of the upland resources. There is some debate as to whether upland grazing by cattle and goats in the eleventh to thirteenth centuries was necessarily part of a transhumant regime in which animals were driven into the hills for the summer under the supervision of herders and then brought down to winter pastures as growth slowed on the higher altitude grounds.[22] This is the kind of regime which seems to be implicit in the allocation of summer and winter grazing resources within the *dabhaichean* pattern that had emerged by the end of the tenth century in the districts between the Tay and the Moray Firth. While written evidence and place-name indicators exist, however, there is no hard archaeological evidence as yet identified for the operation of transhumance in any *dabhach*. Where intensive fieldwork has been undertaken in areas like the northern Cheviot Hills, there is likewise no conclusive evidence for the reservation of parts of the uplands exclusively for livestock pasture – grazing seems to have been fairly ubiquitous – yet at least intermittent cultivation, a practice established in later prehistory and through the Early Historic period, also continued in these areas across the Middle Ages. Cultivation and transhumance are not regarded by modern scholarship as compatible cohabitors of the uplands, but the manuring input of the grazing animals was presumably well understood by their herders. The livestock owners are unlikely to have passed up an opportunity to win a crop where and whenever plough or spade cultivation was possible. The documented presence of *scalingas* (shielings) in the Cheviots, however, and place-name evidence in Gaelic and Middle Scots forms ('àirigh' and 'shiel') in locations as dispersed as western Galloway, Carrick, Cunningham and Renfrewshire, the northern Cheviots, Lauderdale, north-eastern Lammermuirs, Strathearn, Angus and the Nairnshire uplands from the mid twelfth century provide evidence of functioning and widespread short-range transhumance systems that were exploiting upland grazing opportunities on a seasonal basis by the twelfth century and probably much earlier.[23] The practice was clearly not related to underlying cultural factors but points instead to common responses in all areas of Scotland to the environmental opportunities for a mixed agricultural regime in which lowland

22 Tipping, *Bowmont*, pp. 191–192.

23 Barrow, *The Charters of David I*, nos 177, 185; *Regesta Regum Scottorum*, vol. 1, no. 217; *Regesta Regum Scottorum*, vol. 2, nos 136, 185, 197, 236, 364, 398; Innes, *Arbroath Liber*, vol. 1, no. 89.

communities moved their herds into the nearby hills to benefit from summer foliage and grass growth. As discussed in previous chapters, the uplands were clearly not an empty zone awaiting 'discovery' but one whose resources had already been exploited routinely for millennia within a complex economic system. What was changing in the MCA was the intensity of that exploitation, as human population levels increased after a long early medieval decline and as external factors that encouraged the production of surpluses for trade extended their reach into remoter districts.

It is likely that patterns of use in upland woods and grassland that had become traditional between the seventh and the tenth century saw change from the eleventh century, as new political and economic systems evolved and the new climatic and weather conditions offered opportunities for expansion. Most record evidence for any expansion of grazing activity in upland areas, however, relates to twelfth-century developments in the central and eastern Southern Uplands, probably connected to the rapid expansion of a marketised and monetised regional economy during the reign of King David I (1124–53).[24] Three areas – upper Clydesdale, the south-eastern Lammermuirs and the northern Cheviots – have been relatively well researched for the period in question and stand as examples for the wider region. Although no cartographic evidence survives from earlier than the late sixteenth century for any of these areas, the oldest surviving map of upper Clydesdale, Timothy Pont's survey of *c.*1596, offers a sense of regional characteristics in the valley systems flowing out of the central Southern Upland massif in earlier centuries. It depicts progressively thinning settlement in Clydesdale south of the Tinto Hills, with small communities flanking the Clyde and its main tributaries gradually diminishing in number towards the centre of the region.[25] Topography partly determines this thinning out: the valley narrows and provides fewer opportunities for cultivation as it rises into the heart of the Southern Uplands and its flanks steepen and heighten. Even during the MCA, the alluvial soils of the floodplain between Crawfordjohn and Biggar, let alone the peat-covered podzols of the hillier moorland expanses, could not sustain a large population of subsistence farmers. Outside the Forest of Ettrick and the other regional royal and baronial forests, this was mainly grassland supporting a primarily pastoral regime and already had been for millennia.

Heavy grazing throughout the Historic era maintained the striking absence of trees recorded by Pont throughout the whole of this region.[26] It is the extensive practices of a pastoral economy that are reflected in the relative emptiness of Clydesdale south of Lanark. Its emptiness, however, is deceptive, for it was a far-from-virgin landscape. This was a zone where settlement-supporting arable agriculture was limited in extent, woodland scarce and most land given over to pasture, but where the physical remains of late prehistoric settlement littered the valley sides and hilltops.[27] The archaeological record indicates that human settlement and exploitation of the valley's resources was more widespread and intensive at times in the deeper past than in the late medieval and early modern periods. As discussed in the intro-

24 R.D. Oram, 'Breaking new ground: the monastic orders and economic development along the northern European periphery *c.*1070 to *c.*1300', in Francesco Ammannati (ed.), *Religion and Religious Institutions in the European Economy, 1000–1800: Atti della 'Quarantatreesima Settimana di Studi', 8–12 maggio 2011* (Florence, 2012), pp. 331–343.

25 NLS, Timothy Pont, [Glasgow and the County of Lanark] – Pont 34, http://maps.nls.uk/view/00002331.

26 NLS, John Blaeu, Glottiana Praefectura Superior, [vel], The Upper Ward of Clyds-dayl / Auct. Timoth. Pont (Amsterdam, 1654), http://maps.nls.uk/view/00000434.

27 Royal Commission on the Ancient and Historical Monuments of Scotland, *Lanarkshire: An Inventory of the Prehistoric and Roman Monuments* (Edinburgh, 1978), provides a detailed view of the pre-medieval settlement of Upper Clydesdale.

Upper Clydesdale from Tinto, South Lanarkshire. The valley of the Clyde south of Biggar had been settled heavily in the later Iron Age but shows little evidence of a significant human population through the Early Historic period before fresh colonisation, spearheaded by families of Flemish descent, commenced in the second half of the twelfth century.

duction and Chapter 1, settlement had already had a profound impact on regional ecologies by the late Iron Age, principally through an acceleration in woodland clearance after *c.*500 BC as a northern European style of agriculture penetrated the low-lying districts and a stock management regime with intermittent cultivation expanded through the uplands.[28] As noted in Chapter 2, regional-level palaeoenvironmental analysis suggests that this central zone of the Southern Uplands experienced a decline in agricultural exploitation in the latter part of the first millennium AD.[29] In upper Clydesdale, this contraction is reflected in a striking dearth of evidence – in terms of both physical remains and place-names – for earlier medieval settlement in contrast to its pre-Roman and Roman state. Despite speculation about how the twelfth-century incomers may have interacted with indigenous communities,[30] there is little evidence for a numerous settled indigenous community with whom they could interact.

Within the palaeoenvironmental record there is no evidence that early medieval population decline and agricultural contraction brought renewed tree growth. The extent of Southern Uplands woodland *c.*500–1100 was little greater than at present and the notion of a densely canopied, oak-dominated forest

28 Armit and Ralston, 'The coming of iron'.

29 Ibid., p. 52 fig.2.

30 For example, C.J. Tabraham, 'Norman settlement in Upper Clydesdale: recent archaeological fieldwork', *Transactions of the Dumfriesshire and Galloway Natural History and Antiquarian Society*, 3rd series, 52 (1977–8), pp. 114–128 at 126–127.

filling its valleys is partly a nineteenth-century romantic construction in which visions of Robin Hood's Sherwood Forest are applied to the use of 'the Forest' as a secure base by the Scots in the Wars of Independence. It is likely that significant stands of trees did remain, but they were in dispersed though locally extensive tracts, mainly in areas like Ettrick proper but also in the headwater valleys of Clydesdale, and with expanses of birch, willow, hazel and alder scrub in the valley floors. Although the valley was no longer settled intensively or permanently by the later first millennium AD, its upland landscape of mixed scrub, wood and grassland was exploited extensively and seasonally by the populations of neighbouring lowland districts for hunting, grazing and materials needs. These trends of partial abandonment began to reverse after 1000 as Scotland's population first stabilised then grew and improving climatic conditions brought a more favourable regime for upland exploitation. In the valley systems feeding into the Solway Firth south of Clydesdale, agricultural readvance is evident in a decline in willow scrub, the reintroduction of hazel coppice practice, the intensification of grazing impacts and a progressive expansion of arable cultivation.[31] A transition from scrub and woodland to grassland was also underway in upper Clydesdale by *c.*1150, probably linked to the establishment here of a settler plantation of Flemish origin which drove a reordering of the structures of local social and economic organisation.[32] Past discussion has focused on the sheep-farming activities of these early twelfth-century colonists and their links to the supply side of the Flemish wool trade rather than their possible military role.[33] It is likely that their introduction of large numbers of sheep into the Clydesdale hills established the grassland and heather environment which remains such a prominent regional characteristic, but which also started the development of the gully erosion that is so marked on the hillsides that plunge from the upland plateaux into the Clyde and Daer valleys.

The development of large-scale sheep husbandry in response to the economic opportunities of the international wool trade also became widespread in Tweeddale, Teviotdale and the Lammermuir Hills from at least the mid twelfth century. Here, palaeoenvironmental and documentary evidence can illustrate the impact and consequences of this change in livestock patterns. In the south-eastern Lammermuirs, a transition from scrub and woodland to grassland occurred during the twelfth century, and that transition has been linked to the creation of extensive livestock ranges within this region.[34] Although these have traditionally been seen as dominated by major monastic interests, like the Cistercians of Melrose and Newbattle and Tironensians of Kelso, and by greater magnates like the earls of Dunbar, their operations accounted for a tiny percentage of sheep numbers in the region, with the vast majority being in the small flocks of individual peasants.[35] Their intrusion, and in number, into grazing areas traditionally available to peasant farmers triggered conflict that revolved around claims of overstocking, failure to adhere to established stocking and grazing norms,

31 Tipping, 'Medieval woodland history', pp. 66–67, 68–71.

32 For an overview of the extent of pre-1150 political organisation in Clydesdale, see Oram, *David I, King of Scots*, pp. 299–302. The position is explored in detail in A. Grant, 'Lordship and society in twelfth-century Clydesdale', in H. Price and J. Watts (eds), *Power and Identity in the Middle Ages* (Oxford, 2007), pp. 98–124, https://doi.org/10.1093/acprof:oso/9780199285464.003.0009.

33 R.L.G. Ritchie, *The Normans in Scotland* (Edinburgh, 1954), p. 374 and note; Tabraham, 'Norman settlement in Upper Clydesdale', p. 119.

34 Tipping, 'Palaeoecology and political history', p. 17.

35 D. Ditchburn, *Scotland and Europe: The Medieval Kingdom and Its Contacts with Christendom, c.1214–1545*, volume 1: *Religion, Culture and Commerce* (East Linton, 2001), p. 164.

Penshiel Grange, Lammermuir Hills, Scottish Borders. Penshiel, whose ruins stand in the green field to left of centre in this view from the west, was one of Melrose Abbey's most valuable sheep farms. In addition to the valley-bottom grazing here at the head of the Whiteadder Valley, the monks had a second grange at Friardykes on the higher ground to the north-east (now occupied by the windfarm) which was a source of conflict with the earl of Dunbar and his tenants, who also used the common grazings on the southern side of the Lammermuirs.

and consequent damage to pasture to the detriment of all. Melrose Abbey, whose aggressive expansion of its grazing interests in the royal forest between Gala and Leader we have already touched on and will discuss further below, attracted particular hostility from lesser lay landowners and peasant communities alike for its rapid development of flocks based on its granges at Penshiel, Hartside and Edmundston in the Lammermuirs.[36] Litigation and arbitration led to limits being set on the monks' flock sizes and overall sheep numbers and tight designation of the areas in which they could graze and where the grange operations could have their outlying folds.[37] But hidden behind this focus on Melrose's possible abuse

36 R. Fawcett and R. Oram, *Melrose Abbey* (Stroud, 2004), pp. 221–225. For a similar dispute at Hassendean, where the king's peasant tenants led the complaints against the abbey, see A.A.M. Duncan, 'Roger of Howden and Scotland, 1187–1201', in B.E. Crawford (ed.), *Church, Chronicle and Learning in Medieval and Early Renaissance Scotland* (Edinburgh, 1999), pp. 135–160 at 138, and Oram, *Domination and Lordship*, pp. 262–263.

37 Innes, *Melrose Liber*, vol. 1, nos 46–64, 76, 77, 80, 81.

of its grazing rights, it must be remembered, the earl of Dunbar, other regional lords and their multiple peasant tenants were all busily expanding their wool-production operations. The restrictions on Melrose might appear to have environmental concerns and sustainable grazing practices at their root, but they were principally about controlling a rival producer.

Although an apparent post-1050 contraction of cultivation in the south-eastern Lammermuirs has been tentatively suggested as a reflection of intensified livestock management in this district, elsewhere in the eastern Borders there is little evidence for such a shift.[38] At Bowmont in the northern Cheviots, palaeoenvironmental analysis of the uplands flanking the valley shows the continuation of possibly intermittent and opportunist cultivation amidst a landscape that had long been largely given over to cattle grazing.[39] Documentary records reveal continuity in cattle management locally but with an increasing number of sheep present, which is possibly reflected in an intensification of grassland management and the development of grassy heathland from the twelfth century.[40] A number of changes were detected in the upland ecology, including the infilling with fen or bog of shallow basins in the leveller upland plateaux that had not seen the blanket peat growth that had swathed the higher summits from the fourth millennium BC. This peat infilling was possibly due to a change in grassland management techniques, as it occurred whilst conditions were generally drier and unfavourable for peat growth during the middle stages of the MCA.

The culprit seems to have been the adoption of muirburn as a technique to increase the volume of green pasture for grazing and improve nutrient availability. Such burning was successful in the short term but led over time to a build-up of charcoal which impeded drainage. Increased waterlogging of the grassy areas – which became significant in the later thirteenth and fourteenth centuries – contributed to the spread of a new dominant vegetation, *Calluna* (heather or ling), to form heathland, but burning alone was not responsible for this change. The spread of *Calluna* also seems to have been opportunistic, perhaps accelerated by a shift in grazing practice which saw livestock removed from upland areas over winter. The expansion of *Calluna* then occurred through competitive advantage with the grasses that were more favoured by sheep and cattle than heathers and were consequently overgrazed. As the MCA declined into a cooler and wetter climate regime, the conditions were right for the spread of *Calluna* to accelerate and change the balance of vegetation cover in this part of the Southern Uplands from species-rich grassland to ecologically degraded heath.

Similar intensification of sheep farming occurred in Galloway and along the southern edge of the Highland Boundary Fault in Perthshire, Angus, the Mearns and the glens that penetrate the Mounth, as well as in upland Banffshire, Moray and Nairn. Here, however, despite the richness of the medieval material from Coupar Angus and Arbroath abbeys especially, we lack the same volume and type of record evidence for the processes involved and the conflicts it created. We also so far lack the palaeoenvironmental studies by which to assess the ecological impact of the new grazing regime. It is unlikely that Coupar Angus's rise to become the most important wool-producing monastery in Scotland was achieved without friction with its neighbours. But its large-scale sheep-farming operations were enabled by major grants of land and rights from the crown rather than built up piecemeal as occurred in the case of Melrose and its multiple small parcels of property and incremental expansion from its core estate. That wood-

38 Tipping, 'Palaeoecology and political history', p. 17

39 Tipping, *Bowmont*, pp. 190–192.

40 Ibid., pp. 194–196.

land and scrub in Glenisla and up the Ericht into Strathardle and Glen Shee came under early pressure seems implicit, for example, in the rapid disappearance of reference to the royal forest of Drimmie, where Coupar Angus had been granted extensive grazing and easement rights at its foundation, and in Drimmie's re-emergence as one of the abbey's granges before 1165.[41] Yet foresters of Drimmie were still on record in 1549 and the tenants there had tight regulations imposed on their access to what were evidently still substantial areas of woodland that were being managed for the abbey's use.[42] This does not speak of indiscriminate clearance to create pasture for profit-yielding flocks. Likewise, within Strathardle there seems to have been a more nuanced pattern of exploitation than the traditional image of large-scale conversion to sheep-ranges. Certainly, at what became Monk's Cally, the abbey pursued development of an exclusive wool-producing grange, but further to the north-west they became landlords of the tenants of the long-established mixed-farming community on the *dabhach* of Tullochcurran and seem not to have attempted to change the economic basis of that property.[43] An award made to the canons of Scone by King David I of the teind of his cain in cheese and hides from Strathardle indicates that this had been cattle country before the 1160s.[44] While its tenants might have introduced more sheep to capitalise on their lords' connections to international wool merchants in the thirteenth century, there is little evidence to suggest that there was a general scramble into sheep husbandry.

Throughout the central, north and western Highlands and extending into the Hebrides, cattle rather than sheep remained the principal livestock type until the later eighteenth century. In these areas, the patterns of seasonal grazing that had developed by the tenth century and become fixed within the mosaic of *dabhaichean* by the eleventh (see above, pp. 148–150) persisted. In western Glen Affric, palaeoenvironmental studies suggest that until the second quarter of the thirteenth century there was no significant change to local agricultural regimes that had exploited the fragmented suite of environmental niches on the valley floor away from the increasingly unproductive blanket-peat-covered hillsides. Large-scale change to the way this landscape of scattered components was exploited was a later fourteenth- and fifteenth-century phenomenon.[45] Cattle were grazed on the higher and poorer grassland areas, including the peat, but the richest pasture remained that on the alluvial soils of the valley floor that had been the focus of grazing since the Bronze Age.

As in previous centuries, it was bovids that provided one of the principal markers of wealth throughout the Highlands and Islands, as exemplified by the payments in cows and silver pennies granted in the late twelfth century to the monks of Paisley from the mainly Islay and Kintyre territories of Ranald son of Somerled.[46] There is, however, little indication that numbers of animals increased significantly across the twelfth and thirteenth centuries. Certainly, there is no evidence from within the Highlands for a shift in emphasis from cattle to sheep to capitalise on the trade in wool that helped to drive some of the major expansion of open grassland graz-

41 *Regesta Regum Scottorum*, vol. 1, nos 226, 282.

42 C. Rogers (ed.), *Rental Book of the Cistercian Abbey of Cupar-Angus*, 2 volumes (London, 1879), volume 2, p. 55 (hereafter *Cupar-Angus Rental*). The foresters may have been operating in that part of Drimmie that the monks only leased rather than owned: Easson, *Coupar Angus Charters*, vol. 1, p. 34.

43 Easson, *Coupar Angus Charters*, vol. 1, pp. 84–87.

44 Barrow, *The Charters of David I*, no. 225 (4).

45 Davies, 'Upland agriculture and environmental risk', pp. 2057–2058.

46 Innes, *Paisley Registrum*, p. 125.

West Glen Affric looking towards Camban in Fionngleann, Highland. Cleared of shepherds and sheep in the mid nineteenth century, who had themselves replaced the settled mixed-farming communities who had worked this land since prehistory, even this elevated location is shown by pollen analysis to have been exploited as cattle pasture with patches of arable.

ing that occurred in more southerly and easterly districts. Here, regardless of any better opportunities presented by climatic amelioration, management practices that had evolved by the eleventh century and aligned with the environmental limitations of the wetter, windier Atlantic-facing West changed little. Intensification of pressure from grazing livestock, it seems, was not a phenomenon experienced everywhere in Scotland during the MCA.

From 'waste' to productivity

Amongst the records relating to the creation and then subsequent break-up of royal forests there are many references to 'waste' in respect of land. Unlike the modern usage of this word, which has connotations of dereliction and lack of value, 'waste' in the Middle Ages, derived from the Latin 'vastum', meaning an undeveloped or uncultivated place, suggested as yet unrealised economic potential and value. It did not denote 'wilderness', for records suggest the presence of limited and possibly seasonal settlement or the activity of graziers, charcoal burners and other foragers in waste territory. Lords could allocate portions of their income from waste to support monastic foundations, which reveals not only routine, regular exploitation but that it was regulated and paid for.[47] Nor yet was it simply a synonym for forest or heather moorland, for the terms occur

47 See, for example, King Alexander II's 1236 grant of the teind of his waste at Bavelaw in the Pentland Hills to the canons of Holyrood (Innes, *Holyrood Liber*, no. 58) and Walter son of Alan's *c.*1200 grant of the teind 'from all my wastes' to the monks of Paisley (Innes, *Paisley Registrum*, p. 6).

Ednam, Scottish Borders. The nineteenth-century successor to Thor Longus's late eleventh-century church now stands amidst a landscape of cultivation and settlement, where he had found 'a desert place' that he had occupied and started to convert from 'waste' to productivity.

alongside each other in the same documents, implying difference in characteristics or modes of exploitation and legal status as property.[48] Indeed, land that had been previously designated as forest could be redesignated as waste, possibly to allow for future development for alternative economic purposes. In England and throughout much of western Europe, all undeveloped areas lying as waste were viewed in legal theory as the king's land and could be taken into his demesne or disposed of by him at will. It is perhaps this principle that enabled Scottish kings to create so many royal forests in the twelfth century and as easily subsequently to alienate substantial parts of them.

It was perhaps this ability of the king to direct development of waste that we can glimpse in the later 1090s in King Edgar's grant of land at Ednam in Berwickshire to Thor Longus. According to Thor's charter to the monks of Durham granting the church that he had built there and one carucate of land for the support of its priest, the king had given him 'Aednaham desertam' and, with the king's aid and his own money, he 'inhabited' ('inhabitavi') that desert place in the sense that he made it a productive agricultural community sufficiently well populated to sustain a parish.[49] The inclusion of a carucate, or ploughgate, is hugely significant, for it reveals that this *desertum* (an abandoned place or waste) had been broken into plough cultivation over a large area in a short period of time. A similar position can be seen

48 Gilbert, *Hunting and Hunting Reserves*, pp. 94–95.

49 Lawrie, *Early Scottish Charters*, no. 24.

around a century later in the king's 'waste place of Selkirk', into which he had moved some of his own peasant dependents from a nearby community with the intention that they should develop that land and that it should become a productive component within the parish of Selkirk.[50] In both cases, although the agency for change was the labour of peasant tenants, the development was initiated by the king with the conscious intention to create new agricultural communities on land that was previously relatively empty and perhaps only extensively (as opposed to intensively) exploited.

This change dynamic is reflected in references to waste from throughout southern, eastern and north-eastern Scotland, usually at a transitional point where it was being opened for development. Interestingly, the surviving references to large-scale inroads into wastes begin in the third quarter of the twelfth century and start to diminish in number in the second quarter of the thirteenth century, possibly marking the high-tide mark of agricultural expansion during the MCA. On Coldingham Priory's Berwickshire estate, expansion of its agricultural operation through its peasant tenants was marked in a confirmation by the king of the monks' rights to control the use of all 'nemora et gwastina' (woods and wastes) on their land, enabling them to direct clearance or enforce reservation.[51] At Campsie in eastern Perthshire, the monks of Coupar Angus were given the king's hunting chase 'and all of my waste which pertains to it' in free forest, empowering them to develop that land as they chose.[52] Here, this property on the escarpment along the east side of the Tay was retained largely as woodland and pasture, carefully managed by the abbey into the later fifteenth century. Several grants of waste relate to west-central Scotland, mainly within the upland district of southern Renfrewshire. Around 1200, for example, on the expanding southern edge of the developed land in the Stewart lordship of Renfrew, Alan son of Walter gave one of his tenant knights, Robert Croc of Crookston, the lands of Kilbride with its 'waste'.[53] The charter's terms indicate that Robert was intending to introduce cattle onto hitherto lightly exploited upland. It was similarly with a view to enabling Melrose to develop its sheep-runs within the heart of Ettrick Forest that in 1236 King Alexander II gave the abbey the whole of his waste within a carefully defined area.[54] At Bavelaw on the north-western flank of the Pentland Hills, King Alexander II had given the whole of his waste to Henry of Braid, who in 1236 was able to allocate the teind of his profits from that property to the canons of Holyrood.[55] This elevated property lies mainly above the 300m contour but saw extensive arable development alongside more extensive sheep pastures intensify through the Middle Ages.

The potential value in such land designated as 'waste' is further illustrated by the exchange made around 1237 between the earl of Dunbar and Sir David Graham, whereby the latter gave up lands in Galloway in return for the earl's waste in the former royal forest of Dundaff and Strathcarron in southern Stirlingshire, where Graham was already building up a substantial landholding on the southern side of the Campsie Fells.[56] Graham received a confirmation of the lands in free forest from the king, giving him the ability to develop the property without constraint.

50 *Regesta Regum Scottorum*, vol. 2, p. 363.
51 Ibid., no. 46.
52 Ibid., no. 154.
53 NRS GD220/2/1/2.
54 Innes, *Melrose Liber*, vol. 1, no. 264.
55 Innes, *Holyrood Liber*, no. 57.
56 NRS GD220/1/A/1/1/2; Gilbert, *Hunting and Hunting Reserves*, pp. 23, 94.

The site of the Kirk o' Muir, Strathcarron, Stirlingshire. Nowadays swathed by Sitka plantation or drowned beneath the waters of the Carron Reservoir, the upland grazing around Dundaff formed the basis of the valuable estate developed in the mid thirteenth century by the Graham family.

A similar situation appears to obtain in the waste associated with the king's forest of Banff in the hills flanking the Deveron Valley, where in 1242 Alexander II granted five named properties to Waleran de Normanville. These were clearly well-defined components with carefully delimited bounds in a landscape that was demarcated into zones controlled by other royal tenants and lessees.[57] What seems to be happening here was not clearance of previously unexploited land but reoccupation after a period of abandonment, which had enabled the king to declare the district 'waste' and take possession of it into his own hands. Normanville was then given a free hand to undertake the recolonisation of his new holding.

Within all of these awards of waste, the driving motive was conversion of previously loosely, lightly or unexploited areas into productive pasture or arable. It was a step change in intensity of use that was envisaged. The new occupiers were given wide-ranging powers to effect change, introduce tenants and prevent intrusion onto the waste by others who possibly included those who had previously enjoyed forage and seasonal access to the natural resources it contained. Failure to develop the property in some way, however, was actively discouraged. In 1221, for example, a lease of land by the canons of Holyrood at Dalgarnock on the gravel and scrub flats of the Nith's floodplain in central Nithsdale set their tenant a low rent of one silver merk per annum for the first four years, presumably to encourage her to set about the introduction of peasant farmers onto it and to give time for the new community to become established. After four years, a period itself indicative of the possible rapidity of transformation, the annual rent was to rise to twenty shillings per annum, apparently assuming that the former waste was now

57 J. Stuart (ed.), *Illustrations of the Topography and Antiquities of the Shire of Banff*, volume 2 (Aberdeen, 1847), p. 109.

Sifting the parchment record

Charters, the parchment records of property transactions, are amongst our principal sources of evidence for the general appearance of parts of the Scottish landscape and its progressive transformation between the later eleventh and the fifteenth century. Although these documents are very formulaic and legalistic in structure, and the Latin in which they are written can be difficult to interpret, they often contain nuanced detail about the property in question, in respect of topography, land-use, economic practices and changing patterns of social organisation. The document quoted below, for example, provides a snapshot of the district around Craigmillar, south of Edinburgh, in the mid thirteenth century, revealing a mix of arable land (expressed as acres), pasture (and the mix of livestock on it) and hay-producing seasonal meadow. Strikingly, it also presents a landscape in which single, veteran trees are named features used for fixing locations. The named tree mentioned in the charter was not necessarily the only tree in a land otherwise stripped of timber, but it was perhaps a significant and prominent standard amongst woodland dominated by coppice:

> Charter of William son of Henry de Craigmiller [*sic*], granting for his soul's health to God and the Church of the Holy Trinity of Dunfermline, all the toft that Henry de Edmistoune held of his father, and that which he holds on the South of the way which leads from the vill of Nodris [Niddrie], to the Church of Libbertoun, to the oak called Lockyans and one acre on the North, and 2 acres lying between the King's meadow and Craigmillar and common of pasture for 40 sheep and 4 cows or one horse (summary translation from NRS GD122/2/365, dated 'the Feast of St Benedict in Lent' 1250).

Other charters provide unique records of activities that took place within a specific area, which help us to identify when transformative processes were underway. The following example offers a clue as to the emergence of an activity whose development had profound and continuing environmental consequences for central Scotland:

> To all Christ's faithful seeing or hearing this present writ, greetings from William de Vieuxpont, firstborn of the three sons of Lady Emma of St Hilary. Every one of you knows that I, with a view to charity, having given, granted and by this present charter confirmed, to the church of the Holy Rood of Edinburgh and to the canons there serving God, the whole teind of my carbonarium of Carriden, and the tenth penny of all ships and boats loaded and unloaded in my land of Blackness (extract from Innes, *Holyrood Liber*, no. 42, my translation).

This charter, granted to the canons of Holyrood Abbey in *c.*1210 by a member of one of the most important members of landholding nobility of West Lothian, may be one of our earliest records for the extraction of coal from the shallow seams exposed on either side of the Firth of Forth. The Latin noun 'carbonarium' can also mean a charcoal-burning pit, but although 'carbo' means charcoal in classical Latin, in medieval Latin in the British Isles it also acquired the meaning of coal. From around the same date, the Cistercian monks of Newbattle near Dalkeith in Midlothian had acquired possession of a colliery near Tranent in East Lothian. Before the end of the thirteenth century, coal was being extracted on both sides of the Firth of Forth and was quickly overtaking peat as the principal bulk fuel used in processing – chiefly the boiling of brine to produce salt – on the Lothian and southern Fife coasts. The quest for reliable sources of bulk fuel had seen the granting of exclusive rights to peat mosses become of increasing importance for large consumers in the second half of the twelfth century. Newbattle, for example, had sought and secured the peatery called 'Wluestrother' (Wolfstruther) in the parish of Borthwick, Midlothian, from the local lord, David of Lyne (Innes, *Registrum de Neubotle*, no. 13), to help to satisfy its fuel needs. Access to coal saw such grants of peat rights almost disappear in areas close to mine locations or where coal could be transported easily. The transition to coal use, however, was not rapid and there is little evidence for domestic users adopting coal for their household hearths before the later Middle Ages. Instead, its principal consumption seems at first to have been in the saltpans of the outer Firth of Forth, to which this abundant and seemingly inexhaustible source of fuel gave a competitive edge over the peat-fired pans of the inner Forth estuary.

A particular form of record was the perambulation, in which boundaries of property that had been disputed were agreed through a formal process of walking round them with a group of agreed witnesses, who would mark points around the perimeter, all of which were then set out in a formally agreed document. Such records are valuable on multiple levels, not only preserving an account of how, why and when formally prescribed boundaries were settled, but also offering insights on the physical process of creating marches between properties, the character of the landscape at the time of the demarcation and often 'lost' names for topographic features. The early thirteenth-century settlement of a dispute over the boundaries of the kirklands of Kirkconstantine and Kirkbride (Blaiket) in Urr parish in Dumfries and Galloway, between Enguerrand Balliol, lord of Urr, and the canons of Holyrood in Edinburgh, is a very good example. The perambulation, which Elias, abbot of Holyrood (1227–36), and his canons travelled to Galloway to undertake in the company of Enguerrand's knights and 'worthy' (*probus*) men, ran as follows:

> from the spring that is called *tibereba* and in English *Cuwelle* going up to the cross near the spring, and thus rising from that cross by the stones placed in a mound as far as the road which comes from *Kirkpatric* [Kirkpatrick Durham] to *Kirkbride*, and thus crossing that road through the mounds and placed stones as far as the cross positioned next to the road on the south of *anhanscherch*, and thus from that cross by the gathered stones next to the road as far as that other cross that is next to the boundaries about the slope of the hill on the south side, and thus dropping directly by the stones placed upon the little rocks and mounds as far as the cross that is positioned on the east side of *Cloy mac Gilernanth*, and thus straight by the ditch as far as the other cross that is placed over the stream that falls from *Blaraltan* and *kirkehil*, and thus going up by the stream as far as the rock and stone heap beside the ditch which falls from the red moss, and thus between the red moss and the moor going down the ditch which falls from the red moss on the south side and trends to the east as far as the rock beside the road and the bog where the cross is placed, and thus from that rock and cross straight as far as the stream in the bog and going down by that same stream as far as the ditch that comes from the spring of *tibereba* and in English *cuwelle*, and thus rising by that same ditch as far as *tibereba* (Innes, *Holyrood Liber*, no. 70).

Here, we see a landscape partitioned by ditches, bounded by a network of roads, cairns and crosses, running through peat mosses and bogs and up steep hillsides, using both natural and man-made markers. In this case, it is an open, treeless landscape in which those cairns and crosses were prominent and intervisible, where nothing obscured the lines of sight.

productive arable land yielding rent and other profits. A final clause stipulated that if the land was left to lie uncultivated, then the annual rent would be raised to a punitive two merks per annum for as long as it remained in that state.[58]

Breaking new ground

While much of the focus so far has been on the use of uplands and woodlands for pasture, clearance for arable purposes – such as was intended at Dalgarnock – was perhaps the bigger driver of rapid change in land-use across this era. Some areas being brought into cultivation across the MCA had been exploited for this purpose in previous centuries, sometimes extending back to the fourth millennium BC, but had experienced contraction of the tilled land or complete abandonment as population levels fluctuated and climate regimes shifted.[59] More often, however, previously intractable land was being made accessible for cultivation by new agricultural technology,

58 Innes, *Holyrood Liber*, no. 68.

59 For an overview, see Watson and Dixon, *A History of Scotland's Landscapes*, pp. 13–17.

Ox-drawn mould-board plough and horse-drawn harrow, two of the technological advances that revolutionised medieval agriculture in the eleventh and twelfth centuries. The images are from a late nineteenth-century reproduction of the Bayeux Tapestry.

principally the heavier clay and stonier soils that had been unworkable by the older and lighter forms of plough and where the effort of spade cultivation was impractical. The change had come with the introduction of the heavy mould-board plough, or carruca, which was developed in Europe towards the end of the first millennium AD and reached England well before 1000. Better availability of iron tools and new styles of horse harness also speeded and improved the removal of tree stumps, roots and boulders. The carruca was assumed traditionally to have been a twelfth-century importation to Scotland in the train of incoming knightly and monastic colonists.[60] That its northwards spread was perhaps more rapid, more widely distributed and deeper than previously believed, however, is suggested by strong links with Anglo-Saxon England from the early eleventh century and possibly the plantation of English peasants into the land around the head of the Solway Firth in the 1090s, internal colonisation through which settlers were moved from the south-east of Scotland into more northerly districts, and archaeological evidence for carruca-style cultivation on the Angus coast before the end of that century.[61] Whatever the exact date of its introduction, the carruca was a revolutionary development with profound implications for Scotland's environment. Its iron shoe could be pulled through formerly impenetrable turf and root layers, with the broken soil cut by an iron coulter that followed and the sod turned by the mould board to form ridges or rigs.[62] Pulled by a team of eight oxen, the plough was passed several times on either side of a primary cut, piling earth inwards with each pass to form the broad, clearly defined rigs whose corduroy appearance remains a prominent feature of many now once more uncultivated areas of southern and eastern Scotland.

As numerous commentators have observed, this kind of rig-forming cultivation brought multiple advantages that included increasing the extent of land that one community could cultivate, increased

60 Ibid., p. 15.

61 Pollock, 'The Lunan Valley Project', pp. 389, 392–393.

62 P. Dixon, 'Mukked and folded land: the evidence of field data for medieval cultivation techniques in Scotland', in J. Klápšte (ed.), *Agrarian Technology in the Medieval Landscape*, Ruralia 10 (Turnhout, 2016), pp. 107–123 at 121.

nutrient release or manuring opportunity, weed suppression and improved drainage.[63] It also required new forms of social organisation, since the eight-ox teams and expensive iron plough elements were usually beyond the reach of a single peasant family. Cost led to the formation of new peasant communities in townships whose individual members contributed a part of the team or the equipment and who shared proportionally in the commonly cultivated territory. Collectively, these developments represented a major saving in labour and in the calories expended on farming activities by human populations, enabling both the extension of cultivation to new ground – further increasing nutrition levels and creating income-generating opportunities through the sale of surpluses – and the application of saved time to new agricultural or craft activities. What is less well recognised, however, is that these human benefits came at an environmental cost most immediately visible in the radical shift in the physical character of the land brought into cultivation. The first step was the clearance of existing ground cover, possibly with pig grazing to suppress vegetation and prevent woodland regeneration, and the systematic felling of woodland and scrub cover, followed by the removal of stumps and roots. Annual ploughing thereafter prevented the regeneration of the unwanted vegetation or trees. Ploughing with a carruca then disturbed underlying soils to a greater depth than spade or light plough cultivation, permanently altering their composition and changing their content when manure and other fertilisers were added or through mineral and chemical removal as a result of crop growth or leaching. With the added impact of freer drainage, these changes were beneficial to some plant species but largely detrimental to those that had once predominated on the now cultivated land. New domesticated crop species – grain types like rye and 'useful' plants like dyer's rocket – and weeds of cultivation were introduced but at the expense of other condition-sensitive species. The result was a further profound and irreversible ecological shift on top of the many that had already occurred since the end of the last ice age.

As Thor Longus's development of Ednam in the 1090s indicates, arable expansion was well underway

63 Watson and Dixon, *A History of Scotland's Landscape*, p. 15. The mould-board ploughing at Inverkeilor identified by Pollock had not formed rigs, however, suggesting that the practice was not initially ubiquitous: Pollock, 'The Lunan Valley Project', pp. 389, 392–393.

Looking north up Lauderdale towards Kedslie, Scottish Borders. Agriculture here today covers the valley bottom, but in the twelfth and thirteenth centuries it was assarts like Kedslie, pushing upslope westwards into the rolling plateau of the 'forest between Gala and Leader', that were the focus of clearance for cultivation of ground that had lain abandoned since the Early Historic period.

in south-eastern Scotland in the eleventh century. Although record evidence of the process is scanty, where it survives it illustrates that woodland clearance and wetland drainage for plough cultivation was pushing out from the zone of earlier medieval settlement into areas that had been less intensively exploited for several centuries.[64] Older histories have tended to focus on the monastic contribution to this expansion, since the bulk of our surviving evidence is derived from the records of the abbeys founded in the eastern Borders region by David I. Even these records make it clear, however, that the labour through which the biggest advances were made came from peasant farmers. The extent of agricultural expansion is often implicit rather than explicit in the surviving sources, especially where the subject matter was not so much the land itself but the fruits it yielded, expressed most often in ecclesiastical records in terms of the teinds due from that land. A case in point is to be found in the records of Dunfermline Abbey in the third quarter of the twelfth century, concerning teinds due from Stirling and reflecting the rapid expansion of clearance and cultivation operations through new assarts made on the king's demesne lands there or the breaking into cultivation of 'old' land that had formerly lain uncultivated.[65] The agreement determined to which ecclesiastical institution, Dunfermline or the king's chapel in Stirling Castle, the teinds due from these new arable areas were to go. What it tells us is that although no specific locations were named, the district around Stirling was seeing new land being broken into cultivation by peasant or burgess tenants on a scale that made it worth Dunfermline's while to ensure that this land was brought into the formal process of teinding.

Far more locally specific detail of a landscape in which peasant cultivators had already made substantial advances by the early 1100s can be seen in the accounts of the processes by which Melrose and

64 Oram, *Domination and Lordship*, pp. 239–240.

65 C. Innes (ed.), *Registrum de Dunfermelyn* (Edinburgh, 1842), no. 4.

Ben Vorlich and Stuc a' Chroin from Rossie Law, Perth and Kinross. The band of light and shadow across the middle of the view is the Muir of Orchill, a zone now largely composed of commercial conifer plantation, rough hill pasture and grouse muir, which in the thirteenth century was a contested zone between peasant communities with traditional common grazing rights and would-be developers who saw it as ripe for settlement and cultivation.

Dryburgh abbeys expanded their landholdings in the land flanking Lauderdale.[66] At the middle of the twelfth century, activity was focused on *landi* (leas – arable or grassy areas or lawns) and *planities* (cleared areas) that had been pushed up from the older cultivation on the valley's lower slopes into the higher, wooded district to the west. Place-names like 'Blainslie' and 'Kedslie' mark woodland clearings or leas, while 'Sorrowlessfield' and 'Alwin's assart' are even more explicit in their reference to land cleared for cultivation. Blainslie had been given by the king probably before 1150 to his forester in the area, who had cleared and developed the land with a cultivated toft – the *planities* of Blainslie – and house at its heart, ploughing and sowing the area outside the woodland that he was otherwise charged to protect for hunting purposes.[67] Kedslie was described explicitly as a *landus* in *c.*1148–50 when it was given to Dryburgh and lay adjacent to the assart made by an otherwise unknown individual called Alwin, which was likewise given to the abbey.[68] The canons of Dryburgh were clearly expecting to continue to develop the assart beyond the extent it had reached through Alwin's efforts. They also received the land that extended west from the head of the cleared area to 'the traces of a certain old fortification' and then along a road line that ran north from there.[69] Also at Kedslie, they worked a second *landus* or lea called Herdesley.[70] Clearance at the southern end of Lauderdale was being undertaken at what became Sorrowlessfield by William Sorules, who is on record in the 1150s in a charter concerning the adjacent property of

66 These processes are explored in detail in Gilbert, 'The monastic record of a Borders landscape'.

67 Innes, *Melrose Liber*, vol. 1, nos 93, 94.

68 Innes, *Dryburgh Liber*, nos 109, 110.

69 For the use of roads and tracks as boundaries and as starting points from which assarts were pushed out in landscapes of clearance, see Oram, 'Trackless, impenetrable and under-developed?'.

70 Innes, *Dryburgh Liber*, no. 112.

Fauchope (the land flanking Packman's Burn).[71] Sorules appears to have been a tenant of the earls of Dunbar at Earlston in Lauderdale and cleared the assart to which his name was given on the valley slopes opposite that village and which the earl subsequently gave to Melrose.[72] What is most remarkable about these clearances is the rapidity with which large-scale change was effected. Before 1164, Dryburgh and Melrose had clashed over the boundaries between their clearance operations at Kedslie in Lauderdale and Colmslie in the Ale Valley to the west.[73] The record of their dispute refers to various *planities* then divided only by strips of surviving woodland and cut through by paths and roadways in a district that a generation earlier had been largely trackless, wooded hunting grounds.[74] It is easy to understand the despair of the keeper of the forest between Gala and Leader, Richard de Morville, at the end of the twelfth century as he struggled to contain further peasant and monastic inroads into what remained of the forest district.

Conflicting interests in land undergoing development extended far beyond the approximately 80km^2 that Melrose held between the Gala Water and Lauderdale. A classic case dates from the probable peak of expansion in the 1230s on the Muir of Orchill between Strathearn and Strathallan in southern Perthshire, a zone ideal for breaking into cultivation with the carruca. There, Fergus son of Gillebrigte, earl of Strathearn, had attempted to take an area known as Cotken (from the Gaelic 'coitchionn', common grazing) into cultivation. Fergus, however, had chosen to ignore the rights of those who were not his tenants in the shared use of this land, amongst whose number were dependents of Lindores Abbey. The result was litigation which Fergus clearly lost, for he was forced into a humiliating declaration that he recognised that Cotken had been common pasture for generations for all men who lived around the edge of that district, not just the men of his own lands, and that no one had the right to build houses on it or plough there.[75] As a consequence, it remained a zone of upland grazing and forage throughout the Middle Ages into which major inroads commenced only in the mid eighteenth century.[76] In contrast to Melrose's efforts to break such land into more intensive use, here another abbey saw the need to maintain a traditional exploitation regime to preserve the viability of their tenants' mixed agricultural economy.

Only 12km north-east of the Muir of Orchill, another monastic landlord was actively clearing for cultivation. In the late 1190s, the brethren of Inchaffray Abbey had been given three 'yokes' ('jugera') of land on the rising ground towards Fowlis Wester, north-west of the area of pools and marsh in which their monastery stood on its 'inch' (Gaelic 'innis', island).[77] Twenty years later, the abbey gained full rights over the whole area of these marshes which were encompassed within the *fossae* (ditches) they

71 Innes, *Melrose Liber*, no. 12.

72 *Regesta Regum Scottorum*, vol. 2, no. 482.

73 Innes, *Dryburgh Liber*, no. 113.

74 For discussion of roads and their use to break into this district, see Oram, 'Trackless, impenetrable and underdeveloped?', pp. 315–319.

75 Dowden, *Lindores Charters*, no. 28. This case is discussed in C.J. Neville, *Native Lordship in Medieval Scotland: The Earldoms of Strathearn and Lennox c.1140–1365* (Dublin, 2005), p. 98.

76 See the 1750s mapping of this district by William Roy, with areas of unenclosed rig-and-furrow cultivation amidst the otherwise open rough pasture of 'Orchillmoor': NLS, General Roy Military Survey of Scotland 1747–1755, https://maps.nls.uk/geo/explore/#zoom=13&lat=56.28581&lon=-3.92479&layers=3&b=1.

77 W.A. Lindsay, J. Dowden and J. Maitland Thomson (eds), *Charters, Bulls and Other Documents Relating to the Abbey of Inchaffray* (Edinburgh, 1908), no. 2.

had dug by that year,[78] land which then became the heavily worked demesne closest to their precinct. These ditches probably included the first canalisation of the sluggish Madderty Pow, the slow-moving stream that inadequately drained the western end of the fen-filled basin between the Gask Ridge and the south Glen Almond hills. Here, the decision was taken to increase an arable zone at the expense of the wetland resources, which included seasonal pasture on the wet meadows in the basin and the fishery in the wetland pools and pows.[79]

Inchaffray was not alone in being attracted by the arable potential of alluvial soils. On the Forth carses, for example, Holyrood and Newbattle abbeys – or rather their tenants – were active drainers in the third quarter of the twelfth century.[80] In the 1140s, in the Carse of Callendar between Airth and the mouth of the River Avon, a district referred to as 'Blankeland' – the 'white waste ground' – the estuarine flats saw saltpan development on their northern outer edge and the extension of cultivation from the higher dry ground to their south.[81] Always vulnerable to intermittent flooding, expansion into this territory was perhaps safer during the more benign conditions of the MCA, but by the second half of the thirteenth century, as conditions deteriorated, it had already become less tenable and parts were abandoned to the waters by the early 1500s.[82]

Inundation by sea-flood was a risk that other developers were also prepared to take in their desire to win arable from coastal or estuarine flats. From the Nith estuary, where Holm Cultram Abbey received part of Kirkconnell Merse in the mid thirteenth century and confirmation of their right to take 'as much as [they] can reclaim towards the sea',[83] through the Carse of Gowrie and Strathmore to the Laich of Moray, arable yields on alluvium made any threat worthwhile. Lay developers appear to have been at the spearhead of initial drainage and cultivation operations in the Carse of Gowrie from the mid twelfth century, pushing out from the various 'inches' within the wetland and the drier ground along the south-facing escarpment of the Sidlaw Hills.[84] Perth burgesses as well as important secular landholders were speculative developers of grain-producing land in operations that extended the cultivated areas associated with much older settlements like Errol and Inchture.[85] The conversion of wet pasture into oats-, barley- and wheat-yielding arable was perhaps speedier on the Carse of Gowrie than on the Forth on account of the lower salinity of the Tay estuary, which meant that there was less salt in the alluvium to be flushed as the land dried out than on the Forth, Nith or Moray coastlands.[86] Development was already far advanced by the time the monks of Coupar Angus acquired 'Eddirpolles' (between the pools/pows) to the north-east of Errol around 1180, which they developed into an arable operation known simply as 'Grange'.[87] They were not the pioneers here, however, for it is evident that the Hay

78 Ibid., no. 37.

79 The abbey was still taking eels in the Pow near Dollerie in the fifteenth century: ibid., no. 148.

80 Oram, 'Estuarine environments and resource exploitation', pp. 364–365.

81 Barrow, *The Charters of David I*, no. 98. By the late medieval period, Holyrood's interests here included twenty-five farming tounships: Innes, *Holyrood Liber*, appendix, no. 37.

82 This abandonment is discussed in Oram, *Where Men No More May Reap or Sow*, p. 98.

83 Grainger and Collingwood, *Holm Cultram Register*, nos 116–119.

84 Discussed in some detail in Duncan, *Scotland*, pp. 320–321.

85 The archaeological depth of settlement on the islands of dry ground in the Carse from the Neolithic to the late eleventh century is explored in RCAHMS, *South-East Perth*, pp. 41–75.

86 Oram, 'Estuarine environments and resource exploitation', p. 363.

87 Easson, *Coupar Angus Charters*, vol. 1, nos 3, 4, 25.

The Braes of the Carse from the carse floor between Inchture and Carsegrange. Drainage of the wetlands of the carse, pushing out from the dry islands of the 'inches' and down from the ancient raised beach at the foot of the escarpment, accelerated from the later twelfth to the mid thirteenth century as lay lords and monastic proprietors sought to maximise arable opportunities.

lords of Errol had already expanded the arable potential of their property and provided the infrastructure to process the grain, including a mill.[88] The monks, however, moved rapidly to maximise the potential of their operation, buying up adjacent portions of arable already developed by lay proprietors and building a 2.5km-long causeway across the wetland from Inchture to their grange.[89] That causeway itself provided a platform from which to push drainage and clearance out into the areas of wetland to either side. Lay interests also remained active in winning further land from the marshy carse up to – and far beyond – the waning of the MCA. At Inchture, the abbey's peasant tenants extended their community's arable land out from their *inch* towards the shore (*plaga*) of the Tay.[90] But this was piecemeal expansion and, as mid eighteenth-century maps of the district show, the wetlands flanking the larger of the sluggish pows that oozed down to the river from the north remained undrained, exploited for hay from meadow-grasses, reeds, wildfowl and fish until the major drainage operations of the Age of Improvement.[91] Exploitation of some areas was still beyond the technological abilities of medieval cultivators and their best equipment.

An alleged complaint offered by an apologist for one claimant to the throne of Man as justification for his rebellion against his brother in the early thirteenth century was that the portion of the Western Isles given to him for his support – principally Lewis – was 'cultivated by few inhabitants because it is mountainous and rocky, and almost wholly inarable'.[92] Here, he went on to allege, most men lived by fishing and hunting. It is a powerful image but does not accord with post-medieval statements of the agricultural repertoire of the islands, nor with the archaeological evidence for arable expansion along the entire length of the Atlantic-facing machair and blacklands of the Western Isles from the eleventh century onwards. Since the initial stages of Norse

88 Ibid., vol. 1, no. 3; Rogers, *Cupar-Angus Rental*, vol. 1, p. 337.

89 Easson, *Coupar Angus Charters*, vol. 1, no. 37; Duncan, *Scotland*, pp. 320–321.

90 Easson, *Coupar Angus Charters*, vol. 1, no. 25.

91 The abbey was maintaining an eel fishery at Grange to the Reformation: Rogers, *Cupar-Angus Rental*, vol. 1, p. 189.

92 *Chronica Regum Manniae et Insularum*, undated section following annal for year 1217.

settlement of the islands in the ninth century, there had been intensification of the cultivation of both black oats and hulled six-row barley, the latter previously the principal crop grown by the indigenous late Iron Age population. Alongside these, rye, which favoured poor and dry soil, was also introduced onto the machair. Together, these crops and the apparent increase in production suggest that the incomers expanded the zone of cultivation from the stabler blackland zone into areas of machair that earlier communities had been unable or unwilling to farm or had deemed too laborious for profitable development.[93] This increased emphasis on arable farming and especially on grain production was coupled with the introduction of some new technologies to maximise the yields – notably corn-drying kilns and barns[94] – pointing to a farming culture in which cereals were significantly more important than they had been for the indigenous population. This was not a carruca-enabled expansion, however – none of the islands' soils needed that heavy implement – but one that was driven by the cultural expectations of the incomers and especially by the heavy emphasis on cereal products in their diets.

The reorientation of effort evident at sites like Bornais and Cille Pheadair in South Uist underscores how human choice as much as environmental opportunity or climatic constraint played an active role in the shaping of exploitation regimes. There, the arrival of Norse settlers triggered a major shift in

93 N. Sharples, 'The exploitation of the landscape', in N. Sharples (ed.), *The Economy of a Norse Settlement in the Outer Hebrides: Excavations at Mounds 2 and 2A, Bornais, South Uist* (Oxford, 2021), pp. 418–424 at 418–420; Sharples, *A Norse Farmstead*, pp. 187–188; H. Smith, 'Plants', in N. Sharples (ed.), *A Norse Farmstead in the Outer Hebrides: Excavations at Mound 3, Bornais, South Uist* (Oxford, 2005), pp. 189–190; H. Smith, S. College and P. Austin, 'The carbonized plant remains and wood charcoal', in M. Parker Pearson et al. (eds), *Cille Pheadair: A Norse Farmstead and Pictish Burial Cairn in South Uist* (Oxford, 2018), pp. 516–540; H. Smith and J. Mulville, 'Resource management in the Outer Hebrides: an assessment of the faunal and floral evidence from the archaeological investigations', in R.A. Housley and G. Coles (eds), *Atlantic Connections and Adaptation: Economics, Environments and Subsistence in the North Atlantic* (Oxford, 2004), pp. 48–64 at 55–56, 57.

94 Sharples, *A Norse Farmstead*, pp. 89–94, 178, 187–188.

The blackland/machair interface, looking west from Ruabhal, South Uist. The Norse period in the Western Isles saw intensification of arable agriculture and an expansion of cultivation onto the calcareous shell sands of the machair. While these practices were viable during the stable years of the MCA, increasing storminess brought wind and wave erosion in the less stable conditions of the later Middle Ages.

resource exploitation emphasis across the entire marine, maritime and terrestrial environmental range. Some of the change was opportunistic, benefiting from the milder conditions of the MCA, or stemmed from the different technology brought by the new settlers. Collectively, it illustrates how the introduction of new dynamics could overturn the limitations that adherence to the conventions of traditional environmental knowledge imposed on old and culturally conservative established populations. Importantly, the archaeological analysis at sites like Bornais has shown that rather than being locked into the constraints of the regime they brought with them from their Scandinavian homelands, the settlers continued to adapt to opportunities in the Outer Hebrides and to the benefits offered by the MCA. There, cereal cultivation continued to grow in significance, with barley favoured over oats, as the stabler weather of the era became entrenched across the eleventh to thirteenth centuries.[95] The colonists' Scandinavian-style agriculture and new toolkit both required and enabled not only an expansion of the zone of cultivation but also a broader suite of cultivated plants. Flax, whose nutrient requirements made it a competitor for areas suited to the better-quality cereals, was one of the most significant introductions into all areas of Scandinavian settlement in Scotland. Although usually considered more important for the fibrous stems from which linen was produced, it was the oil yielded by its seeds – cold-pressed flaxseed and linseed oil – that was perhaps more in demand for both food preparation and in treatments for wood and other perishable materials.[96] While some seeds could be obtained through foraging for wild plants, this is unlikely to have yielded the volume necessary for regular domestic use. If intro-

95 Sharples, *The Economy of a Norse Settlement*, pp. 428–430.

96 Smith, 'Plants', p. 189; C.A. Dickson and J.H. Dickson, *Plants and People in Ancient Scotland* (Stroud, 2000); J.M. Bond and J.R. Hunter, 'Flax-growing in Orkney from the Norse period to the 18th century', *Proceedings of the Society of Antiquaries of Scotland*, 117 (1987), pp. 175–181.

duced on a large scale as a cultivated crop, the maintenance of nutrient-hungry plants like flax required new approaches to and new levels of manuring, essential on zones like the machair from which nutrients leached easily. In short, the expansion of Norse cultivation practices in the Western Isles across the MCA contributed to a further ecological shift, facilitating the entrenchment of practices – especially machair cultivation – that contributed to devastating losses when the stable weather regime began to break down in the later thirteenth century.

But was the expansion of cultivation that we have seen taking place from the northern Cheviot Hills to Uist, and from the Solway coast to Moray, ubiquitous and uniform? Our conventional understanding has been that the interplay of benign climatic conditions, new agricultural technologies, a growing population and commercial opportunity drove expansion to the limits of viability. It is clear from archaeological and palaeoenvironmental analyses, however, that not everywhere in Scotland conformed to that model, or at least not to the drive to the maximum of cultivable potential that occurred in many parts of England and northern Europe. Yes, the patches of broad-rig cultivation characteristic of the heavy mould-board ploughs employed from the eleventh century that can be seen in the Cheviot uplands are suggestive of opportunistic maximisation, but they do not seem to be a late development indicative of desperation to find new land to sustain an overexpanded population. Indeed, there is evidence that rather than witnessing the beginning of aggressive expansion, in some places the eleventh century experienced a contraction or withdrawal from areas that had been brought into cultivation as early as the first half of the seventh century. At Lair in Glen Shee in north-eastern Perthshire, discussed in Chapter 4, expansion into the uplands saw cultivation reach close to 400m OD before AD 660, well above the limit of modern arable farming in the district, but the settlement disappeared soon after 1000.[97] This was not unviable marginal land, for both the scale of the

97 Strachan, Sneddon and Tipping, *Early Medieval Settlement*, p. 140.

buildings and the artefacts recovered from the sites indicate that the farms in this landscape were occupied by prosperous agriculturalists who grew substantial quantities of barley and grazed their herds on the upland pastures. Why, then, were Lair and similar settlements throughout this district abandoned around the end of the tenth century and beginning of the eleventh? It is unlikely to have been driven by climate change, for conditions were, if anything, warming, albeit with increased rainfall.[98] Wetter conditions were not likely to have forced a change either, for the areas of cultivation were on the freer-draining soils. Indeed, it is likely that the patches of cultivated ground had always been sited to maximise the benefits of such conditions. Similar concentration of arable on the patches of better-quality land has been identified in Glen Affric, suggested there to be a development that was not maximised even during the height of the MCA.[99] Again, a conscious choice was being made to exploit the land and the resources it offered in a different way to the conventional model of eleventh- to thirteenth-century maximising of arable productivity. Such a choice is suggestive of both controlled population pressure and an absence of dependence on cereal cultivation. Socio-economic reorganisation seems a more likely explanation for the abandonment of Lair's high-altitude settlements, with a possible later eleventh- or twelfth-century shift to more nucleated communities at lower altitude.[100] Yet although there is evidence to show later use of some of the abandoned sites, possibly as shielings associated with short-range transhumance, why farmers would choose to abandon viable cultivated ground into which so much energy and soil-enrichment had been invested remains unanswered. All we can say is that despite the opportunities presented by the MCA, population growth locally was clearly not a factor affecting the (non-)exploitation of cultivable land at Lair, even during the pre-plague peak of Scotland's population in the late thirteenth and early fourteenth centuries.[101]

Withdrawal from cereal cultivation, or at least minimal investment in it, appears to have been a conscious decision in some areas during the MCA. Always bearing in mind the limitation of local soils or topography, such a decision was taken at Freswick Links in Caithness. There, marine exploitation rather than cultivation provided the economic underpinning of the settlement during its later Norse phases.[102] The development of the fishing economy at Freswick Links will be discussed below, but it is important to recognise that while a terrestrial, grain-producing regime had existed here in the pre-Norse period and limited quantities of wheat and barley or bere were being grown by its later Scandinavian inhabitants, this investment in agriculture was secondary to specialisation in sea-fishing. One place where widespread withdrawal on this basis seems to have occurred is Shetland, where alternative pastoral exploitation regimes, expansion of deep-water fishing and an ability to obtain grain from external sources encouraged a shift in specialisation away from attempts at cereal cultivation in what even during the best years of the MCA was a hostile environment for cultivators.[103] This regime, described as a 'mixed

98 Ibid., p. 147.

99 Davies, 'Upland agriculture and environmental risk'.

100 Strachan, Sneddon and Tipping, *Early Medieval Settlement*, p. 147.

101 Excavation of the type-site for these forms of upland settlement at Pitcarmick in the moorland above Strathardle revealed later eleventh- and twelfth-century reoccupation there but abandonment again by 1220: ibid., p. 148.

102 Morris, Batey and Rackham, *Freswick Links*, pp. 269–270.

103 G.F. Bigelow, 'Issues and prospects in Shetland Norse archaeology', in C.D. Morris and D.J. Rackham (eds), *Norse and Later Settlement and Subsistence in the North Atlantic*, University of Glasgow Department of Archaeology Occasional Paper 1 (Glasgow, 1992), pp. 9–32 at 20.

Lower Glen Shee looking south from Lair, Perth and Kinross. We do not know what factors led to the abandonment of the viable upland arable community at Lair in the later eleventh century after generations of intensive labour investment in soil-enrichment, but it was not climate-related.

subsistence/surplus-for-exchange pattern', is what potentially evolved at Freswick Links.[104] The trend was not ubiquitous, for on the possibly high-status site at Kebister arable was not only maintained but barley cultivation expanded in the twelfth and thirteenth centuries.[105] This adherence to the local production of cereals was perhaps linked to status in a culture to which grain-derived food and drink were central, as seen in the Norse colony in the Western Isles.

It is, then, a far more complex picture that we see in Scotland than the scramble to expand cultivation postulated for southern Britain and the near continent. Where environmental opportunity, social organisation and access to new agricultural technologies encouraged and permitted that trend, it was largely followed, with expansion already underway in the eleventh century. The result was inroads into areas of woodland and scrub that had regenerated in the earlier Middle Ages, upland grassy heath and coastal or estuarine wetland, pushing back out from the zone of settlement and cultivation that had persisted through the previous era of depopulation and socio-political reconfiguration. Some expansion may have been linked to cultural practices, especially in zones of Scandinavian settlement where cereal

104 Morris, Batey and Rackham, *Freswick Links*, p. 275.

105 O. Owen and C. Lowe (eds), *Kebister: The Four-Thousand-Year-Old Story of One Shetland Township*, Society of Antiquaries of Scotland Monograph Series 14 (Edinburgh, 1999), p. 289. At the Biggings on Papa Stour, evidence for oat and barley cultivation was identified, but not in quantities indicative of widespread arable activity around this important farm: B.E. Crawford and B. Ballin Smith, *The Biggings, Papa Stour, Shetland: The History and Archaeology of a Royal Norwegian Farm*, Society of Antiquaries of Scotland Monograph Series 15 (Edinburgh, 2000), p. 113.

Common seal, West Wemyss, Fife. Despite their name, common or harbour seals are less common than grey seals in the waters around Scotland. Charters from the first half of the twelfth century onwards record the rights to hunt seals on the shores of the Firth of Forth and to receive renders of seal meat and blubber being granted to the monks of Dunfermline by Scottish kings.

cultivation was introduced on a more intensive level than had been practised by previous inhabitants. Extension of cultivation may have been a consequence of different cultural expectations and dietary preferences rather than of having ever more mouths to feed. Such differences perhaps explain why in other districts, including large parts of the Highlands, alternative models of social organisation and agricultural practice did not lead to any drive to the margins. Indeed, in some areas there was a conscious retreat, even when environmental conditions were favourable for further expansion. Such trends drive home the prosaic truth contained in the biblical reminder that 'Man does not live by bread alone.'

Harvesting the waters

Alongside the intensification of agricultural activity and its extension into new territories there was an equivalent expansion of activity into both fresh and salt waters. Archaeological, palaeoenvironmental and historical record evidence all indicate that the MCA saw a step change in the exploitation of aquatic resources in and around Scotland. The seas, especially to the north and west, were fished with what was likely a previously unseen intensity, and the estuaries, rivers and lochs likewise saw rapid growth in their fisheries. Some of this expansion can be linked to the calmer ocean conditions delivered by the MCA or to the influence of warmer ocean surface and deep-water temperatures on the preferred feeding or breeding conditions for pelagic (open-water) and demersal (bottom-feeding) fish, but new technologies (particularly boat-building), cultural food preferences amongst new colonists arriving in Scotland and increasing opportunities for internal and external trade in catches all stimulated a growth in fishing. The sheer richness and diversity of Scotland's fishery at this time is not only evident in the bone assemblages recovered from archaeological sites but is illustrated by the documentary record both of species caught and of places where large-scale fishing operations were based which have long since ceased to yield significant – if any – catches.

Perhaps nowhere in Scotland is the story of riches to dearth clearest than in the Firth of Forth. The scale and range of exploitation here, which was already in operation by the late eleventh century, is revealed by the grants made by kings of Scots to their newly founded monasteries of shares of the catches that were due to them as food renders. It must be emphasised, however, that these were not innovatory awards that established new fishing bases or communities but rights to a share in already functioning operations which the king and his greater nobles possessed as part of their personal property. They reveal, too, already established practices whereby certain of the choicest species or portions of a fish or sea mammal were reserved for the lord. They also show that it was not just fish, seal or whale flesh that was sought but also the blubber, oil and fat obtained from it. For example, the monks of Dunfermline received from Kinghorn in Fife every seventh seal caught (after the Church had received its teind), the heads (except the tongues) of every whale or large fish beached or caught on the Fife side of the firth, and half the oil from all whales and large fish caught between the Forth and the Tay (for use in altar lamps in the abbey church).[106] Kelso Abbey received another half of the whale blubber and fish oil, while Holyrood received every tenth whale caught or beached on the south side of the Forth estuary.[107] Eighteenth-century observers confirm that these whales and 'large fish' were still abundant in later centuries, enabling us to identify them as orca, pilot and even blue whales and also dolphins and porpoises.[108]

These special delicacies were few in number in comparison to the abundance of fish species caught in the outer firth and the tidal reach of the river as far inland as Stirling. Fishbones recovered in excavations in other major estuarine population centres like Dundee on the Firth of Tay or Perth at the head of its tidal reach show that fish from both salt and fresh waters were brought to the burghs for trade, with marine species including herring, cod, ling, haddock, pollack, dab, plaice and ray, which suggests that both netting and line-fishing with baited hooks took place.[109] In

106 Barrow, *The Charters of David I*, no. 33; *Regesta Regum Scottorum*, vol. 1, nos 117, 214.

107 Barrow, *The Charters of David I*, no. 147; *Regesta Regum Scottorum*, vol. 1, nos 131, 177.

108 T.C. Smout and M. Stewart, *The Firth of Forth: An Environmental History* (Edinburgh, 2012), pp. 24–25.

109 D. Perry, *Dundee Rediscovered: The Archaeology of Dundee Reconsidered*, Tayside and Fife Archaeology Committee Monograph 4 (Perth, 2005), p. 17.

the Forth, herring were by far the most plentiful fish taken, but salmon, which were also netted on the freshwater river and its main tributaries upstream of Stirling, was also of great and growing commercial importance.[110] The value of the Forth sea-fishery, mainly for herring but also for cod and haddock, is evident in early awards of general fishing rights and, from the early thirteenth century, in the resolution of disputes over payment of fish teinds due to monasteries from parishes along both sides of the estuary.[111] For salmon, the mid twelfth-century award of fisheries at Kersie and Tullibody to Cambuskenneth Abbey constituted a key component in its revenue base, one that it not only guarded jealously but also sought to expand by licit and illicit means in later centuries.[112] Cambuskenneth was not the only monastic beneficiary, with Scone and Holyrood also receiving salmon fisheries on the Forth's tidal reach.[113] Collectively, these awards reveal that the firth and tidal reach of the river were alive with fishing activity, supporting fisheries that, by the mid twelfth century, were controlled by the king, nobles and ecclesiastical institutions but still supplied an enormous volume of food for the wider population and for trade.

Early records of herring fisheries are not restricted to the Forth alone. Perhaps surprisingly, there is no clear evidence that herring figured in the commercial fish trade of Aberdeen in this period, where operations were dominated by gadid species such as cod, ling and haddock.[114] The absence of herring and other small fishbones from the bone assemblages from major excavations in Aberdeen, however, may relate more to the sampling strategies of the time of the excavations than any dietary aversion to the species in the urban population. This certainly seems to be the case in respect of the absence of herring bones from the 1970s excavations in Perth.[115] While the low salinity of the Tay estuary made it an unsuitable environment for herring shoals and probably stimulated the development of fishing communities on the Angus coast from the river mouth to Arbroath that supplied the Tay burghs,[116]

110 For the salmon fishery, see R.C. Hoffmann and A. Ross, 'This belongs to us! Competition between the royal burgh of Stirling and the Augustinian abbey of Cambuskenneth over salmon fishing rights on the River Forth,' in E. Bhreathnach, M. Krasnodębska D'Aughton and K. Smith (eds), *Monastic Europe: Communities, Landscape and Settlement* (Turnhout, 2020), pp. 451–476.

111 Smout and Stewart, *The Firth of Forth*, pp. 27–28; R. Fawcett and R. Oram, *Dryburgh Abbey* (Stroud, 2005), pp. 164–165; Innes, *Dryburgh Liber*, nos 98, 192. For Holyrood's fishing rights through the port of Leith, see Barrow, *The Charters of David I*, no. 147.

112 Barrow, *The Charters of David I*, nos 159, 213; Hoffmann and Ross, 'This belongs to us!'.

113 Barrow, *The Charters of David I*, no. 147; C. Innes (ed.), *Liber Ecclesie de Scon* (Edinburgh, 1843), no. 5 (hereafter *Scone Liber*).

114 Cameron and Stones, *Aberdeen*, pp. 277–278.

115 A.K.G. Jones, 'The fish bone', in *Perth High Street Archaeological Excavation 1975–1977*, fascicule 4: *Living and Working in a Medieval Scottish Burgh: Environmental Remains and Miscellaneous Finds* (Perth, 2010), pp. 53–67.

116 It was probably for this reason that the monks of Balmerino on the Tay, where they were well provided for with salmon fishing rights, sought to obtain a location for a herring station at the main east coast herring port, Crail, in the fourteenth century: R.D. Oram, 'A fit and ample endowment? The Balmerino estate 1228–1606', in R.D. Oram et al. (eds), *Life on the Edge: The Cistercian Abbey of Balmerino, Fife (Scotland)*, Comentarii Cistercienses 59 parts 1–2 (Pontigny, 2008), pp. 61–80 at p. 73; W.B.B.D. Turnbull (ed.), *Liber Sancte Marie de Balmorinach* (Edinburgh, 1841), no. 49. Similarly, Lindores Abbey, a few kilometres further upstream from Balmerino, possessed extensive salmon fisheries but no sea-fishing and secured an annual gift of 5 hundredweight of dried herring at Crail: Dowden, *Lindores Charters*, no. 75. Herring bone recovered in excavation contexts from Dundee probably came from fish brought to the burgh for sale by boats operating in the North Sea along the Angus and north-east Fife coasts: Perry, *Dundee Rediscovered*, p. 17. For the Crail fishery, see above, Chapter 4.

Crail, Fife. By the early decades of the twelfth century, communities along the south coast of Fife were already home to specialist sea-fishers, who exploited the bounty of the annual herring migrations. Crail, located at the outermost end of the Firth of Forth, was established as one of Scotland's premier fishing ports for herring by the early thirteenth century.

the seawater condition of the Firth of Clyde drew the shoals into the shallow inner tidal stretch of the river where they were netted from boats. Amongst grants by David I to Holyrood was one net and the right to take herring at Renfrew.[117] Its single net, however, was only a fraction of herring-fishing operations there, with the monks of Paisley alone having six nets and one boat for herring based at Renfrew.[118] These few records of twelfth-century herring fisheries on the Clyde, however, barely represent the significance of what by the late fifteenth century was one of the most important commercial operations on the west coast.[119]

As discussed in Chapter 5, herring had become the principal species caught in Western Isles fisheries in the ninth century, when the new ship technologies of the Scandinavian colonists enabled expansion from inshore fisheries to the deeper Atlantic waters.[120] At excavated sites down the west side of Lewis, Harris and Uist, including Bornais, Bostadh, Cille Pheadair, Rosinish and the Udal, later ninth- to thirteenth-century middens reveal a preponderance of offshore

117 Barrow, *The Charters of David I*, no. 147.

118 Innes, *Paisley Registrum*, p. 2.

119 Oram, *Where Men No More May Reap or Sow*, pp. 71–74.

120 R.D. Oram and P. Adderley, 'Re Innse Gall: a Norse colony in the Irish Sea and Hebrides?', in S. Imsen (ed.), *The Norwegian Domination and the Norse World c.1100–c.1400* (Trondheim, 2010), pp. 125–148 at 134–135.

Caithness coast near Keiss, Highland. Whilst Wick/Pultneytown and the coastal villages southwards towards Helmsdale became the centres of the regional fishing fleet in the later eighteenth and nineteenth centuries, it was from locations in the north-east of Caithness, from Keiss, through Freswick and round Duncansby Head to Robert's Haven, that the medieval cod fishing and processing communities operated.

species, principally herring, in their bone assemblages.[121] The nature of the skeletal remains – often lacking head and tail-tip bones – indicates that large-scale processing of the herring catches was being undertaken away from where they were consumed. The possibility has been raised that the Western Isles herring fishery was processed in volume for trade, perhaps with the Scandinavian colonies in Ireland,[122] but the abundance of fish remains from domestic settlement contexts suggests that they were consumed in great quantities in the islands; processing and trade was perhaps largely local.[123] No processing site has yet been identified in the Western Isles, with many locations probably already lost to coastal erosion, but as on Scotland's east coast we should understand these as probably specialist operations, functioning seasonally and dependent on the large, high-status farmsteads that were located further inland. Norse saga evidence, indeed, albeit in an Orcadian context, emphasises the close integration of farming and fishing in Scandinavian culture in the Atlantic islands and the active participation of even the topmost rank of Norse society in fishing alongside their dependents.[124] The integration, however, went beyond personnel, for the by-products of fish processing, principally the guts and heads, could be used as fertiliser on the fields or included in feed for dogs and pigs.[125] Marine protein and mineral content, therefore, enriched cultivation on the otherwise nutrient-poor machair which lay closest to processing

121 C. Ingrem, 'Fish', in N. Sharples (ed.), *A Norse Farmstead in the Outer Hebrides: Excavations at Mound 3, Bornais, South Uist* (Oxford, 2005), pp. 192–194; Parker Pearson et al., *Cille Pheadair*, pp. 478–511.

122 Sharples, *A Norse Farmstead*, p. 196.

123 C. Ingrem, 'Fish', in N. Sharples (ed.), *The Economy of a Norse Settlement in the Outer Hebrides: Excavations at Mounds 2 and 2A, Bornais, South Uist* (Oxford, 2021), pp. 439–441.

124 *Orkneyinga Saga*, chapter 85.

125 Parker Pearson et al., *Cille Pheadair*, pp. 458, 464, 509.

sites and added to the nutrition of important working and food animals.

While gadids – cod, ling, whiting, haddock, saithe and pollack – were also being fished off the Western Isles, greater numbers were caught down the east coast, where Aberdeen had emerged as a major trading port for cod by the thirteenth century.[126] Fishing operations, however, were based especially in Caithness, Orkney and Shetland. The abrupt shift towards deep-water demersal species and principally gadids being caught in these northern locations, like the sudden transition to herring in the Hebrides, has been seen as an indicator of rapid 'Scandinavianisation'.[127] Some research has suggested, moreover, that the rapid development of the Caithness, Orkney and Shetland cod fisheries was related to the integration of the Norse colonies there into the international trade in Atlantic cod that was dominated by Icelandic and northern Norwegian fisheries.[128] While this view of a 'new' fishery developing to capitalise on Scandinavian trade with northern Europe and eastern mainland Britain is contested, what is not in question is the scale of fishing and processing operations at locations like Freswick Links and Robert's Haven in Caithness.[129] These were specialist centres producing a volume of fish that perhaps satisfied the demand of local higher-status farms and regional lay and ecclesiastical elites and possibly delivered a catch surplus that was available for trade beyond the locality. As mentioned above in respect of trends in grain production, it has been suggested for Shetland that the productivity of the catches encouraged and sustained a shift from cereal cultivation in some communities, where fish surpluses could be traded for barley and other grain, perhaps with international merchants.[130] In this way, the burgeoning of trade in foodstuffs to sustain the rapidly expanding European population during the MCA possibly facilitated a major transition in subsistence regimes in the northernmost parts of the British Isles. The shift to exploitation of the sea-fishery was itself enabled by the relatively calmer North Atlantic conditions of the time.

Operations at more than subsistence level are also likely in respect of Scotland's salmon fisheries. Although there are likely to have been some sea catches of this migratory species, the vast bulk of salmon were caught in coastal netting operations and in fisheries on the main tidal estuaries and freshwater streams. As discussed in previous chapters, salmon fishing is likely to have been a millennia-old activity in Scotland before our first documentary records of it commence in the twelfth century, by which time there were evidently operations on every major river around Scotland's coast. Our earliest records of the Tay salmon fishery date from the reign of Alexander I (1107–24), who granted his Augustinian foundation at Scone two nets in the river at Perth, two further nets above the tidal reach within Scone parish and the teind of all salmon catches between Scone, Kinfauns and Craigie on the north and south banks of the river.[131] Alexander's brother and successor David I gave the rights to two other salmon netting operations on the Tay at Perth to St Andrews Priory and Dunfermline Abbey respectively.[132] The first grant to St Andrews explicitly mentions a location

126 Cameron and Stones, *Aberdeen*, pp. 277–278 and table 25.

127 Barrett, Nicholson and Céron-Carrasco, 'Archaeo-ichthyological evidence'; J.H. Barrett, R.P. Beukens and R.A. Nicholson, 'Diet and ethnicity during the Viking colonization of northern Scotland: evidence from fishbones and stable carbon isotopes', *Antiquity*, 75 (2001), 145–154; Oram and Adderley, 'Re Innse Gall', p. 134.

128 J.H. Barrett, 'Fish trade in Norse Orkney and Shetland: a zooarchaeological approach', *Antiquity*, 71 (1997), pp. 616–638.

129 Morris, Batey and Rackham, *Freswick Links*, pp. 179–182, 189–191, 273–275.

130 Bigelow, 'Issues and prospects in Shetland Norse archaeology', p. 20.

131 Innes, *Scone Liber*, nos 5, 12, 20, 85, 107.

132 Barrow, *The Charters of David I*, nos 88, 126, 172.

Mugdrum Island and the Tay sandbanks, looking west from near Flisk, Fife. The tidal channels between the shifting sandbanks of the inner Firth of Tay were ideal for the siting of fixed fish traps and for sweep-net operations, designed especially for the regular migratory runs of salmon up and down the river.

'where my own demesne nets' were located. Grants of fishing on the upstream river and its tributaries proliferated through the later twelfth and thirteenth centuries, with awards to most of the major monastic institutions in the region. Later in the twelfth century, Lindores received fishing rights from its founder, Earl David of Huntingdon, around Mugdrum Island in the river's tidal estuary – excluding a built fish trap, or yair, in the channel between the island and the south shore which he retained for his own household needs – and further net locations on the north side of the river in the same stretch at Clashbenny and 'Red Sands' from the Hay lords of Errol.[133] It was Balmerino Abbey, founded in 1228, however, that became the biggest monastic controller of salmon fisheries on the Tay between Perth and the mouth of the estuary, with salmon exports becoming its principal source of income in the later Middle Ages.[134] As for sea-fisheries, however, it must be borne in mind that lay interests probably predominated along the river's entire course, illustrated by the Hays' retention of a major operation on the north bank of the river and active acquisition of further fisheries there and Earl David's exclusion of his own pre-existing traps from the award to Lindores.[135]

These early records reveal that there were several different methods of catching fish in operation on the Tay by the twelfth century. The first records from the time of Alexander I and David I mainly relate to sweep netting, where a net fixed to a shore point is drawn round in an arc by men in a small rowing boat or coble and back to the shore.[136] Sweep netting was also in operation on the lower reaches of the Tweed

133 Dowden, *Lindores Charters*, nos 6, 76–79.

134 Oram, 'A fit and ample endowment?', pp. 71–73.

135 For the late thirteenth-century acquisition of a fishery at Inchyra by the Hays, see NRS RH1/6/20.

136 The award of the *tractus*, or draught, of a single net is mentioned in Barrow, *The Charters of David I*, no. 88. For a description of these operations, which only ceased on the Tay in 1996, see G. Ogilvy, *The River Tay and Its People* (Edinburgh, 1993), pp. 65–67.

upstream from Berwick early in the twelfth century, on the Clyde at Renfrew, on the Forth at Stirling and probably on most of the larger rivers where cobles and nets could function effectively.[137] Where there was a significant tidal range and large expanses of sandbank, as on the Tay below its confluence with the Earn, stake nets were employed, where the trapped fish were recovered at low tide.[138] Stake nets are likely to have been most often involved in the multiple grants of salmon fisheries located on the sand- and mudbanks in the river estuaries feeding into the Solway from the north, where such traps still survive in small numbers. It was likely a stake net located 'upon the sands' at the mouth of the Doon in Ayrshire that Roger of Skelbrooke granted to Melrose in the late twelfth century.[139] Yairs were also used in tidal zones, usually consisting of stone-built or wooden loops that ran out from the shore into deeper water, forming a pool in which fish were trapped as the tide fell.[140] Such devices are likely to have been used since prehistory and continued to be constructed into the eighteenth and nineteenth centuries in parts of the northern and western Highlands and Hebrides.[141] Yairs were employed on the Tay in the twelfth century, as evidenced by Earl David's retention of his *ihara* (yair) at Colcrike between Mugdrum Island and the shore near Newburgh, and were being constructed on the Leven near Dumbarton in the 1220s.[142] A massive yair can still be seen at low tide on the sands at the mouth of the River Luce on Luce Bay in Galloway, which local tradition identifies as a property of Glenluce Abbey. Beyond such tidal reaches on rivers, cruives were employed. These were large wicker baskets – presumably woven from the withies provided by coppice wood – placed in mid-stream gaps in stone-built weirs, which fish could swim into but not exit.[143] This, too, is an ancient technology but our earliest documented references to a cruive dates from the twelfth century. Probably in the early 1190s, King William permitted the bishop of Moray to construct a mill on the River Lossie at Elgin above the *crohae* (cruives) in the river directly below the king's castle there.[144]

Legislation surviving from 1318 but referencing older statutes indicates that there was an attempt to

137 As in the grant to St Andrews on the Tay, a sweep- or draw-net operation is clearly the subject of David I's grant of the *tractus* of 'Aldstelle' on the Tweed: Barrow, *The Charters of David I*, no. 39. For the Clyde and Forth there are multiple references, but see ibid, no. 147, as an example.

138 David Hay retained his stake nets at Clashbenny and Red Sands in his grant of draw-net operations there to Lindores, described as 'piscariis meis stantibus cum palo et sepi' (my standing fisheries / fishing stances with stake and fence/net): Dowden, *Lindores Charters*, no. 77.

139 See, for example, William de Bruce's grant of a fishery at Rainpatrick on the Esk sands at the head of the Solway to the monks of Melrose: Innes, *Melrose Liber*, vol. 2, appendix, no. 3. For the Doon mouth stake net, see ibid., vol. 1, no. 34. Roland of Galloway granted Holm Cultram a fishery on the Mersehead Sands off the coast of Colvend and Southwick in eastern Galloway, while a Galloway tenant granted them a fishery on the Nith off Kirkconnell: Grainger and Collingwood, *Holm Cultram Register*, nos 128, 149.

140 T.D. Bathgate, 'Ancient fish-traps or yairs in Scotland', *Proceedings of the Society of Antiquaries of Scotland*, 83 (1949), pp. 98–102.

141 C. Martin, 'The silent shores speak: maritime landscapes in North Argyll', *Current Archaeology*, 219 (2008), pp. 26–32; Bathgate, 'Ancient fish-traps', pp. 100–102, for traps on the Beauly and Dornoch firths and in Loch Broom.

142 Dowden, *Lindores Charters*, no. 6; Innes, *Paisley Registrum*, p. 213.

143 Legislation of 1424 restated earlier statutes, no longer surviving, which required the removal of the cruive baskets to open the 'slap' to allow fish to progress freely up- and downstream from Saturday night to Monday morning: *RPS* 1424/11.

144 *Regesta Regum Scottorum*, vol. 2, no. 362.

Above. Stake nets, Cree estuary, Dumfries and Galloway. It was probably fixed nets like these that were recorded on the Tay estuary in the later twelfth century.

Right. Salmon coble, Seggieden, Perth and Kinross. 'Draughts' of nets on the Tay were recorded as gifts from King Alexander I to the canons of Scone. A draught involved a small rowing boat carrying one end of a net in a sweep from a fixed point on the shore, out into the river channel and back to land.

regulate what might otherwise seem to have been a free-for-all in the salmon fishery especially.[145] The legislation concerned cruives placed in tidal zones on river channels and required a specific size of gap to be left in the basket weave to enable salmon smolts and fry to move from the sea into the fresh water and vice versa, with only larger, mature fish being taken. There was clearly an understanding that the movement of these young fish was essential to maintain future stocks, and the value of the fishery was recognised in the forty days' imprisonment and heavy fines that were imposed for contravention of the act. Similarly, it was stated in the same act that the ancient 'prohibited times' for fishing were to be maintained – Sundays still being 'closed' during the salmon season – again with heavy fines imposed for non-observance. Further legislation requiring the destruction of all yairs and cruives in tidal waters for a period of three years was introduced in 1424, the same act restating an old requirement for freshwater cruives to have their baskets removed from Saturday evening through to Monday and to have a minimum weave of three inches in their heck to allow smolt and fry to pass through.[146] Given the volume of fishing that was taking place in Scotland's tidal and fresh waters and the economic importance of the fishery domestically, let alone for trade purposes, such measures were unlikely to have been new even in the later thirteenth century.

Although we can see the roots of the later overfishing of certain species in the intensification of the catches taken in waters close to expanding population centres, on the Firth of Forth especially, it is also clear that there was early recognition of the need for what we would call 'sustainable practice'. While on the one hand the burgeoning record of fishery operations might look like a free-for-all, on the other the close control over those operations exercised by some superior authority and the early evidence for traditional local practices and more all-embracing legal constraints show an awareness of the need to manage catches in order to safeguard future stocks. How effective those measures were in practice is debatable, but it is perhaps significant that the increase in legislation designed to protect the fisheries came after the end of the MCA and at the beginning of the major shifts in oceanic conditions that affected migratory patterns and breeding cycles.[147] It also came, however, at a time when demand for fish was increasing all across western mainland Europe, with Scottish cod, herring and especially salmon becoming major trade commodities commanding good prices and bringing substantial profits to those who controlled the fisheries. Here, we can see the emerging tension between supply and demand, and the temptation of the suppliers to deliver ever more into the voracious European market. That the legislation came to require repeated restatement (with mounting scales of fines) and became more restrictive suggests that sustainable practice and conservation were readily overcome by the lure of financial profit.

• • •

Although not quite the era of endless summers that its former labelling as the 'Medieval Warm Period' might suggest, the Medieval Climate Anomaly had delivered a time of greater stability and prosperity to the people of Scotland. Despite periods of adverse weather and climatic disturbance, it had been overall an era of expansion that generally more positive conditions had stimulated and encouraged. For the first time since the crises at the end of the Roman period and the LALIA, with its episodes of epidemic, crop failure and famine that saw decline in the human population of the northern parts of the British Isles, numbers were rising and people were spreading into

145 *RPS* 1318/13.

146 *RPS* 1424/11.

147 Oram, *Where Men No More May Reap or Sow*, chapter 2.

areas that had perhaps lain abandoned for centuries. Land that had reverted to scrub and woodland, or to lighter exploitation as grazing, was once more seeing clearance, fuelled by the demand for food and materials created by the expanding population and its rulers. Trade, moreover, was opening new horizons and creating export markets with seemingly insatiable appetites for the raw and semi-finished products that Scotland could produce. New technology in agricultural equipment, oxen and horse harness and milling facilities enabled the colonisation of new land and made clearance of woodland and wetland easier and the processing of produce faster and more efficient. Land-hunger and rising demand for commodities like wool or grain introduced competition, however, and led to conflict over access to the land needed to expand arable production, over different values placed on the same land by different social groups and over increasing shortages of staple resources in the zones of densest settlement. Even before the start of the long transition into the cooler and wetter era of the later Middle Ages, the warning signs of overexploitation, resource depletion and increasing marginality in the productive capacity of some land were beginning to emerge. Within a few short years, an already transitioning climate regime worsened by volcanic forcing, coupled with epizootic disease and a descent into one of the most sustained periods of warfare, can have left Scotland's people in no doubt that the best times lay in the past.

CHAPTER 9

Facing into the Coming Storm

Qwhen Alexander our kynge was dede,
That Scotland lede in lauche and le,
Away was sons of alle and brede,
Off wyne and wax, of gamyn and gle.[1]

It is widely recognised on the basis of the historical record and the surviving evidence for lavish spending on building cathedrals, castles, abbeys and churches that the later thirteenth century was a period of great prosperity both for those at the apex of Scotland's social hierarchy and for many lower down in the social order. Indeed, the reign of King Alexander III (1249–86) has long been presented as the apogee of the medieval kingdom, when it achieved almost its territorial maximum with the acquisition of the Western Isles, a level of economic wealth that per capita significantly outstripped England's and a degree of internal peace and stability never before experienced.[2] Much ink has been devoted to discussion of this relative wealth and trade-based prosperity, which underpinned the remarkable architectural efflorescence visible in cathedrals like Elgin and Glasgow or castles like Kildrummy, Dirleton or Caerlaverock.[3] Much of that discussion, however, very consciously looks forward to the traumatic dislocation that followed this time of cultural achievement. Laments for a lost 'golden age' are common at any time, but it was for long accepted that the rhymes bewailing Alexander III's death and the end of the

1 Andrew of Wyntoun, *The Original Chronicle of Andrew of Wyntoun*, ed. F.J. Amours, 6 volumes (Edinburgh, 1903–14), volume 5, p. 145.

2 The most recent assessment of the king's achievement is N.H. Reid, *Alexander III 1249–1286: First among Equals* (Edinburgh, 2019), which presents a more balanced, nuanced account than the older historiography of the reign. Reid is the first historian to give due – if still limited – recognition to the role of climate in the events of Alexander's reign and its aftermath: ibid., pp. 299–303.

3 N. Mayhew, 'Alexander III – a silver age?', in N.H. Reid (ed.), *Scotland in the Reign of Alexander III 1249–1286* (Edinburgh, 1990), pp. 53–73; B.M.S. Campbell, 'Benchmarking medieval economic development: England, Wales, Scotland and Ireland *circa* 1290', *Economic History Review*, 61:4 (2008), pp. 896–945; R. Fawcett, 'Ecclesiastical architecture in the second half of the thirteenth century', in N.H. Reid (ed.), *Scotland in the Reign of Alexander III 1249–1286* (Edinburgh, 1990), pp. 148–180.

good times that his reign had brought reflected a genuine break from that previous era of peace and prosperity and the beginning of one of protracted political crisis and near disaster. The wars, famine and pestilences of the fourteenth century provide context for the second part of that view, but there is also reason to question the idea of Scotland's later thirteenth-century golden age that supposedly died with the king. Although the lines of Andrew of Wyntoun's *Origynale Cronykil of Scotland* (1424) quoted above are commonly regarded as an allegorical reference to the political crises and wars which flowed from the failure of the senior line of Scotland's royal house, there is reason to suggest that it was the horrors of famine and of epizootic and epidemic disease in fourteenth- and early fifteenth-century Scotland that inspired them. Wyntoun perhaps developed his stanza from verses first composed in the uncertain times that followed the king's death on 19 March 1286, although by the time he was writing at the end of the first quarter of the fifteenth century their lament for what had been lost had acquired greater poignancy through the human and environmental shocks that had wracked the country in the preceding century.

In political terms, change – or any sense of impending crisis – might not have been apparent to most of Scotland's population in the immediate aftermath of the events of that March night, but the death only four years later of Alexander's only direct heir, his young granddaughter Margaret 'the Maid of Norway', unleashed a new dynamic.[4] The following hearing of 'the Great Cause of Scotland' before King Edward I of England and the eventual adjudication of the throne to the representative of the most senior collateral line of the Scottish royal family, John Balliol, is one of the most researched episodes in Scottish history and is most often presented as the pivot point between the Alexandrian golden age and the long era of disharmony and struggle for national survival that followed.[5] Later medieval chroniclers such as Walter Bower, whose understandably intensely nationalistic standpoint led him to present the crises of King John's reign as the consequence of Divine displeasure at the illegitimacy of his succession, have been responsible for a view of this Balliol interlude as a period of utter national humiliation before the restoration of Scotland's freedom and honour under the kingship of Robert Bruce. Indeed, Robert I himself was at great pains to stress his place as the legitimate successor to Alexander III and to project himself as the restorer of the good government, peace and stability of that reign. The invasions and devastation of the Wars of Independence were presented as the price exacted for the restoration of liberty from English domination.

It is only relatively recently that the role of climate change in stimulating the political tensions and financial instability that formed an undercurrent to the civil and foreign wars of this era has begun to be recognised.[6] The older historiography is avowedly 'political' in its focus, viewing the unfolding events of the later thirteenth and fourteenth centuries almost exclusively through a lens of high-level human decision-making with the occasional nod towards equally anthropocentric social and economic processes. Indeed, the impression could be given that the only negative environmental experiences of the

4 For a summary of the events of this period, see M. Brown, *The Wars of Scotland 1214–1371* (Edinburgh, 2004), pp. 157–165.

5 For broad discussion of the 'Great Cause' and the reign of King John, see A. Beam, *The Balliol Dynasty 1210–1364* (Edinburgh, 2008), chapters 3–4; G.W.S. Barrow, *Robert Bruce and the Community of the Realm of Scotland*, 3rd edition (Edinburgh, 1988), chapters 1–4; and M.A. Penman, *Robert the Bruce, King of the Scots* (London, 2014), chapters 1–2.

6 Penman, *Robert the Bruce*, index s.v. 'climate'; C. McNamee, *The Wars of the Bruces: Scotland, England and Ireland 1306–1328* (East Linton, 1997).

Glasgow Cathedral, Glasgow. The greatest of Scotland's surviving medieval churches, Glasgow Cathedral is a remarkable monument to the prosperity of the thirteenth-century kingdom founded on the profound environmental transformations delivered across the MCA. Money from burgeoning arable and livestock yields, fish, wool-clips and hides, income from church lands and teind payments and materials from pious donors all helped build much of the structure we see today.

later thirteenth and earlier fourteenth centuries were caused by human actions through warfare. This impression, however, is not simply the product of modern historians' inclinations but is a skewing delivered through the preoccupation of our sources with the high drama of the political turbulence and bloody warfare of the years after 1296 and the beginning of the long conflict with England. There is very little record in our main surviving indigenous chronicle sources of much else beyond very occasional notice of an especially harsh winter or poorer harvest. Amidst the exploits of William Wallace and Robert Bruce, these few seemingly random references to weather events or wider environmental factors can be easily overlooked. Indeed, apart from recognition of the weather's role in the ultimate and catastrophic failure of Edward Bruce's 1315–18 invasion of Ireland,[7] the crop failures, cattle plagues and famine

7 The role of the weather is explored in detail in McNamee, *The Wars of the Bruces*.

Rain squall, North Sea, from Aberdeen to Kirkwall ferry. The disturbed weather conditions of the later thirteenth century and the Dantean Anomaly at the end of the first quarter of the fourteenth century have received scant attention in the traditional narratives of the Wars of Independence.

deaths of the 1310s and 1320s pass almost unnoticed in the celebration of Robert Bruce's military and state-building achievements.

After a very brief interlude of peace (1327–32), war again dominates the main historical narratives into the third quarter of the fourteenth century. These accounts record the conflict over the crown of Scotland between Robert Bruce's son, King David II (reigned 1329–71), and King Edward Balliol, son of the short-reigned King John, further complicated by the efforts of Edward III of England to shore up his Balliol protégé and secure control over Southern Upland Scotland for himself.[8] The widespread devastation of east-central Scotland during the 1330s caused by both sides in the conflict is well represented in modern literature that draws on the surviving chronicle narratives for this period.[9] Seemingly bizarrely, but in fact understandable in terms of its

8 The most detailed modern narrative for this period is M.A. Penman, *David II 1329–71* (East Linton, 2004).

9 R. Nicholson, *Scotland: The Later Middle Ages*, paperback edition (Edinburgh, 1978), chapter 6, provides an overview summary derived for the pre-1335 period from the more detailed treatment of it in R. Nicholson, *Edward III and the Scots: The Formative Years of a Military Career, 1327–35* (Oxford, 1968). Brown, *The Wars of Scotland*, chapter 11, offers an accessible modern overview.

The people of Tournai bury victims of the Black Death: miniature by Piérart dou Tielt illustrating the *Antiquitates Flandriae (Tractatus quartus)* of Gilles li Muisis (Tournai, *c.*1353), Bibliothèque royale de Belgique, ms. 13076–13077, fol. 24v. The seemingly unprecedented human death toll in the Great Mortality, as the first wave of the bubonic plague pandemic that swept west across Europe in the late 1340s and early 1350s was known, changed the dynamic in human/environmental interactions, ushered in a shift in agricultural practices and altered patterns of pressure on natural resources. (Public domain image.)

laconic reporting in the contemporary Scottish chronicle material incorporated into the John of Fordoun compendium, the Black Death which swept across western Europe from 1347/8 and reached Scotland in 1349 has merited far less discussion in general histories than the war damage of the 1330s.[10] Perhaps because the war with England seemed to continue with little pause throughout and beyond the immediate plague years, mainstream discussion into the early 2000s remained focused on the conduct of the conflict until the cessation of hostilities with the Treaty of Berwick and the release of David II from English captivity in 1357.[11]

For the five decades from the Treaty of Berwick until the end of the fourteenth century, most modern historical discussion has continued to dwell on the political struggles for power within the kingdom. The rivalry between David II and his nephew Robert

10 Nicholson, *Scotland*, pp. 148–149, affords only two brief paragraphs, while Brown, *The Wars of Scotland*, p. 250, has a single page, despite Alexander Grant's suggestion that the epidemic was 'possibly the worst disaster' to affect the people of Scotland in the Middle Ages: A. Grant, *Independence and Nationhood: Scotland 1306–1469* (London, 1984), p. 75.

11 K. Jillings, *Scotland's Black Death: The Foul Death of the English* (Stroud, 2003), was the first study to shift the perspective away from the military, political and economic emphases of previous studies of this era.

the Steward has been seen largely in terms of competition over territory and the allegiances of the nobility within the contested land.[12] But apart from recognising the sheer greed, power-hunger and personal animosity that fuelled this competition, there is little consideration given to any wider causal factors that might have driven men into conflict. Yet this was an era of declining noble incomes, in which the release of population pressures everywhere around western Europe was seeing falling rents at a time of rising costs and in which control of resources was increasingly the key to political power. That the struggle for power did not end with David's death in 1371 and the accession of his nephew as King Robert II (1371–90) underscores that there was far more underlying the conflict than traditional political narratives allow. Yes, political factors continued to play a role, especially since the Steward's brood of sons all had territorial ambitions that ensured the extension of factional discord into the early fifteenth century.[13] Although conflict over resources has been recognised as a contributor to instability in this period of ongoing turbulence, the focus has been principally on how men like Alexander Stewart, earl of Buchan, exerted their authority in order to control reserves of manpower and the land on which to sustain it.[14] Again, it has been into a largely political context that these processes have been set, principally in terms of the ebb and flow of regional and national domination between Robert II, his eldest son and successor Robert III (1390–1406) and his younger sons headed by Robert, earl of Fife (later 1st duke of Albany). More recently, however, climate change and a deteriorating weather regime has begun to be explored as a factor, introducing another new dynamic agency whose impact directly affected the land and resources upon which political power was based. This destabilisation of the agricultural foundations of wealth and power threatened the social order and social networks that had been constructed in better times.[15] It is therefore with consideration of that change in North Atlantic or northern hemispheric climate patterns more generally, and the weather they delivered, that the exploration of the 'long fourteenth century' in this chapter commences.

Climate and weather: from 'golden age' to 'little ice age'

We must begin by returning to a question raised briefly in Chapter 6, namely whether the persistent historical view of Alexander III's reign as marking the golden age of medieval Scotland is sustainable in the face of the greater body of data from proxy sources as well as record evidence that is available to modern researchers?[16] Traditionally, it is said that after the political factionalism of his minority (1249–59) the kingdom enjoyed a generation of peace, stability and economic prosperity which, collectively, delivered that famed golden age. By one measure, assessments of the wealth of Scotland in terms of the volume of silver coinage in circulation and the spending on capital-intensive building programmes

12 Penman, *David II*, chapters 7–11.

13 S.I. Boardman, *The Early Stewart Kings: Robert II and Robert III 1371–1406* (East Linton, 1996).

14 R. Oram and P. Adderley, 'Lordship and environmental change in central Highland Scotland c.1300 to c.1450', *Journal of the North Atlantic*, 1 (2008), pp. 74–84.

15 R.D. Oram, 'Between a rock and a hard place: climate, weather and the rise of the lordship of the Isles', in R.D. Oram (ed.), *The Lordship of the Isles* (Leiden, 2014), pp. 40–61.

16 This topic is explored in detail in R.D. Oram, '"Away was sons of alle and brede": the decline of the Medieval Climate Anomaly and the myth of the Alexandrian Golden Age in Scotland', in R.D. Oram (ed.), *'With Our Backs to the Ocean': Land, Lordship, Climate Change, and Environment in the North-West European Past: Essays in Memory of Alasdair Ross*, Environmental Histories of the North Atlantic World 5 (Turnhout, 2023), pp. 83–107.

Ripening oat crop, near Balcormo, Fife. Early interpretations of the MCA as a period of benign weather and abundant harvests in Scotland were coloured by the traditional vision of the reign of King Alexander III as one of bounty, peace and prosperity.

offer *prima facie* evidence for what can be defined as an era of continuous economic growth and confidence that had been triggered and then sustained by rising population and consumer demand.[17] An increase in the financial power of the monarchy and, by extension, of the nobles and burgesses who participated in Alexander III's government has been seen as flowing from a series of fortuitous but interlinked political and fiscal circumstances.[18] Strikingly, that very anthropocentric and traditional view offered no link between the economic expansion that provided the basis upon which the revenues exploited by the crown were founded and the evidence for what at that time (2004) was still regarded as a benign climatic regime that prevailed to *c.*1300. The broad assumption from the early climate models was that the warmer and milder conditions of the MCA increased the carrying capacity of the land and encouraged the expansion of agriculture. This, it was argued, was veritably an age of abundance of ale and bread.

Aspects of that idyllic picture of rosy-cheeked peasants gorging themselves on the fruits of their labour have long been questioned. That questioning, however, has usually been through analogy with conditions in England. There, it has been shown that the seemingly inexorable upward trend in population growth had already begun to slow in the later thirteenth century and that environmental stresses had become increasingly evident as arable agriculture began to reach the maximum of its range. Initially, it was argued that what was happening across a broad region of England must have occurred in parts of Scotland also, though perhaps tempered by the broader mix in Scottish agricultural practices. From these suggestions of initially limited impacts, the traditional, more positive vision began to be challenged in the early 2010s and the critique has continued to develop.[19] Armed with the newly available climate proxy data, scholars began to see the precarity that maintained a firm hold over rural regimes across Scotland from the Northern Isles to Galloway. Far from gorging themselves on Nature's gifts through the extended golden age of the thirteenth century, that more recent scholarship began to suggest that although the better weather conditions of the MCA meant that most Scots were perhaps less hungry than had been their more usual condition in earlier centuries, they were not revelling in abundance and eating to surfeit. In this emerging prescription, hunger was less frequent and harvest failures uncommon. That began to change in the deteriorating weather conditions that came as the climate began to transition out of the MCA. The less benign conditions that the documentary record and climate proxies reveal for the later thirteenth century, however, did not necessarily lead to an abrupt reversion to earlier medieval marginality. Except in a few notable instances, explored below, there were few occurrences of nationwide harvest failure, but there were more frequent poor yields. This new pattern meant that peasants were hungrier but not necessarily starving. While in the main agricultural areas food shortage was perhaps more common than earlier in the thirteenth century and dearth an occasional visitor, true famine was rare.

Much as discussion in Chapter 6 was driven by exploration of the experience of living through the relatively benign conditions of the MCA, this chapter will largely be set against a backdrop of changeable weather patterns. We will examine evidence for the commencement of more unstable conditions in the mid 1250s and trace the main trends that persisted

17 Mayhew, 'Alexander III'.

18 Brown, *The Wars of Scotland*, pp. 56–57.

19 R.D. Oram, '"The worst disaster suffered by the people of Scotland in recorded history": climate change, dearth and pathogens in the long fourteenth century', *Proceedings of the Society of Antiquaries of Scotland*, 144 (2014), pp. 223–244; Oram, 'Away was sons of alle and brede'.

Sunset over St Andrews, 1981, a legacy of the 1980 eruption of Mount St Helens on the Pacific seaboard of the United States. The effects of the volcanic particulates circulating in the high atmosphere give some idea of the impact of the far greater 1257 Samalas eruption of Gunung Rinjani on Lombok in Indonesia.

until a new colder and wetter pattern became established in the 1400s. There are, however, some caveats to be made before we begin. Interpretation of the wider climatic situation from documentary sources alone is deeply problematical as a methodology. Only proxy environmental measures can offer a long-term perspective that lifts our view above the particular and the subjective offered by the medieval narratives and provide a clearer context for recorded events. These proxies, however, as explained in the introduction, are regionally skewed towards the north and west of Scotland or are found in datasets obtained in locations remote to Scotland. All we can take from them is modelled likely trends, not absolute statements about conditions on any given day across the whole country. Bearing that in mind, what we see is a general trend towards less stable conditions, as mentioned in Chapter 6, already becoming established in the 1250s. Profound and lasting change is signalled in the data as occurring from around 1258. A high concentration of volcanic sulphates found in ice cores from Greenland and Antarctica towards the polar extremes indicates that another major volcanic eruption, now identified as the 1257 Samalas eruption in the Gunung Rinjani volcanic complex on Lombok in Indonesia,[20] expelled vast amounts of material into the atmosphere. As aerosol particulates, this volcanic dust circulated the globe and is argued to have led to significantly reduced temperatures in both northern and southern hemispheres to 1261.[21]

20 F. Lavigne et al., 'Source of the great a.d. 1257 mystery eruption unveiled, Samalas volcano, Rinjani Volcanic Complex, Indonesia', *Proceedings of the National Academy of Sciences*, 110:42 (2013), pp. 16742–16747.

21 S. Guillet et al., 'Toward a more realistic assessment of the climatic impacts of the 1257 eruption', EGU General Assembly Conference Abstracts, April 2015, https://ui.adsabs.harvard.edu/abs/2015EGUGA..17.1268G/abstract; S. Guillet et al., 'Climate response to the Samalas volcanic eruption in 1257 revealed by proxy records', *Nature Geoscience*, 10 (2017), pp. 123–129.

That volcanic forcing was sufficient to entrench a downward spiral of climatic cooling globally. Norwegian and Swedish dendrochronological sequences, for example, show a sudden narrowing in annual growth rings between 1258 and 1265, which provides a good indication of the impact of the Samalas eruption across north-western Europe generally, to which Scotland would not have been immune.[22] Further eruptions in *c.*1269, 1275/6 and 1286 maintained a stratospheric veil of volcanic aerosols that suppressed the amount of solar energy reaching the Earth's surface across those decades.[23] We can, almost literally, see a smoking gun towards which some quickly pointed as an explanation for the climatic volatility that was emerging in the later thirteenth century. By the end of the twentieth century, however, further scientific research was already suggesting that volcanism was not the only agency at work in forcing down global temperatures.

Around 1270, the sun had entered a new phase of declining and then, from *c.*1282 until 1342, minimal sunspot activity that has been labelled the Wolf Solar Minimum.[24] Coincidence of the climate-forcing cooling impact of the volcanic eruptions from 1257 onwards with the further cooling associated with a solar minimum closed the door on the more stable climatic conditions of the MCA and introduced a new phase of global climatic reorganisation that delivered a generally cooler temperature regime.[25] A rapid southward extension of sea ice off Greenland recorded in the sagas, despite the persistence across this period of the high summer and winter NAO usually associated with warmer conditions, was the first visible indicator of a gathering instability in the temperature of the surface waters of the North Atlantic. It marked the locally observed manifestation of what was an oscillation between the warming and cooling of oceanic surface water through the first decade of the fourteenth century.[26] There was then a prolonged warming between 1314/15 and 1318, a deep plunge in the early 1320s and an even higher warming spike through the mid 1320s, followed by a shorter drop and a final climb in 1330–2 to the highest temperature level since before 1100.[27] These movements and this temperature spike, it should be remembered, occurred decades after the end of the MCA as a recognisable phenomenon. And the key point is the rapidity of these swings against a generally downward trend. This last high, for example, was followed immediately by a fall below the mean into a sustained period of cold surface temperatures that lasted until the end of the Wolf Minimum in the 1340s. And there surface temperatures remained for over a decade, only beginning to warm again in the early 1350s.

Such shifts matter greatly, for one of the most significant characteristics of instability in ocean surface temperatures is greater year-to-year variance

22 Thun, T. and Svarva, H., 'Tree-ring growth shows that the significant population decline in Norway began decades before the Black Death', *Dendrochronologia*, 47 (2018), pp. 23–29 at fig.5.

23 Campbell, *The Great Transition*, p. 198. For the 1275/6 eruption, see F. Lavigne and S. Guillet, 'The unknown AD 1275 stratospheric eruption: climatic impacts in Europe and tentative volcanic source', in F. Riede (ed.), *Past Vulnerability – Volcanic Eruptions and Human Vulnerability in Traditional Societies Past and Present* (Aarhus, 2015), pp. 63–74, which identifies the culprit as Mt Tarawera in the Okataina Volcanic Centre in New Zealand.

24 L.E.A. Vieira et al., 'Evolution of the solar irradiance during the Holocene', *Astronomy and Astrophysics*, 531 (July 2011), https://doi.org/10.1051/0004-6361/201015843. The 1282–1342 date is from M. Stuiver and P.D. Quay, 'Changes in atmospheric carbon-14 attributed to a variable sun', *Science*, 207:4426 (1980), pp. 11–19.

25 Campbell, *The Great Transition*, pp. 198–199.

26 For these trends, see ibid., p. 200 fig. 3.19 (A) and (B).

27 A.G. Dawson et al., 'Greenland (GISP2) ice core and historical indicators of complex North Atlantic climate changes during the fourteenth century', *The Holocene*, 17:4 (2007), pp. 427–434.

Nith estuary, Dumfries and Galloway, looking east from the slopes of Criffel towards Caerlaverock and the Rainpatrick Sands. The narrowing funnel of the Solway Firth saw a succession of major sea-floods in the later thirteenth century on both Scottish and English sides as storms drove the tidal waters over the low-lying coastal districts. These events contributed to the decision to abandon the old castle at Caerlaverock.

in the summer and winter patterns of the NAO. It is that which, in turn, determines the principal weather trends. As discussed in earlier chapters, in winter a westerly airflow off the Atlantic with higher temperatures and rainfall is associated with positive/high NAO. A negative/low NAO is associated with lower temperatures and rainfall and more northerly and north-easterly airflow. In summer, positive NAO usually means higher temperatures and lower rainfall, while a negative NAO normally results in lower temperatures and higher rainfall. So, ocean surface warming increases atmospheric moisture levels – creating rain-laden cloud masses – and triggers increases in storminess. On the other hand, rapid cooling events bring high surface air pressure associated with anticyclonic circulation. Such high-pressure cells in winter draw cold air further south and into collision with the warmer moist air moving east across the Atlantic. The result is stormy winters and cold, late springs, adverse summer growing conditions and wet and delayed harvests.

Alongside the Greenland ice core and Traligill precipitation data, it has been established that the later thirteenth century witnessed an abrupt reversal in a millennial declining trend in North Atlantic storminess. Modelling of oceanic and atmospheric patterns has shown that the volatility that had been easing almost continuously across prehistory and into the Historic Era returned with a sudden surge in dynamism. This abrupt movement heralded the beginning of a new age of oceanic storm-generating activity that dealt massive and recurrent blows to the west-facing coastlands of the Hebrides and Northern Isles especially but also to what we might consider more sheltered western mainland coasts, particularly down the Clyde coast and into the inner reaches of the Solway Firth.[28] These changes were experienced by the human populations of these districts as

28 L.D. Meeker and P.A. Mayewski, 'A 1,400-year high resolution record of atmospheric circulation over the North Atlantic and Asia', *The Holocene*, 12 (2002), pp. 27–66; P.A. Mayewski et al., 'Holocene climate variability', *Quaternary Research*, 62 (2004), pp. 243–255.

extreme summer wind and rain events, or extremely hot and wet summers and prolonged snowy or icy late winters. Our chronicles for this time are punctuated by examples of such conditions. The converse to this trend was consequent on an extension of the southward range of sea ice off Greenland and Iceland, attested in Icelandic records. Although geographically remote from Scotland, it was what happened in those northern North Atlantic waters which delivered the cold ocean surface temperatures that brought the stormy, wet winters and cold, delayed springs of the late thirteenth century. That trend towards disturbed weather patterns was already evident from the mid 1250s, with the rainfall-producing high North Atlantic surface temperatures dropping sharply through the 1260s and early 1270s before stabilising briefly. This stable episode was then followed first by a sharp plunge and an almost immediate rapid climb in the early 1280s before a renewed slide into prolonged general cooling for the remainder of the century.[29] Greenland ice core data for the same period underscores the depth of the cooling by the late 1270s and the abrupt reversal that occurred in the early 1280s. What this meant for the northern British Isles was more cloud and rain spreading east, before a return to generally colder surface temperatures that persisted through the 1290s.[30]

These swings are understood to have been a symptom of the opening stages of a reordering of global climatic patterns that spanned several centuries, punctuated by episodes of varying length of rapid transitioning or prolonged consistency in weather. Some of these changes were to prove more localised and to affect mainly northern Europe and the northern North Atlantic through a period lasting from the end of the fourteenth to the middle of the nineteenth century, referred to nowadays as the 'little ice age' or LIA. (That episode is the subject of the volume *Where Men No More May Reap or Sow*.) It is the later thirteenth- and fourteenth-century preface to the LIA, however, with which we are concerned here. The record of changeability across that 150-year period charts climate trends and weather phenomena with serious consequences for those parts of the north and west of the British Isles most exposed to Atlantic weather systems. For Irish oak trees and Highland Scots pines whose annual growth responds well to summer rains, we can see robust growth for much of this period, but what was good for trees was not necessarily as good for arable farmers. Most cereal crops, especially where wheat predominates, suffer badly in such conditions. Commenting on the data for temperature variance from the Siberian larch tree-ring sequence for 1290–1340, Bruce Campbell observed that these years were the least stable fifty-year period of the previous 2,000 years.[31] Multiple climate proxies from ice cores and tree-rings obtained across a zone extending from Greenland to central Europe confirm that view, showing that the second quarter of the fourteenth century especially saw an upward trend of year-to-year temperature variability that was sustained through the middle years of the century and reached a peak in the late 1370s/early 1380s before falling rapidly through the 1390s and into the new century as volatility gave way to more sustained trends in the early stages of the LIA.[32]

Analysis of the annual growth of the speleothems in the Traligill cave system in western Sutherland illustrates the main weather trends delivered to Scotland by ocean surface temperature change and NAO

29 Campbell, *The Great Transition*, p. 200 fig. 3.19 (A) and (B).

30 Ibid.

31 Ibid., p. 202.

32 M.E. Mann et al., 'Proxy-based reconstructions of hemispheric and global surface temperature variations over the past two millennia', *Proceedings of the National Academy of Sciences*, 105:36 (2008), pp. 13252–13257.

variation. Speleothem growth indicates that high winter NAO with high rainfall persisted through the second half of the thirteenth century. It shows that the region experienced a peak of precipitation in the 1310s and early 1320s, corresponding with the profoundly disturbed climate episode identified by Neville Brown and labelled as the Dantean Anomaly on account of what he believed to be its co-occurrence with the death of the Italian poet Dante Alighieri (d.1321).[33] It then dropped sharply in the mid fourteenth century into lower NAO conditions before climbing back towards high NAO and higher winter rainfall later in the century.[34] This trend suggests that northern Highland winters were generally milder and wetter until the 1320s, when they became appreciably colder but less wet, then returned to the mild–wet regime albeit at a lower level of rainfall than during the MCA. For summer conditions, speleothems are less forthcoming, and apart from regions like Galloway and southern Argyll, to which the Northern Irish oak dendrochronological sequences can be stretched, we have only the Cairngorm pine chronologies to illuminate weather trends in central and north-east Highland districts. No other area of Scotland has the long data indices with which to construct robust climate models. The northern Cairngorm tree-ring data indicates higher temperatures that peaked around 1284, dipped through the 1290s and early 1300s before climbing to new peaks between 1300 and 1320. These peaks suggest that recurring episodes of high summer NAO affected the central Highland uplands,[35] which correlates strongly with growth trends identified in dendrochronological studies from across the northern hemisphere. In broad terms, reduced precipitation from April to June and cooler late summers delivered reduced tree growth.[36] Lower summer temperatures persisted across the Cairngorm region through the period 1320–50 before climbing to the highest peak since around 1310 in the decade after 1370. After that high point, there was a return to cooler summers with a continued downward trend into the early 1400s as the 'little ice age' manifested itself.

It was, of course, not solely a matter of the NAO and ocean surface temperature trends, for summer and winter conditions in the British Isles are also affected by the Scandinavian and Siberian atmospheric high-pressure cells. In winter, especially, these can channel strong easterly and north-easterly airflow which normally delivers cold and dry conditions. In conjunction with a high NAO, however, which brings moisture-laden air eastwards from the Atlantic, the result can be heavy snowfall or, during a low NAO, protracted periods of frost and ice in eastern districts. Analysis of documentary accounts for the late thirteenth century from England and western Europe accords well with the general temperature trends suggested by climate proxies that were being experienced in Scotland. It has been remarked that 'the 1290s were a harbinger of worse to come',[37] with the high temperatures achieved in eastern France, the Low Countries and Germany through the 1280s and into the early 1290s not being repeated. The trend instead was towards increasing severity.[38]

The new century was one of extremes, starting

33 N. Brown, *History and Climate Change: A Eurocentric Perspective* (London, 2001), pp. 251–258. The choice of Dante as the label for this episode has been criticised, not least because the poet's homeland was little affected by the phenomenon and its ramifications.

34 Proctor et al., 'A thousand year speleothem proxy record', p. 818 fig.4.

35 Rydval et al., 'Reconstructing 800 years of summer temperatures', pp. 2957–2560 and fig.5.

36 Campbell, *The Great Transition*, pp. 198–201 and fig. 3.19.

37 Ibid., p. 202.

38 Pfister et al., 'Winter air temperature variations', pp. 546, 547, 548 fig.10.

Written on the wind: Alexander III's Judgement Day

An unprepossessing stone pillar at the roadside between Burntisland and Kinghorn in Fife, easily missed if you are not deliberately looking for it, is the national monument to an event that moved the trajectory of Scotland's history onto a new path, the death on 19 March 1286 of King Alexander III. As discussed in this chapter and in Chapter 6, Alexander has secured a questionable historical reputation as the king who presided over Scotland's medieval 'golden age', represented in medieval and post-medieval tradition as an era of peace, stability, bountiful harvests and abundance. An increasing volume of climate and environmental historical research, however, is calling that traditional image into question (Oram, 'Away was sons of alle and brede') and highlighting the deteriorating trend in climate and weather patterns across the second half of the thirteenth century and the increasing precarity of the traditional agricultural regimes.

The circumstances of Alexander's death provide an illustration of the extreme events that punctuated the closing years of the century. The earliest account, preserved in the northern English record we know as the Lanercost Chronicle, presents a much-embroidered narrative whereby the king had spent the day in council at Edinburgh and then sat eating and drinking until, late in the evening and despite the horrendous weather conditions, he had decided to return to his new wife, Yolande de Dreux, who was at the royal manor house of Kinghorn in Fife. The weather had deteriorated throughout the daylight hours, with a tempestuous wind blowing so wildly that Alexander was reported to have joked that it was Judgement Day. Arriving at South Queensferry in darkness and in the teeth of the worsening storm, the king had insisted on crossing the Firth of Forth against the advice of the overseer of the ferry. Despite the overseer's certainty that they would all be drowned, the royal party made it safely across the Queensferry Narrows. On the other side, the master of the royal saltpans at Inverkeithing had attempted to persuade him to stay at his house rather than continue in the darkness and storm towards Kinghorn. Alexander, sure of his knowledge of the route – and, as the clerical chronicler snidely alleged, of his expertise in sneaking out in darkness and all weathers to have his wicked way with maidens and matrons alike and returning unscathed – pressed on with a small escort. Between Burntisland and Kinghorn he was separated from them on account of the poor visibility and, in uncertain circumstances, was thrown from his horse and died as the result of a skull fracture.

As long ago as 1886, William F. Skene argued that the Lanercost narrative was contrived to present Alexander in a poor moral light and his death as God's judgement for his manifold and manifest sins. Skene also contended that the distances involved made it highly unlikely that the king could have travelled from Edinburgh to Queensferry (9 miles), crossed the firth (1.5 miles) and then ridden almost to Kinghorn (9 miles) after a full day of business at Edinburgh and post-council feasting, and in such reportedly poor conditions (W.F. Skene, 'On the traditionary accounts of the death of Alexander III', *Proceedings of the Society of Antiquaries of Scotland*, 20 (1885–6), pp. 177–185). What does not seem to be in dispute, however, is that the weather on that fateful day was extreme and that the king's joke about 'Judgement Day', although deemed to be in poor taste, reflected the degree of its unprecedented severity. Here, in the record of its ending, we perhaps have the clearest reflection of the realities of the mythical golden age, offering insight on the physical impact of the changing weather after the waning of the MCA and the sense of human helplessness in the face of storminess such as had not been experienced within living memory. It was, however, but a foretaste of what the next century would deliver.

Alexander III monument, Kingwoodend, Kinghorn, Fife. Erected in 1886, the monument commemorates the death of the king on the foreshore near this place.

Pinewoods by the Allt na Feith Duibhe in Glenmore, Highland. The currently available tree-ring data from the northern Cairngorm pine sequences indicates that a rollercoaster in temperature and precipitation affecting pine growth persisted from the later thirteenth century into the last quarter of the fourteenth.

for west-central Europe with a run of extremely cold winters from 1303 to 1328, that last year marked by unusually narrow growth rings in southern Scandinavian trees.[39] The cold episode commenced with the deep entrenchment of anticyclonic circulation, which established a seasonal pattern of harsh and bitterly cold winters, delayed and relatively cool springs, hot and dry summers and warm, wet autumns for a brief period to 1315. That year saw Brown's Dantean Anomaly bring a dramatic change in Britain and Ireland, and across western Europe too, with cold summers and a westerly airflow which delivered a succession of frontal systems to drop unprecedented rainfall on the British Isles. Greenland ice core data points to this episode as being the result of a rapid and sustained rise in North Atlantic ocean surface temperatures in the summer months.[40] This overheating extended through the winter period to sustain an upward spiral in temperatures and a rise in atmospheric moisture. These conditions peaked in 1318–19, but only in 1325 did a further change become clear, with a return to summer droughts.[41] That dry episode was short-lived, however, with a reappearance in the early 1330s of the high NAO wet summer conditions of the previous decade. This wetness yielded swiftly after 1349 to a plunge to the lowest temperatures experienced across the North Atlantic region since before 1000.[42] Although climate scientists have described this cooling as 'minor by glacial standards',[43] its impact throughout the wider North Atlantic region was profound. This was the period of most rapid southward expansion of the sea ice off Greenland and Iceland, which brought cold, polar waters much further south. It injected a new dynamic into winter storminess to join the already

St John's Kirk, Ayr. The western tower is all that survives of the medieval parish church that stood outside the burgh on the links that extended west of the town. The stripping of turf from the dune systems for building needs and soil-deepening, combined with increased storminess, led to sand inundation of the backlands of properties on the west side of the town and deflation of the sandy area around the church, exposing burials in its graveyard.

39 Thun and Svarva, 'Tree-ring growth', fig. 5.

40 Campbell, *The Great Transition*, p. 200 fig. 3.19 (A) and (B).

41 Dawson, *So Foul and Fair a Day*, pp. 101–102.

42 C. Pfister, G. Schwarz-Zanetti and M. Wegmann, 'Winter severity in Europe: the fourteenth century', *Climatic Change*, 34 (1996), pp. 91–108.

43 Mayewski et al., 'Holocene climate variability', 252.

rising trend that had resulted from high ocean surface temperatures. For north-western Europe, this cold episode was followed by a return to what have been described as 'average' winters that lasted until 1354.[44] Thereafter, the third quarter of the century was characterised by strong year-to-year fluctuations and then a slow return to 'average' winter conditions to the end of the 1390s. Such broad-brush labelling, however, masks great variability in annual weather patterns and signalled still poorer times ahead in the fifteenth century.

It is from this late fourteenth-century phase of climatic instability that some of our most graphic record evidence for weather impacts across Scotland dates. At Ayr, westerly gales and dry conditions in 1380 brought a major episode of erosion of the sandy district between the burgh and the shore. It is likely that cultivation and turf-stripping had worsened a natural cycle of dune deflation and reformation. Blown sand overwhelmed burgage plots in the town, but the erosion had also exhumed burials in the cemetery of St John's Kirk on the links to the west of the town and was undermining the church itself. To encourage efforts to stabilise the sands, King Robert II granted any reclaimed land to those who laboured to halt the erosion.[45] On the opposite side of the country, in 1381, the bishop and canons of St Andrews appealed to Pope Clement VII for financial aid to save their cathedral church 'from the incessant pounding of the sea, which has come well beyond its normal limits'.[46] The appeal was successful, and while the claim may have been overstated in terms of the imminence of the danger,[47] it suggests that recent storm events and losses of land between the cathedral precinct and the shore had been sufficient to cause alarm. Neither of these events, however, is fingerprinted in the climate proxy data, meaning that it is only the surviving written record of their occurrence that bears witness to their impact. This chance visibility is important as a reminder that the broad labels of 'normal' or 'average' conditions which have been applied to the general weather trends of the later fourteenth century mask such one-off annual events that wrought havoc on the lives of those unfortunate enough to experience them.

Noteworthy changes

To what extent is the evidence offered by the climate models and proxy data for less than favourable weather conditions affecting Scotland in the half century to 1300 reinforced by the record evidence? Events such as those described for Ayr and St Andrews above underscore how extreme weather affected our ancestors and how some sought to take advantage of its effects. But the 'insurance fraud' of the St Andrews clergy should not distract us from the value such sources have for illustrating the wider experience of social dislocation and personal loss delivered by the extreme conditions. Across Europe, chronicle accounts are nearly unanimous in narrating the major change in climate conditions traced by our proxies from the 1250s, but starting in 1256. As in earlier centuries, there is a limited suite of surviving narrative and record accounts from Scotland through which to explore this climatic transition and its environmental consequences. The most important surviving sources for this period are the contemporary sections of the Melrose Chronicle and the gath-

44 Pfister, Schwarz-Zanetti and Wegmann, 'Winter severity in Europe'.

45 Cooper, *Charters of the Royal Burgh of Ayr*, pp. xxxv–xxxvi.

46 C. Burns (ed.), *Calendar of Papal Letters to Scotland of Clement VII of Avignon 1378–1394* (Edinburgh, 1976), p. 58.

47 In September 1419, a supplication to the pope challenged the claim that the cathedral had been at risk from storm and tempest: E.R. Lindsay and A.I. Cameron (eds), *Calendar of Scottish Supplications to Rome 1418–1422* (Edinburgh, 1934), pp. 125–126.

erings of thirteenth- and fourteenth-century material in the *Gesta Annalia I* and *Gesta Annalia II*, the latter continuing to the 1360s, which are themselves incorporated into the larger chronicle gathering attributed to John of Fordoun. The 'Fordoun' material itself became the principal source for Walter Bower's early fifteenth-century *Scotichronicon*, which continued the narrative from the early 1380s to the late 1430s. Bower also had access to other, now lost chronicle accounts for late thirteenth-century events. Similar material but supplemented by additional data from sources unused in the Fordoun collection also formed the basis of the chronicle composed in (often excruciatingly bad) verse by Andrew of Wyntoun in the early 1400s. All of these chronicles have a strongly south-eastern Lowlands focus for their reporting. A subtly different perspective is offered by the northern English Lanercost Chronicle, which drew on a Franciscan network that included Haddington and Dumfries for events in southern Scotland.[48] Lanercost is a key source for weather impacts through the second half of the thirteenth century. Finally, the important Irish annal records from Ulster and Connacht provide important context for weather trends affecting the Atlantic-facing seaboard of the north-western British Isles.

For Scotland, the Lanercost chronicler's account of weather events in 1256 is our earliest record of the general impact of the climatic downturn on northern British agriculture:

> In this year there was so great corruption of the air, and inundation of rain, throughout the whole of England and Scotland, that both crops and hay were nearly all lost. And some men's corn rotted in the fields from the day of harvest; some men's corn, shaken out by the wind, grew again under the straw; some men's harvest was so late that they did not reap it until about the festival of St Martin [11 November] or later.[49]

Unsurprisingly, the same chronicler observed the following year that there was a dearth of grain and near famine conditions, with peasants resorting to fighting like dogs over the carcases of horses, or eating food waste normally reserved for pigs, just to enable their families to survive.[50] It was the 1260s, however, that saw regular recurrence of damaging weather impacts, with wide year-to-year variations in general weather and, as will be discussed below, a succession of consequent poor harvests, food shortages and episodes of epidemic and epizootic disease.[51] The decade opened with a year of dearth

48 A.G. Little, 'The authorship of the Lanercost chronicle', *English Historical Review*, 21 (1916), pp. 269–279.

49 *Chronicon de Lanercost*, ed. J. Stevenson (Edinburgh, 1839), p. 64. Independent corroboration of Lanercost's claims of widespread devastation is provided by Matthew Paris, who records extreme wind- and rainstorms beginning on 19 June and running on through the summer, noting especially catastrophic flooding of the Ouse around Bedford and extensive structural damage affecting mills, bridges and houses. In his summary for the year, Matthew claimed that it rained incessantly from the Feast of the Assumption of the BVM (15 August) to the Feast of the Purification (2 February), rendering roads impassable and fields sterile, and at the end of autumn grain rotted in the ears: Matthew Paris, *Chronica Majora*, vol. 5, pp. 561, 600.

50 *Chronicon de Lanercost*, p. 65. Matthew Paris also records poor weather in south-eastern England from February to May 1257, with rain, strong winds and coastal flooding, noting that as a consequence many farmers were obliged to resow their fields: Matthew Paris, *Chronica Majora*, vol. 5, p. 630. Heavy rains continued to affect southern England into late summer, prompting the monks of St Albans to stage a religious procession to secure saintly intercession: Matthew Paris, *Chronica Majora*, vol. 5, pp. 644–645.

51 Matthew Paris, however, recorded prevailing conditions in England in 1258 in equally bleak terms to those of the preceding years: Matthew Paris, *Chronica Majora*, vol. 5, p. 728.

Cooler conditions in the 1260s affected grass growth in even the normally milder Western Isles, accompanied by higher rainfall that waterlogged pastures and stronger winds that delivered greater wind chill. Such weather caused livestock to expend more calories on maintaining body heat rather than on gaining body mass or producing milk.

and with violent storms throughout a dry summer leading into a wet harvest season peppered with destructive gales through the late autumn and early winter.[52] The northern and west-central Irish Annals of Ulster and Annals of Loch Cé record that Ireland experienced a summer of extreme heat and drought in 1263, pointing to a high summer NAO.[53] Scottish sources for that year are preoccupied with the events of the Scoto-Norwegian war and pay little heed to annual weather conditions; the Melrose chronicler, however, attributed the Norwegian defeat on 2 October to heaven-sent pestilence and intense autumn storms.[54] A sharp deterioration affected mainland Britain from autumn 1267, which suffered a series of extreme weather events in late October. The most serious of these was a major sea-flood reported by Bower, probably caused by a combination of high tides, spate conditions in the major river systems feeding into the North Sea, and storm-force north-easterly winds which drove surges that affected the east coast of Scotland from the Tweed to the Tay.[55] Bower continued with a description of winter 1267/8 as 'very harsh with continual whirlwinds of hail, snows and rain', leading into a wet and windy spring and an unusually cold, wet and windswept summer.[56] His thirteenth-century source also recorded that the conditions resulted in abnormally high livestock mortality that year, especially of sheep, but with wild animals also badly affected. Late autumn 1268 and winter 1268/9 offered no relief, with Bower's sources recording extreme conditions from late November to early February that prevented ploughing and preparation for the new season's crop.[57] English

52 Bower, *Scotichronicon*, vol. 5, p. 325.

53 Annals of Ulster, U1263.7 (the same annal's 1263.11 account of mortality through famine and pestilence belongs to 1265 rather than 1263); Annals of Loch Cé, LC1263.15.

54 *The Chronicle of Melrose*, sub anno 1263; Bower, *Scotichronicon*, vol. 5, p. 341.

55 Bower, *Scotichronicon*, vol. 5, pp. 359–361. This event is not recorded in Fordoun. Bower's observation that 'the traces [of the flooding] still show clearly' was carried over from his thirteenth-century source.

56 Ibid., p. 369.

57 Ibid., p. 373.

accounts speak of a spoiled harvest in 1268 followed by a wet, tempestuous and pestilential winter, of which Irish reports of severe storms in January 1269 might be an echo.[58] The decade ended as badly as it had begun, with a weather-related catastrophically bad harvest in 1269, for Scottish and Irish sources identify 1270 as a year of 'dearth and much rain'.[59]

Unfavourable weather conditions throughout the north-east Atlantic region continued in the new decade, a trend which we can identify as a consequence of the colder North Atlantic surface temperatures of the time. Winter 1270/1 seems to have been particularly severe and cold for Scotland,[60] but a new low was reached in 1272, for which Bower's thirteenth-century source recorded

> a great lack of productivity on the land and unfruitfulness at sea, as well as turbulence of the air, as a result of which many people fell ill and many animals died . . . a great famine hit France, England, Scotland and many areas, for the cattle mostly died, the crops failed, and the poor died of their poverty.[61]

The impact of 1272 might have marked the nadir in this run of poor weather but it was not the end, for in 1273 or shortly afterwards a severe windstorm inflicted major damage on the western end of the nave of the cathedral at St Andrews.[62] After this low point, there are no further references to extreme weather events in either Irish or Scottish records until the beginning of the 1280s, probably as a result of a hiatus in the wildly fluctuating ocean surface temperatures until the end of the 1270s. This lull does not mean that conditions had returned to a pre-1256 calm. They were simply less extreme than they had been in the previous twenty years.

A new phase commenced in 1281, with Irish sources recording an extremely poor winter of 1281/2 and the Annals of Innisfallen characterising the whole of 1281 as a year of generally bad weather, beginning with heavy snow in February and early March causing widespread cattle deaths as well as high mortality and general hardship for the human population.[63] A similar gloss is provided by the Annals of Ulster, which start the episode of harsh weather from around Christmas 1281.[64] Most of the main Irish annals present winter 1282/3 as similarly poor, in particular noting extreme weather through December and January and observing that continuing poor conditions thereafter prevented 'any useful work being done'.[65] The summer months of 1283 were characterised by unusually strong winds which resulted in widespread damage to essential winter fodder supplies in hayricks as well as to buildings, but these windstorms do not appear to have been accompanied by rainfall, which suggests anticyclonic, easterly airflow resulting from a high summer NAO. Early 1284 again saw heavy snowfall while spring

58 *Continuation of the Chronicle of William of Newburgh*, in R. Howlett (ed.), *Chronicles of the Reigns of Stephen, Henry II and Richard I*, volume 2 (London, 1884), p. 555; Annals of Innisfallen, AI1269.2.

59 Bower, *Scotichronicon*, vol. 5, p. 381; Annals of Loch Cé, LC1270.6.

60 *The Chronicle of Melrose*, sub anno 1271.

61 Bower, *Scotichronicon*, vol. 5, p. 385.

62 Bower, *Scotichronicon*, vol. 3, p. 399. The change in the design of the nave resulting from the destruction of its western bays by the storm is evident in the surviving south wall. The decision was taken at this time to shorten the building rather than rebuild it to its original full extent.

63 Annals of Innisfallen, AI1281.2, AI1281.8.

64 Annals of Ulster, U1279.7 (there is a three-year error in the dating of the annals at this point).

65 Annals of Connacht, http://www.ucc.ie/celt/published/T100011/index.html, AC1282.9; Annals of Innisfallen, AI1282.2; Annals of Loch Cé, LC1282.8.

Saltmarsh grazing on the Nith estuary, north of Caerlaverock, Dumfries and Galloway. The low-lying pastures around the head of the Solway Firth and the estuaries of the rivers feeding into it became increasingly vulnerable through their exposure to the combined effects of spate run-off from the Southern Uplands and Cumbrian hills and strong westerly winds at the end of the thirteenth century.

1285 was noted for extreme storminess.[66] English records suggest a continuation of these conditions into the second half of the 1280s, with harsh winters and late springs helping to deliver delayed harvests. The resulting risks of reduced yields in wet and windy autumns continued to 1287.[67] *Prima facie*, this poor record seems to confirm the high winter NAO Traligill speleothem evidence for significant precipitation but to contradict the Cairngorm tree-ring suggestion of good summer and autumn conditions through the 1280s and into the 1290s. This view, however, needs to be tempered by indications that even on the north-eastern rim of the Atlantic the 'good' summers tended to be hot and dry, with three – 1284–6 – possibly delivering near drought conditions that were poor for both tree growth and cereal cultivation.

Echoing the proxy climate data, English chronicle and Irish annal sources indicate that the 1290s were a troubled decade for weather. Lanercost, for example, gives details of storms, floods and high winds commencing at the end of September 1291, arriving at the worst possible time for the harvest.[68] Late the following May, the same chronicler recounted a terrible sea-flood around the head of the Solway Firth, driven by a violent tempest, which overwhelmed grazing and arable in areas north and west of Carlisle, probably affecting the carseland on the Scottish side of the firth round to the mouth of the River Nith.[69] It is to events such as this that the medieval storm beaches identified at Caerlaverock probably belong.[70] Bower, perhaps reflecting a central-eastern Lowlands provenance for his late thirteenth-century source, makes no reference to these south-western events

66 Annals of Innisfallen, AI1284.2, AI1285.2.

67 *Chronicon de Lanercost*, p. 122.

68 Ibid., pp. 141–142.

69 Ibid., p. 145.

70 R. Tipping and J. Adams, 'Structure, composition and significance of Medieval storm beach ridges at Caerlaverock, Dumfries and Galloway', *Scottish Journal of Geology*, 43:2 (2007), pp. 115–123.

but reported a prolonged episode of snow and northerly gales that affected much of Britain.[71] This event possibly occurred late in spring 1292, for Bower's source included a verse which spoke of devastation of woodland, frost damage to growing plants and widespread property damage. That these storms caused considerable losses to growing crops is perhaps corroborated by Lanercost, which recorded that the weather brought 'such a lack of produce throughout the whole of England and Scotland that in some parts a quarter of grain sold for thirty shillings'.[72] It was Lanercost that then recorded late and heavy snowfall in East Lothian in mid April 1294, starting on Palm Sunday (10 April), and violent rainfall across the Southern Uplands in early August.[73] The account of this latter episode suggests high precipitation in the Teviotdale hills, for it brought a rapid and violent spate of the River Teviot which swept away mills and flooded ripening crops in the arable lowland east of the district and in the fertile Merse on the north bank of the River Tweed between Roxburgh and Berwick. This brief phase of more northerly- and easterly-driven weather perhaps accounts for an apparent lull in the reporting of poor weather in Ireland in 1294 and 1295. It was a brief respite, for the Innisfallen annals describe 1296 as particularly bad, with stormy conditions, heavy snow and cold weather persisting into the spring and returning in early winter.[74] At this point, Scottish and northern English sources become preoccupied with reporting the outbreak of war between Scotland and England. The last weather-related report of the thirteenth century is provided by Abbot Bower's unnamed source, which recorded a threat of major dearth in autumn 1297 'since there was shortage of grain resulting from inclement weather'.[75]

Here, let us pause for a moment. It was in 1296 that the long era of stable relations with England that had prevailed since 1237, when the Treaty of York had settled the century-long struggle for control of the English northern counties, broke down disastrously.[76] The descent from long-running amity to open warfare across the period 1286–96 has been rehearsed many times, as has the disastrous outcome of the first hostilities in April 1296, the collapse of Scottish resistance in the following weeks and the humiliating submission of the entire Scottish leadership to the overlordship of Edward I of England in June and July.[77] From that point the narratives focus largely on the air of false calm that followed, as Edward slotted his own men, and especially the able and efficient clerk Hugh Cressingham as Treasurer, into controlling positions within the Scottish administrative apparatus. Depending on one's perspective, Cressingham was either effective or ruthless in exploiting the revenue-producing capabilities of his administration, raising some £5,000 (nearly £4.5 million in modern terms) to transmit to Edward for his French war, much of it derived from traditional sources of Scottish royal income but also including the profits of forfeited estates and confiscated wool receipts. We are further told that it was rumours of a plan to send the 'middling folk' ('le menzane de Escoce') to fight overseas in Edward's wars that caused grievances over these financial exactions to coalesce into rebellion.[78] Nowhere in these accounts,

71 Bower, *Scotichronicon*, vol. 6, p. 41.

72 *Chronicon de Lanercost*, p. 154.

73 Ibid., pp. 155, 157.

74 Annals of Innisfallen, AI1296.2, AI1296.7.

75 Bower, *Scotichronicon*, vol. 6, p. 87.

76 Oram, *Alexander II*, pp. 144–145.

77 For an excellent summary, see Brown, *The Wars of Scotland*, chapter 8.

78 Barrow, *Robert Bruce*, pp. 80–83; Brown, *The Wars of Scotland*, pp. 179, 181.

Sheep carcase, Ben Lawers, Perth and Kinross. A combination of adverse weather affecting livestock health and increasing the incidence of wolf predation, seizure of wool crops, increased taxation and the rumour of military conscription for service overseas triggered outbreaks of insurrection across Scotland through the winter of 1296/7.

however, is there reference to the poor weather conditions that had prevailed in the early 1290s and which in 1297 were anticipated as threatening a general failure of crops and consequent dearth of grain. It was the small freeholders, free peasants and burgesses who were staring into the void of a failed harvest, seized wool profits and rigorous revenue raising, further spurred by widespread belief in an English scheme to send them abroad to be slaughtered, who formed the backbone of the armies raised by Andrew Murray and William Wallace. Hungry, impoverished and fearful for their lives under English lordship, they might have believed they had nothing more to lose. Climate, weather and their impacts proved to be as effective here in stimulating resistance to an invader as they would later be for Russia in the wars with Napoleon and Hitler.

Famine I: 'Away was sons of alle and brede'

Bower's link between poor weather and hunger introduces one of the clearest consequences for Scotland of the unprecedented climate variability of this era. The 1297 grain shortage he recorded, however, was just the most recent and probably not the worst dearth that had occurred since the 1250s. The first had followed the harvest failure of 1256, with 1257 seeing 'the greatest dearth of grain' throughout the whole of the British Isles.[79] A second dearth occurred

79 *Chronicon de Lanercost*, p. 65.

in 1260, when Bower's source noted widespread food shortages and price inflation again stemming from harvest failures. The result in 1261–2 was that 'all parts of Scotland suffered from a great famine caused by the harvest being ruined'.[80] A further blow came in 1267, when hopes raised by that year's mild, dry summer for a good harvest were dashed first by plant-eating insect infestations and then, as mentioned above, by violent north-easterly winds and east coast sea-floods in late October.[81] The severe conditions of the following winter of 1267/8 were the precursor to a cold, wet and windy spring and summer in 1268. Bower's unnamed contemporary source reported that 'there was a high death-rate amongst animals, that is amongst red deer, fallow deer, forest ponies, but most of all sheep' on account of the inclement winter and spring but concluded his weather observation for the year with a description of a winter so harsh that the ground could not be ploughed between the end of November 1268 and the beginning of February 1269 to prepare for sowing the next year's crop.[82] Almost inevitably, a poor harvest in 1269 led to widespread shortages in 1270. Worse came in 1272, when the chronicle gave that report of 'a great lack of productivity on the land and unfruitfulness at sea, as well as turbulence of the air' quoted above.[83] The impact of this event could not be mitigated through imports of grain, for it was not just Scotland that experienced these disasters. As Bower's source added grimly, 'a great famine hit France, England, Scotland and many areas',[84] meaning that there was no surplus grain to be had from the traditional trade sources. The climate instability was starting to take its toll in terms of human and animal lives.

Too few prices for staple foodstuffs are known from later thirteenth-century Scotland to enable any deep analysis of how these harvest failures and grain shortages affected the Scottish people in general.[85] A few points of comparison can be made with equivalent values in England, rendered meaningful by the parity of Scottish and English currency values at this time. Higher Scottish than English wheat prices in 1263 and 1264 are suggestive of the lingering effects of harvest failures in the early 1260s, but with no figures surviving from before 1263 it is impossible to determine how high any inflationary spike had been unless we see the one surviving – much lower – figure from 1279 as representing a return to a more normal level.[86] Strikingly similar wheat prices in Scotland and England in 1288 probably reflect what was apparently a particularly good harvest.[87] Indeed, in one of our few surviving royal financial accounts from that time, the sheriff of Lanark observed of 1288 that it had seen 'a good market for wheat'.[88] The doubling of the 1288 figure by 1291 and continuing high prices into the mid 1290s seem to confirm the impact of the poor weather and harvest failures recorded by the contemporary chronicle accounts. This handful of price data tells us little beyond the wide fluctuations in the value of wheat between 'good' and 'bad' years. Relative to incomes, however, there can be no doubt that the near doubling of grain values brought real hardship to many families across

80 Bower, *Scotichronicon*, vol. 5, pp. 325, 335.

81 Ibid., pp. 359, 361.

82 Ibid., pp. 369, 373.

83 Ibid., pp. 381, 385, 387.

84 Ibid., pp. 385, 387.

85 For the few known prices of wheat, barley, malt and oats, see E. Gemmill and N. Mayhew, *Changing Values in Medieval Scotland: A Study of Prices, Money, Weights and Measures* (Cambridge, 1995), pp. 151, 166, 182, 192.

86 Ibid., p. 147 table 9.

87 Ibid., pp. 147, 148.

88 Stuart and Burnett, *The Exchequer Rolls of Scotland*, vol. 1, p. 46.

Ripening wheat, Findogask, Perth and Kinross. Wheat and barley yields declined generally across the second half of the thirteenth century, with bountiful harvests becoming less frequent than at the height of the MCA. Complete failures remained rare, however, and whilst shortages occurred, famine does not seem to have threatened most of Scotland's people.

the kingdom. What that translated into in terms of deaths through starvation or increased susceptibility to disease through malnutrition we shall never know.

Taken together with the proxy climate data, the surviving annal, chronicle and price records for the northern and western parts of the British Isles for the half century after 1250 present a picture of an intermittent but consistently downward trajectory in weather conditions. This deteriorating trend, our sources are unanimous in reporting, affected harvests intermittently but potentially disastrously. There were two sustained periods within which conditions were at least passably acceptable, such that the sources make no comment on either weather events or crop yields: 1273–80 and from summer 1286 to the winter of 1290/1. Both episodes correlate with low grain prices in our few surviving records. In total, twenty years across the period can be categorised as average. In only three years, 1253, 1254 and 1288, however, is there consensus that harvests were particularly good and the weather especially favourable. There is also consensus, on the other hand, that there was a general harvest failure in only two years, 1256 and 1260, which caused famine in the years following, again reflected in the higher early 1260s grain prices. In ten of the fifty years, famine or dearth was recorded, with seven of those years falling between 1261 and 1272, the first part of that span correlating with the poor tree-ring growth across the northern hemisphere. In eighteen of the fifty years, extreme weather events are recorded, chiefly periods of severe cold, snowfall or rain, with ten of the worst years again falling in the period 1260–72 but with four of the five years 1281–6 also witnessing a return to the negative pattern of the 1260s. On the basis of the written record and the collective proxy data, therefore, there hardly seems to be a compelling case for the second half of the thirteenth century and Alexander III's reign especially to be characterised as the era of 'wyne and wax, of gamyn and gle' of the Wyntoun tradition.

War

From 1296 through to 1357, while elsewhere in the British Isles and mainland Europe chroniclers narrated the horrors of famine and disease, our record evidence from Scotland is dominated overwhelmingly by wars, both civil and nationally

between Scotland and England. Indeed, warfare characterised this era to an extent that one modern historian has labelled the period 1214–1371 as 'the Wars of Scotland'.[89] Despite the centrality of the conflict in most Scottish historical narratives, few historians have looked beyond its political contexts, military strategies and socio-political consequences to its short- and long-term ecological or environmental effects. Although the scars of conflict on the built environment have long been studied,[90] war's environmental impact became better understood only after the global conflicts of the first half of the twentieth century and colonial wars in the second half, and it became the subject of detailed academic research only in the twenty-first century.[91] This research has placed much stress on the distinction between collateral environmental damage and intentional acts, in recognition of early human understanding, revealed through millennia of examples, that the wilful destruction of land and its resources is a powerful weapon that can cripple a foe's war-waging capability or act as an even more powerful visible deterrent to potential adversaries.

The environmental legacies of warfare throughout the British Isles in the early fourteenth century have become a recent focus of research and within that the Scottish experience has been prominent.[92] Traditional historiography based on contemporary sources narrates a series of high-profile events of spectacular violence which usually stand as examples

89 Brown, *The Wars of Scotland*.

90 The classic studies are of the devastation of parts of Germany and Poland that occurred during the 'Swedish Deluge' of the Thirty Years' War in Europe. In Scotland, the first comprehensive assessment of the building record for war damage was G. Stell, 'Destruction and damage: a reassessment of the historical and architectural evidence', in N. Macdougall (ed.), *Scotland and War AD 79–1918* (Edinburgh, 1991), pp. 23–35.

91 See, for example, C. Pearson, *Mobilizing Nature: The Environmental History of War and Militarization in Modern France* (Manchester, 2012), and D. Zierler, *The Invention of Ecocide: Agent Orange, Vietnam, and the Scientists Who Changed the Way We Think About the Environment* (Athens GA, 2011).

92 The key work on early fourteenth-century Scotland is P. Slavin, 'Warfare and ecological destruction in early fourteenth-century British Isles', *Environmental History*, 19:3 (2014), pp. 528–550.

of the wider impact of the long wars with England. These are headed by the sack of Berwick in March 1296, the burning of Aberdeen in June 1336, the long siege of Perth in 1339 and the 'Burnt Candlemas' campaign of 1356.[93] These headline incidents, however, took place against a broader backdrop of deliberate devastation of the land and displacement of rural populations, the legacies of which were still being felt decades after the event. Some sense of what was to come can be gained from accounts of William Wallace's raids into northern England in early winter 1297, when his army devastated both Northumberland and Cumberland and carried off the food supplies and seed-corn of the rural population there.[94] The recent experience of Edward I's conquest of Gwynedd had revealed the effectiveness of such operations, with the English harvesting of the cereal crop on Anglesey in 1277 forcing the submission of the Welsh, who otherwise faced starvation. In Scotland, however, the full impact of such tactics was not felt until after the failure of Edward's attempted political settlement in Scotland in 1306, when he moved from an attempt at accommodation with the leadership of the defeated Scots to furious determination to utterly destroy his enemies through targeted devastation of their lands and resources.

One example of the application of this targeted action will suffice. Wemyss in Fife had been visited peaceably by Edward I towards the end of his successful 1303–4 campaign that had brought the Scots to the negotiating table. On 19 June 1306, however, following the eruption of a new rebellion against his lordship, he ordered Aymer de Valence to destroy the manor house there, along with its lands, orchards and gardens, as an example to others, in reprisal for its owner's abandonment of his submission to the English crown.[95] De Valence systematically plundered the estate of all its movables including the stored food needed to carry its people through to the next harvest, burned or trampled its growing crops, drove off or slaughtered all its livestock, destroyed its mills, kilns, barns and ponds and cut down its fruit trees. The psychological impact of such comprehensive devastation on both the local population and residents of neighbouring areas was no doubt profound, as intended, and the use of such terror tactics was to become even more refined later in the wars and applied to terrible effect.[96] Wemyss, for a period of years, effectively ceased to function as an economic unit. The longer-term consequences of the deliberate destruction seen at Wemyss, of fields, orchards and woodland, however, had lasting human ecological impacts that would have taken decades to recover from. The wider, non-human ecology, however, seems to have experienced very limited effects. 'Wasting' of land affected the human capacity to subsist upon its resources, but 'wasting' is not total destruction and fundamental elements like grassland and heather would regenerate rapidly, whilst felled trees would begin to throw new growth in the same manner as coppiced woodland. This kind of action removed the resources that enabled an enemy to mount serious resistance but did not bring about permanent loss. It was a lesson that the Scots who

93 Brown, *The Wars of Scotland*, pp. 175, 240, 241, 253.

94 Bower, *Scotichronicon*, vol. 6, pp. 86–87. The systematic wasting of northern England seems to have been pursued by the Scots thereafter, whenever opportunity allowed. See F. Watson, *Under the Hammer: Edward I and Scotland 1286–1307* (East Linton, 1998), pp. 150–151, and Slavin, 'Warfare and ecological destruction', pp. 539–542.

95 Bain, *Calendar of Documents*, vol. 2, no. 1787. For discussion of the deployment of such tactics by the Scots in northern England, see Slavin, 'Warfare and ecological destruction', pp. 534–536.

96 I.A. MacInnes, '"Shock and awe": the use of terror as a psychological weapon during the Bruce-Balliol civil war, 1332–1338', in A. King and M. Penman (eds), *England and Scotland in the Fourteenth Century: New Perspectives* (Woodbridge, 2007), pp. 40–59.

Nineteenth-century photograph of MacDuff's Castle, East Wemyss, Fife. The tower on the left, which has since been demolished, was the replacement for the manor house destroyed on the orders of Edward I of England as part of his conscious ecocide of the estate of the rebellious Scottish lord of Wemyss.

supported Robert Bruce would soon apply in their own waging of war against the supporters of the English crown in Scotland.

The display of utter ruthlessness introduced by the exasperated Edward I in an attempt to crush his enemies' will was taken up by his Scots opponents and turned against the English and their supporters within Scotland. Systematic ravaging of the lands increased through the campaigns and counter-campaigns of Robert Bruce and his opponents from 1307 to 1314. During this phase of the war two areas in particular experienced prolonged programmes of deliberate devastation that had lasting consequences, Buchan in the north-east and Galloway in the south-west. In the first case, Bruce followed up his victory at Oldmeldrum in spring 1308 with a ruthless harrying of the heartland of the defeated Comyn family's north-eastern power base. This event, referred to as 'the herschip (harrying) of Buchan' by the later fourteenth-century poet John Barbour, involved the burning of the agricultural settlements upon which the Comyns' economic strength and military manpower had been founded and the slaughter of all who stood in the victors' path.[97] As at Wemyss in 1306, the campaign included the slaughter or driving away of livestock, the felling of fruit trees, the destruction of mills and the systematic dismantling of the infrastructure of lordly economy throughout the region. According to Barbour, the effects of the herschip were still felt fifty years afterwards and it was so

97 J. Barbour, *The Bruce*, ed. and trans. A.A.M. Duncan (Edinburgh, 1997), pp. 333–335.

effective as to bring a peace worthy of description by Tacitus to the country north of the Mounth. As underscored by the deaths of humans and livestock experienced across Europe from the 1310s onwards (discussed below), it took decades to recover from such losses and from the ravaging of the land. But here we need to pause, for there are few discernible traces of any lasting impact from the herschip recognisable in either the archaeological or the landscape record for Buchan.[98] Indeed, as Bruce Campbell has highlighted, if the trade of the north-east passing through Aberdeen by the 1320s is any measure of local productivity, the recovery of Buchan in terms of at least its non-human resources had been more rapid than Barbour suggested.[99] King Robert, of course, wanted to cripple the Comyn family by removing its capacity to wage war, but he did not want to waste the land to the point where it would permanently lose the capacity to contribute the human and natural capital upon which effective lordship was founded.

In Galloway, the consequences of the intentional ecological destruction of the Bruce campaigns from 1307 to 1312 had a longer-lasting effect that left identifiable traces in the historic record.[100] There is no poetic account of a herschip akin to that in Buchan, but administrative records and place-name sources make clear the scale of dislocation. Here, accounts of the displacement of whole communities who fled into Cumbria with their herds, or of the wholesale seizure of the livestock of others by Bruce partisans, speak of significant social breakdown.[101] We have a major indication of how profound that dislocation was in the reordering of the parish structures of Galloway's northern Glenkens district and the disappearance of one of the rural deaneries into which the diocese of Whithorn was organised. The disappearance of a long-established component of diocesan government and the parishes within it signals the scale of population loss and associated economic collapse. In essence, these units existed to provide for the spiritual wellbeing of the human population and were sustained by the teinds paid annually. Without people, these structures would quickly wither away. Population loss seems to be confirmed by the loss of settlement names, which appears to mean village abandonment. Evidence for much later re-establishment under new names by peasant colonists later in the fourteenth or early fifteenth century suggests a break of sufficient duration for them to have no cultural memory from which to rebuild the old settlement pattern.[102] This reordering of such fundamental social institutions and the loss of old, established place-names speak of long-term displacement of the people from their land and depopulation to the point where church administrative structures and agricultural communities became economically unsustainable. A collapse in settlement presumably

98 There is no mention of this event in the most recent discussion of late medieval land and environment in the region, C. Shepherd, *The Late Medieval Landscape of North-East Scotland: Renaissance, Reformation and Revolution* (Oxford, 2021). The clearest likely evidence for the 1308 Bruce campaign comes from the excavated remains of the Phase 3 Comyn castle at Rattray, which appears to have been destroyed in a catastrophic burning event in the early fourteenth century: H.K. Murray and J.C. Murray, 'Excavations at Rattray, Aberdeenshire: a Scottish deserted burgh', *Medieval Archaeology*, 37 (1993), pp. 109–218. Reconstruction of the castle here may have started in the late 1310s or early 1320s.

99 Campbell, 'Benchmarking medieval economic development'. For Aberdeen's customable trade in 1328–33, see ibid., p. 916 table 9.

100 R.D. Oram, 'Bruce, Balliol and the lordship of Galloway: south-western Scotland and the Wars of Independence', *Transactions of the Dumfries and Galloway Natural History and Antiquarian Society*, 67 (1992), pp. 29–47.

101 Barbour, *The Bruce*, pp. 344–348.

102 D. Brooke, 'The Glenkens 1275–1456: snapshots of a medieval countryside', *Transactions of the Dumfries and Galloway Natural History and Antiquarian Society*, 3rd series, 59 (1984), pp. 41–56.

Central Dee Valley north of Glenlochar, Dumfries and Galloway. The upland district north of here into the Glenkens area, a zone of strong support for the Balliol and Comyn families, saw repeated and systematic devastation by the supporters of Robert Bruce after 1308. The result was the flight of the local population and the collapse of the old patterns of settlement and agricultural exploitation.

saw contraction of the zone of cultivation and, potentially, a lifting of grazing pressures that led to reversion to a wilder, less managed state which invited a change in land-use. We can see indications of such a repurposing by the 1320s, when King Robert established a new royal forest in this district of Galloway.[103] Although the forest was short-lived – renewed warfare in the region in the 1330s and 1340s overturned the Bruce landholding order there[104] – and population levels seem to have recovered to a degree by the fifteenth century, the rupture in local patterns of social organisation caused by the Bruce raids remained permanently ingrained in the area's socio-economic landscape.

Similar levels of raiding and destruction of rural settlement were again experienced in Scotland in the renewed Anglo-Scottish warfare after 1332. Galloway once more saw widespread devastation in districts that were the principal centres of support for the enemies of the Bruce regime. Attacks on the region by the leadership of the Bruce party were recurrent and spread over the twenty years to the early 1350s, disrupting the recovery that had started after the earlier harrying by King Robert and extending the impacts into the coastal districts around the main urban centres of Kirkcudbright and Wigtown. Most of our sources, however, emphasise that it was the country north of the Firth of Forth and especially from the River Tay northward that suffered an intensity of destruction unseen since Edward I's 1303–4 campaign. English *chevauchées* (fast-moving raids involving small forces of mounted warriors), of the

103 Gilbert, *Hunting and Hunting Reserves*, p. 34.

104 Oram, 'Bruce, Balliol and the lordship of Galloway'. The devastation of Galloway by the Bruce party is recorded as starting in late summer or early autumn 1332 and was presumably directed against the incoming harvest. See W.R. Childs and J. Taylor (eds), *The Anonimalle Chronicle 1307 to 1334: From Brotherton Collection MS 29* (Cambridge, 1991), p. 153.

kind later used by King Edward III in his French wars, penetrated as far as Lochindorb in upland Moray in 1336 and burned a return path through the Moray lowlands to Aberdeen. That important regional trade centre was plundered and put to the torch, Edward III's intention, like that of his grandfather, being to spread fear and alarm and cause maximum destruction, rather than to conquer and hold territory.[105] But the Bruce-supporting Scots were equally ready to wage a scorched-earth war to deprive their enemies of supply. Over the winter of 1336/7 they mounted a campaign of attrition from secure bases in the Forest of Plater in central Angus, which saw both sides ravage the country from the Tay estuary and Sidlaw Hills to the Mearns. As *Gesta Annalia II* describes it, 'through the ceaseless marauding of both sides, the whole land of Gowrie, Angus and Mearns was, for the most part, almost reduced to a hopeless wilderness and to utter want'.[106] Here and in north-east Fife in February and Lothian in summer 1337,[107] the intention was to strip the land of every means through which enemy garrisons could sustain themselves, forcing the occupiers into precarious dependence on supplies brought overland through hostile territory. It was the consequences of these campaigns that *Gesta Annalia II* described when speaking of the districts through which the warring parties had moved being 'reduced to such desolation and distress, that more perished afterwards through starvation and want, than the sword devoured in time of war'.[108] Andrew of Wyntoun, writing more than half a century after these events but using contemporary sources, also recorded the devastation and hunger caused by the army led by Andrew Murray, lieutenant for the young King David II: 'In gret distres þe commownys ware / Pynyt to ded in hungyr sare; / For withe his ost, as he made rade, / Gret wastynge in þe lande he made'.[109]

Walter Bower recorded the same outcome in the account he composed in the 1440s, when he spoke of 'great scarcity of food nearly everywhere in the kingdom'.[110] He explicitly used the Latin word 'fames' (famine) to describe the result and reported the flight of some Scots from the kingdom as refugees. Despite this picture offered by the abbot of general suffering and want, however, the narratives he presents show a strongly localised pattern of impacts. For example, some of the most concentrated devastation reported by *Gesta Annalia II*, Wyntoun and Bower occurred during the 1339 siege of English-held Perth. Wyntoun and Bower especially described how the land around the burgh was laid waste and satellite settlements almost wholly destroyed to the extent that deer came down from the uplands to feed in the deserted countryside.[111] Food supplies locally were exhausted, with reports of the 'common folk' being reduced to such dire straits that they were forced to 'eat grass like sheep' before dying. Amidst such desperate conditions, rumours of cannibalism were rife, with the chroniclers naming 'a certain peasant called Christy Cleke with his fierce woman', who preyed on women and

105 Brown, *The Wars of Scotland*, p. 240. In his 1978 essay on the differing strategies deployed in the first and second phases of the wars, Geoffrey Barrow highlighted Edward III's 'deliberate ruination' of the land: G.W.S. Barrow, 'The aftermath of war: Scotland and England in the late thirteenth and early fourteenth century', *Transactions of the Royal Historical* Society, 5th series, 27 (1978), pp. 103–126 at 124–125 (quoted phrase on p. 125).

106 John of Fordoun, *Chronica Gentis Scotorum*, vol. 2, p. 353.

107 Bower, *Scotichronicon*, vol. 7, pp. 125, 131.

108 John of Fordoun, *Chronica Gentis Scotorum*, vol. 2, p. 355.

109 Andrew of Wyntoun, *The Original Chronicle*, vol. 6, p. 99.

110 Bower, *Scotichronicon*, vol. 7, p. 127.

111 Andrew of Wyntoun, *The Original Chronicle*, vol. 6, p. 132; Bower, *Scotichronicon*, vol. 7, p. 143.

children.[112] As can be seen in these accounts, intentional human action had generated conditions of utter desperation, want and fear, resulting in the fragmentation of social cohesion and the breaching of social and ethical norms.[113] What we must recognise, however, is that the impacts were felt principally by peasant communities locally and the environmental impacts reported are chiefly due to the temporary cessation of agricultural exploitation in the district. As the tales of deer coming down from the uplands to feed in the now empty lowland districts suggest, natural systems began quickly to reassert themselves through recolonising abandoned fields and pasture. As in the recorded episodes of destruction in the earlier 1300s, it was the humanly ordered landscape of economic structures that was disrupted. The wider environment was quick to revert to something altogether less managed.

A final deployment of intentional economic warfare and ecological destruction in the Wars of Independence occurred in February 1356 during Edward III's so-called 'Burnt Candlemas' campaign.[114] A retaliatory and unmistakably punitive *chevauchée*, it had no real strategic aims other than inflicting maximum suffering at a time of year when supplies laid down the previous autumn were starting to run low. It might have helped to concentrate Scottish minds in negotiations for the peace that would bring the release of the captive King David II in 1357. To Bower, Edward was 'like a bear whose offspring have been carried off venting its rage on the forest, he entered the land of Scotland in a savage fashion'.[115] A native of Haddington, Bower was able to draw on local traditions regarding this devastating event in his narrative of indiscriminate looting, the wanton destruction of churches and monasteries and the (to him) heaven-sent windstorm that destroyed an English fleet in the Firth of Forth.[116] His account provides remarkable insights on the persistence of the trauma of such events for over two generations after the actual physical damage was inflicted. Although similar devastating invasions were conducted in Lothian and Teviotdale by John of Gaunt in 1384 and Richard II in 1385, such terror tactics thereafter ceased to be a characteristic of Anglo-Scottish warfare until their reapplication in the sixteenth century.[117] Wilful ecological destruction, however, continued to be deployed as a weapon in civil wars and private feuds in Scotland for the remainder of the Middle Ages, leading to the wanton devastation of managed environmental resources on a local level from which it would have taken years if not generations to recover.

Famine II: 'mortality of men through hunger and pestilence, unheard of in our times'

In the mid twentieth century, economic and social historians were vaguely aware of the slackening of the pace of population growth towards the end of the thirteenth century and then the beginnings of a

112 Bower, *Scotichronicon*, vol. 7, p. 145. Andrew of Wyntoun, *The Original Chronicle*, vol. 6, p. 132, describes it thus: 'Sa gret a hungire wes neir þat steid / That mony folkis for falt were deid. / [Thai] said als þat a carll þarby / Wald mak nettis and set quayntly / Childer and woman for to ta, / And cum þan on þaim, and þaim sla, / And eit þaim all, þat he get mycht. / Criste Cleke to name he hecht / That laithfull lif lang hantit he, / Quhill waist but folkis wes þe cuntre.'

113 MacInnes, 'Shock and awe'.

114 Brown, *The Wars of Scotland*, p. 253.

115 Bower, *Scotichronicon*, vol. 7, p. 289.

116 Ibid., pp. 289–295.

117 The key episode in this resurgence in intentional ecological destruction was the war of the late 1540s that was intended to force the Scots into negotiations for the marriage of Queen Mary to the Prince of Wales. See M. Merriman, *The Rough Wooings: Mary Queen of Scots 1542–1551* (East Linton, 2000).

South-eastern Angus from Turin Hill, Angus. Still one of the richest agricultural zones of Lowland Scotland, this fertile and densely populated region was a major theatre of warfare in the first half of the fourteenth century, subject to punitive raids by English armies and devastated through the scorched earth tactics of the Scots.

downward trend in European population levels towards the end of the first quarter of the fourteenth century. Heavily influenced by contemporary political concerns about global population growth and food supply, which finally gained widespread publicity beyond the academic community through the 1972 publication of *The Limits to Growth*,[118] researchers began to examine the potential factors that lay behind these trends. Most were quick to see in the record from medieval Europe and England a Malthusian inevitability in historical evidence for overpopulation and overexploitation of the resources of productive land. In line with the thought of early nineteenth-century social and economic theorist the Rev. Thomas Malthus, they saw what they interpreted as the increasing pressure of human population outstripping the ability of the agricultural system of the day to sustain its relentlessly rising numbers. As Malthus had proposed, this had resulted in systemic overload and ultimate failure, leading first to malnutrition and increased mortality, and then to inevitable demographic collapse. That was the theory, but precise dates and causes were the subject of speculation. In one widely read university text, for example, the great French historian Georges Duby mused about a correlation between a series of bad harvests in England between 1317 and 1319 and the beginning of a population decrease there, but he could not definitively establish causation.[119] In another hugely influential study, the English social and economic historian Michael Postan articulated a collapse theory that dominated academic thinking for a generation.[120] He expressed it in terms of a reversal of the seemingly relentless expansion of agri-

118 D.H. Meadows et al., *The Limits to Growth: A Report for the Club of Rome's Project on the Predicament of Mankind* (Falls Church VA, 1972).

119 G. Duby, 'Medieval agriculture 900–1500', in C.M. Cipolla (ed.), *The Fontana Economic History of Europe* (Glasgow, 1972), pp. 175–220 at 182.

120 Postan, *The Medieval Economy*.

culture and population in England, which he saw as possibly dating from the 1320s. In what has since become known as 'the Postan Thesis', he linked the turning of the tide to a combination of overpopulation and overexpansion of settlement and cultivation, worsened by poor manuring techniques. Most historians of medieval Britain and Europe concurred with Duby and Postan, however, that bad though this Malthusian check might have been, the real collapse in population followed the first Black Death epidemic of 1346–53. Indeed, the early fourteenth-century situation was viewed as only a gradual decline or slight check, insufficient to relieve the real pressure within a system which struggled forward on a knife-edge of subsistence until the massive corrective of the mid fourteenth-century pandemic.

Although the Postan Thesis dominated discourse in terms of its broader argument and theories about the causes of the incipient crisis, its more minimalist view of earlier fourteenth-century decline has been greatly modified over the last fifty years. Through the 1970s and 1980s, new research shifted both appreciation of the scale of the crisis in the later 1310s and also its causes.[121] Then, in 1996, William Chester Jordan drew together this newer and more refined, data-driven scholarship and the fruits of his own research in a ground-breaking study of what he began to term the 'Great Famine'. When coupled with the identified weather-related effects of Neville Brown's 'Dantean Anomaly', a smoking gun was finally revealed as both agent and vector of catastrophe. Our understanding of the chaos wrought by

121 The nature of the crisis and its short- and long-term consequences have been highlighted – and questioned – by major studies such as I. Kershaw, *Bolton Priory: The Economy of a Northern Monastery, 1286–1325* (Oxford, 1973), and the advance of a new thesis of harvest failures and supply crises in I. Kershaw, 'The Great Famine and agrarian crisis in England 1315–1322', *Past and Present*, 59 (1973), pp. 3–50, reinforced through close examination of the experience of peasant communities revealed in the records of major estate owners like Christchurch Canterbury or the bishops of Worcester by scholars such as Christopher Dyer, Mavis Mates or Jack Ravensdale.

Graveyard excavation in 1986, Whithorn Priory, Whithorn, Dumfries and Galloway. Inhumation burials in the cemetery revealed detailed evidence for the poor physical state, health and wellbeing of the medieval population, who had lived through the bad times of the fourteenth century.

climate change on the people of medieval Europe was transformed.[122] Jordan's Europe-wide perspective created for the first time an overview of the extent of population decline and the key drivers behind it. Rather than a straightforward Malthusian tragedy in which overpopulation was reduced to levels that the over-colonised and overexploited land could sustain, the finger of blame was pointed at a run of bad weather that delivered successive harvest failures and consequent famine between 1315 and 1322. So much was uncontested, but what was greatly debated was the long-term outcome of the protracted dearth and hunger. Much research suggested a rapid 'bounce back' rather than that rural society remained 'unrelievedly mired in a trough from 1322 to 1348'.[123] Indeed, some historians went so far as to suggest that the population drop and experience of the Great Famine led to an improvement in the diet of agricultural labourers in the quarter century down to the Black Death.[124] Importantly, however, Jordan highlighted that although they and their parents may have enjoyed better diets after 1322, the children of the famine years had suffered nutritional deprivation that left them with compromised immune system development from which they never fully recovered. The children who lived through and survived the famine of 1315–22 became the adults who fell victim to the plague a generation later.[125]

In recent historiography, the scale and impact of the 1315–22 event have been accepted as fundamental contributors to the social and political destabilisation of the major arable-based economies of north-western Europe that occurred at this time. For John Aberth, the Great Famine was 'easily the most severe subsistence crisis of the late Middle Ages – and perhaps in all of medieval history', witnessing mortality rates of between 10 and 15 per cent across England and Europe, some two to three times the number of mortalities in a 'normal' year.[126] It is unsurprising that such protracted hardship delivered an increase in social tensions. Aberth, however, is one of several scholars who have refined the original and quite blunt crop-failure-and-starvation model. He

122 W.C. Jordan, *The Great Famine: Northern Europe in the Early Fourteenth Century* (Princeton, 1996).

123 Ibid., pp. 184–185.

124 C. Dyer, 'Changes in diet in the late Middle Ages: the case of harvest workers', *Agricultural History Review*, 36 (1988), pp. 21–37; Jordan, *The Great Famine*, pp. 185–186.

125 Jordan, *The Great Famine*, p. 186.

126 J. Aberth, *From the Brink of the Apocalypse: Confronting Famine, War, Plague and Death in the Later Middle Ages*, 2nd edition (Abingdon, 2010), pp. 16–17.

has highlighted the likelihood that the diseases which accompany acute famine and the physiological side effects of starvation were more likely killers than starvation itself. Aberth has also brought renewed focus to the steps taken by local and royal administrations to alleviate the crisis, building on older work by scholars like Ian Kershaw. These steps ranged from failed attempts at price control to measures intended to curb the activities of hoarders and speculators.[127] They were far from innovatory, basically repeating mechanisms employed since Greek and Roman times and which continued to be deployed for centuries to come. Identifying scapegoats to blame for shortages provided some release of stoked-up social tension, but, more generally, as the crisis rolled on for successive years it was accompanied by rising disorder, political tensions within the ruling community and an increase in brigandage and vagrancy. As social norms were increasingly strained, it is unsurprising that numerous stories of abnormal behaviour circulated widely across the regions, including the consumption of normally shunned or taboo food, even cannibalism.[128] While the cannibalism stories tend to follow a series of common forms across Europe and are probably largely apocryphal, the unclean or unnatural foods to which they refer seem to have been a more probable and common resort.

The positive gloss given by some scholars to post-famine economic conditions for those at the bottom of the social spectrum has been challenged by Bruce Campbell. His analysis of the 1315–31 period indicates English harvest failures in five out of fifteen years. With net shortfalls in excess of 25 per cent, it was the worst sequence experienced since the failures of the late 1250s.[129] In Campbell's model, agricultural output shrank dramatically and unemployment soared against a backdrop of rapidly rising food prices and contracting wages. Indebtedness, mortgaging and foreclosure rocketed, contributing to the mounting political tensions and social disorder that toppled King Edward II from his throne in 1327.[130] It was not until the good harvests of the mid 1330s that the decline in peasant economic circumstances eased, breaking a cycle of crop failures, crisis-selling of property and descent into the ranks of poorly paid, landless labourers. Philip Slavin has taken this discussion further and emphasised the interconnectedness of anthropogenic and non-anthropogenic factors in creating the social and economic environment in which a crisis triggered by climate change wrought havoc with England's rural economy and society.[131] Although it was touched on by Campbell, it is Slavin who has drawn together the impact of the harvest failures of the 1310s, '20s and early '30s with the great cattle epizootic that struck England in 1319 (discussed in Chapter 10) and palaeopathological discussion of the effects of childhood malnutrition from both the famine and the shortage of milk proteins on the generation who were adults a quarter century later.[132] His assessment of English manorial records is that the south-eastern parts of the British Isles lost between 15 and 20 per cent of their population during this period.[133] When we put this into perspective against modern famines, such that which struck Ethiopia in the early 1980s, the level of suffering and

127 For a contemporary view of Edward II's failed price control measures among other things, see Childs and Taylor, *The Anonimalle Chronicle*, pp. 89–91.

128 Aberth, *From the Brink of Apocalypse*, pp. 16–30. For such actions during the 1330s in Scotland, see above, p. 318.

129 Campbell, *The Great Transition*, p. 191.

130 Ibid., pp. 192–198.

131 P. Slavin, *Experiencing Famine in Fourteenth-Century Britain* (Turnhout, 2019).

132 Ibid., pp. 243–289, 367–369.

133 Ibid., p. 366.

loss behind these numbers becomes apparent and the true horror of this previously little-researched episode is revealed. In Ethiopia, approximately 3 per cent of the population perished from hunger and the diseases of malnourishment, with a further 20 per cent displaced as refugees. Seen in those terms, the 'relatively minor check' of the Postan Thesis becomes a catastrophe of epic proportions.

Did Scotland experience similar suffering in the Great Famine? Slavin has rightly cautioned against simply extending the models for England to Scotland (and Wales and Ireland) or falling into an environmentally deterministic trap of seeing climate as the universal trigger of cataclysm.[134] Simply because Scotland experienced the same climate change and weather impacts as the rest of north-west Europe does not mean that it suffered identical consequences. Except perhaps in Lothian and the Merse, where wheat was an important arable crop, Scotland had avoided the population growth, agricultural expansion and wheat-based arable dependency seen in England. It also had a more varied basis for its rural economies, with a broader spread of more weather-resilient grain types, greater dependency on livestock throughout its upland zones and a greater contribution of fish to both economies and diets.[135] These circumstances did not mean that Scotland was immune to the weather instability of the early fourteenth century, for both arable and livestock regimes, as well as the fishing practices that developed during the MCA, were exposed to the hammer blows of climate change. But did Scotland's food shortages lead inevitably to famine on an English scale? Slavin's answer is an emphatic no: while Scotland certainly suffered food shortages, there is an absence of evidence for famine, except perhaps in the south-east Lowlands.

Slavin's view is based on another critical factor, the interplay of population levels and the rural economic regime. There is widespread consensus still that England was a classic Malthusian society in the sense that it was at or beyond the limits of sustainable population relative to its agricultural productivity. Consequently, it was particularly vulnerable to environmental and biological disasters.[136] Scotland was no more economically advanced than England, but its population numbers and density were lower in most regions outside the south-east, even after the better times of the thirteenth century. For the most part, it may have been non-Malthusian in the sense that its population had not outstripped the capacity of the land to sustain it, even during the crisis years of the early fourteenth century. So, when harvests failed, livestock died and fish shoals changed their migratory patterns as ocean surface temperatures fluctuated and delivered deluge or drought, Scots experienced food shortages – and the hardship and possible deaths those brought – but they avoided the slide into famine that the economic backwardness and fragility of England's rural society delivered there.

That view must be tempered by the fact that we lack Scottish chronicle or record evidence for this period that is not preoccupied with the war with England. That warfare, as discussed above, brought short-term but significant dislocation of the agricultural economy of its main theatres in Scotland and northern England. By 1315, however, northern England and Ireland were the primary conflict zones and there the weather impacts on the military campaigns are well documented.[137] Dire accounts

134 Ibid., pp. 15–16.

135 Oram and Adderley, 'Lordship and environmental change', p. 79; Oram, 'The worst disaster', pp. 231–232.

136 Slavin, *Experiencing Famine*, 19.

137 McNamee, *The Wars of the Bruces*, offers the most comprehensive analysis of the conflict. The essays in S. Duffy (ed.), *Robert the Bruce's Irish Wars* (Stroud, 2002), explore specific aspects and episodes in detail. The period is summarised in Brown, *The Wars of Scotland*, chapter 9.

from Ireland and southern England are echoed in northern parts only by the well-informed Lanercost chronicler, who commented on crop failures and famine in England and Scotland, noting that in 1316 there was 'mortality of men through hunger and pestilence, unheard of in our times', and by Walter Bower, who drew on an otherwise unknown early fourteenth-century Scottish source to report a severe winter in 1321–2, 'which was a sore trial to men and killed off nearly all their animals'.[138] Lanercost's extension of deaths arising from the 1315 harvest failure to Scotland – presumably the southern districts with which there was most contact – is plausible in that the chronicler's information network extended into the Scottish region of greatest social and economic similarity to England. It is important to note of Bower's 1321–2 account that he went on to report Scottish raids into northern England in January 1322 at the beginning of the famine there but did not extend his reference to famine as occurring in Scotland also.[139] Indeed, in his following section, which reported Edward II's retaliatory raid into Lothian in August 1322, Bower described the Scottish king as instructing the removal of all cattle and foodstuffs from the path of the advancing enemy to deprive them of opportunities to replenish supplies through foraging. The implication is that there was adequate food in southern Scotland in contrast to the dire conditions in England. That no subsistence crisis is recorded as affecting the Scots over this whole period is supportive of Slavin's reading of the situation: Scotland faced food shortages but not famine.

• • •

The downbeat picture painted by this review of evidence for the experience of many Scots in the century after the death of King Alexander II in 1249 sits uncomfortably with the traditionally positive projection of the so-called Alexandrian golden age and the standard 'war and recovery' focus of mainstream historiography.[140] It is in many ways somewhat ironic, for this era is one of the most intensely studied in the history of medieval Scotland, and yet most discussion of the period reveals a massive blind spot in our awareness of the complexity of circumstances that contributed to bringing the kingdom of Alexander III so close to dissolution and conquest in the early 1300s. But the mainstream narratives, driven for the most part by political and more especially military history, have followed agendas established in Scottish historical writing as long ago as the later fourteenth century. It is, moreover, an often inspiring narrative of triumph over adversity in the face of seemingly overwhelming odds, stalked by heroic and villainous figures who stand high against a dramatic backdrop of tragedy, duplicity, treachery and bloodshed. With heroes and villains of the proportions of William Wallace, Robert Bruce or Andrew Murray, and of Edward I, John Comyn or Hugh Cressingham, what need is there to further complicate matters by introducing climate change, weather impacts and pathogens as a new set of threats lurking in the shadows? What is more, the impact of these new 'villains of the piece' is less immediately visible in the parchment record of the Scottish experience, especially during the span of the Wars of Independence, and the quite detached – almost abstract – manner in which modern scholars have introduced them into the equation has failed to achieve the impact of yet another retelling of the Bannockburn story. Precipitation and temperature models, percentage mortality rates and estimates of grain yields might produce a far grimmer picture than ever composed by the likes of Barbour, Wyntoun or

138 *Chronicon de Lanercost*, p. 233; Bower, *Scotichronicon*, vol. 7, p. 11.

139 Bower, *Scotichronicon*, vol. 7, p. 11.

140 As critiqued in Oram, 'Away was sons of alle and brede'.

Bower, but they lack the immediacy, the visceral energy, of the epic tales of individual human endeavour to which we, as humans, can relate personally.

Does taking the environmental turn in any way diminish the significance of this era? As the material reviewed above should have illustrated, the environmental historical data and the nuance it adds to the discussion enhances our understanding of the human experience across this critically formative century in Scotland's national development. Awareness of the increased precarity of the period after 1256 makes the remarkable achievements of Alexander III's reign all the more notable as demonstrations of resilience in uncertain times. The investments visible in the great building programmes of Scotland's secular nobility and ecclesiastical leadership were founded on an agricultural regime whose diversity provided some safeguard against the vulnerability engendered by overreliance on one dominant crop type or animal species. Rain might cause a harvest to fail and regions sustained by a primarily arable economy might have faced hardship and occasional subsistence crisis, but grass continues to grow in even the wettest years, provided daytime temperatures rise above the 6°C threshold for grass growth. Sheep and cattle provided security across those years of harvest failure, for their hides, wool and woolfells were in demand internationally, bringing cash to even the smallest of livestock-owning peasant farmers. When Hugh Cressingham seized that wool income in the face of an incipient subsistence crisis in 1296–7, the result was a rebellion that shook loose English control of Scotland.

An environmental turn, then, provides nuance and adds depth and detail to the otherwise superficial list of factors influencing human behaviour in this era. We can see rebellion as a response driven by desperation, where the alternative was accepting at best a drop in living standards and at worst impoverishment, starvation and death either at home or abroad in the service of a foreign king. We can see too how the continued destabilisation of weather patterns played a critical role in affecting the progress of warfare across the whole of the British Isles in this era, from the failure of the Scottish invasion of Ireland in 1318 and the rising tide of hostility to Edward II in England during the famine years of 1315–22, to the scattering of Edward III's fleet in 1356. The targeting of the environmental resources of enemy lords was perhaps more critical to the success of military campaigns than the chancy outcomes of a battle, which might swiftly be overturned. Warfare exposed the superficiality and fragility of human settlement patterns and agricultural regimes, where even brief breakdowns brought the quick beginnings of reversion to something less tamed or controlled. The insights offered by the environmental turn make clear both the vulnerability of a society founded on agricultural exploitation during a period of protracted transition from a stabler past regime, and also the resilience and responsiveness of such a society to both environmental and anthropogenic threats. That resilience, we shall see in the final chapter of this volume, was tested by challenges across many fronts, from livestock murrains to shortages of essential resources such as fuel and timber, and was stretched to the limits by the return of epidemic disease in the second half of the fourteenth century.

CHAPTER 10

The Worst Disaster? War, Famine, Disease and Death in the Long Fourteenth Century

I that in heill wes and glaidnes,
Am trublit now with gret seiknes,
And feblit with infirmitie:
Timor Mortis conturbat me.[1]

Deteriorating weather across the last decades of the thirteenth century and the destabilisation and destruction inflicted by warfare in the first half of the fourteenth were but two dimensions of a multifaceted and interconnected suite of challenges facing Scottish society across the 150 years to 1400. As explored in Chapter 9, traditional ideas about a golden age that had ended with the death of King Alexander III and the re-establishment of a previous equilibrium after the traumatic upheaval of the Wars of Independence have been contested and recalibrated through more recent analysis of the Scottish experience of the waning of the MCA. A further challenge to traditional views has come through reappraisal of the range of consequences for agricultural regimes that the Dantean Anomaly brought to the northern British Isles, and the less dramatic but cumulatively impactful changes in the North Atlantic climate regime either side of that traumatic episode. Although we have no reason to believe that Scotland experienced a sustained crisis such as struck south-eastern England in the period 1315–22, with its projected mortality rates of 15–20 per cent, we can also be sure that it did not escape unscathed. That discussion of the impact of the Dantean Anomaly also introduced us to the probability that the more limited expansion of agriculture that occurred in Scotland during the better times from the later eleventh to the middle of the thirteenth century, and the greater diversity in Scotland's range of staple products or modes of agricultural exploitation, had perhaps helped to avoid a 'push to the margins' akin to that which occurred in England and so had established greater community resilience. More resilient, however, does not mean immune from the consequences of weather extremes, from the impact of cumulative anthropogenic pressure on the Scottish environment or from the effects of biological agency on crops, flocks and herds. Indeed, the periods of increased

1 William Dunbar, 'Lament for the Makaris. Quhen he wes seik', in W. Dunbar, *The Poems of William Dunbar*, ed. J. Small, volume 2: *Poems* (Edinburgh, 1893), p. 48.

precipitation and warmth brought new threats from parasites, fungal infestations and pathogens, to human, animal and plant life.

In this final chapter, we will explore the changing structures and patterns of rural exploitation across Scotland and the imprint made by the different agricultural and fishing practices that were in place by the fourteenth century, from the Scottish Borders to the Northern and Western Isles. Central to this discussion will be consideration of the responses of the population to the failure of the traditional environmental knowledge (or TEK) that had been built up through the more benign era of the MCA. How did communities that had grown in dependence on the sea fisheries confront the challenges of changing oceanic conditions? And how did those dependent on cereal cultivation ensure viable harvests and stored grain yields in a time of increased rainfall and often colder conditions? This, as one early seventeenth-century English ballad put it, describing the social, economic and political conflicts of the day, was a time of 'the world turned upside down'.[2] Old certainties no longer held, methods and practices that had emerged in better times were no longer effective and the confidence and stability that had flowed from the previous trustworthiness of TEK had given way to uncertainty and disturbance. But some things did not change and stresses already evident by the waning of the MCA remained as persistent challenges, worsening as the weather deteriorated. Access to building materials continued to pose problems in areas where substantial woodland had been lost millennia earlier, exacerbated by mounting demands from a still growing population that no amount of careful coppicing could supply. And fuel too could no longer be taken for granted as an inexhaustible abundance, at least in lowland areas where centuries of inroads into peat mosses had depleted accessible reserves, and where spring and summer rainfall hindered the winning of adequate supplies to last through the poorer, colder winters. Life, as we shall see in the following discussion, was far from joyous for the bulk of the Scottish population even before the eruption of war, civil and foreign, at the close of the thirteenth century.

Woodland wasted?

In Chapter 8 we saw that shortage of woodland resources close to many of the larger eastern and south-eastern burghs was already acute by 1200. Wood had given place to peat as the primary fuel in many districts from Shetland to Galloway long before the twelfth century. Some wood, mainly as bundles of brush, broom and heather, continued to be used for firing ovens and kilns, as charcoal in specialist crafts and as kindling for peat and coal fires for the rest of the Middle Ages, leading to specialist management of growing trees.[3] The steadily tightening levels of lordly control over woodland resources left easements (traditionally established rights) to gather dead wood and windfall as the sole rights enjoyed by the bulk of the peasantry in otherwise privatised assets. Where no such easements existed, these commodities were still gleaned and sold to local consumers. It is dead wood and windfall gleanings from the king's land in Jed Forest that we see gathered for sale by his local officers in 1290.[4] The use of such gathered materials for firewood was largely practised by those on the economic margins; primarily wood-

2 T.J., a Well-Wisher to the King, Parliament and the Kingdom, *The World Turn'd Upside Down, or A Briefe Description of the Ridiculous Fashions of These Distracted Times* (London, 1646).

3 Oram, 'Environmental change', pp. 215–216; Gilbert, 'Place-names and managed woods'; J.M. Gilbert, 'Medieval woodland management in southern Scotland', *Transactions of the Dumfries and Galloway Natural History and Antiquarian Society*, 3rd series, 86 (2012), pp. 77–117; Gilbert, 'Woodland management in medieval Scotland'.

4 Stuart and Burnett, *The Exchequer Rolls of Scotland*, vol. 1, p. 43.

fuelled fires were probably reserved for displays of conspicuous consumption in mostly elite contexts long before that date. At the end of the fourteenth century, throughout most of lowland Scotland, the burning of large billets of wood was probably an established marker of social status.

Where wood was steadily declining in availability, peat had become the principal source of thermal energy for households across a broad social spectrum. Even those elite, wood-burning households only burned wood in the visible and highly symbolic hearths in halls and chambers while using peat in other contexts. Increasingly unpredictable summer weather, however, rendered it difficult to dry adequate quantities of peat for winter needs over this same period. That was perhaps a major factor in an increasingly evident shift away from dependence on peat in areas where coal was available. Lords with access to coal on their property, even of relatively poor quality, moved to mining it for their own use. Much was used in estate-controlled saltpans or sold to nearby pan operators, who were bulk consumers.[5] Although mining in the later thirteenth and fourteenth centuries was restricted to the accessible coal seams on either side of the Forth basin and possibly on Melrose Abbey's properties in Kyle, it was at a scale sufficient in those areas to reduce dependence on a fuel type whose availability was subject to weather conditions.[6] At St Andrews, it was both the shortage of peat locally and the deteriorating weather conditions at the turn from the fourteenth to the fifteenth century that led to a decision by the canons of the cathedral priory to commence the mining of the shallow coal seams on Drumcarrow Craig, to the west of the burgh.[7] Except in a few possibly specialist cases, discussed in the volume *Where Men No More May Reap or Sow*, the later Middle Ages in Scotland do not seem to have seen the development of a fuelwood coppicing tradition, as distinct from coppicing for building and wickerwork materials, in response to these supply pressures.

Double fireplace in the Duke's Hall, Doune Castle, Stirling. This remarkable double-hearthed fireplace is designed for the burning of large billets of timber held in place by firedogs or andirons. Such conspicuous consumption of fuel was a status indicator in the later fourteenth century, when this fireplace was built in the new castle of Robert Stewart, duke of Albany, at Doune.

It was not only firewood that was becoming scarce in the hinterlands of Scotland's east coast and south-eastern burghs. Building timber, as we have seen in the case of Perth and other expanding towns, was in short supply even before 1200, with shortfalls increasingly made up through timber imports from the Baltic region. What managed timber there was on the burghs' common land was reserved for the use of full burgesses. Those who did not hold that status probably had to resort to scavenging, recycling and gleaning from informal sources. If the use of multiple tree and scrub species – including apple,

5 Oram, 'The salt industry in medieval Scotland'; Oram and Whatley, 'A brief history of Scottish salt'.

6 Oram, 'Environmental change', pp. 220–221.

7 Lindsay and Cameron, *Calendar of Scottish Supplications*, pp. 222–223.

Multi-trunked, out-grown coppiced oak, Dalkeith Park, Midlothian. Although this oak was itself probably last harvested as coppice in the sixteenth century, it grows from a stool several centuries older. Such trees illustrate the careful reservation of growing timber in the managed parks possessed by members of the greater nobility in the later Middle Ages.

hawthorn, pear and rowan, as well as the more common hazel, willow, alder and birch – for house building at Kirk Close in Perth is a measure of what was locally available to non-burgesses, it seems that hedgerow cuttings and gleaning from informally managed woodland were necessary to make up what they were denied from the burgh's managed willow and hazel coppices.[8] Oak is noticeable by its scarcity in both urban and rural excavations, possibly because its high price, exacerbated by its limited availability, made it accessible to only a select few. Most oak timbers recovered from urban archaeological sites had been re-used, indicating a high salvage value for this durable wood.[9] Rural contexts from eastern and south-eastern Scotland also show signs of the limited availability of large timbers and a reliance on probably coppiced hazel, willow, birch and alder, but also on probably foraged wood like wild cherry and rowan. In the late thirteenth- and early fourteenth-century houses at Springwood Park near Kelso, ash

8 Crone and Watson, 'Sufficiency to scarcity', pp. 66–67; H. Murray, 'The buildings', in *Perth High Street Archaeological Excavation 1975–1977*, fascicule 1: *The Excavations at 75–95 High Street and 5–10 Mill Street, Perth* (Perth, 2012), pp. 127–198 at 132–133. For evidence of coppice management, see A. Curteis, C.A. Morris and N.Q. Bogdan, 'The worked wood', in *Perth High Street Archaeological Excavation 1975–1977*, fascicule 2: *The Ceramics, the Metalwork and the Wood* (Perth, 2012), pp. 223–322 at 223.

9 For evidence of wood re-use, see Curteis, Morris and Bogdan, 'The worked wood', pp. 295–298.

Glen Lichorn, Perth and Kinross. The mixed woodland in the valley at the edge of the Braco estate is a remnant of the resource contested in the mid thirteenth century between the monks of Lindores and two local knights. Records of wood use elsewhere on the Lindores estate indicate that it was the birch, alder and willow in the wetter areas rather than the pines on the drier ground that were most in demand from their peasant tenants.

seems to have been used for the main structural timbers, possibly with some elm and a limited amount of oak; the walling was most probably of hazel, willow and alder wattling.[10] Even aristocratic housing in late thirteenth- to early fourteenth-century rural contexts seems to have made use of what was available locally. At Castlehill of Strachan in northern Kincardineshire, large structural timbers were of oak but the main walls were of alder wattle, probably from coppiced trees along the nearby Water of Feugh.[11] Here, however, it does not seem to have been a matter of shortage, for oak woodland was still extensive in the Dee Valley just north of Strachan,[12] and should alert us to the likelihood that the use of coppiced wood in wattle-and-daub construction was a matter of choice rather than necessity and, indeed, a question of cultural preference.

Oak, however, was clearly considered a prestige material in districts where large numbers of substantial oak standards were rare outside managed royal

10 For ash use, see P. Dixon, 'A rural medieval settlement in Roxburghshire: excavations at Springwood Park, Kelso, 1985–6', *Proceedings of the Society of Antiquaries of Scotland*, 128 (1998), pp. 671–751 at 704.

11 H. Murray, 'Appendix 1: The building reconstruction', in P. Yeoman, 'Excavations at Castlehill of Strachan, 1980–1981', *Proceedings of the Society of Antiquaries of Scotland*, 114 (1984), pp. 315–364 at 346–347.

12 The oakwood at Trustach on the Dee only 4km north of Strachan was a source of timber for building work at Arbroath Abbey, nearly 60km away, in the second quarter of the thirteenth century: Innes, *Arbroath Liber*, vol. 1, p. 91.

and baronial forests and woods. Grants of oak from these preserves were used both as gifts and as compensation for losses, especially during the English occupation of Scotland to 1307. In 1291, for example, Edward I ordered four oaks from the forest of Clackmannan to be given to the Knights Templar for their preceptory at Balantrodach in Midlothian.[13] The abbot of Jedburgh petitioned in 1303/4 for twenty oaks from the Forest of Plater in Angus to repair damage inflicted by the English army on his abbey's dependent cell at Restenneth near Forfar, with a further twenty from the forest of Clackmannan going to the canons of St Andrews for similar reasons.[14] A princely award of 200 oaks from the royal forests of Darnaway and Longmorn in Moray was made to Sir Reginald Cheyne of Duffus for his service in Edward's administration in northern Scotland.[15] But it was not only large structural timbers for which access to wood from royal forests was sought, as illustrated by Robert I's grant to Restenneth of the right to take wood (*meremium*) and understorey brushwood (*subboscum*) from Plater for making carts, ploughs, harrows and other wooden tools and equipment, not for building timber or firewood.[16] Access to woods for such purposes could be contested between those looking to safeguard the resource and those who believed they had an easement or a gifted right to exploit it. In 1256/7, for example, the monks of Lindores Abbey complained that their tenants at Feddal in Strathallan had been denied access to the woods in nearby Glen Lichorn and at 'Curelundyn' to cut timber for their building and agricultural needs. Unable to get a negotiated settlement, the monks resorted to calling down the full weight of excommunication on Robert, brother of the steward of the earl of Strathearn, and another Strathearn tenant, Joachim of Kinbuck, who were obstructing them.[17] Robert and Joachim seem to have been fulfilling their responsibility to prevent their lord's woods from being wasted but were possibly acting beyond their authority in the way they went about it.

What such grants, disputes and royal directives issued to foresters locally help to demonstrate is that woodland, perhaps especially in royal preserves throughout southern, eastern and north-eastern Scotland from Galloway to Moray, was still being managed at the turn from the thirteenth to the fourteenth century. Management, however, does not necessarily mean that they were being exploited in what we would nowadays regard as a sustainable manner. The one-sided record of the disputes tends to present the complainers as unjustifiably denied access to a resource to which they had legitimate rights, but the defenders can be seen as fighting – and often losing – a rearguard action to limit their inroads. It is in the context of the attempted preservation of a truncated reserve that we should see some of the enclosures that appear in the record in the earlier 1300s. In some royal forests in the first quarter of the fourteenth century, a decision was made to create smaller, protected areas within the afforested part and give over the zone outside that to other uses. Thus, in 1319 Robert I gave the burgesses of Aberdeen freedom to exploit the lands of the forest of Stocket to the west of the burgh for building materials and fuel and for cultivation. His generous grant specifically excluded 'the wood of Stocket' ('boscum de Stoket') and also reserved his rights of vert and venison in the forest generally.[18] Here, we can see

13 Bain, *Calendar of Documents*, vol. 2, no. 519.

14 Ibid., nos 1428, 1704.

15 Ibid., no. 1736.

16 J. Maitland Thomson (ed.), *Registrum Magni Sigilli Regum Scotorum: The Register of the Great Seal of Scotland*, new edition, volume 1: *a.d. 1306–1424* (Edinburgh, 1882), appendix 1, no. 30.

17 Dowden, *Lindores Charters*, nos 111, 112.

18 Gilbert, *Hunting and Hunting Reserves*, p. 130; *Regesta Regum Scottorum*, vol. 5, no. 158.

that there was already a recognised *boscum* within the forest before 1319, suggesting that burgess and peasant inroads into the afforested area had already significantly reduced the woodland area but that this core block was considered worth reserving for its hunting and wood-timber value. Similar reservation of specific areas, often of enclosed parks, occurred at most of the royal forests in the wider northeast, from Cowie around Stonehaven to Boyne and Enzie between Banff and the Spey.[19] An opposite course of action was taken in the Forest of Plater in Angus, which was still sufficiently wooded in the 1320s for the king to make his award of materials from there to the canons of Restenneth and in the early 1330s to provide cover for Andrew Murray's warriors during the winter campaign against English garrisons in the region. Although in 1360 King David II confirmed Scone Abbey's rights to cut wood in Plater, granted to them in the twelfth century by King Máel Coluim IV, and in 1366 he granted one Murdac Rhynd over fifty acres of the forest to be broken by him into arable cultivation,[20] there was an effort to retain the wider forest area as a royal preserve. Thus, when King Robert II granted the office of forester of Plater to Sir Alexander Lindsay in 1375, he stipulated that the forest area was to be preserved 'in its nature as now is' and forbade further assarts for arable or alienations to other lords.[21] He met with little long-term success, but it was only in the later fifteenth century that the crown finally granted away what remained of Plater Forest.[22]

David II's 1366 grant of assarting rights in Plater appears odd, given that the depopulation effects of the first and second plague pandemics, discussed later in this chapter, should have reduced the pressure for arable expansion. Elsewhere, there are signs that the third quarter of the fourteenth century saw some recovery in lowland woods that had been subject to heavy inroads in earlier decades. In the Forest of Clackmannan, for example, which had been exploited intensively as a source of oak since at least the twelfth century, new oak growth during that period possibly saw the regeneration of trees of sufficient girth for use as roof timbers in nearby Alloa Tower in the early 1400s.[23] These, however, were young, fast-growth trees no more than fifty years old, not the great oak standards needed for major construction work, and their use perhaps indicates the level of depletion at Clackmannan after three centuries of felling. But this apparent record of ravaged woodland should not be applied across the board in Scotland, for elsewhere it is clear that old-growth trees remained in substantial numbers through the fourteenth century and beyond. Darnaway in Moray, for example, from which Reginald Cheyne had received his 200 trees before 1306, and nearby Longmorn Forest, which was probably the source of the 358-year-old tree used for building in Elgin in 1301, provided the oaks used in the magnificent hammerbeam roof of the great hall at Darnaway Castle, erected shortly after 1387. One of the trees used at Darnaway was over 418 years old when felled.[24]

Darnaway was on the edge of the Highlands,

19 Gilbert, *Hunting and Hunting Reserves*, p. 130; *Regesta Regum Scottorum*, vol. 5, p. 46 and nos 220, 223, 248, 261.

20 *Regesta Regum Scottorum*, vol. 6, nos 238, 325; Maitland Thomson, *Registrum Magni Sigilli*, vol. 1, no. 257.

21 Maitland Thomson, *Registrum Magni Sigilli*, vol. 1, nos 618, 712.

22 Paul, *Registrum Magni Sigilli*, vol. 2, no. 1191.

23 T. Ruddock, 'Repair of two important early Scottish roof structures', *Proceedings of the Institute of Civil Engineers*, 110 (1995), pp. 296–307 (the date of the roof structure in question is in fact less clear-cut than suggested here); Crone and Watson, 'Sufficiency to scarcity', p. 81.

24 G. Stell and M. Baillie, 'The great hall and roof of Darnaway Castle, Moray', in W.D.H. Sellar (ed.), *Moray: Province and People* (Edinburgh, 1993), pp. 162–186; Crone and Watson, 'Sufficiency to scarcity', pp. 71, 72 plate 3.5, 76 plate 3.6, 80.

Great hall roof, Darnaway Castle, Moray. Some of the trees used in the construction of the roof in the late fourteenth century were between 400 and 500 years old when felled, reflecting centuries of careful conservation of major trees in what had become the royal forest of Darnaway.
(Photograph by Ian Keillar, late 1970s.)

within which, it has long been argued, extensive oak and pine woodlands have remained relatively untouched since prehistory. As established in earlier chapters, however, that view has little palaeoenvironmental or historical record evidence behind it. Doubt has been cast on the accuracy of the main sources cited in support of the vision of a medieval 'Great Wood' blanketing the Highlands. The principal source is the description of Highland forests contained in the *Chronica* section of the compendium of earlier materials gathered in John of Fordoun's late fourteenth-century chronicle. Much of that early material is derived from generic descriptions of uncivilised lands in works such as the seventh-century *Etymologia* of Isidore of Seville, glossed by rumours and reports of remote districts that Fordoun heard as a priest in Aberdeen.[25] What was being offered here was a standard Roman and Late Antique representation of the stereotypical abode of barbarian peoples, applied to a region inhabited by people who were culturally different from the lowland Scots of Fordoun's day. But we need to be cautious about utterly rejecting the notion of more densely forested Highland areas as just a literary device deployed to culturally 'other' the inhabitants of that region. Other forms of record evidence suggest that while there was no blanket of closed canopy forest spanning the region, large areas of woodland were present throughout the central and west Highlands.[26] Place-names imply that even the intensively settled, cultivated and grazed central Highland basin of the Spey contained large islands of woodland at least as extensive and widespread as the woods recorded and mapped there in the seventeenth and eighteenth centuries.[27] Those islands, however, were interspersed through areas of intensively managed arable and pasture, exploited as wood pasture and managed as reserves for the substantial timbers that were in demand for the roofing and flooring of the elite residences and churches of the region.

Given that multifaceted management, it is unlikely that any of those Highland woods were in any sense 'wildwood', pristine, naturally regenerating successors of the forests of the post-glacial maximum. Their timber resources, as already noted for the tenth and eleventh centuries, were providing materials for substantial building programmes in the Moray and Easter Ross lowlands. That earlier medieval use continued in the thirteenth century, illustrated by the reconstruction of the royal hall and

25 Smout, MacDonald and Watson, *A History of the Native Woodlands*, p. 41; John of Fordoun, *Chronica Gentis Scotorum*, vol. 2, p. 36.
26 This evidence includes the trade in deerskins (red and roe deer) and the pelts of pine marten and polecats, exported mainly through Inverness: Ditchburn, *Scotland and Europe*, p. 161.
27 Smout, MacDonald and Watson, *A History of the Native Woodlands*, p. 43.

West Highland galley, Keils church, Lochaline, Highland. The ability of chieftains throughout the Atlantic west of Scotland to continue to build vessels such as this underscores their reservation of suitable woodland to provide strakes, decking, frame and masts at a scale that reflected their status.

associated buildings at 'Rothenet' on the lower Spey, which involved what were probably locally sourced oak or pine boards and planks.[28] As in later centuries, the wooded valleys of north-eastern Inverness-shire and Easter Ross also continued to provide timber for royal building projects in the region. We see this in the enclosure around Inverness Castle and the construction of a *domus Scoticana* (Scottish-style house) and a wardrobe building there in 1264.[29] In the south-west Highlands, the galleys that provided the basis of military power in the MacDonald lordship of the Isles to the end of the fifteenth century are strong evidence for the availability of suitable timber sources in Argyll and the Inner Hebrides.[30] Much of the oak used in these vessels appears to have been obtained from the still extensive Atlantic oakwood that fringes mainland areas from Ardnamurchan round to Knapdale as well as inland waters like Loch Awe, which was exploited for galley-building purposes into the late seventeenth century.[31] Some of these were ships of great size, like the forty-oar galley to be provided in time of war by Colin Campbell from his barony of Ardskeodnish and Lochawe by the terms of his crown charter of 1315.[32] The trees for this vessel likely came from woods in southern and eastern Lorn, bounded by Loch Crinan and the basin of Loch Awe. A 1354 agreement between John MacDonald of Islay and John MacDougall of Lorn included permission from the former for the latter to build eight twelve- or sixteen-bench galleys, presumably from oakwoods in MacDougall's territory in northern Lorn.[33] These were, of course, high-end construction projects and the necessary timber supplies were controlled by the greater lords throughout the region. But the ordinary inhabitants of the Highlands and Inner Hebrides had wood needs for their own houses and equipment, much of which continued to be met, as in earlier periods, through coppice management of the same suite of hazel, willow, birch and alder as supplied their lowland neighbours. The widespread tradition of basketwork or wattle for buildings and fences and the rich evidence for a diverse treen material culture is indicative of a land where wood was managed, conserved and exploited, and was not necessarily yet in short supply.[34]

Despite this substantial evidence for what we might regard as sustainable practices and the conservation of woodland resources, there remains a vision of this period as witnessing the final acceleration of clearance towards the nadir of Scotland's woodland coverage. Indeed, there are strong and persistent narratives according to which the fourteenth century witnessed major inroads into Scotland's forests, driven not solely by material needs or clearance for agriculture but by the whim and caprice of their privileged owners. Twentieth-century novels drawing on fantastic post-medieval accounts of woodland set deliberately aflame for the pleasure of infamous late fourteenth-century characters like Alexander Stew-

28 Stuart and Burnett, *The Exchequer Rolls of Scotland*, vol. 1, p. 14.

29 Ibid., pp. 13, 14. For *domus Scoticanae* or houses in *alpinatus mos* (Highland fashion), see R.D. Oram, 'Prelatical builders: a preliminary study, c.1124–c.1500', in R.D. Oram and G.P. Stell, *Lordship and Architecture in Medieval and Renaissance Scotland* (East Linton, 2005), pp. 1–25 at 8–9, where the pre-1400 palace of the bishops of Dunkeld is described.

30 C. Martin, 'A maritime dominion – sea-power and the lordship', in R.D. Oram (ed.), *The Lordship of the Isles* (Leiden, 2014), pp. 176–199.

31 McWhannell, 'The galleys of Argyll'; D.C. McWhannell, 'Campbell of Breadalbane and Campbell of Argyll boatbuilding accounts 1600 to 1700', *Mariner's Mirror*, 89:4 (2003), pp. 405–424.

32 *Regesta Regum Scottorum*, vol. 5, no. 46.

33 J. Munro and R.W. Munro (eds), *Acts of the Lords of the Isles 1336–1493* (Edinburgh, 1986), no. 5.

34 For Highland vernacular building, see Cheape, 'Every timber in the forest for MacRae's house'. For woodland produce and treen, see Smout, MacDonald and Watson, *A History of the Native Woodlands*, chapter 3.

Wind-thrown trees from Storm Arwen (November 2021), Camustone Hill, Monikie, Angus. Chronicle references suggest that scenes like this became more common as the storms gained in strength and frequency through the fourteenth century.

art, the 'Wolf of Badenoch', have been used to explain central Highland deforestation.[35] There is also unsubstantiated projection of recorded small-scale episodes of woodland destruction as an instrument of terror in times of war onto far larger canvases. There is a deeply ingrained popular view that orchestrated attacks on woodland – not just the felling of orchard trees as occurred at Wemyss, discussed in Chapter 9 – were deployed to inflict deliberate ecological damage during English invasions. Edward II, for example, is claimed to have sent 10,000 men to fell the woods of Jed Forest in the Borders in 1316, to prevent its use as a hideout by Scottish guerrilla fighters, while John of Gaunt's 1384 campaign into Scotland is said to have been provided with up to 80,000 axes to destroy the Borders' woodland.[36] These are fanciful stories but they are fixed deeply in local tradition as explanations for the creation of the largely treeless grassland landscapes that existed across most of the Borders region until the later twentieth century. Even if true, a single widespread felling event would not have destroyed the woodland permanently, for the largely oak broadleaf forests of the region would regenerate from seed and sprout from the stumps, unless the regrowth was subject to intensive grazing.

War, as discussed in the previous chapter, could be devastating in the short term but is unlikely to have been a significant factor in loss of woodland cover before the era of modern warfare from the First World War onwards. For widespread destruction that had wider ramifications, we should not underestimate the damage to woodland by natural agency recorded in chronicle accounts, such as the windstorm of 6 December 1372 'which demolished and flattened churches and houses, towers and oak

35 See, for example, Nigel Tranter's novel *The Lords of Misrule* (London, 1976).

36 Smout, MacDonald and Watson, *A History of the Native Woodlands*, p. 40.

trees, to the ground'.[37] Just as the devastating storms of winter 2021/2 have shown us in recent times, millions of trees could be damaged or destroyed in single major events,[38] with splintered and weakened timbers left unsuitable for the uses for which they were being grown and conserved. As in early centuries, therefore, we must understand climate change as contributing to continuing woodland decline, deepening, extending and worsening the impact of human action. But even extensive storm damage is reversible, provided the woodland is given the chance to regenerate without additional extractive pressures or unregulated grazing adding to the loss. Nothing affecting the forests in the later thirteenth or fourteenth centuries was irreversible. It would be some centuries before, in the depths of the 'little ice age', the interplay of climate change and human demands wreaked real havoc on Scotland's Highland woodlands.[39]

Fish

In contrast to the growing body of research that has shed light on the state of Scottish woodland in the period after the 1250s, the effects of the waning of the MCA and the slide into unpredictable climatic conditions on Scotland's fisheries in this same period are not as yet well understood. While there is clear evidence for 'aquatic systems under stress' that have been analysed for much of Europe by Richard Hoffmann, indications that the seas around Scotland or the rivers and freshwater lochs inland were experiencing equivalent levels of pressure are less unambiguous .[40] It is not that the Scottish fisheries had not seen an intensification of human exploitation, for as we have observed for the twelfth and earlier thirteenth centuries, there is an upswing in record evidence for both sea and freshwater operations that speaks of the assertion of control over what was becoming a valuable economic resource.[41] It would be wrong to assume that the small scale of Scotland's medieval urban centres and their tendency to be coastal or estuarine meant that there was little impact on aquatic ecosystems, akin to that inflicted on the mainly riverine fisheries of mainland Europe. Burghs like Berwick, Stirling, Dundee, Perth, Aberdeen, Banff and Inverness on the North Sea coast, Glasgow, Renfrew, Dumbarton and Ayr on the Clyde and its estuary, or Kirkcudbright and Dumfries on the Solway, all located at river mouths or at the head of the tidal reach of important waterways, are recorded as centres of fishing for anadromous species, especially salmon, from their earliest foundation. Whilst local consumption perhaps remained low on account of the relatively small size of the population, the impact of river fishing on migratory species in Europe by the fourteenth century was stimulating demand for imported Scottish-caught salmon to compensate for depletion locally.[42] As a consequence, the salmon runs up Scotland's major rivers were subjected to mounting pressure which, as we shall see, led to a steady ratcheting up of legislative controls over their operations.

There is little likelihood that Scottish rivers saw the levels of aquatic habitat destruction witnessed

37 Bower, *Scotichronicon*, vol. 7, p. 375.

38 BBC News, Science and Environment, 'More than eight million trees lost this winter in the UK', 16 February 2022, https://www.bbc.co.uk/news/science-environment-60348947.

39 For the main trends in these processes, see Oram, *Where Men No More May Reap or Sow*, chapters 2 and 3.

40 Hoffmann, *The Catch*, chapter 5.

41 For the development of specialist fishing communities and the growth of a market for fish, see ibid., pp. 77–88 and chapter 4.

42 Ibid., pp. 194–198.

on European and southern British waterways. That is not to say that they were unaffected by the same pollution, watercourse modification or sedimentation issues as are recorded elsewhere in the medieval West. The Tay at and below the leather-production and wool-processing centre at Perth, for example, is likely already to have seen some consequences of the discharge of trade wastes into the river, although the rate of natural discharge through its basin would have flushed this rapidly through the system. A bigger issue here is likely to have been the increase in the river's sediment load attributable to the deforestation of its lower basin and the intensification of cultivation in those districts. Woodland clearance on most river catchments is likely to have increased run-off and associated riverbed turbation or erosion during spates, affecting the small invertebrates on which fish feed and disturbing spawning grounds. Increased nutrient loads and possibly the decreased limpidity of waters caused by sediment in suspension would also have changed the diatom, algal and plant life in and beside the rivers, affecting the ecologies in which habitat-sensitive species thrived. And there was a further by-product of arable expansion in the proliferation of mills on smaller watercourses or the diversion and modification of larger channels to provide perennial supplies to power mills or flush the latrines of monastic complexes. Perth's probably later twelfth- or thirteenth-century lade diversion from the River Almond and Melrose Abbey's cauld and lade diversion of the River Tweed stand as key illustrations of the scale and effect of medieval waterway modification.

Going beyond the purely anthropogenic, a number of impacts on fisheries can be suggested using modelling of changes in northern hemispheric climate patterns and their intimate linkage to atmospheric and oceanic circulation. Ocean surface and deep-water temperatures are now better understood as affecting the presence and movement of both pelagic and demersal fish species, but perhaps more importantly affected the weather regimes over the ocean areas, with significant knock-on effects attributable to the limitations of TEK that had evolved over the preceding centuries. The climatic shift that became evident post-1256 is likely to have brought two major changes in sea conditions that affected maritime communities across Scotland. Most visibly, stormier conditions with rougher seas and a deepening North Atlantic swell presented a new range of hazards to those engaged in deep-water fishing. It was not simply a case of fewer days at sea than had been enjoyed by previous generations but also a question of the limitations of the ship technology available. The small vessels of the time were at great risk of swamping or capsize. In most areas around Scotland's coasts and islands, the boat crews consisted of farmers who fished rather than specialist fishermen, so the loss of even one fishing boat could bring the deaths of a substantial percentage of the adult male population – the main labouring workforce – of a small agricultural community, affecting its wider social and economic viability. Risk of loss had been a present but less frequent threat during the better conditions of the MCA, but the increasing storminess meant that new evaluations of risk were needed and the trend seems to have been towards caution and withdrawal. This was not the only factor affecting fishing strategies, however. A second and less immediately visible change was the effect of fluctuating ocean surface temperatures on migratory pelagic species like herring and deeper-water cooling on demersal species like cod. This factor is still not well understood for the fourteenth century, but it seems that both herring shoals and cod changed behaviour and moved to different ocean areas in response to temperatures, bringing bust to traditional fisheries in some areas but possibly boom to others.

For communities dependent on deep-water fishing, such as had evolved in the Outer Hebrides by the eleventh century, climate change brought both technological and environmental challenges which, coupled with new trading trends, proved insurmountable. Palaeoenvironmental evidence from sites

Although first mentioned in 1494, Low's Work, the name of the weir and sluice on the River Almond at Almondbank near Perth, was already described as 'old' and probably dated from the late twelfth or thirteenth century when the mill lade and burgh mills were first constructed. The weir and sluice (covered by the modern railing and grille) and the 7.2km channel of the lade running past the City Mills and debouching into the Tay below Smeaton's Bridge are monuments to medieval water management at scale.

like Bornais in South Uist reveals that the possibly commercialised exploitation of marine fish resources implicit in the fishbone assemblages there from the twelfth and thirteenth centuries collapsed in the fourteenth century.[43] The collapse has in the past been linked to political upheaval in the region post-1266 (when control over the Hebrides was ceded formally to the Scots by the king of Norway) but more likely relates to changes in ocean temperatures affecting the migratory behaviour of the herring upon which the Hebridean fish trade was based. In the northern mainland and in Orkney and Shetland, the already intensive exploitation of the cod, saithe and ling stocks for both local subsistence consumption and longer-range trade seems to have increased into the fourteenth century. That increase occurred possibly in response to poorer crop-growing conditions, leading to more fish being traded to secure grain needs from external sources. It is also possible that these species may have become more abundant in southern waters as the deep seas around Iceland became too cold to support temperature-sensitive gadids.[44] Sea conditions around the Northern Isles and the Caithness and Sutherland coasts were no better than in the Outer Hebrides, however, meaning that there was also greater risk associated with capitalising on the enhanced catch opportunities.

While the Caithness–Orkney–Shetland trade in dried, processed fish that had developed during the MCA continued into the fourteenth century, excavation evidence suggests that it entered a decline in that century and collapsed in the fifteenth.[45] A key factor in that process may have been the beginnings of the growing dominance of the North Sea fishery by the Dutch in particular, whose superior shipbuilding and processing technologies pushed out the smaller domestic ventures of Orkney and Caithness from significant participation in the international trade. Certainly, there is evidence for Flemish involvement in the herring fishery off the eastern Scottish coast as early as 1274.[46] Knowledge of the existence of this trade, as discussed in Chapters 6 and 8, is based on archaeological evidence for fish processing rather than on historical records, and the scale and economic significance of the trade is contested.[47] James Barrett has proposed that the invisibility of a commercial fish trade in Caithness and the Northern Isles should not be taken as evidence of absence, pointing both to the political arrangements which required the exports of Norwegian dependencies – then including Orkney and Shetland – to be channelled through Norway, especially the port of Bergen, and to the absence of custom on fish including cod in Scotland until the fifteenth century, indicating that none of Caithness's fish trade was registered.[48] There was evidently an established trade from the northern mainland to eastern Scottish ports into the second quarter of the fourteenth century, as illustrated by the 1328/9 exchequer returns which recorded an account with Simon Falconer of

43 Sharples, *A Norse Farmstead*, p. 196.

44 Morris, Batey and Rackham, *Freswick Links*, pp. 271–276; Bigelow, 'Issues and prospects'; Barrett, 'Fish trade in Norse Orkney'; Barrett, Nicholson and Céron-Carrasco, 'Archaeo-ichthyological evidence'.

45 Peak catch and consumption in Orkney seems to have been reached in the thirteenth century, declining thereafter: J.H. Barrett, 'The pirate fishermen: the political economy of a medieval maritime society', in B. Ballin Smith, S. Taylor and G. Williams (eds), *West over Sea: Studies in Scandinavian Sea-Borne Expansion and Settlement before 1300* (Leiden, 2007), pp. 299–340 at 323, 324.

46 Ditchburn, *Scotland and Europe*, p. 144.

47 Compare the discussion in Morris, Batey and Rackham, *Freswick Links*, pp. 271–276, with that in Barrett, 'The pirate fishermen'.

48 Barrett, 'The pirate fishermen', p. 325.

Caithness for delivery of 15,000 'hard' – presumably dried – fish for the royal household.[49] While this is the only surviving specific reference to fish from Caithness in this period, we can assume that where the provenance of the 'hard' fish purchased for the household is not identified, at least some of it came from the northern fishery.[50] Interestingly, the decline in 'hard' fish from Caithness is matched by a disappearance of references to dried fish processed at more southerly east coast ports. Crail, for example, was a source of dried herring in the middle of the thirteenth century, but although a herring fishery and trade continued there through the fourteenth century, its later trade was in fresh or salted fish.[51] This disappearance could reflect changing dietary preferences or the increasing availability of bulk salt for processing, but it could also point to the difficulties of drying fish effectively in the damper conditions of the later Middle Ages.

Exchequer accounts for the later thirteenth and fourteenth centuries record purchases of fish for the royal household in various locations around the country. As fish was not yet a customable commodity, however, we do not know how significant these fisheries were at this date, nor do we know how much was being produced for local consumption or for sale in locations away from the key markets. For example, twenty lasts of herring – equivalent to 240,000 fish – were purchased at Inverness in 1266.[52] But this was presumably the catch from various communities around the inner Moray Firth, gathered into the burgh by royal officers to meet demand, rather than the product of an Inverness-based fishery alone. Crail, where Scotland's main trade in herring seems to have been conducted, likewise saw fish caught elsewhere brought to that port for processing. In 1330 the crown purchased three and a quarter lasts – 39,000 fish – in Crail, with two of the lasts and salt for processing being carried there from Dysart.[53] By the later fourteenth century, Crail seems to have been the main outport for what has been described as 'a regular, if modest' export trade in herring, given that the numbers of fish purchased there for the royal household represented only a percentage of a much larger available catch.[54] It should also be borne in mind, however, that Scottish merchants were active importers of herring from the Scanian fairs of the southern Baltic and these so-called 'Estland' fish were being bought at Crail for royal use as early as 1329.[55] Ayr was a source of haddock and herring from the Clyde fishery, along with Dumbarton and Rutherglen,[56] again presumably representing catches brought from around the outer firth to the markets in the royal burghs. The numbers are miniscule in comparison to the volumes being caught in the later fifteenth century, but that did not prevent Rutherglen merchants from exporting catches to eastern Irish ports in the early fourteenth century.[57] That does not seem to have been the case for Aberdeen and its early involvement in the overseas fish trade, especially of

49 Stuart and Burnett, *The Exchequer Rolls of Scotland*, vol. 1, p. 239.

50 Ibid., pp. 121, 139, 194, 199, 219, 225, 240, 283, 293, 331, 342, 346, 378, 400, 406. References to purchases of 'hard' fish end in the middle of the fourteenth century.

51 Dowden, *Lindores Charters*, no. 75; Stuart and Burnett, *The Exchequer Rolls of Scotland*, vol. 1, pp. 4, 159, 254, 265, 266, 293, 305, 331, 363, 378, 494, 521.

52 Stuart and Burnett, *The Exchequer Rolls of Scotland*, vol. 1, p. 19.

53 Ibid., p. 266.

54 Smout and Stewart, *The Firth of Forth*, p. 28. See, for example, G. Burnett (ed.), *The Exchequer Rolls of Scotland*, volume 2: *1359–1379* (Edinburgh, 1878), pp. 31, 62, 156, 495, 540, 575, 596.

55 Stuart and Burnett, *The Exchequer Rolls of Scotland*, vol. 1, pp. 134–135; Ditchburn, *Scotland and Europe*, pp. 144–145.

56 Stuart and Burnett, *The Exchequer Rolls of Scotland*, vol. 1, pp. 69, 70, 89, 161, 162, 198, 199, 268.

57 Ditchburn, *Scotland and Europe*, p. 147.

The Port of Aberdeen, Aberdeen. The sheltered mouth of the River Dee has been the location of one of north-eastern Scotland's most important harbours since the twelfth century. By the thirteenth century, Aberdeen was trading internationally in fish, exporting cod in bulk to mainland European customers.

cod, for the burgh lent its name to a generic Flemish word for that fish in the course of the thirteenth century, 'aberdaan'.[58] That trade declined, however, and seems to have disappeared entirely in the course of the fourteenth century, in line with the general decline of the northern Scottish and Orcadian cod fishery.

Estuarine and freshwater salmon fisheries have left equally little trace in the record for this period, other than in the multiple charter references to fishing rights and fishing stations. Early fourteenth-century legislation to regulate the size and placing of fish traps, cruives and yairs alludes to older legislation and penalties, which emphasises the importance of the probably mainly salmon fishery already in the twelfth and thirteenth centuries. The 1318 act which mentions this older legislation also refers to 'prohibited times', which indicates that the season for salmon was already regulated in Scotland before that date.[59] The punishments for breaches of the laws were harsh – forty days' imprisonment and 'amerced heavily' (potentially fined substantially) – which again points to the economic value of this resource. In 1401 the penalty was raised to 100 shillings, partly to reflect currency debasement, but the heavy amercement of 1318 became possible execution for three successive breaches of the close season legislation.[60] How much was traded as opposed to being consumed within

58 Ibid., p. 144.

59 *RPS* 1318/13. But exemptions from this close season were also given, as in the three days per week in that season allowed to Walter Oliphant and his wife, Elizabeth Bruce, on the River Earn in 1365: *RPS* 1365/1/1.

60 *RPS* 1401/2/11.

Scotland is unknown, but records from the later fourteenth century indicate that a growing trade in salmon was already being being conducted through Aberdeen, Banff and Inverness.[61] The importance of this trade was recognised in 1398, when a custom of one penny for every salmon or two grilse was introduced.[62] This fixing of custom per fish rather than by volume, weight or multiple number points to the scale of the export trade as still being relatively limited. That all changed towards the end of the fourteenth century, however, with the first surviving reference to barrelled salmon occurring in 1380,[63] heralding the explosion in the fifteenth century in the volume traded.

In the absence of substantial quantitative evidence for volumes of salmon caught and traded before the middle of the fourteenth century, we can only conjecture as to how oceanic variability might have affected the numbers of such anadromous fish and influenced their migratory behaviour.[64] As mentioned above, however, there was an increase in legislative efforts to control the salmon fishery starting from the end of the twelfth century and gaining pace through the thirteenth and fourteenth centuries.[65] This earlier legislation was restated with enhanced penalties in King Robert I's December 1318 parliament, where, 'Concerning the method of fishing in waters where the sea flows, and the penalty of transgressors in a matter of this sort',

> it was ordained and assented that all those who have cruives, fisheries, ponds or watermills where the sea climbs and draws itself back and where young salmon, smolts or the fry of other kinds of fish of the sea or fresh water descend and ascend, such cruives and machines placed below should be at least the measure of two inches in length and three inches in breadth, so that no fry of fish are impeded from ascending and descending, according [to the way] that they can freely ascend and descend everywhere. And if anyone does the contrary and should be convicted or attainted concerning this, he should have imprisonment for forty days, and notwithstanding should be amerced heavily. And the king forbids that anyone presume to fish for salmon or young salmon at prohibited times upon the ancient penalty. And that each sheriff in his sheriffdom should cause this statute to be kept in all manner of waters upon the king's heavy forfeiture and present any defaults before the justiciar in his ayre.[66]

There, in black and white, is a clear concern that the scale of fishing operations and of interference with the waterways was such that even the runs of salmon up the kingdom's main rivers were threatened. Enforcement of this act proved difficult in the long term, but the principle had been established. Whilst it was to be a further century before legislative efforts to extend protection to woodland and growing timber began, here in the first quarter of the fourteenth century we see the first nationwide enactment aimed at protecting a natural resource. Yes, the motivation driving it was economic, but there is in its

61 Burnett, *The Exchequer Rolls of Scotland*, vol. 2, pp. 308, 599.

62 *RPS* 1398/11.

63 Gemmill and Mayhew, *Changing Values in Medieval Scotland*, pp. 306, 311.

64 Hoffmann, *The Catch*, pp. 221–225.

65 R.C. Hoffmann, '*Salmo salar* in late medieval Scotland: competition and conservation for a riverine resource', *Aquatic Science*, 77 (2015), pp. 355–366.

66 *RPS* 1318/13.

Nineteenth-century fish trap near Ratagan, Loch Duich, Kintail, Highland. Although on a grander scale than the fish traps and engines targeted by fourteenth-century legislation, traps like this on the lower reaches of important salmon-spawning rivers were long recognised as posing a threat to the sustainability of the lucrative trade in salmon.

measures a recognition that even a wild species seemingly as prolific as Atlantic salmon was not inexhaustible and that intervention was needed to ensure its future sustainability.

Scab, 'pluk' and murrain

As noted in the previous chapter, a focus on arable agriculture and questions of marginality, subsistence and failure, such as dominate much of the English historiography of rural life in the later thirteenth and fourteenth centuries, has limited application in Scotland. Here, both the more mixed basis of cereal agricultural regimes and the preponderance of pastoralism in many regions have given cause to question the scale of documented harvest failures and references to hunger, dearth and even famine, especially in respect of extending accounts of the English experience of the Dantean Anomaly and the associated 'Great Famine' north of the Border. Although parts of the more arable south-east of the kingdom may have been vulnerable to such subsistence crises, there is an almost unconscious suggestion in older historiography that other regions, especially in the upland districts of the north and west, were more insulated from these events. And yet, as related in the volume *Where Men No More May Reap or Sow*, supposedly more pastoral-based Highland locations experienced some of the worst cereal

Sheep suffering from infestation by *Psoroptes ovis* (non-burrowing sheep mite).
(Image ID: 2PKJ0CW BROKER.com GmbH & Co. KG / Alamy Stock Photo.)

harvest failures and resultant food shortages into the late eighteenth century. As the circumstances of those recorded events underscore, greater dependence on the dairying and meat of a pastoral regime does not mean that there was no large-scale cultivation of grain or cereal component in Highland diets.

Greater dependence on dairying or meat consumption also brought its own vulnerabilities. Impacts on grazing in times of colder weather, when low daytime temperatures led to suppressed grass growth, affecting biomass production and livestock nutrition, or had a direct physical impact on the animals through greater calorific expenditure in maintaining body heat, in part lie behind reports of widespread losses to flocks and herds in the thirteenth and fourteenth centuries (see above, pp. 307, 311).[67] There were also weather-related livestock mortalities, through cold and smothering in snow, such as those reported by Bower's thirteenth-century source for winter 1267/8.[68] But the greatest threat to peasant subsistence came through parasitic and pathogenic agency, which affected not only the direct contribution of sheep and cattle to peasant diets but also their wider contribution to incomes in terms of fleeces and hides. As threats to the health and welfare of Scotland's human population, infestation and infection may have been as significant as the better-known crop failures of the time.

Arguments that the onset of animal and human plagues in the later thirteenth and fourteenth

67 Oram and Adderley, 'Lordship, land and environmental change'.

68 Bower, *Scotichronicon*, vol. 5, p. 369.

centuries was linked to the major climatic events of the period have only recently gained acceptance amongst academics. Epizootics may have been at least partly responsible for unusual levels of mortality amongst wild animals and domesticates recorded as early as 1268 and 1272, but these incidents also coincided with unusually severe winters and the associated suppression of plant growth. Research in Europe and England, however, has yielded evidence for declining livestock health through the earlier thirteenth century. After a sporadic series of non-specific and apparently regional sheep murrains in England between 1201 and 1258, a sheep disease positively identified as scab (a highly acute and transmissible dermatitis caused by mite infestation) spread through ovine populations in north-western Europe. The spread of the mites from mainland Europe into the British Isles is detectable from the late 1270s and lasted into the 1330s, with English flocks being hit for the first time – and ravaged – in 1279/80.[69] It is extremely unlikely that this was the first scab epidemic in the British Isles, given the species-specific preference of the vector – the non-burrowing sheep mite (*Psoroptes ovis*) – but it was the first where symptoms were described. It was also possibly the first to strike England on a national scale, being responsible for the deaths of almost 50 per cent of the national flock. Scotland, however, had already seen high levels of sheep mortality a decade earlier. An unusually high ovine death toll was possibly a consequence of the poor weather throughout spring and summer 1268 referred to above, but records of a cattle disease in northern England and southern Scotland that same year raise the probability of a simultaneous and quite devastating sheep murrain.[70] While 1268's cattle murrain affected a wide region of mainland Britain, we have no record evidence that the catastrophic losses experienced in England in 1279/80 were seen anywhere in Scotland. The high-level Scottish response to outbreaks in the early 1290s, however, suggests that the north of Britain had not escaped the earlier spread of scab and that measures were already in place to contain any recurrence.

That the sheep disease labelled colloquially as 'the pluk' and 'pilsoucht' was some form of scab is confirmed in the early fifteenth-century version of a brieve first developed in the early 1290s and sent to sheriffs around the kingdom commanding them to hold inquests to identify, impound and slaughter infected flocks.[71] The original brieve was likely the product of King John's parliament of August 1293, the summons to which referred to possible infection in the flocks of the men assembling for the session.[72] Its requirement to make inquiries, arrest flocks which were in any way suspected of being scabbed and cull any that were infected might appear draconian, but in a country whose main overseas trade was in wool it was evidently accepted as appropriate to the threat. Whereas in England manorial accounts reveal the level of impact from the efforts to contain scab outbreaks, resulting in tens of thousands of animals being culled, we have no Scottish evidence for the local application of the prescribed responses. Their effectiveness can be judged, however, both from the repeated use of the 1293 brieve into the fifteenth century and beyond and from the fact that no

69 P. Slavin, 'Epizootic landscapes: sheep scab and regional environment in England in 1279–1280', *Landscapes*, 17 (2016), pp. 156–170; P. Slavin, 'Mites and merchants: the crisis of English wool and textile trade revisited, c. 1275–1330', *Economic History Review*, 73:4 (2020), https://doi.org/10.1111/ehr.12969.

70 Bower, *Scotichronicon*, vol. 5, p. 369. For the cattle murrain, see *Chronicon de Lanercost*, p. 85.

71 A.A.M. Duncan (ed.), *Formulary E: Scottish Letters and Brieves 1286–1424*, University of Glasgow Department of Scottish History Occasional Paper (Glasgow, 1976), p. 10 no. 16 ('morbo scabei qui vocatur pilsoucht vel pluk'); T.M. Cooper (ed.), *The Register of Brieves as Contained in the Ayr MS., the Bute MS., and Quoniam Attachiamenta* (Edinburgh, 1946), p. 59 no. 74.

72 *RPS* 1293/8/1.

Modern Highland cattle, Glamis, Angus. The ancestors of cattle like these were a key measure of wealth and status in the cultures of conspicuous consumption in both Highland and Lowland Scotland in the Middle Ages. Providing beef for feasting, milk and dairy products for general sustenance and hides for leather – many of them semi-processed for international trade – they were of vital economic significance.

subsequent outbreak became national.[73] Although scab remained – and remains – a threat to ovine health and to the wool value of flocks, it ceased to be a subject of interest to Scottish chroniclers in the late Middle Ages.

It is likely that Scotland's rulers had seen and learned from England's 1279/80 experience. For landowners with interests in southern and midlands England as well as in Scotland, there had probably been direct experience of that murrain and its financial consequences. Few amongst the kingdom's noble and monastic flock owners would have been unaware of the threat posed by sheep scab in an economy where the profits from wool were fundamental to the incomes of peasants and princes alike. Sheep had been and remained the principal source from which

73 There is evidence for a localised outbreak in the north-east in the late 1320s, when the steward of the household accounted for the slaughter of four sheep 'at the water of Dee' because they were scabby: Stuart and Burnett, *The Exchequer Rolls of Scotland*, vol. 1, p. 148.

the great royal, lordly and monastic building projects of the thirteenth century were funded and from which 'contributions for the peace' and payments of a king's ransom were drawn in the fourteenth century.[74] Even so, the cattle murrains that were recorded intermittently in chronicle accounts across the medieval period were perhaps more deeply damaging at a cultural level, given the central place of cattle in societies – especially in the Scottish and Irish Gàidhealtachd – where lordship revolved around display, gift-giving and exchange fuelled by the conspicuous consumption of beef and dairy products. While the numbers lost in these cattle murrains are unquantified in every surviving source, their wide and prominent chronicle reportage gives a clear sense of the socio-economic distress they unleashed.

Cattle were not only a feature of Highland and Hebridean economic systems, however, and our first recorded later thirteenth-century instance of a bovine murrain occurred in the south-east Lowlands. It may not be purely coincidental that Lothian was struck in 1268 by a possibly pulmonary cattle epizootic, the so-called 'Lungessouth', for spring and summer there that year had been wet, cold and windy and, as mentioned above, also experienced high levels of sheep mortality.[75] A second but this time not specifically localised incidence of unusually high cattle deaths occurred in 1272, a year that was otherwise reported as having seen widespread famine and human deaths in western mainland Europe and the British Isles arising from 'great lack of productivity on the land'. It is unclear if these cattle deaths were related to an epizootic disease or a potential failure of grazing.[76] The first epizootic blows of this period to hit Gaelic Irish society landed as winter ended in 1286, the new season being described in the western Irish Annals of Connacht and Annals of Loch Cé as 'the spring of the cattle-plague' or 'cow-mortality'.[77] It is unlikely that this outbreak was unconnected with a recorded murrain on the bishop of Exeter's manors in Devon in 1286,[78] given the strong mercantile and landownership links between south-west England and south Wales and the south and east of Ireland. Human long-distance networks were the probable medium for spread between south-west Britain and Ireland. With no surviving Scottish chronicle with a Highland or Hebridean locus to record such events, however, we have no evidence for the spread of this murrain within the long-established trade and communication network connecting Ireland and western Scottish regions. Over the next decade, recurrent waves of undefined murrains afflicted herds in north and west Ireland – a 1295/6 outbreak is recorded in Ulster[79] – and it is unlikely that Scottish cattle in regions from Galloway to Argyll, Mull and Islay, which had the most frequent exchanges with Ireland, were unaffected by some of these episodes.

There may have been some optimism that the cattle plagues of the later thirteenth century were passing when no outbreak was reported for over a decade after 1302.[80] This lull, however, ended in a new and even more devastating plague that affected

74 For the importance of the wool economy to Scotland, see Gemmill and Mathew, *Changing Values in Medieval Scotland*, pp. 283–290, 371–373; Mayhew, 'Alexander III', pp. 54, 59–60, 62, 66, 67 note 11; and Grant, *Independence and Nationhood*, pp. 79, 236–237.

75 *Chronicon de Lanercost*, p. 85; Bower, *Scotichronicon*, vol. 5, p. 369.

76 Bower, *Scotichronicon*, vol. 5, p. 385.

77 Annals of Connacht, AC1286.3; Annals of Loch Cé, LC1286.2.

78 P. Slavin, 'The great bovine pestilence and its economic and environmental consequences in England and Wales, 1318–50', *Economic History Review*, 65:4 (2012), pp. 1239–1266 at 1250.

79 Annals of Ulster, *U1298 (the Ulster annals are three years out of synchrony with the other Irish annals at this point).

80 Annals of Loch Cé, LC1302.6, records 'A great destruction of cows, and mortality amongst all cattle, in this year' and then no further episode of murrain until the great epizootic reached Ireland in 1321.

both oxen and cattle across western Europe through the period of the Dantean Anomaly between *c.*1315 and 1325, heaping further misery on a population reeling from the recurrent failures of the grain harvests.[81] Past debate has raged over the nature of this disease, with anthrax, foot-and-mouth or rinderpest seen as potential culprits, but with the most recent assessments supporting the last of these as the most likely.[82] *Rinderpest morbillivirus*, although capable of droplet transmission, is spread most easily through the consumption of grass and water contaminated by the diarrhoea of infected animals. The rainy conditions of the Dantean Anomaly were ideal for the rapid spread of an epizootic that perhaps originated in the Eurasian steppes in the later thirteenth century and spread westwards along the Silk Road and from the Caucasus throughout mainland Europe and, from *c.*1318, the British Isles. The murrain had reached the European North Sea and Channel coasts by 1317. From there, herds in south-eastern England were first affected in 1318–19. The outbreak reached the Scottish border by August 1319, perhaps in infected livestock brought to supply Edward II's army in its siege of Berwick, from where it spread west into Cumberland.[83] Given the high levels of Scottish raids into northern England at this time and the numbers of cattle being driven northwards, it is unlikely that it did not enter Scotland. Again, the absence of Scottish records from this period renders it impossible to calculate its impact, but a stark image of the likely situation is given by references to 'the great cattle slaughter' in Irish annals – it entered that island in 1321 – and by English manorial accounts that indicate a national 62 per cent death rate and that in 45 per cent of estates examined in detail over 71 per cent of the meat and dairy herds and draught oxen perished or were slaughtered.[84]

In the absence of Scottish data, it is important to avoid hyperbole as to the probable impact of rinderpest as it spread around the country, though spread it must have done. Nevertheless, given the prevalence of dairy produce in the Scottish diet, the consequences of a murrain that might have killed off upwards of 62 per cent of the national herd should not be underestimated. As analysis of the short- and long-term outcomes of the murrain in England has shown, it took over a decade for stocking levels to reach even 75 per cent of what they had been pre-murrain, and it was nearer two decades before some classes of bovid – including younger high-milk-producing cows – reached 90 per cent of the 1318 levels.[85] In the Highlands and the Hebrides, where both meat and dairy cattle were central within traditional production and exchange systems, the long-term political and economic ramifications of such a massive drop in numbers followed by a slow recovery were potentially profound. Across the whole country, however, the consequences might have been more serious yet, for the English evidence has been interpreted as suggesting a possible correlation between the deficiency in supplies of meat and more especially in dairy proteins like casein and whey protein in the 1320s and early 1330s and a weakening of the immune systems of children and adolescents growing up in this period.[86] Given also the malnourishment of the Great Famine era, it is likely that the

81 Slavin, 'The great bovine pestilence'; T. Newfield, 'A cattle panzootic in early fourteenth-century Europe', *Agricultural History Review*, 57 (2009), pp. 155–190.

82 Newfield, 'A cattle panzootic', pp. 160–161.

83 *Chronicon de Lanercost*, p. 240.

84 P. Slavin, 'Flogging a dead cow: coping with animal panzootics on the eve of the Black Death', in A. Brown, A. Burn and R. Doherty (eds), *Coping with Crisis: Re-evaluating the Role of Crises in Economic and Social History* (Woodbridge, 2015), pp. 111–135; Slavin, 'The great bovine pestilence'.

85 Slavin, 'The great bovine pestilence', p. 1251.

86 Ibid., p. 1264.

adults of the late 1340s were immunologically ill-prepared to face the epidemics that blazed across Europe in those years.[87] For Scotland's human population, the full price of any post-1319 rinderpest epizootic might not have been evident until thirty years later.

A final recorded fourteenth-century epizootic has passed almost without notice in historical discussion, namely that which afflicted poultry in the early 1340s. Our earliest Scottish source for this is the *Gesta Annalia II*, the chronicle probably composed in the 1360s that was incorporated into the later fourteenth-century compendium attributed to John of Fordoun. It reported that in 1344 fowls throughout Scotland suffered a disease that spoiled their flesh so horribly that no one would eat them.[88] The pathogen at work was likely some variant of the European fowlpox virus, most readily transmitted between flocks by biting mites and insects. It causes scabby or wart-like lesions on non-feathered parts of the body, canker lesions in the mouth and eyelid swelling leading to the lids being sealed by muco-purulent discharge, which accords well with descriptions of folk recoiling from them as they would from a leper. While a plague amongst poultry might not seem a major issue in comparison with the dietary and economic consequences of cattle and sheep epizootics, not only were chickens and eggs important protein sources in the diets of Scots of all social classes but they formed a large segment of rent payments in kind into the seventeenth century and beyond. In 1509, for example, from within the parish of Little Dunkeld alone, the bishop of Dunkeld's household received an annual render of 331 'poultry', 180 chickens, 173 capons, 22 domestic geese and 12 hens.[89] Since these 718 birds would have been but a small portion of the total domestic fowl population within that parish, let alone the diocese, the numbers give us some sense of the volume of poultry attached to most peasant households across Scotland. Interestingly, they also provide some insight on the management of flocks of domestic geese.[90] The sheer scale of the national flock of domestic fowl in the Middle Ages is staggering, hence the notice given to this event by the chronicler of *Gesta Annalia II*. Coming as it did towards the end of one of the most devastating episodes of the second phase of the Wars of Independence, when agriculture in a broad zone from the Mearns in the north-east to Lothian, Stirling and Galloway had been seriously disrupted for several successive years, this epizootic was just a further turn of the subsistence screw.

The foul death of England

Cattle, sheep and poultry epizootics paled into insignificance behind the human pandemic that affected Europe after 1347. Scots called it 'the foul death of England' when it struck their enemies in

87 Ibid.; Slavin, *Experiencing Famine*, chapter 7; P. Slavin and S. DeWitte, 'Between famine and death: physiological stress and dairy deficiency in England on the eve of the Black Death (1315–50): new evidence from paleoepidemiology and manorial accounts', *Journal of Interdisciplinary History*, 44 (2013), pp. 37–60; Jordan, *The Great Famine*, pp. 185–187.

88 John of Fordoun, *Chronica Gentis Scotorum*, vol. 2, p. 358. This episode is dated to 1336 in Walter Bower's fifteenth-century *Scotichronicon*, vol. 7, p. 157, where describes the birds as perceived as 'unclean and riddled with leprosy', a description that tallies with the symptoms of fowlpox. The placement of Bower's reference to the fowlpox, however, indicates that his date is wrong and should have read 1344 (ibid., p. 247).

89 R.K. Hannay (ed.), *Rentale Dunkeldense: Being Accounts of the Bishopric (a.d. 1505–1517) with Myln's 'Lives of the Bishops' (A.D. 1483–1517)* (Edinburgh, 1915), pp. 32–33.

90 Goose farming is not a researched aspect of medieval rural economies in Scotland. For the scale of goose farming and its economic significance, see P. Slavin, 'Late-medieval goose farming: evidence from eastern England, c.1250–1400', *Agricultural History Review*, 58:1 (2010), pp. 1–29.

1348,[91] a taunt that ended when the plague reached Scotland the following year. In contrast to the voluminous accounts by contemporary English and European writers, only one fourteenth-century Scottish account of the epidemic exists, that embedded in *Gesta Annalia II*. Its chronicler is laconic, stating factually that 'there was in the kingdom of Scotland so great a pestilence and plague amongst men . . . as . . . had never been heard of by man', that 'fully a third of the human race was killed', and that its main victims were amongst 'the meaner sort and common people' rather than the magnates.[92] If this class-differentiated impact is true, it is possible that we might be seeing here the legacy of the hungry years of the Dantean Anomaly, when the adults now dying of plague had been young children whose immune systems were compromised through nutritional deficits. The epidemic is reported in only slightly less restrained terms in the early fifteenth-century accounts of Wyntoun and Bower, by the time of whose writing Scotland had suffered at least three further outbreaks. Wyntoun labelled this 'þe fyrst pestilence' and Bower 'the first mortality', following the by then established convention of numbering the recurrent pandemics.[93] It is Bower who provides us with our only plague statistic, reporting that twenty-four of the Augustinian canons of St Andrews cathedral priory had died, approximately two-thirds of the community. If similar levels of loss were experienced generally, the 'fully a third' of the chroniclers becomes a gross underestimate of mortality rates. There are, however, few other direct references to the epidemic's impact on daily life, although notice of a reduced annual rent in a 1354 lease agreement between the bishop of St Andrews and the priory of Restenneth in Angus 'on account of modern times becoming harder and worse than before and by the hostilities and persecutions which thrive in these fevered days' seems to allude to recent warfare and disease.[94]

Bower's 'second mortality' struck in 1362–3 with a strength and virulence that left no class spared, in his view killing 'one third' of those who had survived the first pestilence.[95] This *pestis secunda* might have been more devastating across Europe than the first pandemic, as it affected the nutritionally robust wealthier classes as well as the possibly immunologically compromised poor. Numerous accounts highlighted its greater impact on a younger age group and on men in particular, displaying a gender differentiation that is borne out by modern epidemiological studies, including those of SARS Covid-19. Transmission from England to Scotland might be indicated by King David II's decision to move north from Lothian into Mar and Moray, but there is little other evidence to indicate the nature and direction of its spread.[96] Again, there is extremely limited notice of its wider impact. One contemporary charter to Holyrood Abbey refers to losses through war and pestilence but speaks only in those very general terms.[97] There is no indication that Scotland saw a visitation in 1369/70 of what English sources referred to as a third pestilence.[98] Scotland's 'third plague' instead followed in 1380, Bower again claiming that

91 Extract from the Chronicle of Henry Knighton, in R. Horrox (ed. and trans.), *The Black Death* (Manchester, 1994), p. 78.

92 John of Fordoun, *Chronica Gentis Scotorum*, vol. 2, p. 359.

93 Andrew of Wyntoun, *The Original Chronicle*, vol. 6, p. 197; Bower, *Scotichronicon*, vol. 7, p. 273.

94 Glamis Castle Archive, 251/24. For a wider assessment of the longer-term social and economic consequences of the plague pandemic in Scotland, see Jillings, *Scotland's Black Death*, chapter 2.

95 Bower, *Scotichronicon*, vol. 7, p. 319. For accounts of its impact, see Horrox, *The Black Death*, pp. 85–88.

96 For David II in the north-east parts of his kingdom from September 1362, see *Regesta Regum Scottorum*, volume 6: *The Acts of David II*, ed. B. Webster (Edinburgh, 1982), pp. 309, 317 and nos 276–285.

97 Innes, *Holyrood Liber*, no. 98.

98 Horrox, *The Black Death*, p. 88.

Crouching death figure, Melrose Abbey, Scottish Borders. Fifteenth-century sculpture of this sort, with its emphasis on human mortality, the decay of the flesh and the suddenness of the end, was once a common expression of the heightened preoccupation with preparation for death that arose in the aftermath of the first plague pandemic of the fourteenth century.

one third of Scotland's population died in this outbreak and suggesting that the plague returned with soldiers from a raid into Cumbria.[99] This visitation had struck England in 1374 and was passing by 1379;[100] Scotland had been lucky to avoid it until 1380. A fourth Scottish outbreak perhaps occurred in 1392, possibly spreading north from the epidemic that had begun in England in 1389 and that ravaged York in 1391,[101] but no contemporary Scottish record reports this event. Instead, Scotland's 'fourth plague' struck in 1401, but given the single sentence afforded to it by Walter Bower, who would have been in his

99 Bower, *Scotichronicon*, vol. 7, p. 381.
100 Horrox, *The Black Death*, pp. 88–90.
101 J.F.D. Shrewsbury, *A History of Bubonic Plague in the British Isles* (Cambridge, 1971), pp. 137–138.

Mortality made meaningful: recording and visualising the Black Death

It is quite striking that so little record evidence survives from fourteenth-century Scotland for the impact of the plague that raged throughout the country from 1349 in a succession of devastating waves. A handful of laconic chronicle accounts, commencing with the so-called *Gesta Annalia II* material in the later fourteenth-century compilation attributed to John of Fordoun (John of Fordoun, *Chronica Gentis Scotorum*, vol. 2, p. 359), stand in stark contrast to the voluminous texts composed throughout England and in Ireland. None of the existing Scottish texts provide any hard evidence for the spread of the epidemic, which helped to give rise to a now challenged notion that the Black Death had a limited impact, possibly restricted to mainly Lowland districts and perhaps chiefly in the urban centres. (For an exploration of the older historiography, see R.D. Oram, '"It cannot be decernit quha are clean and quha are foulle": responses to epidemic disease in sixteenth- and seventeenth-century Scotland', *Renaissance and Reformation*, 30:4 (2007), pp. 13–39.) More recent studies have drawn on the more plentiful epidemiological evidence from the late fifteenth-, sixteenth- and seventeenth-century plague events to argue that the network of trade and international travel connections between Scotland and areas that were affected badly in the recurrent plague episodes between 1347 and the end of the century makes it likely that no part of the mainland or islands was unaffected (Oram, 'The worst disaster'). Movement of people and transmission of the *Yersinia pestis* bacterium that causes bubonic plague along seaborne trade routes can be seen in the claim in the Icelandic Lawman's Annal that 'the pestilence visited' the Hebrides, Orkney, Shetland and the Faroe Islands (quoted in O. Benedictow, *The Black Death 1346–1353: A Complete History* (Woodbridge, 2004), p. 154). Although the Gaelic Irish regions were said to have been less severely affected than the Anglo-Irish urban centres (see M. Kelly, *A History of the Black Death in Ireland* (Stroud, 2001)), perhaps because of their more dispersed population, they were not immune, and it was most likely via the long-established trade and mercenary service connections between northern and western Ireland and the Scottish Gàidhealtachd that the disease was transmitted. While we will probably never know the true scale of the death toll across Scotland, we can be quite certain that no region escaped unscathed, and even in places where mortalities were fewer the fear and uncertainty that accompanied this shocking killer delivered a legacy of lasting social and economic change.

There is nothing in the limited fourteenth-century record evidence to give much idea of the psychological impact on the Scottish people of the mass mortalities that occurred. It is not until the mid fifteenth century and the poetry of William Dunbar that we catch a glimpse of the degree of trauma that was experienced and the deep-seated hold it had taken in the Scottish psyche. Across Europe, however, the horrors of the fourteenth century, famine as well as epidemic disease, had seen an outpouring of expressive literature and visual art. The decaying corpses depicted on late fourteenth- and fifteenth-century cadaver and transi tombs are part of a post-Black Death focus on the fragility of human life, mortality, sudden death and the physical corruption of the flesh. The post-Reformation destruction of most of Scotland's medieval funerary monuments perhaps accounts for the absence of such sculptural expressions in Scottish contexts (for a tomb slab at Ardchattan Priory, Argyll and Bute, see Oram, *Where Men No More May Reap or Sow*, p. 32), but a crouching skeletal figure at Melrose Abbey and a remarkable death's head with worms crawling over its jawbone and empty eye sockets in the vault of the Blacader Aisle at Glasgow Cathedral offer tantalising hints of what we have lost.

Glasgow Cathedral, Glasgow. The death's head vault boss in the cathedral's Blacader Aisle, carved probably in the last decade of the fifteenth century, is one of the few such post-plague symbols of mortality to have survived the iconoclasm of the late sixteenth and seventeenth centuries.

mid teens at the time, there was possibly a lower level of mortality on this occasion.[102]

With such limited reference to Scotland's experience of the Black Death in fourteenth- and fifteenth-century narrative accounts, it is unsurprising that plague has until recently been almost invisible in most modern academic histories of later medieval Scotland.[103] Only Alexander Grant, writing in 1984, offered the sober but unsubstantiated opinion that 'plague still appears to have been the worst disaster suffered by the people of Scotland in recorded history'.[104] How great a disaster, however, is still debatable. Apart from Bower's notice of the deaths in the monastic community at St Andrews, there are no exact figures for Scottish mortalities. Instead, contemporary sources from Scotland and the rest of the British Isles and mainland Europe speak of 'fully a third of humanity' or more perishing.[105] At that level, now widely regarded as an underestimate, and assuming a population figure within Bruce Campbell's proposed range for the early fourteenth century of 0.67 million to 0.75 million, by the time the first outbreak ended in 1351–2 between at least 220,000 and 250,000 of Scotland's population would have been dead.[106] Plague mortalities declined in later outbreaks, perhaps only because the starting population in each was smaller, but possibly because different pathogens were at work though not distinguished by chroniclers, who labelled all outbreaks simply as 'plague' or 'pestilence'.[107] The identity of other epidemic diseases present in late medieval Scotland is discussed in the opening chapter of *Where Men No More May Reap or Sow*.

Even if we take the 'fully one third' claim at face value, the loss of something like a quarter of a million people on top of the deaths of the previous half century from warfare and malnourishment would have been truly catastrophic. Contrary to the scepticism regarding numbers of plague deaths expressed in some past studies,[108] an indication of the cumulative effect of mortalities from war, dearth and disease can be recovered from contemporary record evidence. The clearest example is the falling property values recorded in the *verus valor* (true valuation / extent) land assessment of 1366, which took the higher figures of an imprecisely dated *antiqua taxatio* (ancient taxation / extent).[109] Although often

102 Bower, *Scotichronicon*, vol. 8, p. 39.

103 As illustrated by Nicholson, *Scotland*, which affords it just two short paragraphs (pp. 148–149), and Brown, *The Wars of Scotland*, which has a single paragraph (p. 318). Karen Jillings's *Scotland's Black Death*, based on her Aberdeen PhD thesis, was the first monograph-length study of the first outbreak to be published. My own 2014 study, 'The worst disaster', is the first extended exploration of the wider impact of climate, weather and pathogens on Scotland's people.

104 Grant, *Independence and Nationhood*, 75.

105 See, for example, the extracts from contemporary mainland European, English and Irish accounts gathered in Horrox, *The Black Death*. For estimates of overall mortality, see O. Benedictow, *The Black Death 1346–1353: A Complete History* (Woodbridge, 2004), chapter 33.

106 Campbell, 'Benchmarking medieval economic development', pp. 925–933, population figures at 933.

107 As argued in the plague-sceptical discussion in Shrewsbury, *A History of Bubonic Plague*, pp. 127–128, 131, 132. Shrewsbury's minimalist view is no longer tenable as a result of the half century of new research since its publication.

108 Ibid.

109 The legislation for this valuation is in *RPS* 1357 / 11 / 2. For discussion, see A. Stevenson, 'Taxation in medieval Scotland', in P.G.B. McNeill and H.L. MacQueen (eds), *Atlas of Scottish History to 1707* (Edinburgh, 1996), pp. 298–305; Grant, *Independence and Nationhood*, pp. 77–79; and Brown, *The Wars of Scotland*, pp. 318–319. For a comprehensive revision of the data provided in the 1366 valuation, see A. Ross, 'A post-plague golden age? Scotland, *antiqua taxatio*, and thirteenth-century climate change', in R.D. Oram (ed.), *'With Our Backs to the Ocean': Land, Lordship, Climate Change, and Environment in the North-West European Past: Essays in Memory of Alasdair Ross*, Environmental Histories of the North Atlantic World 5 (Turnhout, 2023), pp, 109–134.

presented as a consequence of war damage or monetary contraction, this decline is now understood to relate directly to a sharp drop in population levels and the consequent ending of around a century of upward pressure in rental values.[110] The 1366 assessment reveals that falling values affected all parts of the country, in areas as relatively free from the ravages of war as Caithness, Sutherland and the Isles as well as the most heavily affected districts of southern Perthshire, Lothian, Teviotdale and Galloway: values north of the Tay fell by 44 per cent, those to its south by around 52 per cent. Apart from the climate change impacts of the later thirteenth and earlier fourteenth centuries and the deaths consequent on dearth arising from epizootic losses of flocks and herds and the failure of crops, plague is the only agent whose impact was likely felt throughout the entire kingdom and could have contributed to such a dramatic decline.[111] If the *verus valor* is a proxy for population levels, its figures suggest that the population fell nationwide, which runs against traditional views of Scotland's experience of the Black Death that imply a differential in impact related to dispersed settlement characteristics in the Highlands and Islands and more nucleated settlement in lowland areas, but matches arguments advanced elsewhere for the experience of the country in sixteenth- and seventeenth-century epidemics.[112] In this light, we can begin to sense the true scale of the loss of life in Scotland's experience of the fourteenth-century plague pandemics.

Rural regimes reconfigured

How did those combinations of deteriorating weather and population loss affect the way in which people exploited the environmental resources of the land? Much attention was given in the late 1970s and 1980s to a Scottish equivalent of the 'retreat from the marginal' that was evident in England in the later Middle Ages. This was believed to occur where the least viable settlements and field systems were abandoned as plague survivors gravitated towards better-quality tenancies that had become available as a result of high peasant mortalities. On the basis of the visible evidence of abandoned settlements and high-altitude rig-and-furrow cultivation in the Lammermuir Hills, it was proposed that Scotland had witnessed the same expansion to the technological and environmental limits of medieval agricultural practice in the thirteenth century, and that that expansion had been followed by a withdrawal from the most marginal districts as first climate change struck and then plague mortalities relieved pressure within a cereal-based rural regime.[113] It is in this light that the *verus valor* evidence discussed above is most commonly viewed, with climate change and population collapse seen as delivering the greatest falls in property value.[114] More recent research, however, has suggested alternative responses to deteriorating conditions and that divergent responses occurred in different regions of Scotland. We should, therefore, be cautious in seeing the severe weather of the earlier fourteenth century as

110 Grant, *Independence and Nationhood*, p. 78.

111 Icelandic accounts evidence the spread of the plague along the Atlantic trade routes from Bordeaux, Bristol and Dublin to Bergen, reporting its arrival in the Hebrides, Orkney, Shetland and the Faroe Islands: Benedictow, *The Black Death*, pp. 154–155.

112 Oram, '"It cannot be decernit."'

113 See especially M.L. Parry, 'Secular climatic change and marginal agriculture,' *Transactions of the Institute of British Geographers*, 64 (1975), pp. 1–13; M.L. Parry, *Climatic Changes, Agriculture and Settlement* (Folkestone, 1978); and M.L. Parry and T.R. Carter, 'The effects of climatic variations on agricultural risk', *Climatic Change*, 7 (1985), pp. 95–110.

114 Brown, *The Wars of Scotland*, p. 318.

Turf Law, Lauderdale, Scottish Borders. The ephemeral traces of rig-and-furrow cultivation near the highest point of Dere Street as it climbs out of Lauderdale towards Soutra were once seen as evidence for the 'retreat from the margins' of agriculture after the depopulation effects of the Black Death. They are now thought more likely to be traces of intermittent and opportunistic cultivation from later centuries.

the exclusive or even primary driver of rapid change. Peasant communities would have been reluctant simply to abandon land into which they had made major investments of energy, nutrients and material resources. We should also be cautious in assuming that common responses occurred everywhere across the kingdom, regardless of the different economic regimes practised in zones of different culture and of fundamentally different exploitative potential.

Until the early 2000s, most studies seeking evidence for climate change impacts on medieval Scottish agriculture focused on the zones of intensive arable cultivation in the upland districts of the south and east, where the evidence had not been obliterated by modern farming. There, what might have encouraged people to move was visibly better opportunities in less marginal areas. It is likely to have been the freeing up of tenancies in agriculturally superior districts that persuaded some occupants of those more marginal communities to relocate. But we should not suppose that there was a scramble to abandon less productive holdings for tenancies that were simply lying vacant on the better land. There were survivor families who took the opportunity to increase the size of the holdings they occupied through taking on the lease of lands once held by neighbours who had fallen victim to the plague. Such a strategy is illustrated in the surviving estate rental of 1376–7 of the Douglases of Dalkeith,[115] which shows that most tenants with shares in the arable

115 C. Innes, T. Thomson and A. Macdonald (eds), *Registrum Honoris de Morton*, 2 volumes (Edinburgh, 1853), volume 1, appendix, pp. xlvii–lxxvi.

Ben Lawers from near Wester Tulloch, south Loch Tay-side, Perth and Kinross. High-altitude settlement at the limits of cultivation was abandoned here at the end of the fourteenth century and replaced by seasonal, short-range transhumance use of the uplands as summer pasture for dairying.

land in a tounship held four or more oxgangs (roughly 52 acres in Scotland), which was more than adequate to support a family and yield a comfortable surplus when conditions were good. Pre-plague, with greater competition, tenancies are likely to have been considerably smaller.[116] Thus, the 'slack' created by plague mortalities was likely first taken up by surviving tenants in the same community. That trend created some men of substance with large property holdings, with a much slower drift away from the agriculturally poorer districts. In poorer areas, too, we might expect to see surviving tenants taking up a greater share and diminishing their own economic marginality, which would be visible as abandonment of settlement but not necessarily of the cultivated areas.

Evidence from research in the uplands around the Bowmont Valley in the northern Cheviots lends some support to this interpretation. There, the changing weather patterns brought physical changes with wetter ground conditions by the last quarter of the fourteenth century.[117] There was, however, no evidence that wetness and cold had caused abandonment either of settlement or of upland activities that had become well established from the early twelfth century in what has been described as 'a patchwork multi-tasked landscape'.[118] Every opportunity was still exploited in a busy mosaic of patches of cultiva-

116 Grant, *Independence and Nationhood*, pp. 67, 78–79.

117 Tipping, *Bowmont*, p. 194.

118 Ibid., p. 195.

tion and pasture of differing quality. What might have changed was the density of human presence and the balances of exploitation methods to reflect changes that were as much economic and social as environmental. What projects like that at Bowmont cannot tell us, however, is the year-on-year narrative of agricultural success or failure. They cannot reveal how and why decisions to vary established modes of exploitation were made, or how the local communities responded to the disruption of markets locally and internationally by war or plague. Importantly, they do not record if the evidence for upland cultivation and possibly associated habitation sites represented one-off, temporary/intermittent or perennial activities. More important yet, the detailed findings from Bowmont cannot and should not be generalised beyond the south-eastern Borders into regions with fundamentally different geologies, ecologies, demographic patterns and cultural traditions.

This last point is emphasised by the findings of the large-scale archaeological research project around Ben Lawers and its environs in central Perth and Kinross. There, the response was quite different to what was seen at Bowmont, with an apparent withdrawal from permanent occupation of the highest-altitude and most marginal sites by around 1300. These places saw a possible shift to seasonal occupation thereafter to *c.*1400, probably associated with transhumance pastoralism at some locations, before the development of new seasonally occupied shieling sites during the fifteenth century.[119] No evidence was found of any persistence of even intermittent upland cultivation, with the focus of agriculture shifting firmly to the lower-altitude zone on the slopes and terrace above Loch Tay. The higher ground was developed exclusively for summer grazing and peat-cutting for the lochside settlements. Differing strategies, we can see, were adopted at Bowmont and Ben Lawers, some of which perhaps arose from differing cultural practices. Others, however, perhaps reflect nuanced political and economic drivers and alternative responses to climatic change.

While the Ben Lawers project posited a decisive and permanent shift in upland exploitation post-1300, one of the key propositions from the Bowmont project was that innovation and intensification in both arable and pastoral regimes were processes that continued through the changing conditions after *c.*1250. The biggest difference was in terms of withdrawal from upland cultivation. In contrast to the abandonment of the permanent mixed-farming settlement at Kiltyrie on Ben Lawers,[120] pollen evidence from Bowmont shows that barley and either wheat or oats (or both) were still cultivated in all but the highest upland areas of the valley to at least the early 1300s. This cultivation was probably for subsistence consumption locally rather than disposal at market, but there is some suggestion of a break from then into the later fifteenth century in some districts.[121] It is important not to overread this apparent correlation between the interruption in cultivation and the successive episodes of war and plague through the fourteenth century, especially as elsewhere in this border-zone valley no such break was observed. What can be said, however, is that the continuous cultivation took place on the lower-altitude sites where the maximum efforts at soil-enrichment and conservation had always been concentrated. At upland locations, which probably only benefited from manuring by animals penned overnight in turf-built folds for much of the time, there was perhaps a shift to opportunistic, short-term

119 J.A. Atkinson, *Ben Lawers: An Archaeological Landscape in Time: Results from the Ben Lawers Historic Landscape Project, 1996–2005*, Scottish Archaeological Internet Report 62 (2016), https://archaeologydataservice.ac.uk/library/browse/issue.xhtml?recordId=1137495&recordType=MonographSeries, pp. 94, 269, 271–272.

120 Ibid., chapter 5.

121 Tipping, *Bowmont*, p. 194.

cultivation during episodes of better weather conditions.[122] On Ben Lawers, there is no evidence for this pivot towards opportunistic, short-term cultivation. What was grown at lower altitude apparently sufficed for local subsistence needs.

In the West Highland glens, the response was again different. Research in western Glen Affric suggests that agriculture expanded or contracted opportunistically as local conditions allowed and need demanded.[123] Archaeologically less visible than the expanses of abandoned rig-and-furrow cultivation of the south and east, medieval farming practice in this region was probably mostly dependent on spade or foot-plough cultivation rather than on ox team and carruca. It also tended to be undertaken in smaller blocks on patches of better land dispersed amidst a patchwork of rough grazing, rockier or wetter ground and scrub- or woodland.[124] Pollen data from Affric has brought a major advance in the understanding of this physically dispersed cultivation practice and of pulses of intensification or contraction of arable that do not necessarily align with climate trends. Surprisingly perhaps, not every area of the best available land in the glen was in permanent cultivation. One fertile block at Camban seems to have been brought into cultivation for the first time since the Early Historic era only towards the end of the MCA.[125] We have no way of knowing if this intake was driven by population pressure or by generally declining yields against a backdrop of climate change, or even by some unrelated socio-political decision. By *c.*1300, at sites spread through the glen, there was a shift to intensified cultivation and a diversified cereal range, including oats, barley and wheat, accompanied by an episode of significant woodland clearance. Peak agricultural activity was reached only in the second quarter of the fifteenth century, and it was well into the 'little ice age' before identifiable contraction began here.[126] This is a record that does not accord well with the traditional models of retreat from the margins in the face of deteriorating weather conditions.

Looking further afield, we can see that continuity of established agricultural practices in the face of deteriorating weather is also evident in what might seem areas of the most hostile conditions for arable cultivation. Although it has been argued that intensification of deep-water fishing and dairying had seen an equivalent decline in arable cultivation in the Shetland Islands before 1300, with the islands' grain needs being secured through their external fish trade instead,[127] there is evidence for continuation and possible intensification of cereal cultivation on some sites there. At Kebister on Mainland Shetland, for example, despite the islands' exposure to the increasing storminess of the North Atlantic from the mid thirteenth century, the surrounding field systems continued to evolve through the fourteenth century and arable cultivation was still pursued despite the worsening conditions.[128] Persistence with cereal cultivation here might be linked to the association of Kebister with high-status ownership as part of the archdeacon of Shetland's properties by the early sixteenth century at least.[129] Crop growing here was

122 Dixon, 'Mukked and folded land'.

123 Davies, 'Upland agriculture and environmental risk', section 5.1, pp. 2057–2058.

124 Ibid., p. 2058.

125 Ibid., section 4, p. 2057.

126 Ibid., section 4, p. 2057, and section 5.2.

127 Bigelow, 'Issues and prospects', p. 20.

128 Owen and Lowe, *Kebister*, pp. 289–290.

129 B. Smith, 'A short history of Kebister', in O. Owen and C. Lowe (eds), *Kebister: The Four-Thousand-Year-Old Story of One Shetland Township*, Society of Antiquaries of Scotland Monograph Series 14 (Edinburgh, 1999), pp. 17–21 at 19.

The excavated remains of the parish kirk at Forvie, Aberdeenshire. The remains are now surrounded by the dune systems at the mouth of the River Ythan, while the sands have overwhelmed the small village and its complex of fields that extended over the headland on the east side of the estuary.

perhaps sustained by tenant labour as a display of lordship and the ability of its owner to have a scarce commodity produced from his own land. As with the Bowmont valley's patches of upland cultivation, however, there is nothing to provide insights on the quality or volume of the yields. The cessation of arable cultivation here in the fifteenth century perhaps evidences a long, steady decline before eventual abandonment.

At lowland east coast sites, from Eldbotle in East Lothian to Forvie in Aberdeenshire, there are also indications of intensification of effort on the core infield areas of cereal cultivation and the tofts attached to the houses, where the manuring, deepening and enriching efforts were concentrated. Evidence for these measures is visible in the archaeological and historical record of soil-deepening using turf stripped from the nearby dunes and waste from middens to create anthrosols.[130] In those coastal cases, however, the long-term outcome was negative, as

130 J. Morrison, R. Oram and F. Oliver, 'Ancient Eldbottle unearthed: archaeological and historical evidence for a long-lost early medieval East Lothian village', *Transactions of the East Lothian Antiquarian and Field Naturalists' Society*, 27 (2008), pp. 21–45; E. Hindmarch et al., 'Eldbotle: the archaeology and environmental history of a medieval rural settlement in East Lothian', *Proceedings of the Society of Antiquaries of Scotland*, 142 (2012), pp. 245–300; W. Kirk, 'Prehistoric sites at the sands of Forvie, Aberdeenshire: a preliminary examination', *Aberdeen University Review*, 35 (1953), pp. 150–171; W. Kirk, *'Sands of Forvie', Discovery and Excavation in Scotland, 1957* (Dundee, 1957), p. 4; W. Kirk, *'Sands of Forvie', Discovery and Excavation in Scotland, 1958* (Dundee, 1958), p. 2; W. Kirk, *'Forvie', Discovery and Excavation in Scotland, 1960* (Dundee, 1960), p. 2.

continuation and intensification of cultivation and soil-enriching practices developed during the MCA exacerbated dune destabilisation leading to deflation, sand-blow and the 'drowning' of fields into the later Middle Ages. New settlement and permanent intakes from outfield or muirland are less evident in these areas. The assarting of the twelfth and thirteenth centuries through which agriculture and settlement expanded, recorded throughout south-east Scotland, petered out rather than coming to an abrupt halt, before resuming in the later Middle Ages.[131]

The Outer Hebrides experienced a similar trajectory to the east coast sites, probably due to the concentration of settlement on the sandy machair district down the islands' ocean-facing west coasts. Here, however, the deteriorating conditions through the later thirteenth and fourteenth centuries dealt a series of blows to all niches of the established economic regime and not simply to agricultural practices. Cereal cultivation declined and some of the more exposed settlements on the coastal strip began to contract or be abandoned in this period. At the Udal in North Uist, dune deflation and erosion of the machair, which was always problematical in dry and windy conditions once the land had been ploughed and before the crops had started to grow, worsened in the stormier conditions of the fourteenth century. It was, nevertheless, after 1400 that the greatest losses of cultivable land were suffered there and the site was finally abandoned only in the seventeenth century.[132] The high-status farm site at Mound 3 at Bornais in South Uist revealed a more nuanced picture of changes in arable practice on the machair, with the probably mid thirteenth-century construction of a grain-drying kiln and winnowing barn indicative of a need to improve the condition of grain, harvested either unripe or wet, prior to milling.[133] At Bornais, the corn-drying kiln was functional until the abandonment of the site in the early 1400s, when erosion of the machair locally forced the relocation of its inhabitants inland to a higher and rockier location at Beinn na Mhic Aongheis.[134]

Corn-drying kilns were not new elements in the Scottish agricultural landscape in the mid thirteenth and fourteenth centuries but they became almost ubiquitous through the later Middle Ages and to the eighteenth century.[135] The deteriorating weather conditions from the late 1250s and in particular the increased precipitation experienced after 1315 made the use of such kilns essential for preventing the ruining of the seed-corn to be stored for the next year's sowing through mildew or other fungal infestation. There were also changes in threshing practice, as indicated by the association of the kiln with a winnowing barn at Bornais, and kiln-drying became essential for malting, milling or a combination of these functions. Each formed an indispensable component in the preparation of staple elements of Scottish peasant diets, from bread, bannocks or porridge to ale.[136] Grain may still have been grown

131 Oram, *Where Men No More May Reap or Sow*, chapters 1–3.

132 Serjeantson, *Farming and Fishing*, p. 32.

133 Sharples, *A Norse Farmstead*, pp. 86–128, 155–156.

134 Ibid., p. 195.

135 Shepherd, *The Late Medieval Landscape*, pp. 102–105. For tenth-/eleventh-century corn drying in south-east Sutherland, see R. Coleman et al., 'Early medieval settlement and ironworking in Dornoch, Sutherland: excavations at the Meadows Business Park', Scottish Archaeological Internet Report 28 (2008), http://journals.socantscot.org/index.php/sair/article/view/1354/1343, p. 17. For two thirteenth-century kilns at Capo in the Mearns, see A. Gibson, 'Medieval corn-drying kilns at Capo, Kincardineshire, and Abercairney, Perthshire', *Proceedings of the Society of Antiquaries of Scotland*, 118 (1989), pp. 219–229.

136 P. Dixon, 'Of bannocks and ale: cereal processing in Scotland c.1100–1750', in J. Klápšte and P. Somner (eds), *Food in the Medieval Rural Environment*, Ruralia 8 (Turnhout, 2011), pp. 155–172.

Post-medieval corn-drying kiln, Glen Tilt, Perth and Kinross. Later and much larger than the kilns used in the fourteenth century, structures like this underscore the persistence of communities throughout rural Scotland in pursuing a cereals-based arable regime through an era of colder and wetter conditions. Such kilns emphasise resilience and adaptation in the face of change.

in seemingly unfavourable locations like the Cheviot uplands right through this era of climatic deterioration, but significant and increasing levels of resource had to be committed to ensuring that it was fit for human consumption. The above cases suggest that peasant farmers were not immediately responsive to the changing climate in their crop-growing practices and preferences. It may have taken several seasons of year-on-year deterioration before it was recognised that a shift had taken place that affected their growing crops, and the responses then were widely divergent. In many upland areas, the most common reaction seems to have been to try to grow more to make up any shortfall in yields from the established arable zones, but nowhere can we be certain of how effective that strategy was in the longer term. The expansion or intensification of cultivation even in zones of increased marginality should warn us against seeing all areas of abandoned rig as relating to fourteenth-century events.

Affric's record of a fragmented or dispersed agricultural landscape in which the arable and better-quality pasture formed a mosaic of isolated blocks amidst rockier, poorer and wetter ground has produced indications of judicious management of resources over a long period of time. Islands of good land, as on the alluvial fans of tributary streams, supported good quality grazing (and stands of woodland) as well as having the best arable potential. That some were not taken into cultivation during the MCA suggests that arable expansion was not necessarily the overriding objective everywhere and that the alternative yield in terms of grass, hay, leaf fodder or wood was seen as more valuable in the mixed-farming regime of the West Highlands.[137] Indeed, as has been observed in the Affric study, valley-bottom pasture was probably vital to the survival of the herds upon which the economy of such upland regions was founded, providing both the 'early bite' needed for the spring recovery of overwintered cattle before the higher-altitude summer grazing areas were capable of receiving them and the hay on which they were in part sustained during the winter months.[138] Breaking such areas into cultivation thus signals a major shift in thinking and a reassessment of the balance of resources needed to sustain the communities in the glen.

The Bowmont Valley project suggested that shifts in practice generally were subtle and unlikely to have delivered significant changes in the local ecology, if even intermittent cultivation continued in the marginal areas. It is in the pastures of the Cheviot uplands that more visible change occurred, where what were species-rich grazed grasslands began to undergo a gradual transformation spanning several centuries from *c.*1100.[139] Muirburn had been employed for at least a century before 1250 and, subject to legal controls first introduced in the early 1400s,[140] continues to be used as a technique to stimulate new vegetation growth on which nowadays mainly game fowl and sheep feed. Its unforeseen consequences of contributing to peat growth in shallow natural basins by creating an impervious charcoal pan and increasing the acidification of surface soils, however, probably worsened as the fourteenth century progressed. The spread of *Calluna* heathland, discussed in Chapter 6, appears to have accelerated through the thirteenth and fourteenth centuries, with the high-quality grazing on species-rich grassland being gradually displaced by nutrient-poor heath. There is a need to better understand the processes underlying this deterioration in the quality of the upland pastures and to research other areas of intensifying grazing, such as on the Glenisla and Glen Brichty sheep-ranges of the monks of Coupar Angus Abbey, but an important driver of such change might have been new livestock management strategies and new balances in sheep-to-cattle ratios.

Around Ben Lawers, that above-mentioned shift to seasonal occupation post-1300 at what had previously been permanently occupied settlements might be linked to new short-range transhumance practices, probably involving cattle.[141] It is difficult not to see this development as reflective of wider socio-economic changes but possibly responding more to population contraction in the fourteenth century than to the immediate impact of climate change after 1256. Summertime use of high-altitude pastures, however, was hardly a new phenomenon of the post-1300 period. What changed was the location and nature of the temporary habitation provided for the herders. The provision in the later twelfth and thirteenth centuries of cattle sheds and sheep houses in upland areas of the Melrose Abbey estate shows that livestock were being taken off elevated districts in the southern Lammermuirs for at least part of the winter during the MCA. There is evidence from Bowmont for similar seasonal removal of grazing

137 Davies, 'Upland agriculture and environmental risk', p. 2058.

138 Ibid.

139 Tipping, *Bowmont*, pp. 194–196.

140 Oram, *Where Men No More May Reap or Sow*, chapter 1.

141 Atkinson, *Ben Lawers*, pp. 271–272.

animals from the high-level pastures in the pre-1400 period.[142] There is also a suggestion from there of variation in strategies relating to sheep and cattle, possibly responding to overgrazing, with the monks of Kelso and their tenants in the valley being required to remove their sheep from the high pastures between 10 June and 1 August annually from around 1250 onwards, and only cattle allowed during this period.[143] In Bowmont, intensive summer grazing of sheep in the thirteenth century was possibly a factor in accelerating *Calluna* growth, especially since animals were taken off the pasture in winter when the heath would otherwise have been grazed. The seven-week break from sheep was possibly designed to relieve pressure on the grassland from these preferential grazers. It is improbable that the higher and more exposed Highland grazings were managed much differently.

The intensification of sheep farming that is evident in the parchment record from Mow in the Bowmont Valley was replicated throughout much of Southern Upland Scotland but also in the eastern Perthshire and Angus glens, Strathisla in Banffshire and through the Moray lowlands. As discussed in Chapter 6, the expansion of sheep farming has its origins in the twelfth-century growth in the international demand for wool, but it was in the later thirteenth and fourteenth centuries that Scotland's wool output reached its medieval peak.[144] It is estimated that by the 1370s wool and woolfells from 2.1 million sheep were being exported annually.[145] Total headcount would have been significantly higher. Although discussion of sheep and wool is usually set in the context of the major monastic, especially Cistercian estates, its significance to Scots of all social classes should not be underestimated. Over 50 per cent of the annual wool-clip probably came from peasant-owned flocks in the later fourteenth century, and it has been famously described as 'perhaps the medieval equivalent of North Sea oil' in terms of its contribution to the thirteenth-century monetisation of the Scottish economy through the massive injection of foreign silver bullion it brought.[146] A significant portion of that silver came into peasant hands and would have made those with significant shares in common pastures rich in relative terms. It probably encouraged further investment in sheep at the expense of arable cultivation. Some sheep-farming operations were on an almost industrial scale. Melrose Abbey, with flocks of some 17,000 head by the 1390s, and the earl of March, with around 7,000, were major players whose sheep probably outstripped in number those of any other Scottish landowner.[147] The cottar tenancy granted to the canons of Holyrood, however, with the right to graze sixteen sheep on the common land at Twynholm in Galloway, perhaps gives a better sense of the scale of individual peasant flocks,[148] with a tounship perhaps having a couple of hundred animals collectively. Even for a peasant of lowly cottar status, who had no share in the arable land, the silver earned through sale of the wool from these few animals would have made a considerable difference to his family's economic security. The importance of this income makes the draconian steps taken to eradicate any threats to it from outbreaks of scab more understandable.

Elsewhere in Britain and in Europe an intensifi-

142 Tipping, *Bowmont*, p. 195.
143 Innes, *Kelso Liber*, vol. 1, pp. 117–118, 121; Tipping, *Bowmont*, p. 195.
144 For detailed discussion, see Ditchburn, *Scotland and Europe*, pp. 163–176.
145 Grant, *Independence and Nationhood*, pp. 62–63.
146 Ibid., p. 72; Mayhew, 'Alexander III', pp. 61–62.
147 Ditchburn, *Scotland and Europe*, p. 164.
148 Innes, *Holyrood Liber*, p. 60.

Hownam Law and the hills overlooking Mow in the Bowmont Valley, Cheviot Hills, from Cessford, Scottish Borders. Increased pressure from pasturing in the thirteenth century perhaps led to changes in the vegetation character of these hills, from open grassland to *Calluna*-dominated upland, but the landscape we see today is a product mainly of the intensification of sheep farming in the eighteenth and nineteenth centuries.

cation in livestock farming followed the Black Death, as marginal zones formerly in cultivation were given over to pasture. Those parts of southern Scotland with a pre-plague rural regime closest to the English model might have seen a similar trend, but the evidence is ambiguous. Livestock generally had always had a greater presence within the Scottish rural regime than in England, and since we lack reliable data for late thirteenth- and fourteenth-century Scotland, we cannot be certain that the high numbers of sheep indicated by volumes of traded wool in the last quarter of the fourteenth century represented a significant increase over already high pre-plague levels.[149] Mid thirteenth-century efforts to regulate flock sizes and periods of access to upland sheep pastures, such as those at Mow in the Cheviots, are unlikely to have been enforced rigorously when the profits from wool were potentially so great, but they are also unlikely to have been abandoned altogether. In the Bowmont Valley, there is no evidence palynologically or geomorphologically that grazing pressure intensified through overstocking. Detectable changes in upland vegetation there arose rather from deteriorating soil quality exacerbated by damaging

149 Ditchburn, *Scotland and Europe*, p. 163.

Eaval from Baleshare, North Uist, looking across the machair to the blackland with the upland summer pastures in the mountain area beyond. Extended periods of livestock grazing on the machair to compensate for shorter growing seasons on summer pastures were likely important contributors to the accelerating erosion of fragile coastal districts.

grassland management techniques, such as muirburn.[150] The degraded grassland landscapes of much of upland Scotland, to which sheep grazing is a major contributor, are a consequence of the intensification of stocking in the later eighteenth and nineteenth centuries, not in the Middle Ages.

While sheep and wool have tended to dominate discussion of Scotland's medieval rural economy, cattle were equally important and, in some parts of the country, economically and culturally more significant.[151] Apart from some areas of sheep specialisation on the estates of Kinloss Abbey and other major landholders in the Laich of Moray,[152] much of northern and western Scotland appears to have been dominated by cattle rearing throughout the Middle Ages. As the stocking regulations from the northern Cheviots demonstrate, however, even in areas of more intensive sheep farming, cattle constituted a substantial component of the livestock mix at sheep-to-cattle ratios averaging between 5.8:1 and 7:1. At a crude calculation based on the estimated national flock size relative to volume of wool being exported,

150 Tipping, *Bowmont*, pp. 195–196.

151 Highland and Hebridean cattle culture is discussed in Oram and Adderley, 'Lordship and environmental change', and Oram and Adderley, 'Lordship, land and environmental change'.

152 The just over 7 tonnes of wool seized at Pluscarden Priory outside Elgin in 1304/5, representing wool held in storage there for a foreign merchant, points to the clip of around 12,000 sheep. This is likely to have been gathered from multiple sources throughout the region, however, rather than from a single owner's flock. See Bain, *Calendar of Documents*, vol. 2, no. 1646 (p. 440).

that would suggest a national herd of around 400,000 head in the late fourteenth century. Certainly, it is estimated that by the 1370s, in addition to cattle skins being processed for Scottish domestic use, the hides of 50,000 cattle were being exported annually.[153] Such volumes indicate that livestock numbers, if affected in the same way as those documented for fourteenth-century England, had recovered from any effects of the cattle epizootic of the early 1320s and perhaps exceeded thirteenth-century levels.[154]

Chronicle evidence, as discussed above, indicates that high livestock mortalities also occurred during episodes of more extreme weather, especially in the severe winters of the later thirteenth and earlier fourteenth centuries. The impact of winter and spring losses of the fertile, milk-producing animals that formed the core herds would have been more damaging than higher mortality of calves or greater incidence of miscarriage by cold and undernourished cows. In districts where dairying was the dominant regime, the climatic downturn and its shortening of the grass-growing season perhaps saw an increase in the numbers of calves being slaughtered, but the aim would still have been to maintain stable herd numbers rather than deliver an overall reduction.[155] Delayed springs, where new grass growth started later, were especially damaging to herds that were normally taken to upland pastures for summer

153 Grant, *Independence and Nationhood*, pp. 62–63.

154 Ibid., pp. 62–33, 70, 76, 79–80.

155 F. McCormick, 'Calf slaughter as a response to marginality', in C.M. Mills and G. Coles (eds), *Life on the Edge: Human Settlement and Marginality* (Oxford, 1998), pp. 49–51.

grazing, and it is unlikely that the fourteenth-century expansion of transhumance pastoralism in areas like north Loch Tay-side identified in the Ben Lawers landscape project was an unregulated free-for-all. The movement of cattle onto shieling grounds where spring grass growth was delayed risked long-term damage to the vegetation, especially where the ground was also waterlogged and trampling pocked the turf and broke up root mats, leading to erosion. That being said, evidence for widespread overgrazing of upland pastures is no more visible throughout the central and north Highlands for this period than it is in the northern Cheviots. A contributor to accelerating erosion of the Hebridean machair, however, was probably extended winter grazing there of herds whose summer pastures could no longer sustain them for the normal period between May and September.

• • •

From climate change to the fish trade, timber supply to plague, the century and a half explored in this chapter saw profound and transformational developments that affected Scottish society at every level. Extreme weather events and their direct impact on crops or flocks and herds in the later thirteenth century perhaps created the headline news that most folk would remember. It was, however, the steadily incremental decline rather than the seemingly abrupt step change of these highly visible episodes that wrought the greatest changes across the century, although some of the most profound effects were achieved through mass human – and animal – mortality in very short periods of time. Despite the deeply rooted belief in a later thirteenth-century 'golden age', we have seen that from the opening of the period a generally downward trajectory in climate and weather had already been established. That deterioration persisted into the middle of the fourteenth century before entering a new phase of instability, then descended in the fifteenth century into the early stages of the 'little ice age'. Production of the dietary staples of the bulk of both the lowland and the upland population – grain and dairy products – was most affected, yet Scotland's smaller population and more diverse agricultural economy perhaps helped it to avoid the worst of the subsistence crises that afflicted England and mainland Europe in the second and third decades of the fourteenth century. It would probably have been little comfort to starving Scots, however, to know that they were experiencing only 'dearth' rather than full-blown 'famine'. And even if Scotland's experience of the Great European Famine and concurrent cattle plague was less severe than that of southern England, their impact on the nutritional intake of the bulk of the population and especially on pre-adolescent children perhaps left a legacy of weakened immune responses to bacterial and viral infection that delivered high mortality during the recurrent epidemics of the later fourteenth century.

Responses to climatic deterioration and resulting declining yields, however, and to the mass mortalities of the plague, were different to those seen further south in Britain and Europe. There, the lifting of population pressure saw a 'retreat from the marginal' with abandonment of the least productive and most labour-intensive zones as peasants took up vacant tenancies on the better land. Instead, across Scotland we see indications of an expansion of agriculture rather than a contraction, and the maintenance of arable cultivation in marginal zones even after the plague mortalities of the later fourteenth century. Such cultivation may have been driven by necessity, as communities sought to make up shortfalls in yields simply by growing more, or perhaps represents opportunistic and intermittent cultivation during better times. Most clearly, however, it reinforces another view that runs against the notion of a thirteenth-century golden age, namely that arable cultivation in Scotland was more often than not poised on that razor-thin edge between subsistence and failure. Scottish peasant cultivators practised a regime which sought always to maximise potential yields through cropping as much viable land as they could without eating too far into pasture or other environmental resources necessary for the functioning of their community. It is an image far removed from Wyntoun's rosy-cheeked peasantry revelling in surplus, but it is also one that is less grim than the new norms that would become established as the 'little ice age' tightened its grip.

Reredorter drain, St Andrews Cathedral-priory, St Andrews, Fife. The reredorter or latrine of the Augustinian priory was flushed by a constant flow of water. As the priory was located at the end of a long sandstone promontory where water was obtained from wells sunk deep into the rock, that constant flow was obtained by diverting water from the Kinness Burn west of St Andrews into a lade that ran for 2.5km to the precinct.

Conclusion – Stepping Back from the Brink?

> To take pity on people in distress is a human quality which every man and woman should possess, but it is especially requisite in those who have once needed comfort, and found it in others.[1]

These opening words of the preface to Giovanni Boccaccio's bawdily colourful suite of tales set in the plague-ravaged Florence of 1348 are often lost amidst the riotous word images that follow, yet they are a simple yet powerful self-reflective comment on medieval European society as it stared into the abyss. Having experienced the fear of the unknown, the isolation and the long-term physiological and psychological shockwaves of Covid-19, we perhaps have a better understanding of these words and of the sense of standing on the brink of apocalypse that inspired them than any recent generation. Just as before, during and after lockdown have become common chronological parameters in conversation, so before, during and after the Great Mortality became a measure of our ancestors' time. They had lived through one of the most profoundly transformative episodes in recorded human history, and although they might not instantly have understood it, they had emerged into a new and very unstable world, in which past practice, the increasingly challenged TEK built up over the MCA, no longer held firm. Behind them, the already fluid 'certainties' of the twelfth and thirteenth centuries became swiftly fading memories of a lost 'golden age'. Nothing would ever be quite the same again.

Yet we need to pause on that thought and reflect on what we have covered in the course of the millennium and a half that is the main span of this volume, and on the millennia before that which we moved through rapidly in the introduction. Except possibly for the plagues of the sixth century AD, there was no single agent of human mass mortality to rival the *Yersinia pestis* bacterium. But we have seen that there were other agencies at work throughout this era, whose devastating impact on humans, animal life and plants had punctuated the centuries and marked out moments of transformation. Of these agencies,

1 G. Boccaccio, *The Decameron*, trans. G.H. McWilliam (Harmondsworth, 1972), p. 45.

humans themselves can easily and with justification be presented as the worst perpetrators, for humans make conscious choices and those choices result in actions, at least some of the outcomes of which are intentional. In any exploration of our ancestors' relationship with the world around them, our eyes are drawn to the evidence of large-scale, at least semi-planned change, and in that evidence we can see the repercussions of ancient decisions still playing out today. The nature of those transformational impacts leaves little room for doubt that our predecessors in prehistory and well into the Historic era – indeed, into the relatively recent past – had an adversarial perspective on Nature. They saw it as something chaotic and raw, to be confronted and controlled or tamed, converted to something productive to satisfy their needs and thereby turned to optimal human advantage. We can see this in the record of felling of woodland, draining of wetland and breaking of ground into arable production, evident from the Neolithic onwards. It is, after all, a fundamental Old Testament view of the world as a gift from God to be exploited for the benefit of humanity.[2] And the adversarial aspect is not hard to comprehend when we consider that communities teetering on the knife-edge of subsistence saw their crops destroyed in a single storm event, consumed beneath a blanket of wind-blown sand or devoured by insect infestation. Uncontrolled Nature was, to them, the greatest threat to their survival, greater even than the Black Death, and as explored in *Where Men No More May Reap or Sow* and *Standing on the Edge of Being*, it continued to be seen this way to the twentieth century.

Given what we know of our ancestors' approach to Nature, it is easy to form a view of most of the environmental changes for which there is visible evidence around us as anthropogenic in origin. Misanthropy runs deep in much modern environmental discourse, and in the new environmentalism that has emerged since the 1960s we see the impulse to put right the wrongs inflicted by the generations who came before us. As argued throughout the preceding chapters, however, the traditional and still commonly expressed view of Scotland's environment to be found in popular and populist histories, of a relentless downward trajectory impelled by human action, needs a radical overhaul in the light of more recent historical and scientific research. Let us be clear: there is no doubt that human agency was a significant and often horrifyingly destructive factor in delivering change across the millennia. Damage and devastation were inflicted wittingly through deliberate choices and actions – from the conversion of 'waste' into agricultural land to the extermination of wolves – as well as unconsciously through the 'fellow travellers' that our ancestors imported, from invasive weeds brought with our preferred food plants to tree-killing fungi, pathogens and predatory small mammals. But we must also recognise that natural processes had been at work for a far longer time, delivering a succession of transitions from one climatic state to another and significant shifts in weather regime with each step. Species extinctions, woodland climaxes and declines, the growth of blanket peat bog, changes in migratory patterns, all these things were already in train before human populations had established a foothold in what was to become Scotland. From that point, however, the pace of change began to accelerate. But it was the interaction of natural and anthropogenic agency that shaped future trends, with human choices and responses delivering unexpected outcomes, at least some of which were beneficial to both the human population and the wider environment with which they were engaged. A greater number, however, entrenched and in some cases speeded movements already in progress, particularly woodland loss, that had commenced centuries earlier as the post-glacial

2 White, 'The historical roots of our ecologic crisis', is the classic presentation of this view.

climax peaked then began to decline.

Only by taking the *longue durée* approach adopted in this volume can we see that both the human relationship with Scotland's environment and the influence of the climate upon it are punctuated by episodes of intensifying or slackening pressure, and occasionally by phases of temporary reversal in the direction of dominant trends. We find this pattern of intensification and slackness in respect of changes in temperature, precipitation and storminess, which affected geographical and altitudinal ranges of successful plant growth and animal habitat as well as cycles of fish migration, but which also determined the viability of human subsistence strategies. In anthropogenic terms, the pattern is clear in the spread of inroads made by early agriculture into expanses of the most easily tillable soils, and in the reversion to scrub or grassland, and the regeneration of woodland, following abandonment of the cleared ground in the Neolithic and Bronze Ages, again in the depths of the LALIA, and more subtly in the later thirteenth and fourteenth centuries when weather regimes shifted or population numbers dropped. Most discussion, as noted in this volume and as seen in the almost daily media coverage of the subject, concentrates on the form and fate of Scotland's woodland. There is an abundance of literature and web-based discussion or polemic on the topic, much of it aspirational and inspirational in tone, but some of it hopelessly romantic, ill-informed or simply dangerous. There is no doubt that the unremitting contraction of the climax forest is everywhere in evidence, but the palaeoenvironmental and historical evidence for the causes of that decline are too often ignored when they run contrary to current agendas. Never let fact and reality get in the way of a good story. Rather than any attempt to understand why woodland was failing even in areas where human pressure was absent, there is fantasising about the wilful human destruction of the 'Great Wood' by an ever-shifting panoply of historical pantomime villains. Instead of exploring the processes driving formation of the successor land forms and their character, there is a quixotic search for relict areas of purity or 'primeval' survival. Rarely outside academic discourse is there meaningful engagement with what pre-modern woodland management looked like or with the character and composition of the woodland and its resident biota that evolved through the interplay of human and natural agency.

When we look back at the evidence for the millennia of evolutionary change in our interaction with Scotland's environment, we realise that there is something understandable but also deeply questionable and possibly fundamentally flawed in the rationales that underlie our current drive to create new forests through 'rewilding' to replace what has gone. There is little that is logical in the insistence that these modern anthropogenic creations are somehow 'natural', while at the same time the human role in the shaping of woodland designated as semi-natural and celebrated for its antiquity and its biodiversity is excised or at best denigrated as destructive intervention. Throughout this volume, to obtain the broader context that is needed when making decisions about our future environment that will have far-reaching and potentially irreversible consequences, the interrelationships and interdependencies between human and non-human populations and plants have been stressed. Through those inseparable past engagements, we have explored the multifaceted nature of the still unfolding relationship between humans and their environment. A clear theme has emerged in the course of that exploration, namely that this relationship can rarely be seen in starkly black-and-white terms. For example, those semi-natural woods are, conversely, semi-anthropogenic and took their form through the woodland's capacity to respond to the ever-changing patterns of wood use introduced by our ancestors, from selective felling and coppicing, conservation and clearance, to charcoal burning, pig grazing and wood pasture for cattle, with both human and natural factors operating within the constraints of geology, climate trends and weather effects. Similarly, it is that multi-factor interrelationship that left our uplands already largely devoid of extensive tree cover by around three millennia before intensive sheep farming was introduced in parts of the Southern Uplands and four millennia before the infamous sheep-walks of the Highland Clearances were formed. Wetlands – coastal and landward – have come and gone as sea levels have changed, precipitation levels have fluctuated, ground cover has shifted and the human exploitation of such landscapes has varied in purpose and intensity. To better understand the factors that have led to the present state of our environment, we need to look beyond current preoccupations and fashions to those interrelationships across the *longue durée* that expose the simple and 'quick-fix' solutions as facile, superficial and methodologically unbalanced.

All too often, the value of the deeper historical and prehistoric past as a source of detailed, high-quality information for the analysis of environmental trends over long periods is overlooked in modern discourse. Instead, much of the current debate is founded on relatively shallow data and works from a baseline of assumptions that seems to have been fixed arbitrarily. Perhaps even more concerning is the repeated failure of many well-resourced multi- or interdisciplinary environmental projects to work with the environmental historians who have the depth of knowledge of the archives, the experience of ground-truthing their evidence and the skills in interdisciplinary methodologies to offer unique insights on the human processes that contributed to a given situation. It is only by looking at problems across that deeper chronological range that we can properly approach the analysis of our current circumstances. As we shall discover in the volumes *Where Men No More May Reap or Sow* and *Standing on the Edge of Being*, many of our modern analyses look no further than the arbitrary benchmark of an often vaguely defined Improvement era, without reflecting on the biases in much of the Improvement literature or asking how we arrived at the position that has been used as the threshold from which subsequent change is measured. Most critically, they usually fail to recognise that there were processes within the 'new' developments that had already been in progress for centuries. If we consider just the *dabhaichean* of the kingdom of Alba, we see a system whose presence affected the exploitation of the land, water and their resources across a millennium, putting in place socio-economic and political boundaries that constrained patterns of resource use over broad expanses of lowland and upland districts, and producing widely varying outcomes to either side of shared boundaries. That those *dabhaichean* functioned effectively into the early nineteenth century in districts that contain many of our most prized areas of ancient woodland, species-diverse grassland or water meadow rarely figures in current environmental strategies. Surely we need to understand what it was in the management practices of past populations in those places that enabled the survival of landscape character and biodiversity that we so value today. It is those overlooked processes that we have begun to explore in this volume. In doing so, we have made clear that there was nothing that was truly 'pristine' by the beginning of the Historic era and shown how different degrees of anthropogenic and non-anthropogenic pressures working with variations in geography, geology, geomorphology and climate, cultural preferences and technological capacities delivered widely diverse outcomes. Our distant ancestors had the capacity to make profound changes to the environment, to shape it such that it would better deliver those things upon which their subsistence depended. They could overexploit and exhaust, and persecute perceived competitors to extinction, but they could also recognise depletion and take steps to regulate overuse and bad practice. Above all, however, they developed a fundamental social resilience in the face of adversity and crisis, and it is from that characteristic that we have most to learn in these uncertain times.

Bibliography

Unpublished primary sources

Glamis Castle Archive

GCA 251/24

National Library of Scotland (NLS)

NLS, General Roy Military Survey of Scotland 1747–1755, https://maps.nls.uk/geo/explore/#zoom=13&lat=56.28581&lon=-3.92479&layers=3&b=1.

NLS, John Blaeu, Glottiana Praefectura Superior, [vel], The Upper Ward of Clyds-dayl / Auct. Timoth. Pont (Amsterdam, 1654), http://maps.nls.uk/view/00000434.

NLS, Timothy Pont, [Glasgow and the County of Lanark] – Pont 34, http://maps.nls.uk/view/00002331.

National Records of Scotland (NRS)

GD40 Papers of the Kerr family, marquises of Lothian
GD40/1/15

GD122 Papers of the Gilmour family of Craigmillar and Liberton, Midlothian
GD122/2/365

GD220 Papers of the Graham family, dukes of Montrose (Graham Muniments)
GD220/1/A/1/1/2
GD220/2/1/2

RH1 Miscellaneous transcripts, etc. Individual documents
RH1/6/20

Printed and digital primary sources and editions

Adomnan's Life of Columba, ed. and trans. A.O. Anderson and M.O. Anderson (Edinburgh: Thomas Nelson and Sons, 1961).

Amos, A. (ed.), *Martial and the Moderns* (Cambridge: Deighton, Bell, 1858).

Anderson, A.O. (ed.), *Early Sources of Scottish History to a.d. 1286*, 2 volumes (Edinburgh: Oliver and Boyd, 1922).

Andrew of Wyntoun, *The Original Chronicle of Andrew of Wyntoun*, ed. F.J. Amours, 6 volumes (Edinburgh: Scottish Text Society, 1903–14).

The Anglo-Saxon Chronicle, ed. and trans. G.N. Garmonsway, new edition (London: J.M. Dent, 1972).

Annales Monastici, ed. H.R. Luard, 5 volumes (London: Longman, 1864–69).

Annals of Connacht, http://www.ucc.ie/celt/published/T100011/index.html.

Annals of Innisfallen, https://celt.ucc.ie/published/T100004/index.html.

Annals of Loch Cé, http://www.ucc.ie/celt/published/T100010A/index.html.

Annals of the Four Masters, http://www.ucc.ie/celt/published/T100005B/index.html.

Annals of Tigernach, https://celt.ucc.ie/published/T100002A/index.html.

Annals of Ulster, https://celt.ucc.ie/published/T100001A/index.html (431–1201); https://celt.ucc.ie/published/T100001B/index.html (1202–1378).

Arrian on Coursing: The Cynegeticus of the Younger Xenophon, Translated from the Greek, with Classical and Practical Annotations, and a Brief Sketch of the Life and Writings of the Author: To Which Is Added an Appendix, Containing Some Account of the Canes venatici of Classical Antiquity, ed. and trans. W. Dansey (London: J. Bohn, 1831).

Bain, J. (ed.), *Calendar of Documents Relating to Scotland*, volume 2: *1272–1307* (Edinburgh: H. M. General Register House, 1884).

Barbour, J., *The Bruce*, ed. and trans. A.A.M. Duncan (Edinburgh: Canongate Books, 1997).

Barrow, G.W.S. (ed.), *The Charters of David I: The Written Acts of David I King of Scots, 1124–53, and of His Son Henry, Earl of Northumberland, 1139–52* (Woodbridge: Boydell Press, 1999).

Boccaccio, G., *The Decameron*, trans. G.H. McWilliam (Harmondsworth: Penguin Books, 1972).

Boece, H., *The Chronicles of Scotland by Hector Boece, Translated into Scots by John Bellenden, 1531*, ed. E.C. Batho and H.W. Husbands, volume 2 (Edinburgh: Scottish Text Society, 1941).

Boece, H., *The Chronicles of Scotland by Hector Boece, Translated into Scots by John Bellenden, 1531*, Oxford Text Archive, https://ota.bodleian.ox.ac.uk/repository/xmlui/bitstream/handle/20.500.12024/A16292/A16292.html.

Bower, W., *Scotichronicon*, ed. D.E.R. Watt et al., 9 volumes (Aberdeen: Aberdeen University Press, 1987–98).

Burnett, G. (ed.), *The Exchequer Rolls of Scotland*, volume 2: *1359–1379* (Edinburgh: H. M. General Register House, 1878).

Burnett, G. (ed.), *The Exchequer Rolls of Scotland*, volume 4: *1406–1436* (Edinburgh: H. M. General Register House, 1880).

Burnett, G. and Mackay, A.J.G. (eds), *The Exchequer Rolls of Scotland*, volume 13: *1508–1513* (Edinburgh: H. M. General Register House, 1891).

Burns, C. (ed.), *Calendar of Papal Letters to Scotland of Clement VII of Avignon 1378–1394* (Edinburgh: Scottish History Society, 1976).

Childs, W.R. and Taylor, J. (eds), *The Anonimalle Chronicle 1307 to 1334: From Brotherton Collection MS 29* (Cambridge: Cambridge University Press, 1991).

Chronica Regum Manniae et Insularum: The Chronicle of Man and the Isles: A Facsimile of the Manuscript Codex Julius A. VIII in the British Museum (Douglas, 1924).

The Chronicle of Melrose (facsimile edition), ed. A.O. Anderson et al. (London: Percy Lund Humphreys, 1936).

Chronicon de Lanercost, ed. J. Stevenson (Edinburgh: Maitland Club, 1839).

Chronicon Scotorum, http://research.ucc.ie/celt/document/T100016.

Claudius Claudianus, *Panegyricus de Quarto Consulatu Honorii Augusti*, ed. and trans. W. Barr (Liverpool: Liverpool University Press, 1981).

Continuation of the Chronicle of William of Newburgh, in R. Howlett (ed.), *Chronicles of the Reigns of Stephen, Henry II and Richard I*, volume 2 (London: Rolls Series, 1884).

Cooper, T.M., Baron Cooper of Culross (ed.), *The Register of Brieves as Contained in the Ayr MS., the Bute MS., and Quoniam Attachiamenta* (Edinburgh: Stair Society, 1946).

Cooper, W.S. (ed.), *Charters of the Royal Burgh of Ayr* (Edinburgh: Ayr and Wigton Archaeological Association, 1883).

Dowden, J. (ed.), *Chartulary of the Abbey of Lindores 1195–1479* (Edinburgh: Scottish History Society, 1903).

Dunbar, W., *The Poems of William Dunbar*, ed. J. Small, volume 2: *Poems* (Edinburgh: Scottish Text Society / William Blackwood and Sons, 1893).

Duncan, A.A.M. (ed.), *Formulary E: Scottish Letters and Brieves 1286–1424*, University of Glasgow Department of Scottish History Occasional Paper (Glasgow: University of Glasgow, 1976).

Easson, D.E. (ed.), *Charters of the Abbey of Coupar Angus*, 2 volumes (Edinburgh: T. & A. Constable for the Scottish History Society, 1947).

Easson, D.E. and MacDonald, A. (eds), *Charters of the Abbey of Inchcolm* (Edinburgh: Scottish History Society, 1938).

Fraser, W. (ed.), *Registrum Monasterii S. Marie de Cambuskenneth* (London: Grampian Club, 1872).

Fraser, W. (ed.), *The Annandale Family Book of the Johnstones, Earls and Marquises of Annandale*, 2 volumes (Edinburgh: David Douglas, 1894).

Gerald of Wales, *The History and Topography of Ireland*, trans. J. O'Meara (Harmondsworth: Penguin Books, 1982).

Y Gododin, trans. J. Williams (London: Longman, 1852).

Grainger, F. and Collingwood, W.G. (eds), *The Register and Records of Holm Cultram* (Kendal: Cumberland and Westmorland Antiquarian and Archaeological Society, 1929).

Hannay, R.K. (ed.), *Rentale Dunkeldense: Being Accounts of the Bishopric (a.d. 1505–1517) with Myln's 'Lives of the Bishops' (A.D. 1483–1517)* (Edinburgh: Scottish History Society, 1915).

Harvey, C.C. (ed.), *Calendar of Writs Preserved at Yester House 1166–1503* (Edinburgh: Scottish Record Society, 1916).

Henry of Huntingdon, *The History of the English People 1000–1154*, ed. D. Greenway (Oxford: Oxford University Press, 1996).

Historia Brittonvm cvm Additamentis Nennii, in *Monumenta Germaniae Historica, Auctorum Antiquissimi* 13: *Chronica minora*, volume 3, ed. T. Mommsen (Berlin, 1898), pp. 111–222.

Horrox, R. (ed. and trans.), *The Black Death* (Manchester: Manchester University Press, 1994).

In Praise of Later Roman Emperors: The Panegyrici Latini, ed. and trans. C.E.V. Nixon and B.S. Rodgers (Berkeley: University of California Press, 1994).

Innes, C. (ed.), *Registrum Monasterii de Passelet 1163–1529* (Edinburgh: Maitland Club, 1832).Innes, C. (ed.), *Liber Sancte Marie de Melros*, 2 volumes (Edinburgh: Bannatyne Club, 1837).

Innes, C. (ed.), *Registrum Episcopatus Moraviensis* (Edinburgh: Bannatyne Club, 1837).

Innes, C. (ed.), *Liber Cartarum Sancte Crucis* (Edinburgh: Bannatyne Club, 1840).

Innes, C. (ed.), *Registrum de Dunfermelyn* (Edinburgh: Bannatyne Club, 1842).

Innes, C. (ed.), *Liber Ecclesie de Scon* (Edinburgh: Bannatyne Club, 1843).

Innes, C. (ed.), *Liber S. Marie de Calchou*, 2 volumes (Edinburgh: Bannatyne Club, 1846).

Innes, C. (ed.), *Liber S. Marie de Dryburgh* (Edinburgh: Bannatyne Club, 1847).

Innes, C. (ed.), *Liber S. Thome de Aberbrothoc*, 2 volumes (Edinburgh: Bannatyne Club, 1848–56).

Innes, C. (ed.), *Registrum S. Marie de Neubotle* (Edinburgh: Bannatyne Club, 1849).

Innes, C., Thomson, T. and Macdonald, A. (eds), *Registrum Honoris de Morton*, 2 volumes (Edinburgh: Bannatyne Club, 1853).

Jackson, K.C., *The Gododdin: The Oldest Scottish Poem* (Edinburgh: Edinburgh University Press, 1969).

John of Fordoun, *Chronica Gentis Scotorum*, ed. W.F. Skene, 2 volumes (Edinburgh: William Paterson, 1871–2).

Lawrie, A.C. (ed.), *Early Scottish Charters prior to 1153* (Glasgow: James Maclehose and Sons, 1905).

Laxdaela Saga, trans. M.A.C. Press (London: J.M. Dent , 1899).

Leslie, J., *The Historie of Scotland, Written First in Latin by the Most Reuerend and Worthy Jhone Leslie Bishop of Rosse and Translated in Scottish by Father James Dalrymple*, ed. E.G. Cody and W. Murison, 2 volumes (Edinburgh: Scottish Text Society, 1888–95).

Lindsay, E.R. and Cameron, A.I. (eds), *Calendar of Scottish Supplications to Rome 1418–1422* (Edinburgh: Scottish History Society, 1934).

Lindsay, W.A., Dowden, J. and Maitland Thomson, J. (eds), *Charters, Bulls and Other Documents Relating to the Abbey of Inchaffray* (Edinburgh: Scottish History Society, 1908).

MacPhail, S.R. (ed.), *History of the Religious House of Pluscardyn* (Edinburgh: Oliphant, Anderson and Ferrier, 1881).

Maitland Thomson, J. (ed.), *Registrum Magni Sigilli Regum Scotorum: The Register of the Great Seal of Scotland*, new edition, volume 1: *A.D. 1306–1424* (Edinburgh: H. M. General Register House, 1882).

Martin, M., *A Description of the Western Islands of Scotland circa 1695, by Martin Martin, Gent., Including A Voyage to St. Kilda by the Same Author and A Description of the Western Isles of Scotland by Sir Donald Monro*, ed. D.J. Macleod (Stirling: Eneas Mackay, 1934).

Matthew Paris, *Chronica Majora*, ed. H. Luard, volume 5 (London: Rolls Series, 1880).

Munro, J. and Munro, R.W. (eds), *Acts of the Lords of the Isles 1336–1493* (Edinburgh: Scottish History Society, 1986).

Nennius, *British History and the Welsh Annals*, ed. J. Morris (London: Philimore, 1980).

Orkneyinga Saga: The History of the Earls of Orkney, trans. H. Pálsson and P. Edwards (Harmondsworth: Penguin Books, 1981).

Paul, J.B. (ed.) *Registrum Magni Sigilli Regum Scotorum*, volume 2: *1424–1513* (Edinburgh: H. M. General Register House, 1882).

Pliny the Elder, *The Natural History of Pliny*, ed. and trans. J. Bostock and H.T. Riley, volume 1 (London: Henry G. Bohn, 1855).

The Records of the Parliaments of Scotland to 1707, ed. K.M. Brown et al. (St Andrews, 2007–24), http://www.rps.ac.uk/ (*RPS*).

Regesta Regum Scottorum, volume 1: *The Acts of Malcolm IV 1153–1165*, ed. G.W.S. Barrow (Edinburgh: Edinburgh University Press, 1960).

Regesta Regum Scottorum, volume 2: *The Acts of William I 1165–1214*, ed. G.W.S. Barrow (Edinburgh: Edinburgh University Press, 1971).

Regesta Regum Scottorum, volume 4 part 1: *The Acts of Alexander III 1249–1286*, ed. C.J. Neville and G. Simpson (Edinburgh: Edinburgh University Press, 2013).

Regesta Regum Scottorum, volume 5: *The Acts of Robert I 1306–1329*, ed. A.A.M. Duncan (Edinburgh: Edinburgh University Press, 1988).

Regesta Regum Scottorum, volume 6: *The Acts of David II 1329–1371*, ed. B. Webster (Edinburgh: Edinburgh University Press, 1982).

Reginald of Durham, *Vita Sancti Oswaldi Regis et Martyris*, in Symeon of Durham, *Symeonis Monachi Opera Omnia*, ed. T. Arnold, 2 volumes (London: Longman, 1882–85), volume 1, pp. 326–385.

Robert of Torigni, *Chronicle of Robert of Torigni, Abbot of the Monastery of St Michael-in-Peril-of-the-Sea*, in R. Howlett (ed.), *Chronicles of the Reigns of Stephen, Henry II and Richard I*, volume 4 (London: Longman, 1889).

Rogers, C. (ed.), *Rental Book of the Cistercian Abbey of Cupar-Angus*, 2 volumes (London: Grampian Club, 1879).

A Scottish Chronicle Known as the Chronicle of Holyrood, ed. M.O. Anderson with notes by A.O. Anderson (Edinburgh: Scottish History Society, 1938).

The Scottish Collection of Verse to 1800, ed. E. Dunlop and A. Kamm (Glasgow: Richard Drew, 1985).

Stuart, J. (ed.), *Illustrations of the Topography and Antiquities of the Shire of Banff*, volume 2 (Aberdeen: Spalding Club, 1847).

Stuart, J. (ed.), *Records of the Monastery of Kinloss* (Edinburgh: Society of Antiquaries of Scotland, 1872).

Stuart, J. and Burnett, G. (eds), *The Exchequer Rolls of Scotland*, volume 1: *1264–1359* (Edinburgh: H. M. General Register House, 1878).

Symeon of Durham, *Historia Regum*, in *Symeonis Monachi Opera Omnia*, ed. T. Arnold, 2 volumes (London: Longman, 1882–85), volume 2.

T.J., a Well-Wisher to the King, Parliament and the Kingdom, *The World Turn'd Upside Down, or A Briefe Description of the Ridiculous Fashions of These Distracted Times* (London: John Smith, 1646).

Tacitus, *The Agricola and the Germania*, trans. H. Mattingly, rev. S.A. Handford (Harmondsworth: Penguin Books, 1970).

The Táin: Translated from the Irish Epic Tain Bo Cuailnge, ed. and trans. F. Kinsella (Dublin: Dolmen Press, 1969).

Turnbull, W.B.B.D. (ed.), *Liber Sancte Marie de Balmorinach* (Edinburgh: Abbotsford Club, 1841).

William of Newburgh, *Historia Rerum Anglicarum*, in R. Howlett (ed.), *Chronicles of the Reigns of Stephen, Henry II and Richard I*, volume 1: *Containing the First Four Books of the Historia Rerum Anglicarum of William of Newburgh* (London: Longman, 1881).

William of Newburgh, *Historia Rerum Anglicarum*, book 5, in R. Howlett (ed.), *Chronicles of the Reigns of Stephen, Henry II and Richard I*, volume 2 (London: Longman, 1885).

Secondary works

Aberdeenshire Historic Environment Record (HER), https://online.aberdeenshire.gov.uk/smrpub/master/default.aspx

Aberth, J., *From the Brink of the Apocalypse: Confronting Famine, War, Plague and Death in the Later Middle Ages*, 2nd edition (Abingdon: Routledge, 2010).

Allen, J.R. and Anderson, J., *The Early Christian Monuments of Scotland: A Classified Illustrated Descriptive List of the Monuments with an Analysis of Their Symbolism and Ornamentation* (Edinburgh: Neil, 1903).

Anderson, M.L., *A History of Scottish Forestry*, 2 volumes (Edinburgh: Nelson and Sons, 1967).

Anderson, M.O., 'The Celtic church in Kinrimund', in D. McRoberts (ed.), *The Medieval Church of St Andrews* (Glasgow: John S. Burns and Sons, 1976), pp. 1–10.

Anderson, S. et al., 'The excavation of a large double-chambered souterrain at Ardownie Farm Cottages, Monifieth, Angus', *Tayside and Fife Archaeological Journal*, 12 (2006), pp. 14–60.

Andrades Valtueña, A. et al., 'Stone Age *Yersinia pestis* genomes shed light on the early evolution, diversity, and ecology of plague', *Proceedings of the National Academy of Sciences*, 119:17 (2022), https://doi.org/10.1073/pnas.2116722119.

Anklin, J. et al., 'Climate instability during the last interglacial period recorded in the Greenland Ice-core Project', *Nature*, 364:6434 (1993), pp. 203–207.

Arjava, A., 'The mystery cloud of 536 CE in the Mediterranean sources', *Dumbarton Oaks Papers*, 59 (2005), pp. 73–94.

Armit, I., 'The abandonment of souterrains: evolution, catastrophe or dislocation?', *Proceedings of the Society of Antiquaries of Scotland*, 129 (1999), pp. 577–596.

Armit, I., 'Life after Hownam: the Iron Age in south-east Scotland', in W. Bevan (ed.), *Northern Exposure: Interpretive Devolution and the Iron Ages in Britain*, Leicester Archaeology Monographs 4 (Leicester: Leicester University Press, 1999), pp. 65–80.

Armit, I., *Celtic Scotland*, new edition (London: B.T. Batsford, 2005).

Armit, I. and Ralston, I., 'The coming of iron', in T.C. Smout (ed.), *People and Woods in Scotland: A History* (Edinburgh: Edinburgh University Press, 2003), pp. 40–59.

Atkinson, J.A., *Ben Lawers: An Archaeological Landscape in Time: Results from the Ben Lawers Historic Landscape Project, 1996–2005*, Scottish Archaeological Internet Report 62 (2016), https://archaeologydataservice.ac.uk/library/browse/issue.xhtml?recordId=1137495&recordType=MonographSeries.

Baillie, M.G.L., *Tree-Ring Dating and Archaeology* (London: Croom Helm, 1982).

Baillie, M.G.L., 'Do Irish bog oaks date the Shang dynasty', *Current Archaeology*, 117 (1989), pp. 310–313.

Baillie, M.G.L., 'Dendrochronology raises questions about the nature of the ad 536 dust-veil event', *The Holocene*, 4 (1994), pp. 212–217.
Baillie, M.G.L., *A Slice through Time* (London: B.T. Batsford, 1995).
Baillie, M.G.L., 'Bad for trees – bad for humans?', in C.M. Mills and G. Coles (eds), *Life on the Edge: Human Settlement and Marginality* (Oxford: Oxbow Books, 1998), pp. 13–19.
Baker, A. et al., 'A composite annual-resolution stalagmite record of North Atlantic climate over the last three millennia', *Scientific Reports*, 5 (2015), https://doi.org/10.1038/srep10307.
Ballin, T.B., 'ARO48: A likely late Upper Palaeolithic assemblage from Lunanhead in Angus: re-examination and discussion', Archaeology Reports Online (2021), https://www.archaeologyreportsonline.com/PDF/ARO48_Lunanhead.pdf.
Ballin, T.B. and Wickham-Jones, C., 'Searching for the Scottish Late Upper Palaeolithic: a case study from Nethermills Farm, Aberdeenshire', *Journal of Lithic Studies*, 4 (2017), https://doi.org/10.2218/jls.v4i1.1907.
Ballin, T.B., Saville, A., Tipping, R. and Ward, T., 'An Upper Palaeolithic flint and chert assemblage from Howburn Farm, South Lanarkshire, Scotland: first results', *Oxford Journal of Archaeology*, 29:4 (2010), pp. 323–360.
Ballin, T.B. et al., *Reindeer Hunters at Howburn Farm, South Lanarkshire: A Late Hamburgian Settlement in Southern Scotland – Its Lithic Artefacts and Natural Environment* (Oxford: Archaeopress, 2018).
Ballin Smith, B. (ed.), *Howe: Four Millennia of Orkney Prehistory*, Society of Antiquaries of Scotland Monograph Series 9 (Edinburgh: Society of Antiquaries of Scotland, 1994).
Barber, J.W. et al., 'Excavations on Iona, 1979', *Proceedings of the Society of Antiquaries of Scotland*, 111 (1981), pp. 282–380.
Barclay, G.J., 'The Neolithic', in K.J. Edwards and I.B.M. Ralston (eds), *Scotland after the Ice Age: Environment, Archaeology and History 8000 BC–AD 1000*, 2nd edition (Edinburgh: Edinburgh University Press, 2002), pp. 127–149.
Barrett, J.H., 'Fish trade in Norse Orkney and Shetland: a zooarchaeological approach', *Antiquity*, 71 (1997), pp. 616–638.
Barrett, J.H., 'The pirate fishermen: the political economy of a medieval maritime society', in B. Ballin Smith, S. Taylor and G. Williams (eds), *West over Sea: Studies in Scandinavian Sea-Borne Expansion and Settlement before 1300* (Leiden: Brill, 2007), pp. 299–340.
Barrett, J.H., Nicholson, R. A. and Céron-Carrasco, R., 'Archaeo-ichthyological evidence for long-term socioeconomic trends in Northern Scotland: 3500 bc to ad 1500', *Journal of Archaeological Science*, 26 (1999), pp. 353–388.
Barrett, J.H., Beukens, R.P. and Nicholson, R.A., 'Diet and ethnicity during the Viking colonization of northern Scotland: evidence from fishbones and stable carbon isotopes', *Antiquity*, 75 (2001), 145–154.
Barrow, G.W.S., 'The aftermath of war: Scotland and England in the late thirteenth and early fourteenth century', *Transactions of the Royal Historical* Society, 5th series, 27 (1978), pp. 103–126.
Barrow, G.W.S., *The Anglo-Norman Era in Scottish History* (Oxford: Oxford University Press, 1980).
Barrow, G.W.S., *Robert Bruce and the Community of the Realm of Scotland*, 3rd edition (Edinburgh: Edinburgh University Press, 1988).
Bartlett, R., *The Making of Europe: Colonisation, Conquest and Cultural Change 950–1350* (London: BCA, 1993).
Bartley, D.D., Chambers, C. and Hart-Jones, B., 'The vegetational history of parts of south and east Durham', *New Phytologist*, 77 (1976), pp. 437–468.
Bathgate, T.D., 'Ancient fish-traps or yairs in Scotland', *Proceedings of the Society of Antiquaries of Scotland*, 83 (1949), pp. 98–102.
Beam, A., *The Balliol Dynasty 1210–1364* (Edinburgh: John Donald, 2008).
Benedictow, O., *The Black Death 1346–1353: A Complete History* (Woodbridge: Boydell Press, 2004).
Bennett, K.D., 'Late Quaternary vegetation dynamics of the Cairngorms', *Botanical Journal of Scotland*, 48 (1996), pp. 51–63.
Bennett, K.D., 'Post-glacial dynamics of pine', in J.R. Aldhous (ed.), *Our Pinewood Heritage: Proceedings of a Conference at Culloden Academy, Inverness, 20th–22nd October, 1994* (Royal Society for the Protection of Birds, 1996), pp. 23–39.
Berman, C., *The Cistercian Evolution: The Invention of a Religious Order in Twelfth-Century Europe*, illustrated edition (Pennsylvania: University of Pennsylvania Press, 2010).
Bibby, J.S., Douglas, H.A., Thomasson, A.J. and Robertson, J.S., *Land Capability Classification for Agriculture* (Aberdeen: Macaulay Institute for Soil Research, 1982).
Bigelow, G.F., 'Issues and prospects in Shetland Norse archaeology', in C.D. Morris and D.J. Rackham (eds), *Norse and Later Settlement and Subsistence in the North Atlantic*, University of Glasgow Department of Archaeology Occasional Paper 1 (Glasgow: University of Glasgow, 1992), pp. 9–32.
Boardman, S.I., *The Early Stewart Kings: Robert II and Robert III 1371–1406* (East Linton: Tuckwell Press, 1996).
Bond, J.M., 'Beyond the fringe? Recognising change and adaptation in Pictish and Norse Orkney', in C.M. Mills and G. Coles (eds), *Life on the Edge: Human Settlement and Marginality* (Oxford: Oxbow Books, 1998), pp. 81–90.
Bond, J.M., 'The mammal bone', in J. Hunter with J.M. Bond and A.N. Smith, *Investigations in Sanday, Orkney*, volume 1: *Excavations at Pool, Sanday: A Multi-period Settlement from Neolithic to Late Norse Times* (Kirkwall: The Orcadian / Historic Scotland, 2007).
Bond, J.M., 'The plant remains', in J. Hunter with J.M. Bond and A.N. Smith, *Investigations in Sanday, Orkney*, volume 1: *Excavations at Pool, Sanday: A Multi-period Settlement from Neolithic to Late Norse Times* (Kirkwall: The Orcadian / Historic Scotland, 2007).
Bond, J.M. and Hunter, J.R., 'Flax-growing in Orkney from the Norse period to the 18th century', *Proceedings of the Society of Antiquaries of Scotland*, 117 (1987), pp. 175–181.
Bond, J.M., Guttmann, E. and Simpson, I.A., 'Bringing in the sheaves: farming intensification in the post-broch Iron Age', in R.A. Housley and G. Coles (eds), *Atlantic Connections and Adaptations: Economies, Environments and Subsistence in Lands Bordering the North Atlantic* (Oxford: Oxbow Books, 2004), pp. 138–145.
Bondeson, L., and Bondeson, T., 'On the mystery cloud of ad 536, a crisis in dispute and epidemic ergotism: a linking hypothesis', *Danish Journal of Archaeology*, 3:1 (2014), pp. 61–67.
Bowman, A. and Thomas, D., *Vindolanda: The Latin Writing Tablets* (London: Society for the Promotion of Roman Studies, 1983).
Boyd, W.E., 'Environmental change and Iron Age land management in the area of the Antonine Wall, Central Scotland: a summary', *Glasgow Archaeological Journal*, 11 (1984), pp. 75–81.

Boyd, W.E., 'Palaeobotanical report', in L.J.F. Keppie et al., 'Excavations at the Roman fort of Bar Hill, 1978–82', *Glasgow Archaeological Journal*, 12 (1985), pp. 79–81.

Breeze, D.J., 'The Great Myth of Caledon', in T.C. Smout (ed.), *Scottish Woodland History* (Edinburgh: Scottish Cultural Press, 1997), pp. 47–51.

Breeze, D.J., *Roman Scotland: Frontier Country*, new edition (London: B.T. Batsford and Historic Scotland, 2006).

Brooke, D., 'The Glenkens 1275–1456: snapshots of a medieval countryside', *Transactions of the Dumfries and Galloway Natural History and Antiquarian Society*, 3rd series, 59 (1984), pp. 41–56.

Broun, D., *Scottish Independence and the Idea of Britain: From the Picts to Alexander III* (Edinburgh: Edinburgh University Press, 2007).

Brown, M., *The Wars of Scotland 1214–1371* (Edinburgh: Edinburgh University Press, 2004).

Brown, N., *History and Climate Change: A Eurocentric Perspective* (London: Routledge, 2001).

Bruford, A., 'What happened to the Caledonians?', in E.J. Cowan and R.A. McDonald (eds), *Alba: Celtic Scotland in the Medieval Era* (East Linton: Tuckwell Press, 2000), pp. 43–68.

Büntgen, U. et al., 'Cooling and societal change during the Late Antique Little Ice Age from 536 to around 660 ad', *Nature Geoscience*, 9 (2016), pp. 231–236.

Burgess, C., 'Volcanoes, catastrophe and the global crisis of the late second millennium bc', *Current Archaeology*, 117 (1989), pp. 325–329.

Buteux, S. et al., *Settlements at Skaill, Deerness, Orkney: Excavations by Peter Gelling of the Prehistoric, Pictish, Viking and Later Periods, 1963–1981*, British Archaeological Reports, British Series 260 (Oxford: BAR Publishing, 1997).

Butler, S., 'The environmental history of Shetland', in O. Owen and C. Lowe (eds), *Kebister: The Four-Thousand-Year-Old Story of One Shetland Township*, Society of Antiquaries of Scotland Monograph Series 14 (Edinburgh: Society of Antiquaries of Scotland, 1999), pp. 3–7.

Cage, A.G. and Austin, W.E.N., 'Marine climate variability during the last millennium: the Loch Sunart record, Scotland, UK', *Quaternary Science Reviews*, 29 (2010), pp. 1633–1647.

Caldwell, D., *Islay: The Land of the Lordship*, paperback edition (Edinburgh: Birlinn, 2017).

Cameron, A.S. and Stones, J.A. (eds), *Aberdeen: An In-Depth View of the City's Past*, Society of Antiquaries of Scotland Monograph Series 19 (Edinburgh: Society of Antiquaries of Scotland, 2001).

Campbell, B.M.S., 'Benchmarking medieval economic development: England, Wales, Scotland and Ireland *circa* 1290', *Economic History Review*, 61:4 (2008), pp. 896–945.

Campbell, B.M.S., *The Great Transition: Climate, Disease and Society in the Late-Medieval World* (Cambridge: Cambridge University Press, 2016).

Cartledge, J., Ingrem, C., Mulville, J. and Powell, A., 'Animal bone', in N. Sharples (ed.), *A Late Iron Age Farmstead in the Outer Hebrides: Excavations at Mound 1, Bornais, South Uist* (Oxford: Oxbow Books, 2012), pp. 119–121.

Carver, M., *Portmahomack: Monastery of the Picts* (Edinburgh: Edinburgh University Press, 2008).

Carver, M., *The Tarbat Discovery Programme*, Archaeology Data Service, 2017, https://doi.org/10.5284/1031216.

Cerón-Carrasco, R. and Parker Pearson, M., 'The fish bones', in M. Parker Pearson and N. Sharples (eds), *Between Land and Sea: Excavations at Dun Vulan, South Uist* (Sheffield: Sheffield Academic Press, 1999), pp. 274–282.

Charman, D.J., 'Late-glacial and Holocene vegetation history of the Flow Country, northern Scotland', *New Phytologist*, 127 (1994), pp. 155–168.

Charman, D.J. et al., 'Paleohydrological records from peat profiles and speleothems in Sutherland, northwest Scotland', *Quaternary Research*, 55:2 (2001), pp. 223–234.

Cheape, H., '"Every timber in the forest for MacRae's house": creel houses in the Highlands', *Vernacular Building*, 37 (2013–14), pp. 31–50.

Coleman, R. et al., 'The excavation of a souterrain at Shanzie Farm, Alyth, Perthshire', *Tayside and Fife Archaeological Journal*, 8 (2002), pp. 77–101.

Coleman, R. et al., 'Early medieval settlement and ironworking in Dornoch, Sutherland: excavations at the Meadows Business Park', Scottish Archaeological Internet Report 28 (2008), http://journals.socantscot.org/index.php/sair/article/view/1354/1343.

Coles, B., *Beavers in Britain's Past* (Oxford: Oxbow Books / WARP, 2006).

Coles, B., 'The European beaver', in B. Sykes and T. O'Connor (eds), *Extinctions and Invasions: A Social History of British Fauna* (Oxford: Windgather Press, 2010), pp. 104–115.

Cook, E.R. et al., 'Old world megadroughts and pluvials during the Common Era', *Science Advances*, 1 (2015), e1500561.

Cowie, T.G. and Shepherd, I.A.G., 'The Bronze Age', in K.J. Edwards and I.B.M. Ralston (eds), *Scotland after the Ice Age: Environment, Archaeology and History 8000 bc– ad 1000*, 2nd edition (Edinburgh: Edinburgh University Press, 2002), pp. 151–168.

Crawford, B.E., *Scandinavian Scotland* (Leicester: Leicester University Press, 1987).

Crawford, B.E., *The Northern Earldoms: Orkney and Caithness from ad 870 to 1470* (Edinburgh: John Donald, 2013).

Crawford, B.E. and Ballin Smith, B., *The Biggings, Papa Stour, Shetland: The History and Archaeology of a Royal Norwegian Farm*, Society of Antiquaries of Scotland Monograph Series 15 (Edinburgh: Society of Antiquaries of Scotland, 2000).

Crawford, B.E. and Taylor, S., 'The southern frontier of Norse settlement in north Scotland: place-names and history', *Northern Scotland*, 23 (2003), pp. 1–76.

Crone, A., *The History of a Scottish Lowland Crannog: Excavations at Buiston, Ayrshire 1989–90*, Scottish Trust for Archaeological Research Monograph 4 (Edinburgh: Scottish Trust for Archaeological Research, 2000).

Crone, A., 'A tenth-century tree-ring event in Scotland: proxy evidence for social and economic change?', in A. Woolf (ed.), *Landscape and Environment in Dark Age Scotland* (St Andrews: Committee for Dark Age Studies, University of St Andrews, 2006), pp. 49–56.

Crone, A. and Brunning, R., 'Woodland reconstruction', in A. Crone, *The History of a Scottish Lowland Crannog: Excavations at Buiston, Ayrshire 1989–90*, Scottish Trust for Archaeological Research Monograph 4 (Edinburgh: Scottish Trust for Archaeological Research, 2000), pp. 78–82.

Crone, A. and Mills, C.M., 'Seeing the wood and the trees: dendrochronological studies in Scotland,' *Antiquity*, 76 (2002), pp. 788–794.

Crone, A. and Watson, F., 'Sufficiency to scarcity, 500–1600', in T.C. Smout (ed.), *People and Woods in Scotland* (Edinburgh: Edinburgh University Press, 2003), pp. 60–81.

Crumley, J., *The Great Wood: The Ancient Forest of Caledon* (Edinburgh: Birlinn, 2011).

Curteis, A., Morris, C.A. and Bogdan, N.Q., 'The worked wood', in *Perth High Street Archaeological Excavation 1975–1977*, fascicule 2: *The Ceramics, the Metalwork and the Wood* (Perth: Tayside and Fife Archaeological Committee, 2012), pp. 223–322.

Curtis-Summers, S., Montgomery, J. and Carver, M., 'Food for the body, sustenance for the soul: dietary contrast between Pictish and medieval populations at Portmahomack, Scotland, using stable isotope analysis', in M. Carver, *The Tarbat Discovery Programme*, Archaeology Data Service, 2017, https://doi.org/10.5284/1031216.

Cussans, J.E. and Bond, J.M., 'Mammal bone', in S.J. Dockrill et al., *Excavations at Old Scatness, Shetland*, volume 1: *The Pictish Village and Viking Settlement* (Lerwick: Shetland Heritage Publications, 2010), pp. 132–156.

Dark, K., *Britain and the End of the Roman Empire* (Stroud: Tempus, 2000).

Davidson, D.A. and Simpson, I.A., 'Soils and landscape history: case studies from the Northern Isles of Scotland', in S. Foster and T.C. Smout (eds), *The History of Soils and Field Systems* (Aberdeen: Scottish Cultural Press, 1994), pp. 66–74.

Davies, A.L., 'Upland agriculture and environmental risk: a new model of upland land-use based on high spatial-resolution palynological data from West Affric, NW Scotland', *Journal of Archaeological Science*, 34 (2007), pp. 2053–2063.

Davies, R.R., *Domination and Conquest: The Experience of Ireland, Scotland and Wales 1100–1300* (Cambridge: Cambridge University Press, 1990).

Dawson, A.G., *So Foul and Fair a Day: A History of Scotland's Climate and Weather* (Edinburgh: Birlinn, 2009).

Dawson, A.G., Hickey, K., Mayewski, P.A. and Nesje, A., 'Greenland (GISP2) ice core and historical indicators of complex North Atlantic climate changes during the fourteenth century', *The Holocene*, 17:4 (2007), pp. 427–434.

De Paor, L., 'A survey of Scelig Mhichil', *Proceedings of the Royal Society of Antiquaries of Ireland*, 85 (1955), pp. 174–187.

Deary, H., 'Restoring wildness to the Scottish Highlands: a landscape of legacies', in M. Hourdequin and D.G. Havlick (eds), *Restoring Layered Landscapes: History, Ecology, and Culture* (Oxford: Oxford University Press, 2016), pp. 95–111.

Desprat, S., Goñi, M.F.S. and Loutre, M.-F., 'Revealing climatic variability of the last three millennia in northwestern Iberia using pollen influx data', *Earth and Planetary Science Letters*, 213:1–2 (2003), pp. 63–78.

Diamond, J., *Collapse: How Societies Choose to Fail or Succeed* (New York: Viking, 2005).

Dickson, C., 'Plant remains', in B. Ballin Smith (ed.), *Howe: Four Millennia of Orkney Prehistory*, Society of Antiquaries of Scotland Monograph Series 9 (Edinburgh: Society of Antiquaries of Scotland, 1994), pp. 125–139.

Dickson, C.A. and Dickson, J.H., *Plants and People in Ancient Scotland* (Stroud: Tempus, 2000).

Ditchburn, D., *Scotland and Europe: The Medieval Kingdom and Its Contacts with Christendom, c.1214–1545*, volume 1: *Religion, Culture and Commerce* (East Linton: Tuckwell Press, 2001).

Ditchburn, D. and Macdonald, A.J., 'Medieval Scotland, 1100–1560', in R.A. Houston and W.W.J. Knox (eds), *The New Penguin History of Scotland: From the Earliest Times to the Present Day* (London: Penguin Books, 2001), pp. 96–181.

Dixon, P., 'A rural medieval settlement in Roxburghshire: excavations at Springwood Park, Kelso, 1985–6', *Proceedings of the Society of Antiquaries of Scotland*, 128 (1998), pp. 671–751.

Dixon, P, 'Of bannocks and ale: cereal processing in Scotland c.1100–1750', in J. Klápšte and P. Somner (eds), *Food in the Medieval Rural Environment*, Ruralia 8 (Turnhout: Brepols, 2011), pp. 155–172.

Dixon, P., 'Mukked and folded land: the evidence of field data for medieval cultivation techniques in Scotland', in J. Klápšte (ed.), *Agrarian Technology in the Medieval Landscape*, Ruralia 10 (Turnhout: Brepols, 2016), pp. 107–123.

Dixon, P., 'Rural settlement patterns on the ground', in K.J. Stringer and A.J.L. Winchester (eds), *Northern England and Southern Scotland in the Central Middle Ages* (Woodbridge: Boydell Press, 2017), pp. 237–271.

Dockrill, S.J. et al., *Excavations at Old Scatness, Shetland*, volume 1: *The Pictish Village and Viking Settlement* (Lerwick: Shetland Heritage Publications, 2010).

Duby, G., 'Medieval agriculture 900–1500', in C.M. Cipolla (ed.), *The Fontana Economic History of Europe* (Glasgow: William Collins and Son, 1972), pp. 175–220.

Duffy, S. (ed.), *Robert the Bruce's Irish Wars* (Stroud: Tempus, 2002).

Dugmore, A., Keller, C. and McGovern, T., 'Norse Greenland settlement: reflections on climate change, trade, and the contrasting fates of human settlements in the North Atlantic islands', *Arctic Anthropology*, 44:1 (2007), pp. 12–36.

Dugmore, A. et al., 'Norse Greenland settlement and limits to adaptation', in W.N. Adger, I. Lorenzoni and K. O'Brien (eds), *Adapting to Climate Change: Thresholds, Values, Governance* (Cambridge: Cambridge University Press, 2009), pp. 96–113.

Dumayne, L., 'Invader or native? – Vegetation clearance in northern Britain during Romano-British times', *Vegetation History and Archaeobotany*, 2 (1993), pp. 29–36.

Dumayne, L., 'Iron Age and Roman vegetation clearance in northern Britain: further evidence', *Botanical Journal of Scotland*, 46 (1993), pp. 385–392.

Dumayne, L., 'The effect of the Roman occupation on the environment of Hadrian's Wall: a pollen diagram from Fozy Moss, Northumbria', *Britannia*, 32 (1994), pp. 217–224.

Dumayne-Peaty, L., 'Forest clearance in northern Britain during Roman-British times: re-addressing the palynological evidence', *Britannia*, 26 (1998), pp. 315–322.

Dumayne-Peaty, L., 'Human impact on the environment during the Iron Age and Romano-British times: palynological evidence from three sites near the Antonine Wall', *Journal of Archaeological Science*, 25 (1998), pp. 203–214.

Dumayne-Peaty, L., 'Continuity or discontinuity? Vegetation change in the Hadrianic-Antonine frontier zone of northern Britain at the end of the Roman occupation', *Journal of Biogeography*, 26 (1999), pp. 643–665.

Duncan, A.A.M., *Scotland: The Making of the Kingdom*, paperback edition (Edinburgh: Oliver and Boyd, 1978).

Duncan, A.A.M., 'Roger of Howden and Scotland, 1187–1201', in B.E. Crawford (ed.), *Church, Chronicle and Learning in Medieval and Early Renaissance Scotland* (Edinburgh: Mercat Press, 1999), pp. 135–160.

Dyer, C., 'Changes in diet in the late Middle Ages: the case of harvest workers', *Agricultural History Review*, 36 (1988), pp. 21–37.

Edwards, K.J. and Ralston, I., 'New dating and environmental evidence from Burghead fort, Moray', *Proceedings of the Society of Antiquaries of Scotland*, 109 (1980), pp. 202–210.

Edwards, K.J. and Ralston, I.B.M. (eds), *Scotland after the Ice Age: Environment, Archaeology and History 8000 BC–AD 1000*, 2nd edition (Edinburgh: Edinburgh University Press, 2002).

Edwards, K.J. and Whittington, G., 'Vegetation change', in K.J. Edwards and I.B.M. Ralston (eds), *Scotland after the Ice Age: Environment, Archaeology and History 8000 BC–AD 1000*, 2nd edition (Edinburgh: Edinburgh University Press, 2002), pp. 63–82.

Ervynk, A. and Woollett, J., 'Top-predator or survivor? The Castle of Sugny as seen through its animal remains', *Les cahiers de l'urbanisme*, September 2006, pp. 78–89.

Ervynk, A., Lentaker, A., Müldner, G., Richards, M. and Dobney, K., 'An investigation into the transition from forest dwelling pigs to farm animals in medieval Flanders', in U. Albarella, K. Dobney, A. Ervynk and P. Rowley-Conwy (eds), *Pigs and Humans: 10,000 Years of Interaction* (Oxford: Oxford University Press, 2007), pp. 170–193.

Evans, J.G., *An Introduction to Environmental Archaeology* (St Albans: Granada, 1978).

Evans, N., 'A historical introduction to the northern Picts', in G. Noble and N. Evans (eds), *The King in the North: The Pictish Realms of Fortriu and Ce* (Edinburgh: John Donald, 2019), pp. 10–38.

Fawcett, R., 'Ecclesiastical architecture in the second half of the thirteenth century', in N.H. Reid (ed.), *Scotland in the Reign of Alexander III 1249–1286* (Edinburgh: John Donald, 1990), pp. 148–80.

Fawcett, R. and Oram, R., *Melrose Abbey* (Stroud: Tempus, 2004).

Fawcett, R. and Oram, R., *Dryburgh Abbey* (Stroud: Tempus, 2005).

Fawcett, R. and Oram, R., *Elgin Cathedral and the Diocese of Moray* (Edinburgh: Historic Scotland, 2014).

Fenton, A., *The Northern Isles: Orkney and Shetland* (Edinburgh: John Donald, 1978).

Foster, S., *Picts, Gaels and Scots* (London: B.T. Batsford / Historic Scotland, 1996).

Fowler, P., *Farming in the First Millennium AD: British Agriculture between Julius Caesar and William the Conqueror* (Cambridge: Cambridge University Press, 2002).

Fraser, J.E., *From Caledonia to Pictland: Scotland to 795* (Edinburgh: Edinburgh University Press, 2009).

Fraser Darling, F., 'History of the Scottish forests', *Scottish Geographical Magazine*, 65:3 (1949), pp. 132–137.

Frere, S.S. and Wilkes, J.J., *Strageath: Excavations within the Roman fort, 1973–1986*, Britannia Monograph Series 9 (London: Society for the Promotion of Roman Studies, 1989).

Gemmill, E. and Mayhew, N., *Changing Values in Medieval Scotland: A Study of Prices, Money, Weights and Measures* (Cambridge: Cambridge University Press, 1995).

Gevirtz, S., 'Jericho and Shechem: a religio-literary aspect of city destruction', *Vetus Testamentum*, 13:1 (1963), pp. 52–62.

Gibson, A., 'Medieval corn-drying kilns at Capo, Kincardineshire, and Abercairney, Perthshire', *Proceedings of the Society of Antiquaries of Scotland*, 118 (1989), pp. 219–229.

Gilbert, J., 'The monastic record of a Border landscape 1136 to 1236', *Scottish Geographical Magazine*, 109 (1983), pp. 4–15.

Gilbert, J.M., 'Crossbows on Pictish stones', *Proceedings of the Society of Antiquaries of Scotland*, 107 (1978), pp. 316–317.

Gilbert, J.M., *Hunting and Hunting Reserves in Medieval Scotland* (Edinburgh: John Donald, 1979).

Gilbert, J.M., 'Place-names and managed woods in medieval Scotland', *Journal of Scottish Names Studies*, 5 (2011), pp. 35–56.

Gilbert, J.M., 'Medieval woodland management in southern Scotland', *Transactions of the Dumfries and Galloway Natural History and Antiquarian Society*, 3rd series, 86 (2012), pp. 77–117.

Gilbert, J.M., 'Woodland management in medieval Scotland', *Proceedings of the Society of Antiquaries of Scotland*, 146 (2017), pp. 215–252.

Gilbertson, D.D., Schwenninger, J.-L., Kemp, R.A. and Rhodes, E.J., 'Sand-drift and soil formation along an exposed Atlantic coastline: 14,000 years of diverse geomorphological climatic and human impacts', *Journal of Archaeological Science*, 26 (1999), pp. 439–469.

Girling, M.A. and Greig, J., 'A first fossil record for *Scolytus scolytus* (F.) (elm bark beetle): its occurrence in elm decline deposits from London and the implications for neolithic elm disease', *Journal of Archaeological Science*, 12:5 (1985), pp. 347–351.

Grant, A., *Independence and Nationhood: Scotland 1306–1469* (London: Edward Arnold, 1984).

Grant, A., 'Lordship and society in twelfth-century Clydesdale', in H. Price and J. Watts (eds), *Power and Identity in the Middle Ages* (Oxford: Oxford University Press, 2007), pp. 98–124, https://doi.org/10.1093/acprof:oso/9780199285464.003.0009.

Gräslund, B. and Price, N., 'Twilight of the gods? The "dust veil event" of ad 536 in critical perspective', *Antiquity*, 86:332 (2012), pp. 428–443.

Greenstein, J.M., 'The vision of peace: meaning and representation in Ambrogio Lorenzetti's Sala della Pace cityscapes', *Art History*, 11:4 (1988), pp. 492–510.

Grigson, C. and Mellars, P.A., 'The mammalian remains from the middens', in P.A. Mellars (ed.), *Excavations on Oronsay* (Edinburgh: Edinburgh University Press, 1987), pp. 243–289.

Groucutt, H.S. and Carleton, W.C., 'Mass-kill hunting and Late Quaternary ecology: new insights into the "desert kite" phenomenon in Arabia', *Journal of Archaeological Science*, 37 (2021), https://doi.org/10.1016/j.jasrep.2021.102995.

Guillet, S. et al., 'Toward a more realistic assessment of the climatic impacts of the 1257 eruption', EGU General Assembly Conference Abstracts, April 2015, https://ui.adsabs.harvard.edu/abs/2015EGUGA..17.1268G/abstract.

Guillet, S. et al., 'Climate response to the Samalas volcanic eruption in 1257 revealed by proxy records', *Nature Geoscience*, 10 (2017), pp. 123–129.

Hall, D. and Malloy, K., '"Always chasing deer" – further excavations at the deer parks of Buzzart Dykes and Kincardine Park and new excavations at Kincardine Castle', *Tayside and Fife Archaeological Journal*, 21/22 (2015–16), pp. 25–34.

Hall, D., Oram, R. and Malloy, K., 'A hunting we will go? Stirling University's medieval deer parks project', *Tayside and Fife Archaeological Journal*, 17 (2011), pp. 59–67.

Halliday, S., 'The wider context of 7th century ad agrarian expansion', in D. Strachan, D. Sneddon and R. Tipping, *Early Medieval Settlement in Upland Perthshire: Excavations at Lair, Glen Shee 2012–17* (Oxford: Archaeopress, 2019), pp. 140–143.

Hanson, W.S., 'Forest clearance and the Roman army', *Britannia*, 27 (1996), pp. 354–358.

Hardin, G., 'The tragedy of the commons', *Science*, 162 (1968), pp. 1243–1248.

Hardy, K. and Ballin, T., 'South Cuidrach: a new Late Upper Palaeolithic and Mesolithic site on the Isle of Skye, western Scotland', *Proceedings of the Society of Antiquaries of Scotland*, 153 (2024), https://doi.org/10.9750/PSAS.153.1367.

Hartman, S., Ogilvie, A., Ingimundarson, J., Dugmore, A., Hambrecht, G. and McGovern, T.H., 'Medieval Iceland, Greenland, and the New Human Condition: a case study in integrated environmental humanities', *Global and Planetary Change*, 156 (2017), https://doi.org/10.1016/j.gloplacha.2017.04.007.

Hedges, J.W., *Isbister: A Chambered Tomb in Orkney*, British Archaeological Reports, British Series 115 (Oxford: BAR Publishing, 1983).

Helama, S. et al., 'Volcanic dust veils from sixth century tree-ring isotopes linked to reduced irradiance, primary production and human health', *Scientific Reports*, 8 (article no.1339) (2018), https://doi.org/10.1038/s41598-018-19760-w.

Henshall, A.S. and Ritchie, J.N.G., *The Chambered Cairns of the Central Highlands: An Inventory of Their Structures and Their Contents* (Edinburgh: RCAHMS, 2001).

Hetherington, D., 'A short history of Scotland's lost species 2: the bear', https://www.linkedin.com/pulse/short-history-scotlands-lost-species-2-bear-david-hetherington/.

Highland Historic Environment Record, https://her.highland.gov.uk/home.

Hill, P.H., *Whithorn and St Ninian: The Excavation of a Monastic Town 1984–91* (Stroud: Sutton Publishing, 1997).

Hillman, G., 'Reconstructing crop husbandry practices from charred remains of crops', in R. Mercer (ed.), *Farming Practice in British Prehistory* (Edinburgh: Edinburgh University Press, 1981), pp. 123–162.

Hindmarch, E., Oram, R., Haggarty, G., Hall, D. and Robertson, J., 'Eldbotle: the archaeology and environmental history of a medieval rural settlement in East Lothian', *Proceedings of the Society of Antiquaries of Scotland*, 142 (2012), pp. 245–300.

Hoffmann, R.C., '*Salmo salar* in late medieval Scotland: competition and conservation for a riverine resource', *Aquatic Science*, 77 (2015), pp. 355–366.

Hoffmann, R.C., *The Catch: An Environmental History of Medieval European Fisheries* (Cambridge: Cambridge University Press, 2023).

Hoffmann, R. C. and Ross, A., 'This belongs to us! Competition between the royal burgh of Stirling and the Augustinian abbey of Cambuskenneth over salmon fishing rights on the River Forth,' in E. Bhreathnach, M. Krasnodębska D'Aughton and K. Smith (eds), *Monastic Europe: Communities, Landscape and Settlement* (Turnhout: Brepols, 2020), pp. 451–476.

Holden, T., 'The botanical evidence', in C. Lowe (ed.), *Excavations at Hoddom, Dumfriesshire* (Edinburgh: Society of Antiquaries of Scotland, 2006), pp. 150–155.

Holden, T., 'The corn-drying kilns at Hoddom', C. Lowe (ed.), *Excavations at Hoddom, Dumfriesshire* (Edinburgh: Society of Antiquaries of Scotland, 2006), pp. 102–113.

Holden, T. and Boardman, S., 'Crops', in N. Sharples (ed.), *Scalloway: A Broch, Late Iron Age Settlement and Medieval Cemetery* (Oxford: Oxbow Books, 1998), pp. 99–106.

Holmberg, P. et al., 'The Rök Runestone and the end of the world', *Futhark: International Journal of Runic Studies*, 9:10 (2020), pp. 7–38.

Hunter, F., 'Excavations at Birnie, Moray, 2004', National Museums Scotland Research Repository, 2005, https://nms.iro.bl.uk/concern/reports/308939ee-35f4-40bd-9f90-0390237d0426.

Hunter, J., *On the Other Side of Sorrow: Nature and People in the Scottish Highlands* (Edinburgh: Mainstream Publishing, 1995).

Hunter, J. with Bond, J.M. and Smith, A.N., *Investigations in Sanday, Orkney*, volume 1: *Excavations at Pool, Sanday: A Multi-period Settlement from Neolithic to Late Norse Times* (Kirkwall: The Orcadian / Historic Scotland, 2007).

Ingrem, C., 'Fish', in N. Sharples (ed.), *A Norse Farmstead in the Outer Hebrides: Excavations at Mound 3, Bornais, South Uist* (Oxford: Oxbow Books, 2005), pp. 192–194.

Ingrem, C., 'Fish', in N. Sharples (ed.), *The Economy of a Norse Settlement in the Outer Hebrides: Excavations at Mounds 2 and 2A, Bornais, South Uist* (Oxford: Oxbow Books, 2021), pp. 439–441.

Innes, C., 'Some account of early planting in Scotland', *Transactions of the Royal Highland and Agricultural Society*, new series, 9 (1861), pp. 40–53.

Jackson, K., *The Gododdin: The Oldest Scottish Poem* (Edinburgh: Edinburgh University Press, 1969).

James Hutton Institute, 'Soil data and maps', https://www.hutton.ac.uk/soil-data-and-maps/.

Jepson, P. and Blythe, C., *Rewilding: The Radical New Science of Ecological Recovery* (London: Icon Books, 2020).

Jillings, K., *Scotland's Black Death: The Foul Death of the English* (Stroud: Tempus, 2003).

Jobey, G., 'Between Tyne and Forth: some problems', in P.A.G. Clack and S. Haselgrove (eds), *Rural Settlement in the Rural North* (Durham: Council for British Archaeology, 1981), pp. 7–20.

Jones, A.K.G., 'The fish bone', in *Perth High Street Archaeological Excavation 1975–1977*, fascicule 4: *Living and Working in a Medieval Scottish Burgh: Environmental Remains and Miscellaneous Finds* (Perth: Tayside and Fife Archaeological Committee, 2010), pp. 53–67.

Jones, G.D.B. and Walker, J., 'Towards a minimalist view of Romano-British agricultural settlement in the north west', in J.C. Chapman and H.C. Mytum (eds), *Settlement in North Britain 1000 BC–AD 1000*, British Archaeological Reports, British Series 118 (Oxford: BAR Publishing, 1983), pp. 185–204.

Jordan, W.C., *The Great Famine: Northern Europe in the Early Fourteenth Century* (Princeton: Princeton University Press, 1996).

Kelly, F., 'The old Irish tree-list', *Celtica*, 11 (1976), pp. 107–124.

Kelly, M., *A History of the Black Death in Ireland* (Stroud: Tempus, 2001).

Kenworthy, J.B., 'The prehistory of north east Scotland', in A.M.D. Gemmell (ed.), *Quaternary Studies in North East Scotland* (Aberdeen: Department of Geography, University of Aberdeen, 1975), pp. 74–81.

Kershaw, I., *Bolton Priory: The Economy of a Northern Monastery, 1286–1325* (Oxford: Oxford University Press, 1973).

Kershaw, I., 'The Great Famine and agrarian crisis in England 1315–1322', *Past and Present*, 59 (1973), pp. 3–50.

Kirk, W., 'Prehistoric sites at the sands of Forvie, Aberdeenshire: a preliminary examination', *Aberdeen University Review*, 35 (1953), pp. 150–171.

Kirk, W., *'Sands of Forvie', Discovery and Excavation in Scotland, 1957* (Dundee: Council for British Archaeology, Scottish Regional Group, 1957).

Kirk, W., *'Sands of Forvie', Discovery and Excavation in Scotland, 1958* (Dundee: Council for British Archaeology, Scottish Regional Group, 1958).

Kirk, W., *'Forvie', Discovery and Excavation in Scotland, 1960* (Dundee: Council for British Archaeology, Scottish Regional Group, 1960).

Kitchener, A.C. and Conroy, J.W.H., 'The history of the Eurasian Beaver *Castor fiber* in Scotland', *Mammal Review*, 27:2 (1997), pp. 95–108.

Kitchener, A.C. and Doune, J., 'A record of the aurochs, *Bos primigenius*, from Morayshire', *Glasgow Naturalist*, 25:4 (2012), https://www.glasgownaturalhistory.org.uk/gn25_4/kitchener_1.pdf.

Lacaille, A.D., 'Mesolithic implements from Ayrshire', *Proceedings of the Society of Antiquaries of Scotland*, 64 (1930), pp. 34–48.

Lamb, H., 'The early medieval warm epoch and its sequel', *Palaeogeography, Palaeoclimatology, Palaeoecology*, 1 (1965), pp. 13–37.

Lamb, H., *Climate, History and the Modern World* (London: Methuen, 1982).

Lamb, H., *Weather, Climate and Human Affairs* (London: Taylor and Francis, 1988; reprinted London: Routledge, 2012).

Lane, A. and Campbell, E., *Dunadd: An Early Dalriadic Capital* (Oxford: Oxbow Books, 2000).

Langdon, P.G., Barber, K.E. and Hughes, P.D.M., 'A 7500-year peat-based palaeoclimatic reconstruction and evidence for an 1100-year cyclicity in bog surface wetness from Temple Hill Moss, Pentland Hills, southeast Scotland', *Quaternary Science Reviews*, 22:2–4 (2003), pp. 259–274.

Lavigne, F. et al., 'Source of the great a.d. 1257 mystery eruption unveiled, Samalas volcano, Rinjani Volcanic Complex, Indonesia', *Proceedings of the National Academy of Sciences*, 110:42 (2013), pp. 16742–16747.

Lavigne, F. and Guillet, S., 'The unknown ad 1275 stratospheric eruption: climatic impacts in Europe and tentative volcanic source', in F. Riede (ed.), *Past Vulnerability – Volcanic Eruptions and Human Vulnerability in Traditional Societies Past and Present* (Aarhus: Aarhus University Press, 2015), pp. 63–74.

Lepper, F. and Frere, S.S., *Trajan's Column: A New Edition of the Cichorius Plates: Introduction, Commentary and Notes* (Stroud: Alan Sutton, 1988).

Little, A.G., 'The authorship of the Lanercost chronicle', *English Historical Review*, 21 (1916), pp. 269–279.

Logan, J., *The Scottish Gaël, or Celtic Manners as Preserved among the Highlanders*, 2 volumes (London: Smith, Elder, 1831).

Lowe, C. (ed.), *Excavations at Hoddom, Dumfriesshire* (Edinburgh: Society of Antiquaries of Scotland, 2006).

Ludlow, F. et al., 'Medieval Irish chronicles reveal persistent volcanic forcing of severe winter cold events, 431–1649 CE', *Environmental Research Letters*, 8 (2013), pp. 1–10, http://dx.doi.org/10.1088/1748-9326/8/2/024035.

Lynch, M., *Scotland: A New History* (London: Century, 1991).

Lynn, C.J., 'Ballywee', *Excavations*, 5 (1974), pp. 4–6.

MacDonald, A., 'The *papar* and problems: a brief review', in B.E. Crawford (ed.), *The 'Papar' in the North Atlantic: Environment and History* (St Andrews: Committee for Dark Age Studies, University of St Andrews, 2002), pp. 13–29.

MacInnes, I.A., '"Shock and awe": the use of terror as a psychological weapon during the Bruce-Balliol civil war, 1332–1338', in A. King and M. Penman (eds), *England and Scotland in the Fourteenth Century: New Perspectives* (Woodbridge: Boydell Press, 2007), pp. 40–59.

MacKenzie, N.A. and Callander, R.F., 'Birchwoods in a Deeside parish', in T.C. Smout (ed.), *Scottish Woodland History* (Edinburgh: Scottish Cultural Press, 1997), pp. 135–146.

Malloy, K. and Hall, D., 'Archaeological excavations of the medieval royal Kincardine landscape, Aberdeenshire, Scotland,' *Medieval Archaeology*, 61:1 (2018), pp. 157–176.

Malloy, K. and Hall, D., 'Medieval hunting and wood management in the Buzzart Dykes landscape', *Environment and History*, 25 (2019), pp. 365–390.

Mann, M.E. et al., 'Proxy-based reconstructions of hemispheric and global surface temperature variations over the past two millennia', *Proceedings of the National Academy of Sciences*, 105:36 (2008), pp. 13252–13257.

Margaritelli, G. et al., 'Persistent warm Mediterranean surface waters during the Roman period', *Scientific Reports*, 10:10431 (2020), https://doi.org/10.1038/s41598-020-67281-2.

Martin, C., 'The silent shores speak: maritime landscapes in North Argyll', *Current Archaeology*, 219 (2008), pp. 26–32.

Martin, C., 'A maritime dominion – sea-power and the lordship', in R.D. Oram (ed.), *The Lordship of the Isles* (Leiden: Brill, 2014), pp. 176–199.

Marttila, J., 'Resources, production, and trade in the Norse Shetland', *Journal of the North Atlantic*, 29 (2016), pp. 1–20.

Mayewski, P.A. et al., 'Holocene climate variability', *Quaternary Research*, 62 (2004), pp. 243–255.

Mayhew, N., 'Alexander III – a silver age?', in N.H. Reid (ed.), *Scotland in the Reign of Alexander III 1249–1286* (Edinburgh: John Donald, 1990), pp. 53–73.

McCarthy, D.P. and Breen, A., 'An evaluation of astronomical observations in the Irish annals', *Vistas in Astronomy*, 41 (1997), pp. 117–138.

McCormick, F., 'The dog in prehistoric and early Christian Ireland', *Archaeology Ireland*, 5:4 (1991), pp. 7–9.

McCormick, F., 'Calf slaughter as a response to marginality', n C.M. Mills and G. Coles (eds), *Life on the Edge: Human Settlement and Marginality* (Oxford: Oxbow Books, 1998), pp. 49–51.

McCormick, M., Dutton, P.E. and Mayewski, P., 'Volcanoes and the climate forcing of Carolingian Europe, a.d. 750–950', *Speculum*, 82 (2007), pp. 865–895.

McCormick, M. et al., 'Climate change during and after the Roman Empire: reconstructing the past from scientific and historical evidence', *Journal of Interdisciplinary History*, 43:2 (2012), pp. 169–220.

McGovern, T.H. et al., 'Landscapes of settlement in northern Iceland: historical ecology of human impact and climate fluctuation on the millennial scale', *American Anthropologist*, new series, 109:1 (2007), pp. 27–51.

McGuigan, N., *Máel Coluim III 'Canmore': An Eleventh-Century Scottish King* (Edinburgh: John Donald, 2021).

McHardie, L.S., Mighall, T.M. and Schofield, J.E., 'The landscape and environmental setting for the Pictish monastery at Portmahomack,' in M. Carver, *The Tarbat Discovery Programme*, download 7.4.4, Archaeology Data Service, 2017, https://doi.org/10.5284/1031216.

McNamee, C., *The Wars of the Bruces: Scotland, England and Ireland 1306–1328* (East Linton: Tuckwell Press, 1997).

McNiven, P., 'The *Gart*-names of Clackmannanshire', *Journal of Scottish Name Studies*, 1 (2007), pp. 61–76.

McWhannell, D., 'Ship service and indigenous sea power', *West Highland Notes and Queries*, 3rd series, 1 (August 2000), pp. 3–18.

McWhannell, D.C., 'The galleys of Argyll', *Mariner's Mirror*, 88:1 (2002), pp. 14–32.

McWhannell, D.C., 'Campbell of Breadalbane and Campbell of Argyll boatbuilding accounts 1600 to 1700', *Mariner's Mirror*, 89:4 (2003), pp. 405–424.

Meadows, D.H., Meadows, D.L., Randers, J. and Behrens, W.W., *The Limits to Growth: A Report for the Club of Rome's Project on the Predicament of Mankind* (Falls Church VA: Potomac Associates, 1972).

Meeker, L.D. and Mayewski, P.A., 'A 1,400-year high resolution record of atmospheric circulation over the North Atlantic and Asia', *The Holocene*, 12 (2002), pp. 27–66.

Mercer, R.J. and Tipping, R.M., 'The prehistory of soil erosion in the northern and eastern Cheviot Hills, Anglo-Scottish Borders', in S. Foster and T.C. Smout (eds), *The History of Soils and Field Systems* (Aberdeen: Scottish Cultural Press, 1996), pp. 1–25.

Merriman, M., *The Rough Wooings: Mary Queen of Scots 1542–1551* (East Linton: Tuckwell Press, 2000).

Met Office, 'North Atlantic Oscillation', https://www.metoffice.gov.uk/weather/learn-about/weather/atmosphere/north-atlantic-oscillation.

Met Office, 'What are El Niño and La Niña', https://www.metoffice. gov.uk/weather/learn-about/weather/oceans/el-nino.

Miket, R.F., 'The souterrains of Skye', in B. Ballin Smith and I. Banks (eds), *In the Shadow of the Brochs* (Stroud: Tempus, 2002), pp. 77–110.

Miles, H. and Jackman, B., *The Great Wood of Caledon* (Lanark: Colin Baxter Photography, 1991).

Milles, A., 'Charred plant remains', in A. Lane and E. Campbell, *Dunadd: An Early Dalriadic Capital* (Oxford: Oxbow Press, 2000), pp. 221–226.

Mills, C.M., 'Historic pine and dendrochronology in Scotland', *Scottish Woodland History Discussion Group: Notes*, 13 (2008), pp. 9–14.

Mills, C.M. and Crone, A., 'Dendrochronological evidence for Scotland's native timber resources over the last 1000 years', *Scottish Forestry*, 66:1 (2012), pp. 18–33.

Mitchison, R., *A History of Scotland*, 1st edition (London: Methuen, 1970) and 2nd edition (London: Methuen, 1982).

Moe, D. and Rackham, O., 'Pollarding and a possible explanation of the Neolithic elmfall', *Vegetation History and Archaeobotany*, 1 (1992), pp. 63–68.

Morris, C.D., Batey, C.E. and Rackham, D.J., *Freswick Links, Caithness: Excavation and Survey of a Norse Settlement*, North Atlantic Biocultural Organisation and Highland Archaeology Monograph 1 (Inverness: Highland Libraries, 1995).

Morrison, A., 'Souterrains in Sutherland', in J.R. Baldwin (ed.), *The Province of Strathnaver* (Edinburgh: Scottish Society for Northern Studies, 2000), pp. 215–235.

Morrison, J., Oram, R. and Oliver, F., 'Ancient Eldbottle unearthed: archaeological and historical evidence for a long-lost early medieval East Lothian village', *Transactions of the East Lothian Antiquarian and Field Naturalists' Society*, 27 (2008), pp. 21–45.

Mulholland, H., 'The microlithic industries of the Tweed Valley', *Transactions of the Dumfriesshire and Galloway Natural History and Antiquarian Society*, 47 (1970), pp. 81–110.

Mulville, J., Bond, J.M. and Craig, O., 'The white stuff: milking in the outer islands of Scotland', in J. Mulville and A.K. Outram (eds), *The Zooarchaeology of Fats, Oils, Milk and Dairying: Proceedings of the 9th ICAZ Conference, Durham 2002* (Oxford: Oxbow Books, 2005), pp. 167–182.

Murray, H., *Viking and Early Medieval Buildings in Dublin*, British Archaeological Reports, British Series 119 (Oxford: BAR Publishing, 1983).

Murray, H., 'Appendix 1: The building reconstruction', in P. Yeoman, 'Excavations at Castlehill of Strachan, 1980–1981', *Proceedings of the Society of Antiquaries of Scotland*, 114 (1984), pp. 315–364.

Murray, H., 'The buildings', in *Perth High Street Archaeological Excavation 1975–1977*, fascicule 1: *The Excavations at 75–95 High Street and 5–10 Mill Street, Perth* (Perth: Tayside and Fife Archaeological Committee, 2012), pp. 127–198.

Murray, H.K. and Murray, J.C., 'Excavations at Rattray, Aberdeenshire: a Scottish deserted burgh', *Medieval Archaeology*, 37 (1993), pp. 109–218.

Nairne, D., 'Notes on Highland woods, ancient and modern', *Transactions of the Gaelic Society of Inverness*, 17 (1892), pp. 170–221.

Neville, C.J., *Native Lordship in Medieval Scotland: The Earldoms of Strathearn and Lennox c.1140–1365* (Dublin: Four Courts Press, 2005).

Newfield, T., 'A cattle panzootic in early fourteenth-century Europe', *Agricultural History Review*, 57 (2009), pp. 155–190.

Newfield, T.P., 'The climate downturn of 536–50', in S. White, C. Pfister and F. Mauelshagen (eds), *The Palgrave Handbook of Climate History* (London: Palgrave, 2018), pp. 447–493.

Newfield, T.P., 'Mysterious and mortiferous clouds: the climate cooling and disease burden of late antiquity', in A. Izdebski and M. Mulryan (eds), *Environment and Society in the Long Late Antiquity*, Late Antique Archaeology 12 (Leiden: Brill, 2018), pp. 89–115.

Newfield, T.P. and Labuhn, I., 'Towards a messy history of crisis and climate in Carolingian Europe', 11 November 2016, https://www.historicalclimatology.com/features/archives/11-2016.

Newfield, T.P. with Marciniak, S. and Cameron-Steinke, B., '"Verbalist ingenuity" and the evidential basis for virgin-soil smallpox epidemics in the sixth century: from Iona to Ṣanʿā'', in R.D. Oram (ed.), *'With Our Backs to the Ocean': Land, Lordship, Climate Change, and Environment in the North-West European Past: Essays in Memory of Alasdair Ross*, Environmental Histories of the North Atlantic World 5 (Turnhout: Brepols, 2023), pp. 25–82.

Nicholson, R., *Edward III and the Scots: The Formative Years of a Military Career, 1327–35* (Oxford: Oxford University Press, 1968).

Nicholson, R., *Scotland: The Later Middle Ages*, paperback edition (Edinburgh: Oliver and Boyd, 1978).

Noble, G., 'Fortified settlement and the emergence of kingdoms in Northern Scotland in the first millennium ad', in N. Christie and H. Herold (eds), *Fortified Settlements in Early Medieval Europe: Defended Communities of the 8th–10th Centuries* (Oxford: Oxbow Books, 2016), pp. 26–36.

Noble, G., 'Fortified settlement in northern Pictland', in G. Noble and N. Evans (eds), *The King in the North: The Pictish Realms of Fortriu and Ce* (Edinburgh: John Donald, 2019), pp. 39–57.

Noble, G. and Evans, N. (eds), *The King in the North: The Pictish Realms of Fortriu and Ce* (Edinburgh: John Donald, 2019).

Noble, G. and Evans, N., *Picts: Scourge of Rome, Rulers of the North* (Edinburgh: Birlinn, 2022).

Noble, G., Gondek, M., Campbell, E., Evans, N., Hamilton, D. and Taylor, S., 'Rhynie: a powerful place of Pictland', in G. Noble and N. Evans (eds), *The King in the North: The Pictish Realms of Fortriu and Ce* (Edinburgh: John Donald, 2019), pp. 58–80.

Noble, R., *Castles in the Mist: The Victorian Transformation of the Highlands* (Glasgow: Saraband, 2016).

O'Connor, T. and Evans, J.G., *Environmental Archaeology: Principles and Methods*, 1st edition (Stroud: Sutton Publishing, 1999) and 2nd edition (Stroud: Sutton Publishing, 2005).

O'Regan, H., 'The presence of the brown bear *Ursus arctos* in Holocene Britain: a review of the evidence', *Mammal Review*, 48:4 (2018), pp. 229–244.

Ogilvy, R., *The River Tay and Its People* (Edinburgh: Mainstream Publishing, 1993).

Oppenheimer, C. et al., 'The Eldgjá eruption: timing, long-range impacts and influence on the Christianisation of Iceland', *Climatic Change*, 147 (2018), pp. 369–381.

Oram, R.D., 'Bruce, Balliol and the lordship of Galloway: south-western Scotland and the Wars of Independence', *Transactions of the Dumfries and Galloway Natural History and Antiquarian Society*, 67 (1992), pp. 29–47.

Oram, R.D., 'A family business? Colonisation and settlement in twelfth- and thirteenth-century Galloway', *Scottish Historical Review*, 72 (1993), pp. 111–145.

Oram, R.D., *The Lordship of Galloway* (Edinburgh: John Donald, 2000).

Oram, R.D., 'Prelatical builders: a preliminary study, c.1124–c.1500', in R.D. Oram and G.P. Stell, *Lordship and Architecture in Medieval and Renaissance Scotland* (East Linton: Tuckwell Press, 2005), pp. 1–25.

Oram, R.D., 'Castles and colonists in twelfth- and thirteenth-century Scotland: the case of Moray', in *Chateau Gaillard*, 22 (2006), pp. 289–298.

Oram, R.D., 'Capital tales or Burghead bull?', in S. Arbuthnot and K. Hollo (eds), *Fil súil nglais: A Grey Eye Looks Back: A Festschrift in Honour of Colm Ó Baoill* (Ceann Drochaid: Clann Tuirc Books, 2007), pp. 241–262.

Oram, R.D., '"It cannot be decernit quha are clean and quha are foulle": responses to epidemic disease in sixteenth- and seventeenth-century Scotland,' *Renaissance and Reformation*, 30:4 (2007), pp. 13–39.

Oram, R.D., 'A fit and ample endowment? The Balmerino estate 1228–1606', in R.D. Oram, M. Hammond, J. Kerr, R. Fawcett, G. Markus and P.P. Dixon (eds), *Life on the Edge: The Cistercian Abbey of Balmerino, Fife (Scotland)*, Comentarii Cistercienses 59 parts 1–2 (Pontigny: Cîteaux, 2008), pp. 61–80.

Oram, R.D., *Domination and Lordship: Scotland 1070–1230*, New Edinburgh History of Scotland 3 (Edinburgh: Edinburgh University Press, 2011).

Oram, R.D., 'Ouncelands, quarterlands and pennylands in the Western Isles, Man and Galloway: tribute payments and military levies in the Norse west', in S. Imsen (ed.), *Taxes, Tributes and Tributary Lands in the Making of the Scandinavian Kingdoms in the Middle Ages* (Trondheim: Rostra Books, 2011), pp. 57–75.

Oram, R.D., *Alexander II, King of Scots 1214–1249* (Edinburgh: John Donald, 2012).

Oram, R.D., 'Breaking new ground: the monastic orders and economic development along the northern European periphery c.1070 to c.1300', in Francesco Ammannati (ed.), *Religion and Religious Institutions in the European Economy, 1000–1800: Atti della 'Quarantatreesima Settimana di Studi', 8–12 maggio 2011* (Florence, 2012), pp. 331–343.

Oram, R.D., 'The salt industry in medieval Scotland', *Studies in Medieval and Renaissance History*, 3rd series, 9 (2012), pp. 209–232.

Oram, R.D., 'Estuarine environments and resource exploitation in eastern Scotland c.1125–c.1400: a comparative study of the Forth and Tay estuaries', in E. Thoen et al. (eds), *Landscapes or Seascapes? The History of the Coastal Environment in the North Sea Area Reconsidered*, Comparative Rural History Network 13 (Turnhout: Brepols, 2013), pp. 353–377.

Oram, R.D., 'Between a rock and a hard place: climate, weather and the rise of the Lordship of the Isles', in R.D. Oram (ed.), *The Lordship of the Isles* (Leiden: Brill, 2014), pp. 40–61.

Oram, R.D., '"The worst disaster suffered by the people of Scotland in recorded history": climate change, dearth and pathogens in the long fourteenth century', *Proceedings of the Society of Antiquaries of Scotland*, 144 (2014), pp. 223–244.

Oram, R.D., 'Trackless, impenetrable and under-developed? Roads, colonisation and environmental transformation in the Anglo-Scottish border zone *c.*1100 to *c.*1350', in R. Evans and V. Haynes (eds), *Roadworks: Medieval Britain, Medieval Roads* (Manchester: Manchester University Press, 2015), pp. 303–325.

Oram, R.D., 'Paisley Abbey and its estates in the Middle Ages', *Renfrewshire Local History Forum*, 18 (2016), pp. 5–23.

Oram, R.D., 'Environmental change, resource conflicts and social change in Scotland c.1250–1500', in C-F. Mathis and G. Massard-Gilbaud (eds), *Sous le soleil: Systèmes et transitions énergétiques du Moyen Âge à nos jours* (Paris: Éditions de la Sorbonne, 2019), pp. 211–224.

Oram, R.D., *David I, King of Scots 1124–1153* (Edinburgh: John Donald, 2020).

Oram, R.D., 'Landscape, land use and environment', in B.M. Kaczynski (ed.), *The Oxford Handbook of Christian Monasticism* (Oxford: Oxford University Press, 2020), pp. 418–434.

Oram, R.D., 'Public policy and private practice: protection, management and development of woodland resources from the 12th to 17th centuries', in M. Stewart (ed.), *Policies and Products through the Ages: The Role of the State and Private Interests*, Scottish Woodland History Conference Notes 21 (2016) (Native Woodland Discussion Group, 2022), pp. 1–7.

Oram, R.D., '"Away was sons of alle and brede": the decline of the Medieval Climate Anomaly and the myth of the Alexandrian Golden Age in Scotland', in R.D. Oram (ed.), *'With Our Backs to the Ocean': Land, Lordship, Climate Change, and Environment in the North-West European Past: Essays in Memory of Alasdair Ross*, Environmental Histories of the North Atlantic World 5 (Turnhout: Brepols, 2023), pp. 83–107.

Oram, R.D., *Where Men No More May Reap or Sow: The Little Ice Age: Scotland 1400–1850* (Edinburgh: John Donald, 2024).

Oram, R. and Adderley, P., 'Lordship and environmental change in central Highland Scotland c.1300 to c.1450', *Journal of the North Atlantic*, 1 (2008), pp. 74–84.

Oram, R.D. and Adderley, P., 'Lordship, land and environmental change in West Highland and Hebridean Scotland c.1300–c.1450', in S. Cavaciocchi (ed.), *Economic and Biological Inter actions in Pre-industrial Europe from the 13th to the 18th Centuries* (Florence: University of Florence Press, 2010), pp. 257–268.

Oram, R.D. and Adderley, P., 'Re Innse Gall: a Norse colony in the Irish Sea and Hebrides?', in S. Imsen (ed.), *The Norwegian Domination and the Norse World c.1100–c.1400* (Trondheim: Rostra Books, 2010), pp. 125–148.

Oram, R.D. and Whatley, C.A., 'A brief history of Scottish salt from the eleventh century to the late twentieth century', in S.A. Whatley and J. Hambly (eds), *Salt: Scotland's Newest Oldest Industry* (Edinburgh: John Donald, 2023), pp. 16–33.

Owen, O. and Lowe, C. (eds), *Kebister: The Four-Thousand-Year-Old Story of One Shetland Township*, Society of Antiquaries of Scotland Monograph Series 14 (Edinburgh: Society of Antiquaries of Scotland, 1999).

Parker, A.G., Goudie, A.S., Anderson, D.E., Robinson, M.A. and Bonsal, C., 'A review of mid-Holocene elm decline in the British Isles', *Progress in Physical Geography: Earth and Environment*, 26:1 (2002), pp. 1–45.

Parker Pearson, M. and Sharples, N. (eds), *Between Land and Sea: Excavations at Dun Vulan, South Uist* (Sheffield: Sheffield Academic Press, 1999).

Parker Pearson, M., Brennand, M., Mulville, J. and Smith, H. (eds), *Cille Pheadair: A Norse Farmstead and Pictish Burial Cairn in South Uist* (Oxford: Oxbow Books, 2018).

Parry, M.L., 'Secular climatic change and marginal agriculture,' *Transactions of the Institute of British Geographers*, 64 (1975), pp. 1–13.

Parry, M.L., *Climatic Changes, Agriculture and Settlement* (Folkestone: Dawson & Sons, 1978).

Parry, M.L. and Carter, T.R., 'The effects of climatic variations on agricultural risk', *Climatic Change*, 7 (1985), pp. 95–110.

Patterson, W.P., Dietrich, K.A., Holmden, C. and Andrews, J.T., 'Two millennia of North Atlantic seasonality and implications for Norse colonies', *Proceedings of the National Academy of Sciences*, 107:12 (2010), pp. 5306–5310.

Pearson, C., *Mobilizing Nature: The Environmental History of War and Militarization in Modern France* (Manchester: Manchester University Press, 2012).

Penman, M.A., *David II 1329–71* (East Linton: Tuckwell Press, 2004).

Penman, M.A., *Robert the Bruce, King of the Scots* (London: Yale University Press, 2014).

Peregrine, P., 'Climate and social change at the start of the Late Antique Little Ice Age', *The Holocene*, 30:11 (2020), pp. 1643–1648.

Perry, D., *Dundee Rediscovered: The Archaeology of Dundee Reconsidered*, Tayside and Fife Archaeology Committee Monograph 4 (Perth: Tayside and Fife Archaeology Committee, 2005).

Perry, I. and Moore, P.D., 'Dutch elm disease as an analogue of Neolithic elm decline', *Nature*, 326 (1987), pp. 72–73.

Perth High Street Archaeological Excavation 1975–1977, fascicule 4: *Living and Working in a Medieval Scottish Burgh: Environmental Remains and Miscellaneous Finds* (Perth: Tayside and Fife Archaeological Committee, 2010).

Pfister, C., Schwarz-Zanetti, G. and Wegmann, M., 'Winter severity in Europe: the fourteenth century', *Climatic Change*, 34 (1996), pp. 91–108.

Pfister, C., Luterbacher, J., Schwarz-Zanetti, G. and Wegmann, M., 'Winter air temperature variations in western Europe during the Early and High Middle Ages (ad 750–1300)', *The Holocene*, 8 (1998), pp. 535–552.

Piggott, S., 'Native economies and the Roman occupation of the north', in I.A. Richmond (ed.), *Roman and Native in North Britain* (London: Nelson, 1958), pp. 1–27.

Pitts, L.F. and St Joseph, J.K., *Inchtuthil: The Roman Legionary Fortress: Excavations 1952–65* (London: Society for the Promotion of Roman Studies, 1985).

Platt, M.I., 'The animal bones', in J.R.C. Hamilton (ed.), *Excavations at Jarlshof, Shetland* (Edinburgh: HMSO, 1956), pp. 212–215.

Pollock, D., 'The Lunan Valley Project: medieval rural settlement in Angus', *Proceedings of the Society of Antiquaries of Scotland*, 115 (1985), pp. 357–399.

Postan, M.M., *The Medieval Economy and Society: An Economic History of Britain in the Middle Ages* (London: Weidenfeld and Nicolson, 1972).

Proctor, C.J., Baker, A. and Barnes, W.L., 'A three thousand year record of North Atlantic climate', *Climate Dynamics*, 19 (2002), pp. 449–454.

Proctor, C.J., Baker, A., Barnes, W.L. and Gilmour, M.A., 'A thousand year speleothem proxy record of North Atlantic climate from Scotland', *Climate Dynamics*, 16 (2000), pp. 815–820.

Rackham, O., *Ancient Woodland: Its History, Vegetation and Uses in England* (London, 1980).

Rackham, O., *Trees and Woodland in the British Landscape: The Complete History of Britain's Trees, Woods and Hedgerows*, rev. paperback edition (London: Phoenix Press, 2001).

Ralston, I., 'The Green Castle and the promontory forts of North-East Scotland', *Scottish Archaeological Forum*, 10 (1978), pp. 27–40.

Ralston, I., 'Portknockie: promontory forts and Pictish settlement in the North-East', in A. Small (ed.), *The Picts: A New Look at Old Problems* (Dundee: University of Dundee, 1987), pp. 15–26.

Ramsay, S. and Dickson, J.H., 'Vegetation history of central Scotland', *Botanical Journal of Scotland*, 49 (1997), pp. 141–150.

Rascovan, N. et al., 'Emergence and spread of basal lineages of *Yersinia pestis* during the Neolithic decline', *Cell*, 176:1–2 (2019), pp. 295–305.

Rasmusen, P., 'Leaf-foddering of livestock in the Neolithic: archaeobotanical evidence from Weier, Switzerland', *Journal of Danish Archaeology*, 8:1 (1989), pp. 51–71.

Rasmussen, S. et al., 'Early divergent strains of *Yersinia pestis* in Eurasia 5,000 years ago', *Cell*, 163:3 (2015), pp. 571–582.

Raye, L., 'The early extinction date of the beaver (*Castor fiber*) in Britain', *Historical Biology*, 27:8 (2015), pp. 1029–1041.

Reid, N.H., *Alexander III 1249–1286: First among Equals* (Edinburgh: John Donald, 2019).

Ridley, R.T., 'To be taken with a pinch of salt: the destruction of Carthage', *Classical Philology*, 81:2 (1986), pp. 140–146.

Ritchie, G. and Welfare, H., 'Excavations at Ardnave, Islay', *Proceedings of the Society of Antiquaries of Scotland*, 113 (1983), pp. 302–366.

Ritchie, J.N.G., 'Two new Pictish symbol stones from Orkney', *Proceedings of the Society of Antiquaries of Scotland*, 101 (1968–9), pp. 130–133.

Ritchie, R.L.G., *The Normans in Scotland* (Edinburgh: Edinburgh University Press, 1954).

Rivet, A.L.F. and Smith, C., *The Place Names of Roman Britain* (London: B.T. Batsford, 1979).

Rixson, D., *The Small Isles: Canna, Rum, Eigg and Muck* (Edinburgh: Birlinn, 2001).

Roberts, B.K., Turner, J. and Ward, P.F., 'Recent forest history and land use in Weardale, northern England', in H.J.B. Birks and R.G. West (eds), *Quaternary Plant Ecology* (Oxford: Blackwell Scientific, 1973), pp. 207–221.

Ross, A., 'The *dabhach* in Moray: a new look at an old tub', in A. Woolf (ed.), *Landscape and Environment in Dark Age Scotland* (St Andrews: University of St Andrews, 2006), pp. 57–74.

Ross, A., *The Kings of Alba c.1000–c.1130* (Edinburgh: John Donald, 2011).

Ross, A., *Land Assessment and Lordship in Medieval Northern Scotland* (Turnhout: Brepols, 2015).

Ross, A., 'A post-plague golden age? Scotland, *antiqua taxatio*, and thirteenth-century climate change', in R.D. Oram (ed.), *'With Our Backs to the Ocean': Land, Lordship, Climate Change, and Environment in the North-West European Past: Essays in Memory of Alasdair Ross*, Environmental Histories of the North Atlantic World 5 (Turnhout: Brepols, 2023), pp. 109–134.

Royal Commission on the Ancient and Historical Monuments of Scotland, *Lanarkshire: An Inventory of the Prehistoric and Roman Monuments* (Edinburgh: HMSO, 1978).

Royal Commission on the Ancient and Historical Monuments of Scotland, *North-East Perth: An Archaeological Landscape* (Edinburgh: HMSO, 1990).

Royal Commission on the Ancient and Historical Monuments of Scotland, *South-East Perth: An Archaeological Landscape* (Edinburgh: HMSO, 1994).

Royal Commission on the Ancient and Historical Monuments of Scotland, *Mar Lodge Estate, Grampian: An Archaeological Survey* (Edinburgh: HMSO, 1995).

Royal Commission on the Ancient and Historical Monuments of Scotland, *Early Medieval Sculpture in Angus Council*, RCAHMS Broadsheet 11 (Edinburgh: RCAHMS, 2003).

Royal Commission on the Ancient and Historical Monuments of Scotland, *In the Shadow of Bennachie: A Field Archaeology of Donside* (Edinburgh: RCAHMS / Society of Antiquaries of Scotland, 2007).

Ruddock, T., 'Repair of two important early Scottish roof structures', *Proceedings of the Institute of Civil Engineers*, 110 (1995), pp. 296–307.

Rydval, M. et al., 'Reconstructing 800 years of summer temperatures in Scotland from tree rings', *Climate Dynamics*, 49:2951 (2017), pp. 2951–2974.

Scottish Green Party, *A Rural Manifesto for the Highlands: Creating the Second Great Wood of Caledon* (Inverness, 1989).

Serjeantson, D., *Farming and Fishing in the Outer Hebrides ad 600 to 1700: The Udal, North Uist*, Southampton Monographs in Archaeology, New Series 2 (Chandlers Ford: Highfield Press Southampton, 2013).

Sharples, N., 'The exploitation of the landscape', in N. Sharples (ed.), *The Economy of a Norse Settlement in the Outer Hebrides: Excavations at Mounds 2 and 2A, Bornais, South Uist* (Oxford: Oxbow Books, 2021), pp. 418–424.

Sharples, N. (ed.), *Scalloway: A Broch, Late Iron Age Settlement and Medieval Cemetery in Shetland* (Oxford: Oxbow Books, 1998).

Sharples, N. (ed.), *A Norse Farmstead in the Outer Hebrides: Excavations at Mound 3, Bornais, South Uist* (Oxford: Oxbow Books, 2005).

Sharples, N. (ed.), *A Late Iron Age Farmstead in the Outer Hebrides: Excavations at Mound 1, Bornais, South Uist* (Oxford: Oxbow Books, 2012).

Sharples, N. (ed.), *The Economy of a Norse Settlement in the Outer Hebrides: Excavations at Mounds 2 and 2A, Bornais, South Uist* (Oxford: Oxbow Books, 2021).

Shepherd, C., *The Late Medieval Landscape of North-East Scotland: Renaissance, Reformation and Revolution* (Oxford: Windgather Press, 2021).

Shepherd, I.A.G., 'The Picts in Moray', in W.D.H. Sellar (ed.), *Moray: Province and People* (Edinburgh: Scottish Society for Northern Studies, 1993), pp. 75–90.

Shrewsbury, J.F.D., *A History of Bubonic Plague in the British Isles* (Cambridge: Cambridge University Press, 1971).

Simpson, I.A., 'The chronology of anthropogenic soil formation in Orkney', *Scottish Geographical Magazine*, 109:1 (1993), pp. 4–11.

Simpson, I.A., 'Relict properties of anthropogenic deep top soils as indicators of infield management in Marwick, West Mainland, Orkney', *Journal of Archaeological Science*, 24 (1997), pp. 365–380.

Simpson, I.A. and Guttmann, E.B., 'Transitions in early arable land management in the Northern Isles – the *papar* as agricultural innovators?', in B.E. Crawford (ed.), *The 'Papar' in the North Atlantic: Environment and History*, St John's House Papers 10 (St Andrews: Committee for Dark Ages Studies, University of St Andrews, 2002), pp. 59–67.

Simpson, I.A., Lancaster, S.J. and Dockrill, S.J., 'Making arable soils: anthropogenic soil formation in a multi-period landscape', in R.A. Nicholson and S.J. Dockrill (eds), *Old Scatness Broch, Shetland: Retrospect and Prospect* (New York: North Atlantic Biocultural Organisation, 1998), pp. 111–126.

Simpson, I.A., Crawford, B.E. and Ballin Smith, B., 'Papar place-names in the Northern and Western Isles of Scotland: a preliminary assessment of their association with agricultural land potential', http://www.paparproject.org.uk/agricultural.html.

Simpson, I.A., Dockrill, S.J., Bull, I.D. and Evershed, R.P., 'Early anthropogenic soil formation at Tofts Ness, Sanday, Orkney', *Journal of Archaeological Science*, 25 (1998), pp. 729–746.

Sinclair, J. (ed.), *General Report of the Agricultural State and Political Circumstances of Scotland*, volume 4 (Edinburgh: Archibald Constable, 1814).

Skene, W.F., 'On the traditionary accounts of the death of Alexander III', *Proceedings of the Society of Antiquaries of Scotland*, 20 (1885–6), pp. 177–185.

Slavin, P., 'Late-medieval goose farming: evidence from eastern England, c.1250–1400', *Agricultural History Review*, 58:1 (2010), pp. 1–29.

Slavin, P., 'The great bovine pestilence and its economic and environmental consequences in England and Wales, 1318–50', *Economic History Review*, 65:4 (2012), pp. 1239–1266.

Slavin, P., 'Warfare and ecological destruction in early fourteenth-century British Isles', *Environmental History*, 19:3 (2014), pp. 528–550.

Slavin, P., 'Flogging a dead cow: coping with animal panzootics on the eve of the Black Death', in A. Brown, A. Burn and R. Doherty (eds), *Coping with Crisis: Re-evaluating the Role of Crises in Economic and Social History* (Woodbridge: Boydell and Brewer, 2015), pp. 111–135.

Slavin, P., 'Epizootic landscapes: sheep scab and regional environment in England in 1279–1280', *Landscapes*, 17 (2016), pp. 156–170.

Slavin, P., *Experiencing Famine in Fourteenth-Century Britain* (Turnhout: Brepols, 2019).

Slavin, P., 'Mites and merchants: the crisis of English wool and textile trade revisited, c. 1275–1330', *Economic History Review*, 73:4 (2020), https://doi.org/10.1111/ehr.12969.

Slavin, P. and DeWitte, S., 'Between famine and death: physiological stress and dairy deficiency in England on the eve of the Black Death (1315–50): new evidence from paleoepidemiology and manorial accounts', *Journal of Interdisciplinary History*, 44 (2013), pp. 37–60.

Slavin, P. and Sebbane, F., 'Emergence and spread of ancestral *Yersinia pestis* in Late-Neolithic and Bronze-Age Eurasia, *ca.* 5,000 to 2,500 y B.P.', *Proceedings of the National Academy of Sciences*, 119:21 (2022), https://doi.org/10.1073/pnas.2204044119.

Small, A., 'Burghead', *Scottish Archaeological Forum*, 1 (1969), pp. 61–68.

Smith, A.G. and Pilcher, J.R., 'Radiocarbon dates and vegetational history of the British Isles', *New Phytologist*, 72:4 (1973), pp. 903–914.

Smith, B., 'A short history of Kebister', in O. Owen and C. Lowe (eds), *Kebister: The Four-Thousand-Year-Old Story of One Shetland Township*, Society of Antiquaries of Scotland Monograph Series 14 (Edinburgh: Society of Antiquaries of Scotland, 1999), pp. 17–21.

Smith, H., 'Plants', in N. Sharples (ed.), *A Norse Farmstead in the Outer Hebrides: Excavations at Mound 3, Bornais, South Uist* (Oxford: Oxbow Books, 2005), pp. 189–190.

Smith, H. and Mulville, J., 'Resource management in the Outer Hebrides: an assessment of the faunal and floral evidence from the archaeological investigations', in R.A. Housley and G. Coles (eds), *Atlantic Connections and Adaptation: Economics, Environments and Subsistence in the North Atlantic* (Oxford: Oxbow Books, 2004), pp. 48–64.

Smith, H., College, S. and Austin, P., 'The carbonized plant remains and wood charcoal', in M. Parker Pearson, M. Brennand, J. Mulville and H. Smith (eds), *Cille Pheadair: A Norse Farmstead and Pictish Burial Cairn in South Uist* (Oxford: Oxbow Books, 2018), pp. 516–540.

Smout, T.C., *Nature Contested: Environmental History in Scotland and Northern England since 1600* (Edinburgh: Edinburgh University Press, 2000).

Smout, T.C., 'Woodland in the maps of Pont', in I.C. Cunningham (ed.), *The Nation Survey'd: Timothy Pont's Maps of Scotland* (Edinburgh: John Donald, 2006), pp. 77–92.

Smout, T.C., 'The history and the myth of Scots pine', *Scottish Forestry*, 68:1 (2014), pp. 1–4.

Smout, T.C. and Stewart, M., *The Firth of Forth: An Environmental History* (Edinburgh: Birlinn, 2012).

Smout, T.C., MacDonald, A.R. and Watson, F., *A History of the Native Woodlands of Scotland, 1500–1920* (Edinburgh: Edinburgh University Press, 2005).

Smyth, A.P., *Warlords and Holy Men: Scotland AD 80–1000* (London: Edward Arnold, 1984).

Spence, D., *Shetland's Living Landscape* (Lerwick: Thule Press, 1979).

Stauffer, B., 'The Greenland Ice Core Project', *Science*, 260:5115 (1993), pp. 1766–1767, https://doi.org/10.1126/science.260.5115.1766.

Stell, G., 'Destruction and damage: a reassessment of the historical and architectural evidence', in N. Macdougall (ed.), *Scotland and War AD 79–1918* (Edinburgh: John Donald, 1991), pp. 23–35.

Stell, G. and Baillie, M., 'The great hall and roof of Darnaway Castle, Moray', in W.D.H. Sellar (ed.), *Moray: Province and People* (Edinburgh: Scottish Society for Northern Studies, 1993), pp. 162–186.

Stevens, S.T., 'A legend of the destruction of Carthage', *Classical Philology*, 83:1 (1988), pp. 39–41.

Stevenson, A., 'Trade with the south, 1070–1513', in M. Lynch et al. (eds), *The Scottish Medieval Town* (Edinburgh: John Donald, 1988), pp. 180–206.

Stevenson, A., 'Taxation in medieval Scotland', in P.G.B. McNeill and H.L. MacQueen (eds), *Atlas of Scottish History to 1707* (Edinburgh: Scottish Medievalists / Department of Geography, Edinburgh University, 1996), pp. 298–305.

Strachan, D., Sneddon, D. and Tipping, R., *Early Medieval Settlement in Upland Perthshire: Excavations at Lair, Glen Shee 2012–17* (Oxford: Archaeopress, 2019).

Stringer, K.J., 'Acts of lordship: the records of the lords of Galloway to 1234', in T. Brotherstone and D. Ditchburn (eds), *Freedom and Authority: Historical and Historiographical Essays Presented to Grant G. Simpson* (East Linton: Tuckwell Press, 2000), pp. 203–224.

Stringer, K.J., 'Reform monasticism and Celtic Scotland: Galloway, *c.*1140–*c.*1240', in E.J. Cowan and R.A. McDonald (eds), *Alba: Celtic Scotland in the Medieval Era* (East Linton: Tuckwell Press, 2000), pp. 127–165.

Stuiver, M. and Quay, P.D., 'Changes in atmospheric carbon-14 attributed to a variable sun', *Science*, 207:4426 (1980), pp. 11–19.

Summers, J. and Bond, J., 'The agricultural economy of a late wheelhouse', in N. Sharples (ed.), *A Late Iron Age Farmstead in the Outer Hebrides: Excavations at Mound 1, Bornais, South Uist* (Oxford: Oxbow Books, 2012), p. 339.

Sun, C.Q., 'Ash from Changbaishan Millennium eruption recorded in Greenland ice: implications for determining the eruption's timing and impact', *Geophysical Research Letters*, 41:2 (2013), pp. 694–701.

Syse, K.L., 'The ebb and flow of trees and farmland: symbols of nationhood in Scotland and Norway', *Journal of the North Atlantic*, special issue 4 (2013), pp. 219–228.

Tabraham, C.J., 'Norman settlement in Upper Clydesdale: recent archaeological fieldwork', *Transactions of the Dumfriesshire and Galloway Natural History and Antiquarian Society*, 3rd series, 52 (1977–8), pp. 114–128.

Taylor, A., '*Leges Scocie* and the lawcodes of David I, William the Lion and Alexander II', *Scottish Historical Review*, 88:2, no. 226 (October 2009), pp. 207–288.

Taylor, A., 'The assizes of David I, king of Scots, 1124–53', *Scottish Historical Review*, 91:2, no. 232 (October 2012), pp. 197–238.

Taylor, A., *The Shape of the State in Medieval Scotland 1124–1290* (Oxford: Oxford University Press, 2016).

Taylor, M., 'The wood', in M. Parker Pearson and N. Sharples (eds), *Between Land and Sea: Excavations at Dun Vulan, South Uist* (Sheffield: Sheffield Academic Press, 1999), pp. 188–192.

Taylor, S., 'The coming of the Augustinians to St Andrews and version B of the St Andrews foundation legend', in S. Taylor (ed.), *Kings, Clerics and Chronicles in Scotland, 500–1297* (Dublin: Four Courts Press, 2000), pp. 115–123.

Thun, T. and Svarva, H., 'Tree-ring growth shows that the significant population decline in Norway began decades before the Black Death', *Dendrochronologia*, 47 (2018), pp. 23–29.

Tipping, R., 'The form and fate of Scotland's woodlands', *Proceedings of the Society of Antiquaries of Scotland*, 124 (1994), pp. 1–54.

Tipping, R., 'Holocene evolution of a lowland Scottish landscape: Kirkpatrick Fleming, II, Regional vegetation and land-use change', *The Holocene*, 5 (1995), pp. 83–96.

Tipping, R., 'Environment and environmental change in Eastern Dumfriesshire', in Royal Commission on the Ancient and Historical Monuments of Scotland, *Eastern Dumfriesshire: An Archaeological Landscape* (Edinburgh: HMSO, 1997), pp. 10–25.

Tipping, R., 'Medieval woodland history from the Scottish Southern Uplands: fine spatial-scale pollen data from a small woodland hollow', in T.C. Smout (ed.), *Scottish Woodland History* (Edinburgh: Scottish Cultural Press, 1997), pp. 52–75.

Tipping, R., 'Pollen analysis, late Iron Age and Roman agriculture around Hadrian's Wall', in A. Gwilt and C. Haselgrove (eds), *Reconstructing Iron Age Societies* (Oxford: Oxbow, 1997), pp. 239–247.

Tipping, R., 'Vegetation history of southern Scotland', *Botanical Journal of Scotland*, 49 (1997), pp. 151–162.

Tipping, R., 'Palaeoecology and political history: evaluating driving forces in historic landscape change in southern Scotland', in I.D. Whyte and A.J.L. Winchester (eds), *Society, Landscape and Environment in Upland Britain*, Society for Landscape Studies, supplementary series, 2 (Society for Landscape Studies, 2004), pp. 11–20.

Tipping, R., 'Landscape and land-use change in the early historic period on the northern Solway plain', in C. Lowe (ed.), *Excavations at Hoddom, Dumfriesshire* (Edinburgh: Society of Antiquaries of Scotland, 2006), pp. 146–148.

Tipping, R., 'Blanket peat in the Scottish Highlands: timing, cause, spread and the myth of environmental determinism', *Biodiversity and Conservation*, 17 (2008), pp. 2097–2113.

Tipping, R., *Bowmont: An Environmental History of the Bowmont Valley and the Northern Cheviot Hills, 10,000 BC–AD 2000* (Edinburgh: Society of Antiquaries of Scotland, 2010).

Tipping, R. and Adams, J., 'Structure, composition and significance of Medieval storm beach ridges at Caerlaverock, Dumfries and Galloway', *Scottish Journal of Geology*, 43:2 (2007), pp. 115–123.

Tipping, R., Davies, A.L. and McCulloch, R., 'Introduced oak woodlands in northern Scotland: pollen analytical evidence for early historic plantations', in A. Woolf (ed.), *Landscape and Environment in Dark Age Scotland* (St Andrews: Committee for Dark Age Studies, University of St Andrews, 2006), pp. 29–47.

Toorians, L., 'Twelfth-century Flemish settlement in Scotland', in G.G. Simpson (ed.), *Scotland and the Low Countries 1124–1994* (East Linton: Tuckwell Press, 1996), pp. 1–14.

Tranter, N., *The Lords of Misrule* (London: Corona Books, 1976).

Turner, J., 'The environment of northeast England during Roman times as seen by pollen analysis', *Journal of Archaeological Science*, 6 (1979), pp. 285–290.

van der Veen, M., *Crop Husbandry Regimes: An Archaeological Study of Farming in Northern England 1000 BC–AD 500* (Sheffield: University of Sheffield, 1992).

Vésteinsson, O., McGovern, T.H. and Keller, C., 'Enduring impacts: social and environmental aspects of Viking age settlement in Iceland and Greenland', *Archaeologia Islandica*, 2 (2002), pp. 98–136.

Vieira, L.E.A., Solanki, S.K., Krivova, N.A. and Usoskin, I., 'Evolution of the solar irradiance during the Holocene', *Astronomy and Astrophysics*, 531 (July 2011), https://doi.org/10.1051/0004-6361/201015843.

VisitScotland, 'Scotland's landscapes, scenery and nature spots', https://www.visitscotland.com/things-to-do/landscapes-nature.

Visona, P., 'On the destruction of Carthage again', *Classical Philology*, 83:1 (1988), pp. 41–42.

Wainwright, F.T., *The Souterrains of Southern Pictland* (London: Routledge and Kegan Paul, 1963).

Walker, B. and Ritchie, G., *Fife, Perthshire and Angus*, 2nd edition (Edinburgh: HMSO, 1995).

Ward, T., 'Daer Reservoir (Crawford parish): bastle house, cairns, find-spots, Mesolithic knapping site', *Discovery and Excavation in Scotland*, 1995, p. 87.

Ward, T., 'Daer Reservoir (Crawford parish): Mesolithic sites; burnt mounds; cairns', *Discovery and Excavation in Scotland*, 1997, p. 75.

Ward, T., 'Daer Reservoir (Crawford parish): Mesolithic flint-knapping site and other lithic scatters', *Discovery and Excavation in Scotland*, new series, 1 (2000), p. 86.

Ward, T., 'Weston Farm (Carnwath; Dunsyre parishes): Mesolithic and later prehistoric lithic scatters; medieval brooch', *Discovery and Excavation in Scotland*, new series, 1 (2000), p. 85.

Ward, T., 'Daer Valley Project (Crawford parish): excavation; survey', *Discovery and Excavation in Scotland*, 6 (2005), p. 134.

Ward, T., 'Upper Tweed survey – Burnetland Hill: excavation', *Discovery and Excavation in Scotland*, 8 (2007), p. 170

Warmington, B.H., 'The destruction of Carthage: a retractatio', *Classical Philology*, 83:4 (1988), pp. 308–310.

Waterman, D.M., 'The excavation of a house and souterrain at Craig Hill, Co. Antrim', *Ulster Journal of Archaeology*, 19 (1956), pp. 87–91.

Waterman, D.M., 'The excavation of a house and souterrain at Whitefort, Drumarood, Co. Down', *Ulster Journal of Archaeology*, 19 (1956), pp. 73–86.

Watkins, T., 'Where were the Picts? An essay in settlement archaeology', in J.G.P. Friell and W.G. Watson (eds), *Pictish Studies: Settlement, Burial and Art in Dark Age Northern Britain*, British Archaeological Reports, British Series 125 (Oxford: BAR Publishing, 1984), pp. 63–86.

Watkins, T. and Barclay, G., 'Excavation of a settlement and souterrain at Newmill, near Bankfoot, Perthshire', *Proceedings of the Society of Antiquaries of Scotland*, 110 (1981), pp. 165–208.

Watson, F., *Under the Hammer: Edward I and Scotland 1286–1307* (East Linton: Tuckwell Press, 1998).

Watson, F. and Dixon, P., *A History of Scotland's Landscapes* (Edinburgh: Historic Environment Scotland, 2018).

Watt, H.B., 'Scottish forests and woodland in early historic times', *Annals of the Andersonian Institute*, 2 (1900), pp. 89–107.

White, L., Jnr, 'The historical roots of our ecologic crisis', *Science*, 155 (1967), pp. 1203–1207.

Whittington, G. and Edwards, K.J., '*Ubi solitudinem faciunt pacem appellant*: the Romans in Scotland, a palaeoenvironmental contribution', *Britannia*, 24 (1993), pp. 13–25.

Whittington, G. and McManus, J., 'Dark Age agricultural practices and environmental change: evidence from Tentsmuir, Fife, eastern Scotland', in C.M. Mills and G. Coles (eds), *Life on the Edge: Human Settlement and Marginality* (Oxford: Oxbow Books, 1998), pp. 111–119.

Whittington, G. and Edwards, K.J., 'Climate change', in K.J. Edwards and I.B.M. Ralston (eds), *Scotland after the Ice Age: Environment, Archaeology and History, 8000 BC–AD 1000*, 2nd edition (Edinburgh: Edinburgh University Press, 2002), pp. 11–22.

Whittington, G., Edwards, K.J. and Cundill, P.R., 'Late- and post-glacial vegetational change at Black Loch, Fife, Eastern Scotland – a multiple core approach', *New Phytologist*, 18:1 (1991), pp. 147–166.

Wickham-Jones, C.R., *Scotland's First Settlers* (London: B.T. Batsford, 1994).

Wickham-Jones, C.R., *The Landscape of Scotland: A Hidden History* (Stroud: Tempus, 2001).

Wilson, A., 'Roman penetration in Strathclyde south of the Antonine Wall, Part Two: Romanization', *Glasgow Archaeological Journal*, 20 (1996–7), pp. 1–40.

Wilson, R. et al., 'Reconstructing Holocene climate from tree rings: the potential for a long chronology from the Scottish Highlands', *The Holocene*, 22:1 (2011), pp. 3–11.

Wohletz, K., 'Were the Dark Ages triggered by volcano-related climate changes in the 6th century?', *EOS Transactions of the American Geophysical Union*, 48:81 (2000), https://www.lanl.gov/orgs/ees/geodynamics/Wohletz/Krakatau.htm.

Woolf, A., *From Pictland to Alba 789–1070* (Edinburgh: Edinburgh University Press, 2007).

Woolf, A. (ed.), *Landscape and Environment in Dark Age Scotland* (St Andrews: Committee for Dark Age Studies, University of St Andrews, 2006).

Xu, J.D., 'Climatic impact of the Millennium eruption of Changbaishan volcano in China: new insights from high-precision radiocarbon wiggle-match dating', *Geophysical Research Letters*, 40:1 (2013), pp. 54–59.

Yalden, D., *The History of British Mammals* (London: T. and A.D. Poyser, 1999).

Yalden, D.W. and Kitchener, A.C., 'History of the fauna', in S. Harris and D.W. Yalden (eds), *Mammals of the British Isles: Handbook*, 4th edition (Southampton: Mammal Society, 2008), pp. 17–31.

Zierler, D., *The Invention of Ecocide: Agent Orange, Vietnam, and the Scientists Who Changed the Way We Think About the Environment* (Athens GA: University of Georgia Press, 2011).

Newspaper and online news articles and blogs

BBC Scotland News, 7 April 2021, '"Exciting" Stone Age discoveries in the Cairngorms', https://www.bbc.co.uk/news/uk-scotland-north-east-orkney-shetland-56661834.

BBC Scotland News, 8 November 2021, 'Hundreds of prehistoric tools found along river', https://www.bbc.co.uk/news/articles/cz88vzkdgyeo.

BBC Scotland News, 22 January 2022, 'Grazing cattle boost capercaillie habitat', https://www.bbc.co.uk/news/uk-scotland-highlands-islands-60084433.

BBC News, Science and Environment, 'More than eight million trees lost this winter in the UK', 16 February 2022, https://www.bbc.co.uk/news/science-environment-60348947.

BBC Scotland News, [11 March 2022], 'Thriving wild boar population creating havoc in the Highlands' https://www.bbc.co.uk/programmes/articles/2gLYpzPmBtbH4qTt3VQR9bF/the-thriving-wild-boar-population-creating-havoc-in-the-highlands.

BBC Scotland News, 2 February 2023, 'Formal bid made for Flow Country UNESCO status', https://www.bbc.co.uk/news/uk-scotland-highlands-islands-64497032.

University of Aberdeen News, 14 May 2020, 'Hillfort revealed to be largest Pictish site ever discovered in Scotland', https://www.abdn.ac.uk/news/14019/.

Wessex Archaeology Coastal Archaeological Landscapes: Intertidal & Estuarine, https://blogs.wessex.co.uk/coalie/2014/10/30/coalie-project-report-online/.

Doctoral and master's theses

Miller, J.J., 'An Archaeobotanical Investigation of Oakbank Crannog, a Prehistoric Lake Dwelling in Loch Tay, the Scottish Highlands', unpublished PhD thesis, University of Glasgow (1997).

Unpublished reports and 'grey literature'

Kenworthy, J.B., *Excavation of a Mesolithic Settlement Site at Nethermills Farm, Crathes, near Banchory, Grampian, 1978–80: Interim Statement* (1981), unpublished report.

Music recording

'The Wire', from Runrig, *Heartland*, Ridge Records RR005, 1985, LP, music and lyrics C. and R. Macdonald, published by Storr Music.

Index

aberdaan, 343
Aberdeen, 96, 250, 278, 281, 314, 316, 318, 332, 334, 338, 342, 344
Aberdeenshire, 78, 111, 173, 216, 362
Aberfoyle (Stirling), 225
Aberlemno (Angus), 170
Abernethy (Highland), 79, 237
Adomnán, abbot of Iona, 84, 85, 86, 87, 126, 128, 167
aerosol particulates, 32, 133, 135, 295, 296
Agricola, Cnaeus Julius, Roman general, 11, 37, 52, 73
agriculture, 6, 11, 13, 43, 53, 60, 65, 67, 96, 108, 115, 138, 146, 162, 181, 183, 193, 207, 225, 240, 242, 247, 248, 253, 272, 274, 294, 305, 327, 336, 351, 358, 360–365, 370, 375
  abandonment or contraction, 108, 116, 253
  arable, 28, 31–32, 35, 39, 43, 51, 53–56, 57, 60, 61, 62–63, 66, 67, 70, 72, 73–74, 78, 79, 99, 108, 114, 116, 118, 124, 126–127, 129, 130, 140, 142, 146, 148, 150–151, 162, 193, 220, 224, 226, 227, 234, 237, 241, 245, 246, 247, 250, 252, 254, 260–263, 265, 266, 267, 269–271, 273–274, 275, 286, 294, 298, 308, 309, 322, 324, 326, 333–334, 339, 345, 358, 360–365, 366, 370, 374
  Bronze Age, 32, 34
  expansion of, 9, 44, 56, 57, 67, 73, 138, 181, 207, 242, 260, 266, 294, 324, 327, 370
  Neolithic, 6, 26, 28–29, 31
  slash-and-burn, 26, 69, 100
  subsistence, 28, 63, 79, 252, 275, 321, 341, 345, 360, 361, 370, 374, 375
  surplus (arable), 40, 43, 59, 66, 142, 181, 185, 205, 252, 265, 275, 311, 359, 371
Airth (Stirling), 269
Alba, kingdom of, 67, 96, 98, 102, 148, 150, 152, 180, 181, 376
alder trees, 60, 70, 74, 78, 79, 80, 81, 83, 85, 86, 87–88, 89, 90, 92, 93. 94, 118, 158, 225, 230, 237, 241, 254, 330–331, 336
  auhne, 230
  carr, 72, 74, 78, 88
Ale valley (Scottish Borders), 268
Alexander I, king of Scots, 183, 281, 282
Alexander II, king of Scots, 16, 185, 202, 210, 233, 260, 261, 325
Alexander III, king of Scots (*see also* golden age), 179, 238, 287, 288, 300, 325, 327
alien species, 22
Alloa Tower (Clackmannshire), 333
Almond, river (Perth and Kinross), 199, 339
Alps, 41
Altens (Aberdeenshire), 250
Alves (Moray), 226
'Alwin's assart' (Scottish Borders), 267
Alyth (Perth and Kinross), 59
Amazonia, 29, 70
Aneirin, Brythonic poet, 167
Anglesey (Gwynedd), 314
Anglian people, 87, 116, 118, 167
Angus, 22, 55, 59, 60, 166, 173, 185, 193, 211, 216, 231, 251, 256, 264, 278, 318, 332, 333, 352
  Glens of, 22, 231, 366
animal husbandry, 141, 221
anticyclonic circulation, 16, 191, 197, 202, 203, 297, 303, 307; *see also* cyclonic circulation
Annals of Connacht, 349
Annals of Innisfallen, 189, 307, 309
Annals of Loch Cé, 190, 197, 306, 349
Annals of Tigernach, 119, 121, 122
Annals of Ulster, 119, 121, 133, 138, 187, 188, 197, 198, 306, 307
Annandale (Dumfries and Galloway), 52, 210, 211
Antarctica, 295
anthrax, 350
anthrosols, 127–130, 140, 142, 362
antiqua taxatio (see also *verus valor*), 356–357
Antonine frontier, 60, 76
apple trees (*see also* orchards), 248, 329–330
arable: *see under* agriculture
Arbroath (Angus), 59, 278
  Abbey, 230, 231, 244, 256
Arctic region, 5, 104, 197
  air, 16, 42, 119, 191, 203
ard (plough), 55
Ardchattan Priory (Argyll and Bute), 354
Ardestie (Angus), 59, 60
Ardnamurchan (Highland), 14, 85, 336
Ardnave, Islay (Argyll and Bute), 63
Ardownie (Angus), 59, 60
Ardross (Highland), 169, 173
Ardskeodnish and Lochawe, lordship of, 336
Argyll (Argyll and Bute), 22, 29, 48, 78, 113, 199, 237, 299, 336, 349, 354
Arisaig (Highland), 237
Arran, island of (Argyll and Bute), 240
Arthur, king of the Britons, 52
ash trees, 71, 78, 85, 86, 89, 90, 330
aspen trees, 79, 83, 85
assarts (*see also* intakes), 225, 266, 267–268, 333, 353
Assynt (Highland), 14, 41, 106
Atholl, earl of, 230
Atlantic Meridional Overturning Circulation (AMOC), 11, 12
Atlantic Ocean, 6, 11, 12, 13, 14, 15, 21, 31, 41, 42, 63, 71, 74, 90, 99, 104, 105–106, 108, 114, 119, 121, 122, 123, 130, 132, 134, 135, 138, 151, 186, 187, 191–192, 196, 197, 199, 200, 202, 217, 237, 238, 241, 258, 270, 279, 280, 281, 292, 296–298, 299, 303, 305, 307, 308, 327, 336, 339, 345, 361
atmospheric circulation systems: *see* El Niño Southern Oscillation; North Atlantic Oscillation
Augustinian order, 167, 281, 352
Augustus, Roman emperor, 52
auhne: *see under* alder trees
Auldearn (Highland), 221
aurochs, 156
Avenel family, 214
Ayr (South Ayrshire), 161, 232, 233, 304, 338
  fishery, 342
Ayrshire, 24, 87, 227, 233, 283
Azores, 13, 120, 122
  high pressure cell, 13, 42, 105, 120, 122, 135, 191, 197

Baekdu/Changbai, 135
Balantrodach (Midlothian), 332
Balfieth (Aberdeenshire), 244

Balliol family (*see also* Edward, king of Scots; John, king of Scots), 263, 288, 290
Balmerino Abbey (Fife), 282
Baltic Sea, 197, 329, 342
Bamburgh (Northumberland), 118
Banff (Aberdeenshire), 261, 333, 338, 344
  forest of, 261
Banffshire, 256, 366
Bannockburn (Stirling), 325
Barbour, John, 315, 316, 325
barley cultivation, 29, 57, 60, 62, 63, 78, 113–114, 125, 126, 140, 141, 142, 146, 269, 271, 272, 274–275, 281, 360, 361
Barra, 43, 108, 130
Bavelaw (Midlothian), 260
bear (brown), 36, 154, 158–159, 163, 164, 166, 169, 173, 176, 319
  extinction in England, 159
Beauly Firth, 48
beaver, 36, 158, 164, 248
beech trees, 22, 194
beef (*see also* cattle), 144, 145, 349
Beinn na Mhic Aongheis, South Uist (Western Isles), 363
Bellenden, John, 155, 201
Bemersyde (Scottish Borders), 228
Ben Lawers (Perth and Kinross), 360–361, 365, 370
Bergen, 341
Berkeley, Humphrey de, 230
Bermondsey chronicle, 202
Berwick (Northumberland), 228, 291, 309, 314, 338, 350
  fishery, 283
Berwickshire, 158, 217, 222, 259, 260
Biggar (South Lanarkshire), 218, 252
birch trees, 21, 22, 47, 60, 61, 83, 86, 90, 93, 94, 158, 230, 241, 254, 330, 336
  coppice, 61, 74, 78, 81, 83, 87, 330
  wrawes of bule (birch hoops), 230
birch woodland, 69, 74, 81, 86, 89, 116, 118, 218, 225, 237
Birnie (Moray), 74
Biscay, 41
Black Death (*see also* epidemic disease; Great Mortality; pandemic disease; plague, bubonic), 291, 321–322, 354, 356–357, 367, 374
Blackford (Perth and Kinross), 173
blackland, 63, 146, 270, 271
Blackness (West Lothian), 262
Blainslie (Scottish Borders), 267
Blair Atholl (Perth and Kinross), 230
Blairgowrie (Perth and Kinross), 216, 221, 250
blubber, seal and whale, 161, 162, 277
Boar's Raik (Fife), 167
Boccaccio, Giovanni, 372
Boece, Hector, 155, 157, 186, 190
Bo'Ness (West Lothian), 231
Borders (Scottish), 4, 74, 167, 185, 206, 214, 217, 226, 244, 256, 266, 328, 337, 360
Bornais (Western Isles), 53, 57, 62, 71, 90–91, 93, 113–114, 145–146, 148, 150, 271–272, 279, 341, 363–364
boscum, 332, 333
  subboscum, 332
Bostadh (Western Isles), 279
Bower, Walter, abbot of Inchcolm (*see also Scotichronicon*), 181, 185, 186, 189, 195, 288, 305, 306, 307, 308, 309, 310, 311, 318, 319, 325, 326, 346, 352, 353, 356
Bowmont valley (Scottish Borders), 28, 56–57, 62, 63, 72, 74, 78, 116, 124, 256, 359–360, 362, 365–366, 367
Boyd Watt, Hugh, 45
Boyne and Enzie (Aberdeenshire / Moray), forest of, 333
bread, 109, 110, 276, 294, 363
Breadalbane (Perth and Kinross / Stirling), 22
British Isles, 9, 11, 12, 13, 17, 39, 41, 42, 95, 103, 106, 108, 110, 116, 120, 122, 124, 131, 132, 134, 153, 169, 180, 183, 186, 187, 188, 194, 195, 196, 197, 262, 281, 285, 298, 299, 303, 305, 310, 312, 313, 323, 326, 327, 347, 349, 350, 356
brochs, 71, 116
Bronze Age, 29, 31–34, 62, 71, 72, 74, 76, 79, 80, 89, 99, 156, 159, 161, 207, 218, 244, 257, 375
broom, 244, 247, 328
Brough of Birsay (Orkney), 173
Bruce family and supporters, 316–318
Bruce, Edward, 289
Bruce, Robert: *see* Robert I, king of Scots
Brus, Robert de, 211
brushwood, 246, 332
bubonic plague: *see* plague, bubonic
Buchan, 'herschip' of, 315–316
Buiston crannog (North Ayrshire), 87–88
bulls: *see under* cattle
Burghead (Moray), 96–97
Burgie (Moray), 226, 227
burning, of woodland and moorland (*see also* muirburn), 18–19, 22, 29, 36, 55, 69, 76, 125, 233, 256
Burnt Candlemas, 314, 319
Burntisland (Fife), 300
Burrian (Orkney), 173
butter, 141
Buzzart Dykes park (Perth and Kinross), 216
Byzantine Empire, 110, 133

Cadzow (South Lanarkshire), 157
Caerlaverock (Dumfries and Galloway), 308
  castle, 287
Cairngorm mountains, 13, 16, 21, 25, 30–31, 79, 80, 150, 199, 204, 299
  tree-ring sequences from: *see* tree rings
Caithness (Highland), 16, 18, 25, 29, 30, 70, 185, 202, 237, 274, 281, 341–342, 357
  fishery, 341–342
  Flow Country, 18–19, 25
Caledon, Wood of (myth), 20–21, 44–45, 47–50, 51–52, 67, 69, 76, 156, 239, 241, 334
  battle of, 52
Caledones, 47, 49, 54
Caledonia, 40, 47–50, 52, 158
  Silva: *see* Caledon, Wood of
Caledonian forest (classification), 21, 47
calf slaughter: *see under* cattle
Calgacus, Caledonian chieftain, 64–65
Callendar, Carse of, 269
Callendar Park (Falkirk), 52, 232
Calluna (*see also* heather), 55, 107, 116, 125, 218, 256, 365, 366
Camban, Glen Affric (Highland), 61, 115–116, 361
Cambuskenneth Abbey and estate (Stirling), 231, 278
Campbell, Colin, 336
Campsie (Perth and Kinross), 207, 260
Campsie Fells (East Dunbartonshire / Stirling), 260
Canada, 13
cannibalism, 318, 323
Caracalla, Roman emperor, 59
Carlisle (Cumbria), 308
Carlungie (Angus), 59–60
Carnach Mor, Glen Affric (Highland), 61, 115, 116
carnivorous mammals: *see* bear (brown); fox; lynx; wildcat; wolves
carr, 56, 72, 74, 78, 87, 88, 237, 248
Carrick (South Ayrshire), 232, 251
Carriden (West Lothian), 262
carruca (plough), 151, 183, 264, 265, 268, 271, 361
carrying capacity, 149, 294
carse and carseland, 52, 206, 231, 232, 269–270, 308; *see also under* Forth; Gowrie; Stirling
Carthage, 64
Cassius Dio, Roman writer, 52
Cat Coit Celidon, 52
cattle, 26, 53, 55, 58, 69, 73, 74, 125, 127, 133, 140, 144, 145, 161, 162, 213, 218, 220, 221, 223, 224, 227, 251, 256–257, 260, 325, 326, 346, 365, 366, 368–369, 370, 375
  beef, 144, 145, 349
  bulls, 156, 169
  calf slaughter, 115, 141, 144
  dairy: *see* dairying
  dung or manure (soil fertiliser), 130
  dung (fuel), 63, 93
  epizootic disease: *see main entry*
  hides, 145, 161, 234, 248, 257, 326, 346, 369
  milk, 141, 369
  overwintering, 125
  pasture: *see main entry*; *see also* grazing (land)
  pestilence, murrain or mortality, 122, 138, 187, 189, 195, 201, 202, 289, 307, 323, 326, 347, 349–351, 369, 370
  white, 156–157
  wild, 156–158
'Celtic cowboys', 53, 54, 55, 63
'Celtic fields', 56
Celts, 20, 37, 39, 43–44, 53–58, 63, 65, 73, 100, 180, 218
  culture, 173, 180
centralisation, social and economic, 96, 116
Ceratocystis: *see* Dutch Elm Disease
Cerialis, Quintus Petillius, Roman general, 39, 52, 57, 73
charcoal, 18, 45, 60, 71, 78, 90, 91, 92–93, 256, 365
  manufacture, 45, 94, 95, 220, 231, 248, 258, 262, 328, 375
chariot warfare, 51, 54

charters, 148, 162, 179, 184, 211, 207, 218, 220, 221, 223, 226, 227, 228, 229, 232, 233, 241, 243, 244, 246, 248, 249, 259, 262–263, 267, 336, 343, 352
cheese production, 141, 234, 257
cherry trees, 79, 86, 330
chestnut trees, 80
Cheviot Hills (Scottish Borders), 28–29, 56–57, 72, 73–74, 78, 116, 118, 126, 225, 251–252, 256, 273, 359, 364, 365, 367, 368, 370
Cheyne, Sir Reginald, 237, 332, 333
Chillingham (Northumberland), 157
China, 135, 137
  record sources, 135
Christy Cleke, 318
Cille Pheadair (Western Isles), 145, 146, 148, 271, 279
Cistercian monks, 130, 185, 206, 232, 254, 262, 366; *see also* Balmerino Abbey; Coupar Angus Abbey; Dundrennan Abbey and estate; Glenluce Abbey; Holm Cultram Abbey and estate; Kinloss Abbey and estate; Melrose Abbey and estate; Newbattle Abbey and estate
Cistercian propaganda, 206
Clackmannan (Clackmannanshire), 225
Clackmannan, forest of, 211, 332, 333
Clackmannanshire, 225
Clashbennie (Perth and Kinross), 282
clearance (for agriculture), 26, 29, 31, 43, 55, 56–57, 61, 69–74, 79, 80, 86, 99–100, 118, 158, 162, 206, 207, 208, 213, 220, 224–225, 227, 232, 234, 247, 248, 250–251, 253, 257, 260, 261, 263, 265–267, 268, 270, 286, 336, 339, 361, 375
  fire use / slash-and-burn, 25, 55, 69, 100
Clearances (Highland), 3, 4, 376
Clement VII, pope, 304
climate
  amelioration, 58, 125, 179, 188, 243, 258
  cooling, 9, 16, 25, 32, 83, 90, 104–105, 109–110, 112, 122, 132–133, 135–116, 192, 203, 296–298, 303, 339
  deterioration, 42, 71, 107, 112, 114, 116, 119, 142, 150, 159, 218, 300, 364, 370
  proxy data (*see also* ice; speleothems; tree rings), 10, 13–17, 19, 40–2, 104, 106, 110, 119, 121, 122, 124, 132, 137, 191, 192–4, 196, 202–4, 207, 292, 294, 295, 304, 308, 312
  warming, 6, 8–9, 18, 132–133, 142, 145, 150, 198, 274, 296–297
Clyde, river (South Lanarkshire), 24, 47, 60, 218, 252, 254, 283, 338
Clyde, Firth of, or estuary, 16, 229, 279, 297, 338
  fishery, 279, 283, 338, 342
Clydesdale (North and South Lanarkshire), 206, 218–219, 252–253, 254
Clunie (Perth and Kinross), forest of, 211, 223
coal use and coalmines, 262, 328–329
coitchionn, 268
Cold War, 5
Coldingham Priory and estate (Scottish Borders), 222, 226, 228, 260
Colmslie (Scottish Borders), 227, 268
colonisation (human) (*see also* Norse (or Scandinavian) colonisation and settlement; *see also under* England), 6, 22, 32, 53, 63, 80, 124, 145, 151, 152, 205, 206, 218, 264, 286
  internal, 264
Columba, abbot of Iona, 84–85, 86, 120, 128, 162, 167
  Life of Columba, 120, 162
commons and common rights, 220, 232, 245, 246, 247, 248, 262, 268, 329, 366
Comyn family, 315–316, 325
Connacht, 197, 305
coppice and coppicing, 61, 70–76, 78, 80, 81, 83, 85–87, 88, 90, 93, 95, 99, 145, 216, 229–231, 232, 238, 240, 248, 254, 262, 283, 314, 329–331, 336
Corrieyairack Pass (Highland), 80
Cotken (Perth and Kinross), 268
Coupar Angus Abbey (Perth and Kinross), 130, 207, 221, 223, 250, 256, 257, 260, 269–270, 365
Cowie (Aberdeenshire), forest of, 333
Craig Phadrig (Highland), 74
Craigie (Perth and Kinross), 281
Craigmillar (Edinburgh), 262
Crail (Fife), 342
crannogs, 74, 80, 87–88
Crawford (South Lanarkshire), 218
Crawfordjohn (South Lanarkshire), 252
creel houses, 230
Cressingham, Hugh, 309, 325, 326
Crieff (Perth and Kinross), 78
Croc, Robert, of Crookston, 260
Cromarty Firth (Highland), 89
Cromwell, Oliver, 45
crows, 164, 169
cruives: *see under* fish traps
cultivation (*see also* agriculture, arable), 9, 31, 32, 34, 36, 43, 55–57, 60, 61–63, 69, 73, 78, 79–80, 98, 113–116, 118, 125–126, 127, 140, 141–142, 146, 147, 150, 190, 193–194, 205, 206, 212, 224, 226, 227, 248, 251–254, 256, 259, 263, 265–269, 271–276, 280, 281, 304, 308, 317, 321, 328, 332, 333, 339, 346, 357–358, 360–365, 366, 367, 370
  slash-and-burn, 26, 69, 100
  technology, 28, 32, 55, 56, 62–63, 251, 259, 263–266, 273, 361
  terraces, 56
  upland, 360, 362
  weeds of, 28, 78, 126, 265
cultural landscape, 2, 19, 21, 30, 34, 129
Cumbernauld (East Dunbartonshire), forest of, 157
Cumbria (modern English county), 39, 72, 223, 316, 353
Cumbria, kingdom of, 99
Cunningham (North Ayrshire), 251
cyclonic circulation, 14, 191, 192; *see also* anticyclonic circulation

dabhach (pl. *dabhaichean*), 96, 148–150, 151–152, 234, 237, 251, 257, 376
Dacian War, second, 65
Daer Water and valley (South Lanarkshire), 24, 218, 254
dairying, 115, 141, 144, 145, 146, 346, 361, 369
Dál Riata, kingdom of, 102
Dalgarnock (Dumfries and Galloway), 261, 263
Dalkeith (Midlothian), 262, 358
Dante Alighieri, 299
Dantean Anomaly, 299, 303, 321, 327, 345, 350, 352
Darnaway (Moray), 237, 333
  castle, 96, 333
  forest, 227, 237, 332, 333
davach / davoch: *see dabhach*
David I, king of Scots, 183, 210, 211, 221–222, 228, 229, 231, 246, 252, 257, 266, 279, 281, 282
David II, king of Scots, 290, 291, 318, 319, 333, 352
David, earl of Huntingdon, 199, 282
dearth, 115, 122, 179, 186, 196, 205, 234, 253, 277, 294, 305, 307, 309, 310–312, 322, 345, 356, 357, 370
Dee, river (Aberdeenshire), 24, 25, 79, 331
deer, 4, 161, 164, 170, 173, 207, 208, 210–212, 215, 216, 227, 238, 240, 318, 319
  fallow, 311
  red, 154–155, 169, 170, 311
  reindeer, 18, 22, 36, 154
  roe, 69, 154–155, 158, 210
deer forest: *see* forest (deer, hunting)
deforestation, 45, 56, 83, 100, 199, 337, 339
dendrochronology (*see also* climate, proxy data; tree rings), 96, 296, 299
devastation (deliberate) and wasting, 63–64, 288, 290, 314–319
Deveron valley (Aberdeenshire), 261
Devon, 349
diet, 126, 130, 147, 162, 271, 322, 324, 346, 350, 351, 363
  deficiencies, 126, 351
  preferences, 154, 158, 162, 244, 276, 278, 342
  staples, 205, 245
Dirleton Castle (East Lothian), 287
DNA, 32, 140
dogs, 154, 167, 169, 170, 215, 280, 305
Doggerland, 21
domesticated livestock (*see also* cattle; goats; pigs; sheep and sheep husbandry), 22, 36, 156, 162, 169, 212, 251, 347
domesticated plants (*see also* barley cultivation; flax cultivation; oats; rye; wheat), 28, 265
domus Scoticana, 336
Donnchad mac Crinain, king of Scots, 181
Doon, river (South Ayrshire), 232, 283
Dornoch Firth (Highland), 89, 95
Dorset, 138
Douglas of Dalkeith family, 358–359
drainage: *see under* wetlands
Drimmie (Perth and Kinross), forest of, 211, 221, 223, 257
Drosten Stone, 166–167
drought and dryness, 10, 12, 42, 90, 108, 120–121, 122, 124, 137, 196, 202, 205, 303, 306, 308, 324

Drumcarrow Craig (Fife), 329
Drumochter (Perth and Kinross/Highland), 49
Dryburgh Abbey and estate (Scottish Borders), 221, 227, 228, 267–268
Duan Albanach, 181
Dublin, 230
Duby, George, 320, 321
Duffus (Moray), 237, 332
Dumbarton (West Dunbartonshire), 283, 338, 342
Dumfries (Dumfries and Galloway), 305, 338
Dumfriesshire, 118
Dun da Lamh (Highland), 80
Dun Vulan, South Uist (Western Isles), 53, 71, 89, 90, 93
Dunadd (Argyll and Bute), 113, 166
Dunbar, earl of, 254, 256, 260, 268
Dunbar, William, 354
Duncan of Carrick, 232
Dundaff (Stirling), 260
Dundee, 59, 277, 338
Dundrennan Abbey and estate (Dumfries and Galloway), 223, 225
dune systems, 42–43, 107, 193, 362
  deflation, 43, 107, 108, 114, 193, 304, 363
  in Western Isles, 42–43, 108, 114, 363
Dunfermline Abbey and estate (Fife), 162, 211, 216, 221, 231, 262, 266, 277, 281
dung and faecal matter, 63, 93, 94, 127, 140, 244
Dunkeld (Perth and Kinross), 49, 351
  bishop of, 351
Dunlappie (Angus), forest of, 211
Durham (Co. Durham), 118, 187, 259
  monks of, 259
  Symeon of, 118, 188
Durward's Dyke (Angus), 216
Dutch Elm Disease, 26
Drymen (Stirling), 225
Dysart (Fife), 342

eagles, 164, 169, 173
Earlston (Scottish Borders), 268
Earn, river (Perth and Kinross) (*see also* Strathearn), 78, 283
Easter Ross (Highland), 22, 89, 95, 96, 108, 129, 150, 166, 170, 173, 334, 336
Easterton of Roseisle (Moray), 173
Eden, river (Cumbria), 39
Eden, river (Fife), 55
Edgar, king of Scots, 183, 221, 259
Edinburgh, 247, 262, 263, 300
  fuel supply, 246–248
Edmundston (Scottish Borders), 255
Ednam (Scottish Borders), 259, 265
  moor, 244
Edward, king of Scots, 290
Edward I, king of England, 45, 237, 288, 309, 314, 315, 325, 332
Edward II, king of England, 323, 326, 327
Edward III, king of England, 290, 318, 319, 326
Edzell (Angus), 231
Eigg (Highland), 169
Eildon Hill North (Scottish Borders), 74
eileircg (deer trap), 215, 240
Einar, jarl of Orkney, 93
El Niño / La Niña, 11–12
El Niño Southern Oscillation (ENSO), 12
Eldbotle (East Lothian), 362
elder trees, 86
Eldgjá 'fire gorge', 135–136
Elgin (Moray), 223, 227, 230, 237, 283, 333
  cathedral, 170, 287
elk and giant elk, 154
elm trees, 22, 26, 28, 29, 30, 47, 71, 74, 78, 89, 227, 228, 248, 331
  decline, 26, 28, 29
  wych, 71
elrick: *see eileircg*
England, 22, 57, 74, 101, 190, 195, 197, 208, 259, 264, 273, 289, 291, 294, 299, 305, 307, 309, 311, 313, 320, 321, 324, 326, 327, 357, 367
  Anglo-Saxon, 101, 148, 264
  bear extinction in, 159
  colonisation from, 206, 264
  eastern, 206, 243
  famine in, 189, 190, 194, 311, 322–324, 325, 327, 370
  harvests in, 193, 194–195, 200, 309, 320
  kings of: *see* Edward I; Edward II; Edward III; Henry I; Henry II; Richard II
  northern, 9, 57, 58, 73, 76, 85, 116, 122, 187, 195, 314, 324, 325, 350
  panzootics in, 189, 323, 347–350, 351, 369, 370
  plague in, 351–353, 354
  Roman forts and frontiers in, 58, 76
  southern, 158, 186, 187, 189, 194, 195, 197, 325, 327, 350, 370
  weather in, 187, 189, 193–195, 197, 203
  western, 203, 204
English Channel, 197
Enlightenment, 4
epidemic disease (*see also* Black Death; Great Mortality; pestilence; plague, bubonic; pandemic), 34, 103, 104, 110, 111–112, 130, 180, 189, 285, 288, 305, 321, 326, 347, 351–356, 357, 370
epizootic disease 104, 286, 288, 305, 323, 349–351, 357, 369; *see also* cattle (pestilence, murrain or mortality); fowlpox; *Rinderpest morbillivirus*; sheep, scab
Ericht, river (Perth and Kinross), 257
erosion, 248, 254, 339
  coastal, 17, 43, 280
  soil, 32, 43, 56–57, 63, 124, 254
  wind, 43, 63, 108, 114, 147, 304, 363, 370
Erraid (Argyll and Bute), 162
Errol (Perth and Kinross), 269–270, 282
Eskdale (Dumfries and Galloway), 118, 214
Ettrick, river (Scottish Borders), 24
Ettrick (Scottish Borders), forest, 214, 252, 254, 260
Etymologia, 334
Exeter, bishop of, 349

Falconer, Simon, 341–342
falcons, 170, 176
Falkirk (Falkirk), 52, 232
famine (*see also* dearth), 9, 65, 104, 110, 112, 179, 186, 195, 197, 200, 285, 288, 294, 305, 307, 310–312, 318, 319–325, 345, 370
  English, 186, 188–190, 195, 200, 285, 305, 307, 311, 321–323, 325, 326, 345
  Ethiopia, 323
  European, 110, 186, 189, 195, 197, 198, 307, 311, 321–323, 345, 349, 350, 354, 370
  Irish, 119, 120, 190, 289
Faroe Islands, 354
Fauchope (Scottish Borders), 268
Fearn Abbey (Highland), 129–130
Feddal (Perth and Kinross), 332
fence and fencing, 35, 74, 215, 229, 336
Ferenze Hills (West Renfrewshire), 214
Fergus, son of Gillebrigte earl of Strathearn, 268
fermeisun, 214
Feugh, Water of (Aberdeenshire), 331
fields (arable) (*see also* infield; outfield), 28, 29, 35, 39, 43, 56–57, 60, 63, 69, 72, 99, 112, 120,126, 127, 128, 130, 193, 195, 199, 225, 242, 249, 280, 305, 314, 319, 363
  banks and walls, 31, 56
  'Celtic', 56
  clearance rubble, 31
  home, 114; *see also* infield; outfield
  silt, 193
  systems, 357, 361
Fife, 24, 55, 66, 107, 162, 211, 262, 277, 300, 314, 318
Fife, earl of, 292, 329
Fimbulvetr or Fimbulwinter, 105
Findchan, Mull landowner, 85, 90
fish, 53
  anadromous, 53
  cod, 140, 143, 277, 278, 281, 285, 339, 341, 343
  demersal species, 277, 281, 339
  dried, 341, 342
  gadid species, 278, 281, 341
  haddock, 277, 278, 281 342
  herring, 143, 277–280, 281, 285, 339, 341–342
  ling, 277, 278, 281, 341
  oil, 277
  other species (dab, plaice, pollack, ray, whiting), 277, 281
  pelagic species, 277, 339
  saithe, 143, 281, 341, 140
  salmon, 53, 131, 164, 169, 173, 176, 278, 281–285, 338, 343–345
fish prices, 285
fish traps, 161
  cruives, 282, 283, 285, 343, 344
  stake nets, 283
  yairs, 283, 285, 343
fisheries, 11, 143, 232, 248, 269, 277, 285, 338–339, 342, 344
  cod, 281, 339, 342, 343
  herring, 143, 279, 280, 339, 341, 342
  inshore, 53
  riverine, 281–283, 338–339
  salmon, 53, 278, 281–283, 285, 343–344

sea or deep-water, 143, 277, 278, 279, 281, 282, 328, 339, 341
fishing, 28, 112, 131, 141, 150, 151, 270, 274, 277–285, 324, 328, 338–345
commercial, 142, 278–279, 341
line, 140, 277
nets and creels, 161, 277, 279, 282–283
regulation, 283, 285, 343–345
river, 278–279, 281–283, 285, 338, 341; *see also* Forth; Spey; Tay; Tweed
sea or open-water, 53, 140, 141, 274, 279–281, 339–345, 361
subsistence, 281
Flanders (*see also* Flemish colonists and settlers), 206, 243
count of, 237
fish trade, 341, 343
wool trade, 254
Flanders Moss (Stirling), 250
Flavian dynasty, 47
Flavius Constantius, Roman emperor, 40, 52
flax cultivation, 62–63, 113, 140, 146, 147, 272, 273
flaxseed, 272
Flemish records, 186, 194
Flemish colonists and settlers, 218, 254
floods and flooding, 103, 136, 189, 190, 199, 200, 201, 202, 269, 308
sea-floods, 190, 311
Florence, 373
Flow Country (Highland), 18–19, 25
fodder, 73, 74, 114, 115, 141, 144, 145, 212
leaf, 74, 365
winter, 125, 141, 146, 149, 307
foraging, 25, 28, 30, 43, 272, 325
pig, 220, 223; *see also* pannage
Fordoun, John of, 121, 185, 291, 305, 334, 351, 354
forest (deer, hunting) (*see also* forest law; parks), 45, 207–221, 222–223, 227, 230, 234, 237, 240–241, 252, 258–259, 311, 316, 332–333; *see also individual named forests*; Scottish Borders forest
forest and woodland, 6, 18–22, 24, 25, 26–29, 30, 31, 34–35, 36, 37, 40, 43, 45, 47, 50, 51, 52–53, 55–6, 61, 66, 69–100, 116, 118, 145–146, 147, 150, 154–158, 160, 167, 176, 187, 197, 202, 206, 207–208, 211–214, 216, 217–242, 244–248, 250, 251, 252, 253–254, 260, 262, 265–268, 275, 286, 309, 314, 319, 328, 330–338, 339, 344, 361, 365, 374–375, 376
clearance: *see main entry*
climax, 19, 26, 29, 375
conservation, 70, 95, 116, 207, 237, 238, 240, 285, 336, 360, 375
decline, 55–56
disputed, 228
management (*see also* coppice and coppicing), 80, 83, 87, 96, 99, 207, 229, 375
named, 226–7
plantation, 82, 100, 218
'primeval', 20, 21, 25, 69, 241, 375
regeneration, 20, 22, 26, 28, 29, 31, 34, 52, 55, 66, 69, 71, 72, 74, 80, 82, 86, 88, 89, 95–6, 98, 99, 110, 116, 118, 125, 146, 150, 207, 225, 231, 237, 240, 244, 265, 333, 375
semi-natural woodland, 36, 83, 237, 241, 375
soils, 28
forest law, 208, 210–211, 234, 236, 241
foresters and keepers, 99, 212, 213, 215, 218, 232, 257, 332
Forfar (Angus), 332
Forres (Moray), 223, 226, 227, 230, 237
Forth, river (Stirlingshire), 52, 200, 231, 269, 277, 283
basin, 225, 329
carses, 52, 231, 269
Firth or estuary, 162, 231, 262, 277, 285, 300, 317, 319
fishery, 277–279, 283
Forth–Clyde isthmus, 47, 60
Fortriu, kingdom of, 96, 98, 110–111, 148, 150, 166
Forvie (Aberdeenshire), 362
fowlpox, 351
Fowlis Wester (Perth and Kinross), 268
fox, 158, 161, 163
Freswick Links (Highland), 274–275, 281
frost, 186, 187, 189, 191, 200, 201, 202, 203, 299, 309
fuel (*see also* broom; charcoal; coal use and coalmines; dung and faecal matter; heather; peat (fuel); seaweed; wood), 60, 61, 71, 74, 89, 90, 93, 94, 95, 113, 128, 142, 146, 148, 213, 220, 227, 229–232, 242, 243, 244–251, 262, 326, 328, 329, 332
fungal infestation, 26, 328, 363
Fyvie (Aberdeenshire), 173

galleys and galley-building (*see also* shipbuilding), 95, 237–238, 336
Galloway (Dumfries and Galloway), 16, 53, 55, 70, 87, 95, 151, 161, 185, 215, 217, 223–225, 227, 251, 256, 260, 263, 283, 294, 299, 315, 316, 317, 328, 332, 349, 351, 357, 366
'Gart' place-names, 225
Gask Ridge (Perth and Kinross), 269
geese, 164, 169, 173, 351
Germany, 17, 52, 136, 197, 205, 299
Gesta Annalia (*see also* Fordoun, John of), 185, 238, 239, 240, 241, 305, 318, 351, 352, 354
Gille Chriosd, lord of Tarbet, 233
Gillebrigte, lord of Galloway, 224
glaciers and glaciation, 8, 21, 25, 41, 105, 106, 119, 132, 303
Glasgow, 338
cathedral, 96, 99, 287, 354
fishery, 338
Glen Affric (Highland), 61, 74, 116, 124, 225–226, 230
cultivation and settlement in, 61, 74, 115–116, 257, 274, 361, 364–365
Glen Banchor (Highland), 237
Glen Brichty (Perth and Kinross), 365
Glen Derry (Aberdeenshire), 31, 79
Glen Esk (Angus), 231
Glen Feshie (Highland), 237
Glen Gairn (Aberdeenshire), 79
Glen Isla (Perth and Kinross), 124, 365
Glen Lichorn (Perth and Kinross), 332
Glen Lui (Aberdeenshire), 31, 79
Glen More (Highland), 237
Glen Quoich (Aberdeenshire), 31, 79
Glen Shee (Perth and Kinross), 124, 150, 257, 273
Glen Spean (Highland), 80
Glen Tromie (Highland), 237
Glenelg (Highland), 237
Glenfiddich (Moray), forest of, 234
Glengonnar (South Lanarkshire), 218
Glenkens district (Dumfries and Galloway), 316
Glenluce Abbey (Dumfries and Galloway), 283
Glenmore (Highland), forest of, 234
goats, 145–146, 225, 251
milk, 145
golden age (*see also* Alexander III), 179, 180, 185–186, 287–288, 292, 294, 300, 325, 327, 370, 373
Gowrie, 318
Carse of, 206, 269
Graham, Sir David, 260
grain drying (*see also* kilns), 114, 142, 363
grain prices (*see also* wheat), 312
grange farming (*see also* Kirkgunzeon; Penshiel), 218, 225, 255, 257, 269–270
Grantown-on-Spey (Highland), 170
grass, 55, 78, 107, 116, 144, 145, 150, 160, 218, 220, 251, 318, 350, 365
growth, 141, 145, 251, 252, 326, 346, 369, 370
'lawns', 220, 267
grassland, 28, 34, 36, 39, 56, 61, 78, 99, 176, 237, 252, 254, 256, 257, 314, 337, 365, 366, 368, 375, 376
grazing (animal activity), 36, 55, 63, 69, 76, 81, 89, 116, 124, 145–146, 149–150, 154, 204, 211, 213, 214, 215, 218, 220, 221, 224–225, 228–229, 234, 240, 251–252, 254, 255–256, 257, 258, 286, 360, 365, 366, 368, 370
pigs, 222–223, 224–225, 265, 375
summer, 150, 204, 234, 360, 365, 366
winter, 148, 149, 150, 234, 251, 370
grazing (land) (*see also* pasture), 54, 55, 56, 57, 60, 63, 72, 74, 80, 107, 118, 144, 148, 202, 226, 234, 237, 246, 248, 251, 254, 256, 257, 268, 308, 361, 365, 366
overgrazing, 32, 43, 366, 370
pressure, 22, 26, 29, 55, 61, 71, 80, 116, 221, 225, 251, 254, 317, 337–338, 346, 349, 365, 367, 368
rights to, 221, 256–257, 268
upland, 251, 268, 365
Great Famine (*see also* famine, European), 321–322, 324, 345, 350, 370
Great Glen, 48–9, 61, 80, 124, 158
Great Mortality (*see also* Black Death; epidemic; pandemic; plague, bubonic), 138, 187, 190, 373
'Great Winters', 123, 133, 137, 145, 197, 201
Green Castle (Moray), 97–98
Greenan (South Ayrshire), 227, 232
Greenland, 6, 13, 104, 107, 133, 192, 205, 296, 298, 303

Greenland Ice-core Project (GRIP), 5–6, 8, 11, 13, 107, 119, 121, 132–133, 135, 136, 191, 295, 297, 303
Greenland Ice Sheet Project, Phase 2 (GISP2), 6, 8, 13
Gunung Rinjani (volcano), 295
Gwynedd, 314

habitat loss, 153, 154, 158, 162, 224, 338, 339
Haddington (East Lothian), 248, 305, 319
Hadrian's Wall (Cumbria / Northumberland), 58
Hamburgian culture, 17–18, 22
hares, 212
Harris, 43, 129, 238, 279
Hartside (Scottish Borders), 255
harvests, 9, 13, 63, 65, 110, 112, 120, 137, 138, 179–81, 185–6, 187, 189, 190, 194–5, 196, 197, 200, 202, 203, 204, 205, 289, 294, 297, 300, 305–6, 307, 308, 310–12, 314, 320, 322, 323, 324, 325, 326, 328, 345, 348, 350
  failure, 9, 110, 120, 179, 180, 186, 187, 190, 197, 200, 206, 294, 310, 311, 312, 322, 323, 325, 326, 345, 346
hawks, 164
hawthorn, 71, 90, 330
hay (*see also* meadow), 72, 125, 141, 149, 270, 305, 365
Hay family, 269–270, 282
hazel trees, 21, 22, 28, 30, 47, 60, 70, 71, 74, 78, 81, 83, 86, 88, 89, 92, 93, 158, 225, 254, 330
  coppice, 61, 71, 74, 80, 83, 85, 87, 90, 93, 230, 248, 254, 330, 331, 336
heather (*see also Calluna*), 4, 31, 55, 60, 73, 89, 176, 216, 218, 237, 254, 256, 258, 314
  as fuel and kindling, 60, 89, 90, 91, 93, 146, 244, 328
heath and heathland, 19, 55, 57, 60, 61, 78, 93, 116, 125, 160, 220, 256, 275, 365, 366
Hebrides, 30, 34, 53, 63, 90, 113, 121, 127, 128, 145, 161, 185, 202, 238, 239, 241, 257, 281, 283, 297, 341, 350, 354
  Inner, 24, 70, 84–85, 87, 90, 95, 155, 238, 241, 336
  Outer (*see also* Western Isles), 62, 85, 89, 90, 144, 238, 239, 272, 339, 341, 363
  southern, 22
Hekla (Iceland), 32
Henry I, king of England, 196
Henry II, king of England, 224
Henry of Braid, 260
Herodian, Roman writer, 52
hides: *see under* cattle
Highlands, 1, 4, 14, 16, 30, 48, 52, 65, 69, 70, 74, 76, 78–79, 83, 107, 108, 112, 115, 116, 119, 128, 137, 158–159, 160, 161, 192, 196, 215, 221, 225, 230, 234, 241, 257, 276, 283, 333–334, 336, 350, 357, 365, 370
Highland Boundary Fault, 22, 76, 256
Hill of Keillor (Perth and Kinross), 173
Hilton of Cadboll (Highland), 170
Historia Brittonum, 52, 121
Hoddom (Dumfries and Galloway), 118
Holm Cultram Abbey and estate (Cumbria), 223, 269
Holocene, 8, 17, 18
Holyrood, Chronicle of, 197
Holyrood Abbey and estates (Edinburgh), 223, 231, 246, 260, 261, 262–263, 269, 277, 278, 279, 352, 366
Honorius, Roman emperor, 40
hornbeam, 22
horses and ponies, 54, 133, 169, 221, 262, 300, 305, 311
  grazing, 211
  harness, 264, 286
  wild, 221
Howe (Orkney), 114
Humber, 188
hunter-gatherer culture, 18, 25
hunting (*see also* deer; seals; pigs, wild; wild boar; wolves), 24, 25, 28, 30, 36, 56, 69, 153, 154, 157, 159, 161, 162, 163–76, 217, 221, 223, 238, 240, 241, 251, 254, 267, 270
  chases, 154, 167, 207, 215, 260
  dogs and hounds, 154, 167, 169, 170, 208, 210, 215, 219
  reserves (*see also* forest and woodland), 150, 157, 162, 207–216, 218, 221, 227, 234, 238, 240, 241, 268, 333
  traps, 215, 240
Huntingdon, earldom of, 199, 211, 282

ice (*see also* sea-ice), 17, 18, 21, 22, 24, 36, 133, 186, 195, 196, 197, 299
Iceland, 6, 13, 32, 41, 104, 131, 135, 224–225, 281, 298, 303, 341, 354
  low-pressure cell, 13, 42, 106, 132, 135, 191
Improvement, 4, 83, 100, 149, 270, 376
'inch' (Gaelic 'innis', island), 248, 268, 269, 270
Inchaffray Abbey and estate (Perth and Kinross), 268–269
Inchture (Perth and Kinross), 269, 270
Inchtuthil (Perth and Kinross), 76
Indonesia, 295
infield, 56, 62, 93, 113, 362
Innes, Cosmo, 45
insects and insect remains, 10, 241, 311, 351, 374
intakes (*see also* assarts), 31, 225, 361, 363
Inveravon (Moray), 30, 80
Inverkeilor (Angus), 193
Inverkeithing (Fife), 300
Inverness (Highland), 96, 161, 166, 223, 229, 230, 237
  castle, 336
  fishery, 338, 342, 344
Inverness-shire (Highland), 74, 78, 96, 336
Iona (Argyll and Bute), 85–86, 87, 128
  abbey, 84–86, 87, 126, 128
  abbots: *see* Adomnán; Columba
Ireland, 16, 21, 96, 101, 107, 133, 136, 143, 154, 190, 191, 193, 195, 197, 205, 230, 243, 280, 326, 349, 354
  Annals: *see* Annals of Connacht; Annals of Innisfallen; Annals of Loch Cé; Annals of Tigernach; Annals of Ulster
  crop failure, dearth or famine in, 109, 120, 122, 189, 190, 196, 324–325
  law-codes, 96
  missionaries and monks from, 101, 127
  mortalities in, 122, 138, 189, 190, 196, 349, 354
  oak chronology in, 16, 96, 299
  oak woodland, 298, 299
  weather, 119, 122, 133–134, 135, 186, 187, 188, 196, 197, 201, 202, 205, 289, 303, 306, 309, 324
  wetlands, 107
  woodland history, 96, 98
Iron Age, 39, 41, 49, 53–54, 55, 57–58, 63, 67, 74, 78, 79, 80, 90, 100, 108, 112, 113, 114, 116, 127, 140, 143, 146, 154, 156, 158, 160, 161, 218, 253, 271
  pre-Roman, 58, 60, 72, 80, 158, 253
ironmasters, 45
Isbister (Orkney), 173
Isidore of Seville, 334
Islay (Argyll and Bute), 63, 238, 240, 257, 336, 349
Italy, 64, 136, 205

Jacobites, 45
Japan, 136
  record sources, 135
Jarlshof (Shetland), 92, 161
Jed (Scottish Borders), forest, 328, 337
Jedburgh Abbey and estate (Scottish Borders), 221, 231, 332
Jerusalem, 64
jet stream, 13, 106, 122
Joachim of Kinbuck, 332
John, king of Scots, 288, 290
John of Gaunt, 319, 337
juniper, 21, 22, 79, 90
Jura, island of (Argyll and Bute), 24, 240

Kebister (Shetland), 275, 361
Kedslie (Scottish Borders), 267–268
Kelso (Scottish Borders), 330
  abbey and estate, 231, 244, 253, 277, 366
Kersie (Stirling), 278
Kilbride (North Ayrshire), 260
Kilbuiack (Moray), 226, 227
Kildrummy Castle (Aberdeenshire), 287
kilns, 314, 328
  corn drying (*see also* grain drying), 118, 271, 363
Kiltyrie (Perth and Kinross), 360
Kincardine (Clackmannanshire), 231
Kincardine park (Aberdeenshire), 216
Kinfauns (Perth and Kinross), 281
Kinghorn (Fife), 162, 277, 300
Kingussie (Highland, 237
Kinloch, Rum (Highland), 24
Kinloss Abbey and estate (Moray), 226, 227, 368
Kinnoull (Perth and Kinross), 248
Kintail (Highland), 237
Kintyre (Argyll and Bute), 257
Kirkconnell Merse (Dumfries and Galloway), 269
Kirkcudbright (Dumfries and Galloway), 317, 338

Kirkgunzeon (Dumfries and Galloway), 223, 225, 227
Kirriemuir (Angus), 130, 170
Knapdale (Argyll and Bute), 241, 336
Knights Templar, 332
Knocknagael (Highland), 166
Kyle (East and North Ayrshire), 329

Lacaille, Armand, 24
Laggan (Highland), 80, 237
Lair (Perth and Kinross), 124–125, 150–151, 273, 274
Lamb, Hubert, 9
Lammermuir Hills (East Lothian / Scottish Borders), 251, 252, 254, 255, 256, 357, 365
Lanark (South Lanarkshire), 248, 252
  sheriff of, 311
Lanarkshire, 157, 218
land management, 62, 83, 137, 140, 153, 184, 211
Lanercost Chronicle, 185, 300, 305, 308–309, 325
larch, 71, 90, 92, 93
  Siberian, 298
Late Antique era, 104, 169, 334
Late Antique Little Ice Age (LALIA), 43, 44, 63, 82, 83, 99, 110, 111, 112, 114, 116, 125, 130, 176, 285, 375
Low Countries (*see also* Netherlands), 299
Lauder (Scottish Borders), 227
Lauderdale (Scottish Borders), 227–228, 251, 267–268
Lawman's Annal, 354
lead and lead-mining, 183, 218
leather, 205, 339
Leith Hall (Aberdeenshire), 169, 173
Lennox (East and West Dunbartonshire / Stirling), 99, 185, 230, 231
Leslie, Bishop John, 157
Leven, river (West Dunbartonshire), 283
Lewis, 29, 142, 270, 279
lime (building), 183
lime trees, 28
limestone, 10, 42
Lindisfarne (Northumberland), 138
Lindores Abbey and estate (Fife), 230, 268, 282, 332
lions, 154, 166, 169
Lindsay family, 218, 333
Little Dunkeld, 351
'little ice age' (LIA), 15, 16, 240, 292, 298, 299, 338, 361, 370, 371
Loch Affric (Highland), 61
Loch Awe (Argyll and Bute), 336
Loch Duich (Highland), 81, 241
Loch Garten (Highland), 31, 80
Loch Lomond (West Dunbartonshire), 17, 233
  Stadial or Readvance, 17, 18, 21, 25
Loch Moidart (Highland), 85
Loch Pityoulish (Highland), 80
Loch of Roseisle (Moray), 173
Loch Shiel (Highland), 85
Loch Sunart (Highland), bottom water temperature, 13–14, 16, 192, 196, 198, 204
Loch Tay (Perth and Kinross), 74, 80, 360, 370
Lochindorb (Moray), 318
Lochleven, monastery, 231
Lombok, 295
London, 194
Longmorn (Moray), forest of, 227, 237, 332, 333
Lorn (Argyll and Bute), 336
Lossie, river (Moray), 283
Lothian, 43, 53, 60, 63, 167, 223, 262
  East, 231, 262, 309, 362
  West, 262
Lowlands, 1, 2, 59, 61, 72, 192, 305, 308, 324, 349
Lowther Hills (Dumfries and Galloway / South Lanarkshire), 218
Luce, bay and river (Dumfries and Galloway), 283
Lulach, king of Scots, 181
Lunan Bay (Angus), 193
'lungessouth', 349
Lussa, Jura (Argyll and Bute), 24
lynx, 36, 158, 164, 169

Mabie (Dumfries and Galloway), 227
Macbethad, king of Scots, 181, 187, 193
MacDonald, John, of Islay, 336
MacDougall, John, of Lorn, 336
machair (*see also* sand, shell-sand), 42–43, 90, 145, 146, 270, 271, 273, 280, 363
  cultivation, 43, 113, 146, 270–271, 273, 280, 363
  erosion, 43, 108, 113, 114, 147, 363, 370
  grazing, 43, 145
Madagascar, 29
Madderty Pow (Perth and Kinross), 269
Maeatae, 59, 66
Máel Coluim II mac Cinaeda, king of Scots, 180, 181, 186
Máel Coluim III mac Donnchada, king of Scots, 181, 183, 190
Máel Coluim IV, king of Scots, 185, 333
Magnus Bareleg, king of Norway, 95
malnourishment and malnutrition, 312, 320, 323, 324, 350, 356
Malthus, Thomas, 320
Malthusian crisis, 9, 320, 321, 322, 324
Man, Chronicle of, 95
Man, Isle of, 95, 270
  Kingdom of, 270
Manchuria, 136
manure and fertiliser, 62, 63, 93, 113, 114, 127, 128, 130, 142, 146, 265, 280
Manx chronicle: *see* Man, Chronicle of
Margaret, Maid of Norway, 288
Margaret, Queen and saint, 181–183
marginal land and marginality, 32, 34, 57, 63, 112, 114, 116, 124, 125, 126, 130, 245, 273, 286, 294, 345, 357–358, 359, 360, 364, 365, 367, 370
markets and market economy, 58, 60, 62, 65, 67, 183, 205, 206, 233, 248, 249, 285, 286, 311, 342, 360
Martial, Roman poet, 158
Matilda, empress, 196–197
meadow (*see also* hay), 78, 149, 220, 226, 262, 269
  grasses, 270
  water, 376
Mearns (Aberdeenshire), 216, 230, 244, 256, 318, 351
Medieval Climate Anomaly (MCA) (*see also* Medieval Warm Period), 9, 14, 16, 130, 138, 142, 150, 151, 179, 180, 185, 186, 196, 199, 204, 205, 207, 217, 224, 226, 227, 230, 234, 243, 244, 250, 251–252, 256, 258, 260, 263, 269, 270, 272–4, 277, 281, 285, 294, 296, 299, 300, 324, 327, 328, 338, 339, 341, 361, 363, 365, 373
Medieval Warm Period (*see also* Medieval Climate Anomaly), 138, 179, 285
Mediterranean, 40, 64, 119, 205
Meigle (Perth and Kinross), 166, 169
Melrose Abbey and estate (Scottish Borders), 214–216, 221, 223, 227–228, 232, 254–255, 256, 260, 266–267, 268, 283, 329, 339, 354, 365, 366
  Chronicle of, 181, 185, 201, 304, 306
meltwater (*see also* spates), 198, 199
Merlin, 52
Merovingian Franks, 208
Mesolithic era, 18, 23–28, 30, 69, 154, 158, 161
Meuse estuary, 197
midden waste, 18, 60, 63, 93, 114, 161, 162, 279, 362
Midlothian, 59, 218, 262, 332
migration (fish), 12, 13, 162, 375
migration (human), 102, 103, 243
milk (*see also* cattle; dairying; *see also under* goats; sheep and sheep husbandry), 144, 323, 369
mills and milling, 202, 220, 286, 309, 314, 315, 339, 363
Moidart (Highland), 237
Monk's Cally (Perth and Kinross), 257
Monkwearmouth-Jarrow (Co. Durham), 122
Monro, Donald, archdeacon of the Isles, 239–240, 241
Mons Graupius, battle of, 51, 64
moorland, 55, 63, 73, 91, 145, 176, 216, 218, 240, 252, 258
Morar (Highland), 237
Moray, 22, 74, 95, 96, 110, 150, 170, 173, 181, 190, 210, 226, 227, 230, 256, 269, 273, 318, 332, 333, 334, 352
  bishop of, 223, 283
  diocese of, 234
  floods, 190
  forest in: *see* Longmorn *and under* Darnaway
  Laich of, 97, 226, 231, 269, 368
  lowlands, 22, 318, 366
Moray Firth, 54, 95, 166, 173, 206, 222, 237, 251, 342
Morocco, 104, 107
mortality rates (human), 9, 110, 119, 138, 186, 190, 197, 322, 325, 327, 352, 354, 356, 357, 359, 370
mortality rates (livestock), 141, 187, 346, 369
Morvich (Highland), 81–2
Morville family, 214, 215, 216, 268
Moss of Blair (Perth and Kinross), 250
mossland (*see also* peat and peatland; peateries), 247, 251
Mount Pinatubo, 135

Mow (Scottish Borders), 367
Mugdrum Island (Perth and Kinross), 282, 283
Muir of Orchill (Perth and Kinross), 268
muirburn, 55, 256, 365, 368
Mull (Argyll and Bute), 85, 238, 349
  Ross of, 85, 162
Murdac Rhynd, 333
murrain: *see under* cattle (pestilence, murrain or mortality)
Murray, Andrew (13th century), 310
Murray, Andrew (14th century), 318, 325, 333

Nairnshire (Highland), 221, 251
Neolithic era, 18, 25, 26, 28–31, 32, 49, 70, 71, 79, 80, 93, 127, 154, 173, 207, 244, 374, 375
Netherlands (*see also* Low Countries), 40
Newbattle Abbey and estate (Midlothian), 218, 222, 231, 232, 243, 254, 262, 269
Newbigging Leslie, 160, 173
Newburgh (Fife), 283
Newmill (Perth and Kinross), 60, 74
Newtonmore (Highland), 30, 80
Nith, river (Dumfries and Galloway), 218, 223, 225, 227, 269, 308
Nithsdale (Dumfries and Galloway), 261
Normandy, 196, 208
Normanville, Waleran de, 261
Norse era, 83, 90, 92, 93, 95, 142, 143–145, 146, 161–162, 274
Norse language and cultural practices, 128, 138, 141–146, 271–273, 280
  saga tradition, 93, 94, 102, 104, 138, 280, 296
  sea-faring technology: *see main entry*
  shipbuilding: *see main entry*
Norse mythology, 105
Norse (or Scandinavian) colonisation and settlement, 6, 90, 98, 112, 129, 138–145, 152, 238, 270–271, 275, 279, 281
  settlements, 90–95, 142–145, 281; *see also* Bornais; Brough of Birsay; Cille Pheadair; Freswick Links; Jarlshof; Pool; Robert's Haven; Udal
  settlers, 57, 90–95, 139, 142–143, 145, 147, 150, 238, 271, 280
  Viking raiders, 45, 102, 138, 147
North America, 8, 13, 69, 71
  climate, 13
  driftwood from, 71
  forest cover, 69
North Atlantic Oscillation (NAO), 12, 13–14, 16, 42, 43, 105, 106, 112, 119, 120, 122–124, 132, 135, 188, 192, 196, 198, 200, 201, 202, 203, 296–297, 298, 299, 303, 306, 307, 308
North Korea, 135
North Sea, 21, 40, 190, 306, 338, 341, 350, 366
Northern Isles (*see also* Orkney; Shetland), 22, 47, 91, 94, 95, 108, 112, 114, 116, 127, 138, 140, 141, 143, 173, 217, 238, 241, 294, 297, 341
  'papar' in, 127–129
  woodland in, 47, 91, 94
Northumberland, 39, 40, 72, 157, 181, 314

oak trees and woodland, 16, 20, 22, 25, 28, 30, 47, 60, 69, 74, 78, 80–83, 85–86, 87–89, 90, 96–98, 99, 100, 155, 188, 194, 220, 222, 223, 227, 228, 231, 233, 236, 237, 241, 248, 243, 262, 330–332, 333, 334, 336, 337
  coppice, 80, 83, 331
  cross-marked, 227
  Irish (*see also* Ireland), 298
  mast (*see also* pannage), 194, 196
Oakbank crannog (Perth and Kinross), 74
oats, 56, 57, 62, 112–114, 115, 118, 125, 126, 140, 141, 142, 146, 269, 271, 272, 360, 361
oceanic circulation systems, 11–12
oceanic surface temperatures, 12, 16, 191, 204, 296, 298, 299, 303–304, 307, 324, 339
Ochil Hills (Fife / Perth and Kinross / Stirling), 199, 225
Old Kilpatrick (West Dunbartonshire), 230
Old Scatness (Shetland), 70, 92–93, 114, 140–141, 142, 148, 150, 161, 166, 169, 173
Oldmeldrum (Aberdeenshire), 315
Onuist, son of Urguist, king of the Picts, 170
Oort Solar Minimum, 14, 150, 187, 191, 192, 196
orchards (*see also* apple trees), 314, 337
Orkney, 29, 70, 83, 93, 95, 96, 112, 114, 127, 128, 138, 140, 141–142, 173, 237, 281, 341, 354; *see also* Mainland; Rousay; Sanday; South Ronaldsay; Westray
Oronsay (Argyll and Bute), 24, 161
otters, 161, 169
  skins, 161
Outer Hebrides (*see also* Western Isles), 62, 85, 89, 90, 144, 238, 239, 272, 339, 341, 363
outfield, 56, 62, 63, 363
oxen, 151, 169, 264, 286, 350

Pabbay / Pabay (Hebrides), 128, 129
Pacific Ocean, 11, 12
Paible (Western Isles), 129
Paisley Abbey and estate (West Renfrewshire), 214, 221, 223, 226, 233, 257, 279
palaeoecology, 26, 29
palaeoenvironmental records, 6, 10, 13, 16, 18, 40, 43, 45, 52, 60, 67, 70, 72, 76, 99, 103, 104, 113, 118, 124, 126, 127, 147, 218, 238, 253, 254, 256, 257, 273, 277, 334, 339, 375
Palaeolithic, Late Upper, 18, 22, 24, 36
palynology (*see also* pollen and pollen analysis), 45, 71, 367
pandemic disease (*see also* Black Death; epidemic disease; Great Mortality; plague, bubonic), 9, 28, 34, 321, 333, 351, 352, 357
pannage (*see also* foraging, pigs), 221–224, 242
panzootic, 9; *see also under* cattle (pestilence); England; *see also* fowlpox; *Rinderpest morbillivirus*; sheep, scab
Papa Stour (Shetland), 128
papar, 127–9
Papa / Papay (Orkney / Shetland) place-names, 128
parks, *see* Buzzart Dykes; Durward's Dyke; Kincardine; Stirling, 157, 216, 223, 333
pastoral farming and pastoralism (*see also* transhumance), 28, 53–54, 63, 73, 108, 116, 124, 140, 146, 234, 245, 252, 274, 345, 346, 360, 370
pastoralists (*see also* pasture), 18, 161
pasture (*see also* grazing (land)), 28, 32, 34, 35, 36, 55–56, 72, 78, 79, 80, 107, 108, 112, 113, 116, 126, 162, 212, 214, 220, 227–228, 241, 247, 251, 252, 255, 256, 257, 260, 261, 262, 268, 319, 334, 360, 364, 365, 367, 371
  high-altitude, 145, 204, 251, 274, 365
  upland, 55, 145, 151, 218, 274, 263, 365–366, 367, 369, 370
  wetland (*see also* meadow), 269
  wood, 212, 220, 221–222, 227–228, 229, 234, 242, 263, 334, 375
pathogens, 5, 26, 154, 325, 328, 346, 351, 356, 374
peat and peatland, 14, 18, 19, 20, 31, 32, 34, 36, 42, 60, 63, 76, 83, 94, 106, 108, 116, 227, 244, 246, 249–251, 252, 256, 257, 263, 328, 365, 374
  blanket, 19, 25, 31, 36, 83, 227, 249, 256, 257, 374
  formation and growth, 19, 31, 32, 60, 83, 105, 108, 256, 365
peat (fuel), 89, 90, 92–94, 116, 146, 230, 244, 246, 247, 248–250, 251, 262, 328–329
peateries, 244, 246, 249
Penick (Highland), 221
Penshiel (Scottish Borders), 255
Pentland Hills (Lothian), 43, 260
Penuld (West Renfrewshire), 226
Perth (Perth and Kinross), 60, 96, 199, 201, 233, 246, 248, 249, 277, 278, 281, 282, 314, 318, 329, 330, 338, 339
  bridge, 199, 201
  burgesses, 249, 269
  castle, 199
  floods, 199
  Kirk Close, 330
Perthshire, 59, 60, 78, 124, 150, 158, 166, 173, 211, 221, 223, 248, 256, 260, 268, 273, 357, 366
pestilence (*see also* Black Death; epidemic disease; Great Mortality; pandemic disease; plague, bubonic), 186, 189, 200, 206, 288, 319, 325, 352, 354, 356
  cattle, 122; *see also* epizootic disease; *see also under cattle*
Pictish era, 92, 93, 114, 143, 144, 161, 208
Pictish kingdoms or state (*see also* Fortriu), 96, 98, 102–111, 148, 173
Pictish settlements and estates, 92, 95, 110–111, 114, 124–126, 139–1340, 143, 150, 161
Pictish ships, 122
Pictish sculpture and stones, 53, 154, 159, 164, 166, 169, 170–1, 173, 208, 210; *see also* Aberlemno; Ardross; Blackford;

Brough of Birsay; Burrian; Drosten Stone; Easterton of Roseisle; Fyvie; Grantown-on-Spey; Hill of Keillor; Hilton of Cadboll; Kirriemuir; Leith Hall; Meigle; Newbigging Leslie; Old Scatness; Rhynie; Shandwick; St Vigeans; Strathpeffer; Tyrie
Picts, 67, 90, 102, 138, 140, 142, 146, 167, 170, 173
pigs (*see also* pannage; wild boar), 127, 144, 155, 161, 194, 196, 211, 220–222, 223–225, 265, 280, 305, 375
  wild, 154, 155, 210
'pilsoucht', 347
pine (Scots) trees and pinewoods, 16, 20, 21, 22, 24, 25, 29, 30, 31, 45, 52, 60, 71, 74, 78, 79, 85, 86, 89, 90, 93, 99, 237, 241, 298, 299, 334, 336
  tree rings, *see main entry*; *see also* dendrochronology
pine marten, 161, 163
Pitcarmick (Perth and Kinross), 124
  settlement type, 124–125, 150, 151
Pitcur (Perth and Kinross), 59
plague, bubonic (*see also* Black Death; Great Mortality; epidemic disease; pandemic disease; pestilence), 9, 34, 110, 274, 291, 322, 333, 352–354, 356, 357, 358, 359, 360, 367, 370, 373
  Justinianic, 110
  mortalities, 110, 112, 119, 322, 352, 356, 357, 359, 370
planities, 267–8
plantation, *see* forest and woodland
plantation (settlement), 254, 264
Plater (Angus), forest of, *see* 318, 332–333
ploughs and ploughing (*see also* ard; carruca), 55, 56, 63, 64, 151, 183, 201, 251, 259, 264–266, 267, 268, 306, 361
  mould-board, 55, 263, 273
Pliny the Elder, Roman writer, 47, 50
'pluk', 347
polecats, 161, 163
pollarding, 26
pollen and pollen analysis, 10, 21, 26, 28, 55, 56, 57, 60, 61, 66, 67, 73, 78, 80, 81, 83, 85–86, 88, 89, 90, 108, 112, 115, 116, 126, 147, 217, 220, 225, 227, 234, 245, 360, 361
Pont, Timothy, 218, 233, 252
Pool, Sanday (Orkney), 70, 93, 94, 112, 114–115, 141, 142, 150, 283
population pressure (human), 32, 62, 80, 116, 230, 274, 292, 361, 370
Portknockie (Moray), 97
Portmahomack (Highland), 88–89
  monastery, 108, 126, 129, 130, 147
Postan, Michael, 9, 320–321
'Postan Thesis', 9, 320–321, 324
poultry, 351
precipitation, 103, 119
  acidic, 136
  blizzards, 195, 202
  rain and rainfall, 8, 10, 12, 14, 19, 31, 41–42, 43, 99, 103, 119, 120, 121, 122, 123, 132, 136, 137, 179, 189, 190, 192, 195, 199–200, 201, 202, 274, 297–299, 303, 305, 306, 307, 309, 312, 326, 328
  snow and snowfall, 103, 119, 195
predators, 36, 154, 158, 162, 163, 164, 169, 173, 176
protein, 126, 130, 144, 147, 350, 351
  marine, 126, 280
Psoroptes ovis (*see also* sheep, scab), 347
Ptolemy, Graeco-Roman geographer, 47–51
Pytheas of Massalia, Greek geographer, 47

Queensferry (Fife / West Lothian), 300
  Narrows, 300
'Quikege' (Scottish Borders), 221

rabbits, 212
Ragnarök, 105
rainfall: *see* precipitation
  'bloody', 119, 121
Ranald, son of Somerled, 257
Rannoch (Argyll and Bute / Highland / Perth and Kinross), 199
ravens, 164
Reginald of Durham, 118
Reidchalmai (Highland), 81–82
Renfrew, 226, 260
  fishery, 279, 283, 338
  forest of, 227
Renfrewshire, 214, 221, 251, 260
Restenneth Priory (Angus), 332, 333, 352
'retreat from the margins', 276, 357, 361, 370
rewilding, 21, 29, 100, 375
Rhine, river, 52
Rhineland, 243
Rhynie (Aberdeenshire), 111
ribwort plantain, 78
Richard II, king of England, 319
rig-and-furrow, 264, 273, 357, 361, 364
  broad, 273
Rinderpest morbillivirus, 350
Rink Farm (Scottish Borders), 24
Robert I, king of Scots (Bruce, Robert), 288, 289, 315–317, 332, 344
Robert II, king of Scots, 291–292, 304, 333
Robert III, king of Scots, 292
Robert of Torigni, 196, 197
Robert's Haven (Highland), 281
Robin Hood, 254
Roman culture and influence, 37, 39, 102, 130
Roman army, military and legions, 20, 21, 37, 40, 43, 44, 45, 47, 51, 52, 59, 60, 65, 66, 67, 73, 76, 100
  Ninth Cohort of Batavians, 40
  Ninth Legion, 51
Roman emperors: *see* Augustus; Caracalla; Flavian dynasty; Flavius Constantius; Honorius; Septimius Severus; Titus
Roman Empire, 20, 39, 40, 41, 44, 45, 51, 52, 58, 59, 64, 67, 104
  Imperial generals and officials, 11, 154
  policy, 44, 59
Roman forts, frontier lines and garrisons, 40, 47, 55, 58, 59, 65, 76, 78
Roman invasion and occupation, 35, 52, 53, 56, 57, 58, 60, 64, 66, 70, 72, 76, 100, 158, 160, 218, 285
  Antonine, 59, 60, 76
  Flavian, 47, 59, 65, 76, 78, 158
  Hadrianic, 76
  Severan, 67
Roman law, 210
Roman literature and writers, 37, 39–40, 45, 47, 50–51, 52, 158, 334
Roman province, 44, 67, 80
Roman Warm Period (RWP) or Roman Climatic Optimum, 40, 42, 43, 59, 60, 63, 67, 72, 112, 159
Rome, 37, 40, 53, 64, 65, 158
Rosinish (Western Isles), 279
'Rothenet' (Moray), 336
Rothiemurchus (Highland), 31, 80, 237
Roxburgh (Scottish Borders), 229, 309
rowan trees, 21, 330
Rùm, island of (Highland), 24, 29, 238, 240, 241
Rutherglen (North Lanarkshire), 342
rye, 56–57, 125, 146, 265, 271

SARS Covid-19, 352
Sahara, 121, 122
salmon: *see* fish
salt and salt industry, 231–232, 244, 249–250, 262, 269, 277, 342
saltpans, 231–232, 250, 262, 269, 300, 329
Samalas eruption: *see* Gunung Rinjani
sand, 43, 55, 193
  banks, 283
  blow and drift (*see also* dune deflation), 42–43, 107, 108, 193, 304, 363, 374
  see also shell-sand; machair
Sanday (Orkney), 70, 93, 112, 114, 142
Sanquhar (South Ayrshire), forest of, 233
Scalloway (Shetland), 92, 114
Scandinavia, 104, 105, 131, 133, 135, 197, 199
Scipio Aemilianus, Roman general, 64
Scone Abbey and estate (Perth and Kinross), 257, 278, 281, 333
Scotichronicon (*see also* Bower, Walter), 199, 305
Scoto-Norwegian war, 306
Scottish Borders forest (between Gala and Leader), 211, 214, 215, 221, 223, 227, 255, 268
sea-faring technology (*see also* shipbuilding), 53, 138
sea ice, 104, 119, 133, 192, 197, 296, 298, 303
seals, harbour, grey and common, 161–162, 169, 277
seaweed, 146
  dried as fuel, 93, 94, 244
Selkirk (Scottish Borders), 211, 260
  'waste place of', 260
Selkirk, forest of, 211, 223
Septimius Severus, Roman emperor, 52, 59, 66, 76
Shandwick (Highland), 166
Shannon, river, 205
Shanzie (Perth and Kinross), 59, 60
sheep and sheep husbandry, 4, 45, 55, 140, 144, 145, 146, 161, 162, 201, 218, 224, 227, 254–257, 260, 262, 318, 326, 346, 365–367, 368, 376
  dung, 63, 127
  impacts, 254–257, 260
  lamb / mutton, 145
  milk, 145

mortalities, 201, 306, 311, 347, 349
murrains, 347–348, 351
scab, 347–348, 351, 366
wool and woolfells: *see* wool
shell-sand (*see also* machair), 113, 128
Sherwood Forest (Nottinghamshire), 254
Shetland Islands, 53, 59, 151, 166, 173, 354
agriculture, 55, 63, 112–113, 114, 126, 127–129, 130, 140–142, 144, 147, 161, 274, 361
fishing, 140, 143, 161–162, 281, 341
fuel types, 328
Norse raids and colonisation, 138, 140, 161–162; *see also* Jarlshof; Old Scatness
Pictish settlements in, 92, 112, 140, 161–162, 166
woodland, 70, 83, 91
Shiel, river (Highland), 85
shielings and shieling grounds, 234, 251, 274, 360, 370
'shifting baselines' (concept), 3, 5
shipbuilding, 85, 95, 138, 143, 144, 237, 238, 341
Siberia, 104, 119, 124, 132–133, 188, 191, 197, 298, 299
Sidlaw Hills (Angus/Perth and Kinross), 60, 269, 318
silver coinage, 183, 257, 292, 366
Sinclair, Sir John, of Ulbster, 52
Skaill, Deerness (Orkney), 114
Skelbrooke, Roger of, 232, 283
Skene, William F., 300
skins and pelts, 161, 162, 163, 176, 183, 224, 369, 145, 154
Skye, island of (Highland), 29, 59, 85, 167, 237, 238, 240
snow and snowfall: *see* precipitation
SO_4 and sulphates, 121, 133, 135, 295
soil (*see also* anthrosols), 19, 22, 28, 31, 55, 56, 61, 62, 63, 74, 89, 113, 115, 116, 118, 125, 146, 150, 193, 252, 257, 264, 265, 269, 271, 274, 375
acidification, 36, 365
alluvial, 61, 115, 158, 252, 257, 269
brown forest, 28
deepening: *see* anthrosols
enrichment, 56, 127–130, 140, 142, 146, 274, 360, 362–363
erosion, 32, 43, 55, 56, 57
impoverishment and deterioration, 9, 32, 36, 43, 83, 252, 367
sandy: *see* machair
solar activity (*see also* sun), 136, 191, 296
eclipse, 121
irradiance, 110, 133
minima, 14, 150, 187, 191, 192, 193, 196, 296
Solway Firth and coast, 118, 225, 231, 254, 264, 273, 283, 297, 308, 338
Somhairle mac Gillebrigta, 237
Sorrowlessfield (Scottish Borders), 267
Sorules, William, 267–268
Sound of Harris (Western Isles), 129
souterrains, 59–61, 65, 67, 79; *see also* Ardestie; Ardownie; Carlungie; Newmill; Shanzie
Southwick (Dumfries and Galloway), 225
South Ronaldsay (Orkney), 173
Southern Uplands, 22, 24, 25, 30, 52, 53, 54, 60, 69, 70, 100, 112, 118, 119, 160, 167, 198, 218, 221, 223, 229, 252–253, 256, 398, 376
spades, 56, 63, 251, 264, 265, 361
spates (*see also* flood; meltwater), 198, 199, 200, 201, 202, 249, 306, 309, 339
speleothems (*see also* climate proxy data), 10, 13–14, 16, 41–42, 43, 63, 106–107, 108, 119, 122–123, 132, 133, 137, 181, 192, 196, 197, 200, 202, 203, 298–299, 308
Spey, river (Highland/Moray), 49, 234, 237, 333, 334, 336
Speyside (Highland/Moray), 80
sphagnum moss, 31, 60
Springwood Park (Scottish Borders), 330–331
spruce, 92, 93
St Andrews (Fife), 168, 304, 329
bishop of, 352
cathedral–priory, 167–168, 281, 304, 307, 332, 352, 356
parish, 168
St Quivox (South Ayrsire), 233
St Vigeans (Angus), 166, 167, 170
stake nets: *see* fish traps
Steward of Scotland (*see also* Stewart family), 214, 233, 292
Stewart family, 260, 292
Alexander, earl of Buchan (Wolf of Badenoch), 292, 336–337
Robert II, king of Scots: *see main entry*
Robert, duke of Albany, 292, 329
Stirling (Stirling), 52, 183, 231, 266, 277, 278, 351
Carse of, 231
castle, 266
fishery, 283, 338
forest, 211
King's Park, 157, 216
saltpans, 231
Stirlingshire, 260
Stocket (Aberdeenshire), forest of, 332
stocking levels, 243, 350, 367, 368
Stonehaven (Aberdeenshire), 333
storminess, 25, 99, 108, 240, 297, 300, 303, 308, 339, 361, 375
Strachan, Castlehill of (Aberdeenshire), 331
Strageath (Perth and Kinross), 78
Strath Tummel (Perth and Kinross), 124
Stratha'an (Moray), 150, 234
forest of, 150
Strathallan (Perth and Kinross), 268, 332
Strathardle (Perth and Kinross), 257
Strathcarron (Stirling), 260
Strathdon (Aberdeenshire), 80
Strathearn (Perth and Kinross), 78, 251, 268, 332
earl of, 268
Strathmashie (Highland), 80
Strathmore (Aberdeenshire/Angus/Perth and Kinross), 130, 269
Strathpeffer (Easter Ross), 175
Strathspey (Highland), 30, 80, 150
subsistence (*see also* agriculture), 28, 79, 163, 176, 205, 281, 345, 346, 351, 360, 361, 370, 375, 376
crisis, 110, 112, 122, 137, 190, 322, 325, 326, 345, 370
sun (*see also* solar activity), 136, 296
dimmed or weakened, 109, 121
sunspots, 14, 187, 296
Sutcliff, Rosemary, 39
Sutherland (Highland), 14, 16, 18, 41, 59, 81, 106, 217, 298, 341, 357
swamps, 37, 40, 52
Swinewood (Scottish Borders), 221–222
sycamore, 22
Symmachus, Quintus Aurelius, 154

Tacitus, Publius Cornelius, 11, 39–40, 47, 50, 63–64, 65, 316
Táin bó Cúailnge, 53
tanbark, 45
Tap o' Noth (Aberdeenshire), 111–112
Taransay, 129
Tarbatness (Highland), 89, 108, 129–130
Tay, river (Perth and Kinross), 39, 67, 76, 150, 199, 248, 251, 260, 270, 277, 278, 281–283, 306, 317, 339, 357
estuary of Firth, 55, 269, 277, 278, 318
floods: *see* floods and spates
salmon fishery, 281–283
silting, 248
Tay–Tummel basin (Perth and Kinross), 49, 201
Tees, river (Co. Durham), 118, 188
teinds and teinding, 211, 234, 257, 260, 262, 266, 277, 278, 281, 316
Tentsmuir (Fife), 55, 107–108
territorialisation, 32, 210
Teutoburger Wald, battle of, 52
Teviot, river (Scottish Borders), 309
Teviotdale (Scottish Borders), 221, 254, 309, 319, 357
Thames, river, 194, 197
thermal energy, 13, 89, 94, 244, 245, 247, 248, 329
Thor Longus, 259, 265
Threepword/Threipwood (Scottish Borders), 228
timber: *see under* wood
Tinogad, 176
Tinto Hills (South Lanarkshire), 252
Tironensian monks, 254
Titus, Roman emperor, 64, 158
Torridon Mountains (Highland), 17
Torwood (Stirling), 52
towns, planned, 4, 183, 220, 329
traditional environmental knowledge (TEK), 11, 112, 152, 272, 328, 339, 373
Trajan's Column, 65
Traligill (Highland), cave system and speleothems, 13, 14, 16, 41, 42, 106, 119, 132, 135, 137, 192, 196, 197, 198, 200, 202, 203, 297, 298, 308
Tranent (East Lothian), 262
transhumance (*see also* pastoral farming and pastoralism; shielings), 251, 274, 360, 365, 370
Traquair, forest of, 211, 223
tree growth, 22, 25, 34, 135, 253, 299, 308
tree rings (*see also* climate proxy data; dendrochronology), 10, 16, 96 119, 132, 135, 296, 298–299, 303
tree species: *see individual named species*
Troqueer (Dumfries and Galloway), 225
Trossachs (Stirling), 225

Tullibody (Clackmannanshire), 278
Tullochcurran (Perth and Kinross), 257
turf use, 56, 74, 78, 93, 127, 128, 142, 244, 264, 304, 360, 362, 370
Turnberry (South Ayrshire), 232
Tweed, river (Scottish Borders), 24, 198, 282, 306, 309, 339
Tweeddale (Scottish Borders), 52, 254
Twynholm (Dumfries and Galloway), 366
Tyne, river (Northumberland), 39, 118, 122
Tyne–Solway isthmus and frontier, 39, 76
Tyrie (Aberdeenshire), 173

Uamh an Tartair (Highland) (*see also* Traligill), 14, 41, 42, 43, 63, 106, 107–108, 122, 123, 132, 136
Udal (Western Isles), 144, 145, 146, 150, 279, 363
Uhtred, lord of Galloway, 223–224
Uist: *see under* Western Isles
Ulster, 120, 122, 134, 186–187, 202, 203, 305
  Annals: *see* Annals of Ulster
understorey (woodland), 31, 36, 154, 220, 332
Urr, river (Dumfries and Galloway), 223, 227, 263

Valence, Aymer de, 314
Varus, Publius Quinctilius, Roman general, 52
vert, 212, 213, 214, 216, 251, 332
verus valor (see also *antiqua taxatio*), 356–357
Vieuxpont family, 262
Vindolanda, Roman fort and civilian settlement (Northumberland), 40
volcanic forcing, 32, 106, 107, 109, 110, 135–136, 286, 296
volcanoes (*see also* Gunung Rinjani; Hekla), 135–136
  volcanic eruptions, 105, 109, 121, 133–134, 135, 295, 296

Wallace, William, 289, 310, 325
war impacts (*see also* devastation), 59, 65–67, 157, 291, 312–319, 324–325, 327, 337–338, 352, 356–357, 360
warming trends, 6, 8–9, 12, 18, 132–133, 142, 145, 150, 274, 296
  oceanic, 12, 132, 198, 296, 297
Wars of Independence, 254, 288, 319, 325, 327, 351
waste, 118, 207, 225, 258–261, 269, 374
waterlogging, 18, 19, 31, 36, 99, 150, 237, 256
wattling and wicker, 74, 78, 84, 86–88, 90, 93, 99, 229, 283, 329, 331, 336
  wattle-and-daub, 87, 331
Waverley chronicle, 202
Wear, river (Co. Durham), 122
weather (*see also* drought; precipitation; storminess)
  deterioration, 16, 32, 43, 112, 113, 119, 187, 199, 202, 204, 292, 294, 300, 306, 312, 327, 329, 357, 361, 363
Wemyss (Fife), 314–315, 337
Wester Ross (Highland), 21, 81
Western Isles (*see also* Outer Hebrides), 22, 29, 42, 70, 89, 112, 113, 126, 128, 129, 138, 141, 142–151, 155, 270, 273, 275, 279, 280, 281, 287, 328
  Barra, 43, 108, 130
  dune systems in: *see* dune systems
  Harris, 43, 129, 238, 279
  Lewis, 29, 142, 270, 279
  Pabbay, 129
  'papar' in, 127–129
  Taransay, 129
  Uist (North and South), 53, 57, 62, 63, 71, 89, 90, 91, 92, 93, 108, 112, 112, 142, 144, 145, 146, 142, 147, 161, 238, 271, 273, 279, 341, 363
  woodland in, 22, 70, 83, 90, 99
wetlands, 18, 61, 70, 71, 72, 78, 89, 97, 107, 108, 147, 173, 176, 226, 227, 237, 241, 247, 250, 266, 269–270, 275, 286, 374, 376
  drainage of, 126, 206, 265, 266, 269, 270, 374
whales (*see also* blubber), 131, 277
  oil, 277
wheat, 56, 146, 269, 274, 298, 311, 324, 360, 361
  prices, 311
wheelhouses, 92
Whithorn (Dumfries and Galloway), 88, 230
  diocese, 316
  monastery, 87
Wigtown (Dumfries and Galloway), 317
wild boar (*see also* pigs), 85, 169, 210, 212
wildcats, 158
wilderness (concept), 1, 2, 47, 79, 258, 318
wildfowl, 173, 176, 240, 251, 270
William, king of Scots, 185, 207, 210, 212, 216, 222, 232, 233, 243, 283
William I 'the Conqueror', king of England, 188
William II Rufus, king of England, 190
willow (*see also* wattling and wicker), 21, 70, 71, 74, 78, 79, 83, 85, 86, 87, 88, 90, 92, 93, 94, 158, 225, 241, 248, 254, 330
  coppice, 85, 87, 90, 330, 331, 336
  withies and wands, 87, 90, 99, 230, 248, 283
wind and windstorms, 13, 31, 43, 63, 103, 114, 122–123, 193, 194, 199, 202, 298, 300, 305, 306, 307, 308, 311, 374
Wolf of Badenoch: *see* Stewart family, Alexander
Wolf Solar Minimum (*see also* solar activity), 296
wolves, 36, 158, 159–160, 161, 164, 169, 173, 176, 374
  hunting, 173, 214
  skins and pelts, 163, 176
wood
  dead wood and windfall, 220, 230, 244, 328
  driftwood (*see also* larch), 70, 71, 90, 91–95, 99, 238, 241
  fuel, 71, 72, 74, 89, 91, 93, 94, 95, 146, 213, 220, 227, 229–230, 231–232, 242, 244–246, 247, 248, 328, 329
  imported, 70, 89–90, 92, 94, 248, 329
  poles, 74, 87, 99, 228, 230, 248
  rods, 228, 230
  smallwood, 74, 78, 99, 213, 229, 238, 241, 251
  timber, 61, 71, 73–74, 76, 78, 80, 84, 86–87, 88, 89, 95, 96, 98, 99, 100, 116, 146, 156, 207, 213, 216, 218, 220, 225, 228, 229, 233, 238, 240, 241, 242, 244, 248, 262, 326, 329, 332, 333, 334, 336, 344, 370
  turned, 85, 87, 90
  undressed, 85
  worked, 85, 86, 97
woodland: *see* forest and woodland
wool, 145, 234, 248, 257, 286, 309, 310, 326, 347, 346, 366–367, 368
  trade, 254, 257, 339, 347, 348, 366–367, 368
  woolfells and fleeces, 145, 326, 246, 366
  woollen cloth, 205
Worcester chronicle, 203
wrawes of bule, 230
Wulvestrother (Midlothian), 243
Wyntoun, Andrew of, 185, 288, 305, 312, 318, 325, 352, 371

Y Gododdin, 85, 167, 176
yairs, 283, 285, 343
Yersinia pestis (*see also* Black Death; epidemic disease; Great Mortality; pandemic disease; plague, bubonic), 354, 373
yew trees, 86
yields (crop), 9, 10, 34, 63, 98, 113, 114, 115, 118, 120, 124, 126, 129, 142, 147, 148, 185, 205, 269, 271, 294, 308, 312, 325, 328, 359, 361, 362, 364, 365, 370, 371
Yolande de Dreux, queen of Scots, 300
York, 118, 309, 353
Younger Dryas, 17